Cover Crops and Sustainable Agriculture

Editors

Rafiq Islam

Program Director, Soil, Water and Bioenergy Resources
Ohio State University, Piketon, Ohio, USA

Bradford Sherman

Publications Editor
Ohio State University, Piketon, Ohio, USA

CRC Press
Taylor & Francis Group
Boca Raton   London   New York

CRC Press is an imprint of the
Taylor & Francis Group, an **Informa** business

A SCIENCE PUBLISHERS BOOK

First edition published 2021
by CRC Press
6000 Broken Sound Parkway NW, Suite 300, Boca Raton, FL 33487-2742

and by CRC Press
2 Park Square, Milton Park, Abingdon, Oxon, OX14 4RN

© 2021 Taylor & Francis Group, LLC

CRC Press is an imprint of Taylor & Francis Group, LLC

ISBN: 978-1-032-03440-9 (hbk)
ISBN: 978-1-032-03443-0 (pbk)
ISBN: 978-1-003-18730-1 (ebk)

Typeset in Times New Roman
by Radiant Productions

Preface

Agriculture faces the immense challenge of meeting the needs of global food security. Agricultural production must be doubled in order to meet that demand of increasing population growth, which will make our agriculture increasingly dependent on repeated annual plowing, widespread usage of chemical fertilizers and reactive chemicals, inefficient and/or excessive irrigation, and limited cropping diversity. Such an intensification of farming will have negative impacts on soil health, crop productivity, and ecosystem services. Conventional agriculture substantially contributes to, and is affected by, global climate change. In response to global warming, population growth, land degradation, and water war, the future of agriculture will require proactive management of natural resources to provide global food security with enhanced ecosystem services.

Cover crops, as a biochemical primer, together with conservation tillage, cropping diversity, and proactive soil amendments, are the critical components of sustainable agriculture to enhance ecosystem services. They provide a myriad of organic components to support soil biodiversity and efficiency to perform enzymatic functions, residue decomposition and nutrient recycling, and biocontrol services. Cover crops help soil aggregate formation, alleviate compaction, improve hydrological properties, reduce erosion and leaching, smother and control weeds, provide forage, and rejuvenate soil health.

Globally, producers are looking for evidence-based knowledge on the use of cover crops in crop rotations, and their impact on reducing the cost of chemical fertilization (especially nitrogen) and herbicide applications, increasing crop yield, and improving farm income. Those innovative producers who are successful are proactively incorporating cover crops into their existing cropping systems (such as growing cereal rye in corn-soybean rotation) or adjusting their crop rotations to better fit cover crop blends (such as corn-rye-soybean-wheat-mixed cover crops or soybean-wheat-mixed cover crops-sunflower), for multi-lateral benefits.

This book covers a wide range of topics on characteristics, optimization, and applications of cover crops; soil biological, chemical, and physical properties associated with soil health; carbon sequestration and nutrient cycling; organic farming and community vegetable production; forages and livestock grazing; orchard sustainability; weed, pest, and soil-borne disease pressures; soil erosion and water quality; greenhouse gas emissions; and farm economics. We hope that this book will deliver science-based knowledge and information for professionals within the international sectors to collaborate on addressing major concerns associated with agricultural sustainability and food security.

We extended our thanks to all the 47 authors across Africa, Asia, Europe, North America, and South America, under different disciplines, who have contributed to this book by authoring chapters. While most of the chapters were written to address interdisciplinary complex topics, several of the authors have contributed their own research data and results to provide a broader perspective on cover crop applications. Moreover, we thank them for their vigilance and persistence during the review and revision of the book chapters.

Special thanks to Randall Reeder, Joy Bauman, Wayne Lewis, Sarah Swanson, and Tom Worley for their help to acquire field pictures, develop outlines of the book, and support to accomplish this initiative. We also thank other colleagues at South Centers who helped us in one way or another to make this book a reality.

Finally, we acknowledge and convey our profound thanks to the staff at CRC Press for successful completion of the book. We especially extend our gratitude to Vijay Primlani, who encouraged, guided, and worked closely with us from the very beginning of the project and the publication of the book.

Rafiq Islam
Bradford Sherman
The Ohio State University South Centers
Piketon, Ohio, USA

Contents

Preface   iii

1. **Cover Crops and Agroecosystem Services**   1
   Rafiq Islam, Nataliia Didenko and *Bradford Sherman*

2. **Benefits of Cover Crops on Agronomic Crop Yield**   16
   Kateryna Chorna

3. **Potential and Challenges of Growing Cover Crops in Organic
   Production Systems**   28
   Sutie Xu, Sindhu Jagadamma, Renata Nave Oakes, Song Cui, Erin Byers
   and *Zhou Li*

4. **Cover Crops in Vegetable Production and Urban Farming in
   Sub-Saharan Countries**   41
   *Michael Kwabena Osei, Mavis Akom, Joseph Adjebeng-Danquah,
   Kenneth Fafa Egbadzor, Samuel Oppong Abebrese, Kwabena Asare Bediako*
   and *Richard Agyare*

5. **Algorithms to Optimise Cropping Diversity with Cover Crops**   58
   Romashchenko, M, Matiash, T, Bohaienko, V, Kovalchuk, V, Lukashuk, V and *Saydak, R*

6. **Sustainable Suppression of Weeds through Ecological Use of Cover Crops**   69
   Shawn T Lucas

7. **Cover Crops for Pests and Soil-borne Disease Control and Insect Diversity**   84
   *Nataliia Didenko, Vira Konovalova, Somayyeh Razzaghi, Alimata Bandaogo,
   Sougata Bardhan* and *Alan Sundermeier*

8. **Cover Crops for Forages and Livestock Grazing**   99
   Riti Chatterjee

9. **Cover Crops' Effect on Soil Quality and Soil Health**   124
   MA Rahman

10. **Cover Crops for Orchard Soil Management**   147
    Biswajit Das, BK Kandpal and *H Lembisana Devi*

11. **Cover Crop Mixes for Diversity, Carbon and Conservation Agriculture**   169
    Reicosky, DC, Ademir Calegari, Danilo Rheinheimer dos Santos and *Tales Tiecher*

12. **Cover Crops and Soil Nitrogen Cycling**   209
    Nitu, TT, UM Milu and MMR Jahangir

13. **Effect of Cover Crops on Soil Biology**   227
    Harit K Bal

14. **Cover Cropping Improves Soil Quality and Physical Properties**   254
    Yilmaz Bayhan

15. **Cover Crops Effects on Soil Erosion and Water Quality**   268
    Beenish Saba and *Ann D Christy*

16. **Effects of Cover Crops on Greenhouse Gas Emissions**   280
    Somayyeh Razzaghi

17. **Cover Crops Influence Soil Microbial and Biochemical Properties**   299
    Amoakwah, E

18. **Economics of Cover Crops**   309
    Mohammad S Rahman and *James J Hoorman*

Index   319

1

Cover Crops and Agroecosystem Services

Rafiq Islam,[1,]* *Nataliia Didenko*[2] *and Bradford Sherman*[1]

1. Overview

Agriculture is one of the important sectors contributing to the global economy and regulation of ecosystem services. To produce greater amounts of food, feed, and fibre, current agricultural production methods rely heavily on the use of synthetic fertilisers and reactive chemicals, repeated plowing, inefficient and/or excessive irrigation and continuous mono-cropping that exert negative impacts on soil productivity (Islam and Didenko, 2019). Agricultural lands are consequently becoming susceptible to degradation, loss of soil organic matter (SOM), severe plow-pan compaction, lack of groundwater recycling, prolonged drought and pollution, intermittent water-logging and secondary salinisation and accelerated erosion—making them unsuitable for long-term crop productivity. It is estimated that about 20 per cent of the agricultural lands will undergo desertification by 2050 (Rosenthal, 1990; Zwane, 2019). These intermingled problems result in maintaining crop production with higher irrigation frequencies and chemical inputs, accompanied by increased pest and disease pressures, higher operating costs, and reduced farmer incomes, without considering any environmental compatibilities and social responsibilities.

Agriculture, especially conventional farming practices, contributes to and is affected by global climate change. Currently the agricultural sector contributes around 12 per cent of anthropogenic greenhouse gas (GHG) emissions (IPCC, 2007) and faces growing pressure to reduce emissions to mitigate climate change effects (Zwane, 2019). The climate change effects are becoming increasingly noticeable on global crop productivity and ecosystem services. Where irrigation is a routine key management factor, conventional farming practices have severely impacted the soil quality and crop production by altering soil-water-plant-air ecosystems (Islam and Didenko, 2018).

Agriculture faces the challenge of feeding a world population expected to grow to 9.6 billion by 2050 (FAO, 2010; Zwane, 2019). Global agricultural production must be doubled to meet that demand, which will tend to make existing croplands increasingly dependent on reactive chemicals, fresh water and energy (Delgado et al., 2011; Islam and Didenko, 2019). Such intensification of farming will have long-term consequences on soil quality and is expected to be detrimental to agricultural productivity. Moreover, farmers are being caught in a vicious spiral of unsustainability related to depletion and degradation of land and water resources, increasing labour and input costs and decreasing profit margins (Scholberg et al., 2010). With current concerns related to global warming, population growth, land degradation and freshwater availability, agriculture is required to provide more diverse ecological services and make more efficient use of natural/renewable resources to provide global food security (van der Ploeg, 2008; Cherr et al., 2006).

[1] The Ohio State University, Columbus, Ohio, USA.
[2] Institute of Water Problems and Land Reclamation, Kyiv, Ukraine.
* Corresponding author: islam.27@osu.edu

2. Sustainable Agriculture and Cover Crops

Public concern over the consequences of conventional farming practices, including environmental contamination by fertilisers and reactive chemicals, soil erosion, depletion of natural resources and pesticide residues in foods have prompted shifts to sustainable agricultural production systems. Moreover, the consequences of global climate-change effects need to be addressed for the sustainability of long-term food security worldwide. As the demand for food security has and/or will increase, the looming prospect of reduced agro-ecosystem services demands a solution based on sustainable agricultural management practices to improve and/or sustain soil quality for economic crop productivity (Long et al., 2016; Zwane, 2019).

Crop rotation with legumes was first recognised by the Chinese over 2,000 years ago (Pieters, 1927). Virgil, in pre-Christian Rome, proclaimed in verse 'the virtues of fallowing the land from continuous cropping and of rotating small grains with legumes' (Gladstones, 1976). While lupin was common in southern Europe since the beginning of recorded history, crop rotations were not practiced in northern Europe until about the 16th century (Pearson, 1967). During the 1730's, Lord Townsend of Norfolk County in England, introduced the 'Norfolk rotation', which is four-year rotation of wheat-turnip-barley-red clover (Pearson, 1967). With this rotation, average wheat yields in England increased from 0.54 Mg/ha to 1.35 Mg/ha by the early 19th century. The 'Norfolk rotation', with slight modification, is still practiced in northern Europe today.

It is reported that Europeans, especially English settlers, brought green-manuring knowledge to North America. A rotation of corn-wheat-red clover was described by Thomas Cooper in 1794 as being practiced by the 'best' farmers in Pennsylvania (Pieters, 1917). Partridge pea and cowpea were grown in rotation with agronomic crops in Virginia and Maryland by the end of the 18th century (Pieters, 1927). Cover crops, as green manure, became an integral part of diverse cropping systems during the first part of the last century (Pieters, 1927).

Use of cover crops was gradually reduced due to commercial production and abundance of inexpensive chemical fertilisers during the 1950s, providing growers with concentrated primary nutrient sources (e.g., N, P and K) that could be easily managed (Smil, 2001; Tonitto et al., 2006). As a consequence, soil quality strategies shifted from building SOM and inherent soil fertility via crop rotations and supplementary use of organic amendments, to a discrete system dominated by external inputs used to boast soluble nutrient pools and crop yields (Drinkwater and Snapp, 2007). Moreover, externalities associated with the excessive use of reactive agrochemicals were typically ignored while inherent systems' functions and services were gradually lost and degraded (Cherr et al., 2006). Additionally, the shift towards large-scale and highly specialised operations diminished inherent diversity and resilience of agricultural systems (Cherr et al., 2006; Baligar and Fageria, 2007; Shennan, 2008; van der Ploeg, 2008).

A revival in the adoption of cover crops occurred during the 1980s, as farmers in southern Brazil exponentially increased the use of conservation agriculture technologies (Calegari, 2003; Landers, 2001). Increased adoption of cover crops resulted in a gradual reversal of the degradation of the natural production base as farmers were able to partially restore SOM contents and gradually reduce their dependence on external chemical inputs. Moreover, the successful expansion of no-tillage technology with cover-crops expansion in central and South America was clearly driven by farmers who actively engaged in technology development and transfer. In contrast, the use of no-tillage and/or cover crops in the United States was limited due to high crop value and risk-averse behaviour of conventional producers (Phatak et al., 2002). However, increased concerns related to environmental quality, energy use, land degradation and global warming have resulted in a gradual shift towards resource conservation management with an increased focus on sustainability and/or ecological-based or natural production systems (Lu et al., 2000; Hartwig and Ammon, 2002; Ngouajio et al., 2003; Shennan, 2008). While cover crops were abandoned due to green revolution technologies to address food security, use of cover crops is once more becoming the cornerstone of sustainable agro-ecosystems in response to recent interest in green technologies (Shennan, 1992; Phatak et al., 2002; Sullivan, 2003; Cherr et al., 2006; Baligar and Fageria, 2007).

Maintaining a healthy and productive soil is the foundation of sustainable agriculture to improve ecosystems in response to conventional farming and climate change impacts. Sustainable agriculture is a complex and multifaceted concept based on the integration of novel and holistic approaches of conservation tillage, cropping diversity and cover crops and precision soil and plant amendments for ensuring economic productivity, environmental compatibility and social acceptability to enhance agro-ecosystem services (Fig. 1). Sustainable agriculture can be defined as the management practices that sustain farmers, resources, inputs and communities by promoting best management practices and methods that maintain or enhance environmental quality and conservation of natural resources (Anonymous, 1989; Scholberg et al., 2010).

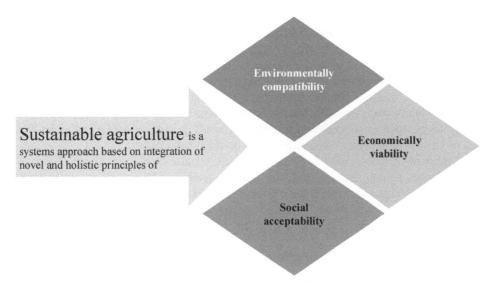

Fig. 1 Sustainable agriculture and its components.

2.1 Cover Crops Terminology

Cover crops, as an integral component of sustainable agriculture, improve soil biodiversity and efficiency, regulate soil balancing and equilibrium and soil physical stability to maintain and rejuvenate soil health. Based on their use, historically cover crops have been expressed in different terminologies, such as allelopathy crops, border crops, break crops, carbon crops, catch crops, companion crops, forage crops, green manure crops, intermediate crops, inter-crops, insectary crops, land cover, mixed crops, nitrogen crops, non-commercial crop, nurse crops, perennial crops, plant cover, pulse crops, revegetation crops, rotational crops, scavenger crops, service crops, siderite crops, smoother crops, summer annual crops, synergy crops, trap crops, and winter annual crops (Hamzaev et al., 2007; Clark, 2007; Magdoff and van Es, 2010; byronseeds.net; powerthesaurus. org/cover_crop).

2.2 Types of Cover Crops

Diverse types of plants have been, and can be, used as cover crops depending on their use in cropping systems (Clark, 2007; Magdoff and van Es, 2010). Over the years, legumes and grasses have been the most extensively used. There are growing interests in use of brassicas and other types of crops as important cover crops. Legumes are often planted as cover crops for biological nitrogen (N) contribution for succeeding agronomic crops. Several legumes as annual cover crops are grown during the summer including cowpea (black-eyed pea), soybeans, moth bean, lima bean, Canavalia, peanuts, rice bean, Tepary bean, lablab bean, winged bean, Sesbania, mung bean, Berseem clover, jack bean, Tephrosia, Sunn hemp, chickpea, and horse gram.

Other legumes that are normally planted in the fall (winter annuals) and counted on to overwinter include crimson clover, chicory vetch, hairy vetch, crown vetch, subterranean clover, field peas, lentils, lupin, sweet clover, moth bean and medic. Hairy vetch can withstand severe winter-weather conditions. Legumes, that are biennial and perennial, include red clover, white clover, sweet clover, birds foot trefoil, kura clover, velvet bean, sainfoin, pigeon pea, faba beans, and alfalfa. Plants usually used as winter annuals can sometimes be grown as summer annuals in cold, short-season regions. Also, Austrian winter peas and field peas, as winter annuals that are easily damaged by severe winter and frost, can be grown as a winter annual in tropical and sub-tropical conditions. Moreover, cowpeas, chickpea, lablab bean, tepary bean, moth bean, and sun hemp can be grown in dry areas as rain-fed crops (Clark, 2007; Magdoff and van Es, 2010).

Grass and cereals used as cover crops include both summer and winter annuals and biennials. While cereal rye, annual ryegrass, annual fescue, wheat, spelt, and triticale are used as winter cover crops, the pearl millet, sorghum, sudangrass, sorghum-sudangrass hybrids, oats, and barley are used as late spring and summer cover crops. Cereal rye is very winter-hardy and easy to establish.

Brassicas, used as cover crops, include mustard, rapeseed, radish, kale, canola, turnip, broccoli, cabbage, cauliflower, carrot, kohlrabi, rutabaga, collard, tatsoi, komatsuna, rapini and Brussels sprouts. They are increasingly used as winter or rotational cover crops in the production of vegetable and specialty crops, such as potatoes and tree fruits. Canola grows well under the moist and cool conditions of late fall when other kinds of plants go dormant for winter.

Several other plant species used as cover crops include buckwheat, beet, chards, flax, spinach, sunflower, safflower, squash, chicory, phacelia, and edible amaranth. Among them, buckwheat is one of the fastest growing cover crops and can grow more than 60 cm tall in the month following planting. It competes well with weeds because it grows so fast and, therefore, is used to suppress weeds following an early spring vegetable crop (Clark, 2007; Magdoff and van Es, 2010).

2.3 Function and Benefits of Cover Crops

Cover crops are generally herbaceous plants that are simultaneously or alternatively grown with agronomic crops or during fallow periods to provide multiple benefits to agronomic, soil health, water quality and ecological and environmental services (Fig. 2). An overview of the benefits and services provided by cover crops (Hoorman et al., 2014; Anonymous, 2019) are described in Fig. 2.

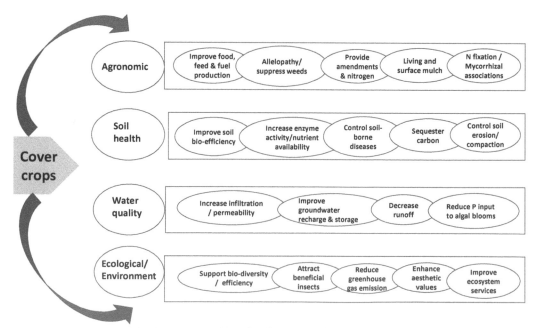

Fig. 2 Functions and benefits of cover crops in agro-ecosystems.

2.3.1 *Agronomic Benefits*

Legumes, as cover crops, fix atmospheric N_2 through symbiosis with Rhizobia bacteria and contribute a substantial amount of nitrogen to fulfil the requirement for agronomic crops, such as corn, sunflower, wheat (Decker et al., 1994), potato, tomato, etc. (Abdul-Baki and Teasdale, 1997). Total biological N contribution by medics and crimson clover biomass ranges from 50 to 150 kg/ha (Murdoch et al., 1985), whereas, berseem, hairy vetch, woolly pod vetch, and subterranean clover biomass may provide or exceed 200 kg/ha. Results conducted by the Ohio State University (Hoorman et al., 2009) showed that cowpea, Austrian winter pea and mung bean can contribute 101, 111, and 124 kg N/ha, respectively. Oilseed radish can recycle more than 80 kg N/ha annually.

Table 1 Average Biomass Production, Nitrogen and Carbon Contents and Biomass Nitrogen Contribution and/or Recycling by Legume and Non-legume Cover Crops (* indicates N recycling), Hoorman et al. (2009).

Cover Crop	Dry Biomass (Mg/ha)	Nitrogen (%)	Carbon (%)	C:N of Biomass	Biomass N (kg/ha)
Legumes					
Alfalfa	4.8	3.72	41.9	11.3	400
Cowpea	3.2	2.97	42.6	14.3	101
Crimson clover	1.8	1.97	41.7	21.2	36
Hairy vetch	2.1	2.33	44.3	19.0	53
Jumbo ladino clover	0.7	3.26	42.8	13.1	55
Medium red clover	0.9	2.77	43.7	15.8	53
Mammoth red clover	0.8	2.00	43.5	21.7	35
Mung bean	2.9	3.92	40.4	10.7	124
White clover	1.3	2.92	43.3	14.8	38
Austrian winter pea	3.1	3.60	44.1	12.3	111
Non-legumes					
Annual ryegrass	1.3	2.11	43.2	20.5	62*
Cereal rye	1.4	0.89	42.6	48.1	26*
Coriander	0.5	1.52	41.4	27.3	18*
Oilseed radish	3.9	2.11	41.1	19.5	82*
Sorghum-Sudan grass	12.2	0.68	43.3	63.3	85*
Spelt	2.1	0.99	42.7	43.3	45*

Other studies reported that cover crops played an important role in contributing and recycling nitrogen for vegetable and agronomic crop growth (Fig. 3; Table 2). Cover crops retrieve nitrogen and other nutrients left in the field after main crops have been harvested, keeping them in biomass and recycled (Abdul-Baki et al., 1998; Sultani, personal communications).

Tropical legume cover crops, such as cluster bean, rice bean, and sesbania contributed a substantial amount of nitrogen for rain-fed wheat production (Sultani, personal communications).

Table 2 Nutrients recycled by cover crops grown in a Date Orchard in Coachella Valley, California (Abdul-Baki et al., 1998).

Cover Crop	Biomass	N	P	K	Ca	Mg	Mn	Fe	Cu	B	Al	Zn
						Kg/ha						
Iron clay cowpea	5.0	123*	17	177	0.10	0.02	0.17	10.8	0.05	2.3	7.5	0.2
Lana vetch	4.7	160*	13	171	0.06	0.02	0.15	8.9	0.06	0.2	9.2	0.1
Seco barley	4.1	73	8	99	0.01	0.01	0.06	4.6	0.05	0.1	—	0.1

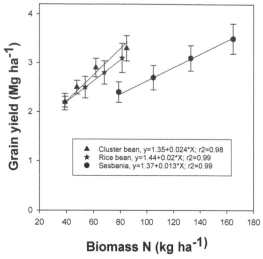

Fig. 3 Tropical legume cover crops' biological nitrogen contribution for rain-fed wheat yield (Sultani (2007)).

Among the tropical legume cover crops, sesbania provided the maximum amounts of biological nitrogen for wheat production (Fig. 3). Several studies reported that grass cover crops enhance mycorrhizal associations to sequester and recycle nutrients for crops growing in poor soils (Murrell et al., 2019). Furthermore, AMF can increase plant nutrient uptake and chemical defence production, both of which can improve plants' ability to resist insect herbivory (Murrell et al., 2019).

Cover crops' impact on agronomic crop yields can be difficult to quantify (Lu et al., 2000; Snapp et al., 2005; Blanco-Canquietal, 2012); however, on-going, long-term, no-till and cover crop studies reported an increase of crop yields due in large part to soil fertility, especially nitrogen contribution, weed suppression and soil quality improvements (Fig. 4). Over a period of 15 years, studies at the Ohio State University South Centres revealed the relative crop yield (integration of corn, soybean and wheat) increased significantly in response to cover crops' synergistic impact (Islam, 2010).

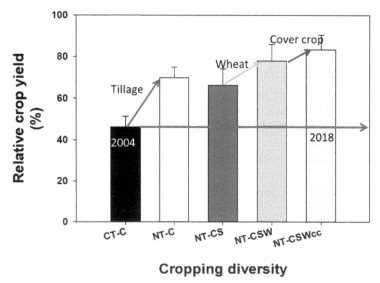

Fig. 4 Tillage and cover crop impact (2004 to 2018) on relative crop yield.

Cover crops, for grazing, are one of the most likely outputs in sustainable agriculture (Hoorman et al., 2009; Anonymous, 2019). A mixture of cover crops can be used for grazing if established immediately after harvesting wheat or established early in the fall, especially with fast-growing cereal or brassica cover-crop mixtures. Cover crops for grazing are not only economically viable and environmentally compatible, they improve long-term soil health, maintain ecological balance and farm income. Combined with reduced cost of feed, higher gain and increased crop yields, cover crops for grazing can pay off economically.

Cover crops provide mulching effect to physically suppressed weeds and subsequently release secondary metabolites (e.g., organic acids and glucosinolates) upon decomposition, to exert allelopathic effects on weeds and control pests, nematodes and other soil-borne diseases (Brown and Morra, 1996; Potter et al., 1998; Lazzeri and Manici, 2001; Wen et al., 2017; Sujan et al., 2019). Brassicas, such as radish, rapeseed, and canola are effective to control nematode and soil-borne pathogens (*Verticillium wilt*). Indian mustard is often used to control *Sclerotium rolfsii, F. oxysporum, R. solani, M. phaseolina*, nematodes and diamond-back moth (Fayzalla et al., 2009).

2.3.2 Soil Health Benefits

Cover crops impact soil health by improving soil biodiversity and efficiency, chemical balancing, sequestering carbon, maintaining C:N stoichiometry and soil physical stability with moisture conservation and soil-erosion control (Islam et al., 2020). A significant increase in the size of the soil microbial biomass and higher metabolic efficiency under cover crops was due to the added effects of rhizo deposition from fine roots production (Bradford et al., 2013; Cotrufo et al., 2013). Cover crop roots exude into the rhizosphere a wide range of diverse labile C and N compounds (sugars, amino acids and organic acids), which are preferentially utilised by microbes and help to promote biological diversity and efficiency as natural diversity closely related to natural growth and efficiency (Canarini et al., 2019).

Cover crops add diverse organic matter to soil that is usually returned to the soil after termination and subsequently, replenish SOM. Several studies have reported an increase in the TOC and TN contents under cover crops (Hessen et al., 2004; Plaza-Bonilla et al., 2016); while others have reported no change or a non-significant increase in TOC and TN contents (Jokela et al., 2009; Frasier et al., 2016). A significantly higher active organic C (a measure of soil quality) and active N content suggests an efficient biochemical response to the cover crop-induced changes that contribute to C:N stoichiometry in SOM.

Bulk density, as a measure of soil compaction, was significantly reduced by cover crop managements (Anonymous, 2019). Brassicas, such as radish, provide large quantities of residue on the surface soil and improve soil porosity and water infiltration properties. Cover crops stabilise soil structures by the physical and biochemical actions of fine roots, particularly when grass species, such as cereal rye, annul ryegrass, wheat, spelt and triticale are grown in the fall. Islam et al. (2020) reported that soil aggregate stability increased by 15 per cent under cover cropping compared to control.

Fig. 5 Effect of radish on soil compaction (photo by Randall Reeder).

A good blend of cover crops increases and conserves soil moisture by allowing more water (rainfall, snowmelt, or irrigation) to infiltrate the soil than the bare fields. Cover-crop roots create soil macropores that hold moisture, keep the soil cooler, decrease irrigation frequency and reduce susceptibility of crops to abiotic stress, such as drought in rain-fed and dry areas. Moreover, the organic mulching effects of cover crops upon termination help to reduce evaporation demand. Karlen et al. (1994) reported that surface residue accumulation plays an important role in improving soil and biological, chemical and physical properties in no-tillage systems and consequently, reduce erosion and improve soil quality (Hargrove, 1991; Karlen et al., 1994).

2.3.3 Water Quality

Generally, compacted soil affects water infiltration with an associated increase in surface runoff and erosion, which are serious problems to long-term soil productivity (Pimentel et al., 1995). Soil erosion from bare fields carries sediment-laden soluble reactive P (SRP) and other nutrients into surface waters, and thus are responsible for algal blooms and degrading fisheries, sports and aquatic habitats (Hoorman and Sundermeier, 2017). Moreover, soil compaction affects plant root growth, water and nutrient availability to plants, and consequently, decreases crop yields.

Cover crops provide support to hold the soil and residue in the field (Fig. 6). They act as a living or dead biological mulch to control soil erosion by increasing water infiltration and permeability, improving groundwater recharge and storage and reducing offsite movement of SRP and other nutrients to freshwater systems to minimise algal blooms. A field study, conducted to evaluate the effect of cover crops on soil erosion and residue movement, showed that living cover crops (cereal rye) act as barriers to slow down soil and residue from conventionally-tilled bare fields. The above-ground biomass of cover crops act as cushions to reduce the impact of rainfall and slows down water movement, while its roots increase soil aggregate stability and form macropores through which water infiltrates, thus reducing surface runoff. Cover crops also reduce nitrate leaching between 40–70 per cent compared with bare soil (Hoorman and Sundermeier, 2017) and reduce the chances of groundwater and drinking water contamination.

Fig. 6 Cereal rye, as a winter cover crop, acts as a barrier to slows down soil and residue loss from the adjacent conventionally-tilled field.

2.3.4 Ecological and Environmental Benefits

Cover crops provide several ecological and environmental benefits, such as increased biodiversity and efficiency by attracting beneficial insects, reduction in greenhouse gas (GHG) emissions, enhanced aesthetic values and improved ecosystem services.

One of the major challenges of sustainable agricultural-management practices is to mitigate agriculture's contribution to GHG emissions, especially N_2O emissions, into the atmosphere (Hu et al., 2015). Globally, agricultural soil ecosystems constitute the largest source of N_2O emissions (estimated at 6.8 Tg N_2O-N year^{-1}), comprising approximately 65 per cent of the total N_2O emitted into the atmosphere, with 4.2 Tg N_2O-N year^{-1} derived from N fertilisation and indirect emissions, 2.1 Tg N_2O-N year^{-1} arising from manure management and 0.5 Tg N_2O-N year^{-1} introduced through biomass burning (IPCC, 2007).

Several species of ammonia oxidiser, nitrifier denitrifiers and heterotrophic denitrifiers bacteria convert excess soil N into the N_2O, which traps 300 times more heat in the atmosphere than CO_2 (Zhu et al., 2013; Shcherbak et al., 2014; Hu et al., 2015). Cover crops, by up-taking N tied up in biomass and SOM, control the pathways of N_2O emissions in the soil. In some cases, cover crops can even increase carbon storage, thereby regulating C:N stoichiometry in SOM, further reducing N_2O emissions in the atmosphere (Mitchell et al., 2013).

The surface deposition and accumulation of winter-killed or herbicide-terminated cover-crop biomass often encourage the growth of diverse predatory insects that help control pests and diseases, thus minimising both herbicide and insecticide applications. Several cover crops can also kill harmful and parasitic disease-causing soil organisms, thus reducing the application of fungicides or nematicides (Snapp et al., 2007; Wen et al., 2017).

Herbicide-resistant weed pressures are increasing day by day. Weed pressure is expected to increase competition with agronomic crops as the global climate change effects continue to accelerate. While cover crops may not eliminate the need for herbicide application, they can suppress the weeds with competition for water, nutrients, space and light and thus, reduce herbicide applications. In some cases, green plantings have been shown to reduce weed density by 90 per cent in corn crops (Teasdale, 1998; Anonymous, 2019).

3. Management of Cover Crops

Cover crops, as a biological primer, have been slowly adopted by producers to build SOM and recycle essential nutrients, supplement N fertilisation, suppress weeds, control soil erosion, develop and improve soil health and ultimately, increase economic crop yield. In addition, cover crops facilitate the transitioning to conservation tillage and improve the functionality of no-till. While transitional no-till often decreases crop yield and profit, the use of cover crops provides a buffer to the expected yield (Islam, 2010; Islam and Didenko, 2018). Cover-cropping practice extends the usefulness of the residues (slower breakdown) and utilises them as mulch to replace plastic mulch (in vegetable or orchard production), conserve and maintain soil moisture and recycle efficient use of organically-bound nutrients.

Cover-crop management is based on two principles and practices: (1) live green covers and (2) dead surface mulch, which depend on winter or summer planting of cover crops. Live green-cover crops grown simultaneously with agronomic crops remain wholly or partly alive during the growing season and provide a living mulch layer throughout the season (Hartwig and Ammon, 2002; Scholberg et al., 2010). Examples include growing cereal rye or radish under matured corn or soybeans at the yellowing-leaf stage (Fig. 7).

In contrast, dead surface mulches are produced from chemically or mechanically terminated cover crops before the row crop is planted (planting of winter cover crops, such as cereal rye, hairy vetch, or crimson clover before planting corn vs. cereal rye and triticale before planting soybeans in

Fig. 7 Living green cover of cereal rye and radish in growing corn and soybean, respectively (*Photo*: Rafiq Islam, 2016).

Fig. 8 Cereal rye mulching effect on suppressing weeds and conserving soil moisture for soybeans (*left*). Winter-killed cowpea residues (*right*) to recycle organically bound N for growing corn without any chemical N fertilisations (*Photo*: Rafiq Islam, 2011).

the spring) or planting of summer cover crop or blends (winter, cowpea, soybeans after harvesting, winter wheat). By rolling, plowing, or disking, the cover-crop residues are often incorporated into the soil one or two weeks prior to planting of agronomic crops (Fig. 8). However, cowpea planted as a summer cover crop after harvesting wheat in July is generally killed by frost (Fig. 8).

4. Typical Crop Rotation with Cover Crops

Adoption of cover-crop-based cropping systems is generally influenced by the perception of stakeholders, the purpose of using cover crops, local conditions, seed availability, cropping systems and intensity and increased awareness of ecosystem services (Anderson et al., 2001). Proper selection and integration of the use of cover crops in diverse cropping systems is expected to be the cornerstone of sustainable agro-ecosystems (Baligar and Fageria, 2007). The development of functional cropping diversity with cover crops requires a more integrated and system-based approach, rather than reinstating traditional production practices. Examples of a few typical cropping diversity systems with cover crops (Hoorman et al., 2009) are as follows:

1. *Corn-rye-soybeans-rye*: This rotation would need to be planted after early maturing soy beans, harvested in early or mid-September. Rye supports weed control, nutrient recycling, erosion control and soil-carbon sequestration.

2. *Corn-rye-soybeans-wheat-cowpea (or cowpea, winter pea, radish, oats, rye, crotalaria, hairy vetch blend)*: In this system, while rye supports weed control, nutrient recycling, erosion control, and soil carbon sequestration, cowpea and/or blend provide a substantial amount of biological N contribution for succeeding corn. Soybeans provide residual biomass N for wheat production. Wheat, in the crop rotation, acts as both a winter and cover crop.

3. *Wheat-brassica (radish/turnips)-corn-rye-soybeans-wheat-cowpea-corn*: Oilseed radish or turnip is planted to reduce soil compaction, recycle nutrients, and suppress oil-borne diseases (by glucosinolates). While rye supports weed and erosion control, nutrient recycling and soil carbon sequestration, cowpea provides substantial amounts of biological N for succeeding corn. Soybeans provide residual biomass N for wheat.

4. *Corn silage-winter pea-corn silage (or legume blends)*: Winter pea or other legume blends are used for biological N contribution to support corn-silage production.

5. *Corn silage-rye or annual ryegrass-corn silage*: Rye as a cover crop used for forage, soil cover to control erosion, or manure amendments.

6. *Soybeans-wheat-winter pea (or legume blends)-sunflower-rye*: While rye supports weed control, nutrient recycling, erosion control and soil carbon sequestration, winter pea or legume blends provide a substantial amount of biological N for sunflowers. Soybeans provide residual biomass N for wheat production. Wheat in crop rotation acts as both a winter and cover crop.

7. *Wheat-sorghum sudangrass-soybeans-Austrian winter pea-corn-cereal rye-soybeans-wheat-Brassica (oilseed or tillage radish or turnip)* back to corn or soybeans.

8. *Wheat-sweet clover (or other legumes)-corn-rye-soybeans-wheat-Brassica (oilseed or tillage radish or turnip)*.

5. Limitations of Cover Crop Adaptation

Despite numerous direct and indirect benefits of cover crops associated with sustainable agriculture, including potential improvements in above- and below-ground biodiversity, agronomic crop productivity, SOM accumulation and nutrient recycling, soil physical stability and soil health, the best is yet to be achieved in routine adoption and management of cover crops in agronomic cropping systems. Several limitations and constraints on cover crop adoption and management include:

Extra costs and financial limitations: To get the most benefits out of cover crops, it is good to start with a cover crop or blend of cover crops to improve the functionality of conservation tillage (no-tillage or reduced tillage) systems. One of the roadblocks for farmers is the extra cost associated with cover-crop seed, planting and termination compared to the time required before economic yields, improved soil health and other benefits (Lichtenberg et al., 1994; Anonymous, 2019). Islam (2010) reported that an increase in crop yield often lags soil-health improvement in response to cover crops impact.

Land lease and ownership: A large percentage of agricultural farmland is leased or under contract, usually on a few years' term. Farmers often lease their land, rendering them less motivated or inclined to make long-term investments on soil quality improvement and ecosystem services. Management practices that increase the improved land's value also increase the associated cost incurred to lease the land. Growers who lease lands use best judgement to get the most out of the farming with the bare minimum investment to land stewardship. Globally, where the agricultural lands are unable to support economic food production, the land is kept under subsistence food production year-round. In such situations, investments on long-term improvements in soil quality are unlikely to occur because cover crops have no space to incorporate into the cropping systems.

Hosts to pests and diseases: Without appropriate selection of cover crops, insects and pathogens may use the cover crops as hosts during the off season, multiply and spread to the main crop. Phytopathogenic nematodes and many pests and insects have diverse hosts. During the off season, these pathogens may survive on cover crops that serve as favourable hosts and subsequently affect the main crop. Under tropical and subtropical conditions with high temperature and humidity conducive to their growth, the pathogen population continues to grow and, by the time the main crop is planted, they can inflict severe economic damages.

Soil moisture limitations: As usual, cover crops require water to grow. Therefore, cover crops may consume water that is much needed by the main crops, especially in dry areas where rainfall is seasonal or limited. Even in humid areas of the world, agronomic crops may suffer from moisture stress during the dry season or under drought if cover crops consume soil moisture in spring. Moreover, competition for moisture, space, light and nutrients between cover crops used as living mulches and the main crop can be intense unless the living mulch is killed or controlled by herbicides; otherwise, main crops will be affected (Anonymous, 2019).

Region-specific considerations: When used without enough experience or knowledge, cover crops can have undesirable effects. In dry regions or during dry years, for example, cover crops (which typically remove water from the soil in late spring) may leave too little moisture for the cash crops that follow. In northern regions, cover crops may not have time to establish themselves after the cash crop has been harvested in the fall.

Plant biology and length of growing season: While cover crops require enough time to produce a substantial amount of biomass, there can be delay in planting of the agronomic crops and consequently, affect the crop yield. Cover crops are best adapted to areas where the suitability of crop-growing season is long enough to support the main crop and establishment of cover crops during the remainder of the year. Planting of main crops in a timely manner may be affected or limited if cover crop growth extends into the main crop growing season, as can happen during a wet season.

Roadblocks and availability of information: Producers are often looking for evidence-based knowledge to operate and manage their farming systems; however, appropriate information on cover crops biology, mixing of cover crops, planting rates of cover crops, N contribution of cover crops, pests and diseases associated with cover crops and economics of cover crops are still inadequate and not routinely available. Data on how cover crops or blends fit into specific cropping systems in each region with economic risks and benefits are limited. Substantial economic and technical roadblocks: are embedded in state and federal government policies and discourage farmers from adopting cover crops (SARE/CTIC, 2016).

6. Summary and Conclusion

Cover crop is one of the important components of sustainable agriculture. By providing and supporting direct and indirect benefits, cover crops can help to achieve the objectives of sustainable agriculture—economic viability, environmentally compatibility and social acceptability. Cover crops, or a blend of cover crops, are useful for providing food and energy to microbes to perform functions associated with soil biodiversity and efficiency, soil chemical balancing and soil physical stability to develop and maintain soil quality to support economic crop production with enhanced ecosystem services. As a living or dead surface mulch, cover crops protect soil against raindrop impact and shearing the force of surface runoff, improve oil aggregation and structural stability, increase water infiltration and minimise soil erosion, provide home-grown N and recycle nutrients, decrease soil compaction and reduce insects and pathogens through increasing predator biodiversity. Fast-growing cover crops can suppress weeds and reduce herbicide uses, thereby reducing risks of widespread environmental contamination. Annual input from biologically fixed N by legume cover crops can greatly reduce dependence on expensive chemical fertilisers. Cover crops have been shown to sequester C and accumulate SOM, thereby, improving soil quality. Developing and improving soil quality by cover crops subsequently increase crop yields with enhanced agro-ecosystem services of sustainable agriculture.

References

Abdul-Baki, A.A. and Teasdale, J.R. (1997). Snap bean production in conventional tillage and in no-till hairy vetch mulch. HortSci., 32: 1191–1193.

Abdul-Baki, A., Aslan, E. Beardsley, Cobb, S. and Shannon, M. (1998). Soil, water and nutritional management of date orchards in Coachella Valley and Bard, California Date Commission, Thermal, California, p. 48.

Anderson, S., Gündel, S., Pound, B. and Triomphe, B. (2001). Cover Crops in Smallholder Agriculture. London: ITDG Publishing,10.3362/9781780442921.

Anonymous. (1989). Sustainable agricultural production: Implications for international agricultural research, Technical Advisory Committee, Consultative Group on International Agriculture Research, Washington, D.C., USA.

Anonymous. (2019). Cover Crops – Good for Crop Yields, Oil Health and Bottom Lines, NSAC'S BLOG (National Sustainable Agriculture Coalition). https://sustainableagriculture.net/.

Baligar, V.C. and Fageria, N.K. (2007). Agronomy and physiology of tropical cover crops. J. Plant Nutrition, 30: 1287–1339.

Blanco-Canqui, H., Claassen, M.M. and Presley, D.R. (2012). Summer cover crops fix nitrogen, increase crop yield, and improve soil-crop relationships. Agronomy J., 104: 137–147.

Bradford, M.A., Strickland, M.S., DeVore, J.L. and Maerz, J.C. (2012). Root carbon flow from an invasive plant to belowground food webs. Plant and Soil, 359: 233–244.

Brown, P.D. and Morra, M.J. (1996). Hydrolysis products of glucosinolates in *Brassica napus* tissues as inhibitors of seed germination. Plant Soil, 181: 307–316.

Calegari, A. (2002). The spread and benefits of no-till agriculture in Paraná State, Brazil. pp. 187–202. *In*: Norman Uphoff (ed.). Agro-ecological Innovations: Increasing Food Production with Participatory Development. London, Earthscan.

Canarini, A., Kaiser, C., Merchant, A., Richter, A. and Wanek, W. (2019). Root exudation of primary metabolites: Mechanisms and their roles in plant responses to environmental stimuli. Review article 157, Front. Plant Sci., 21. https://doi.org/10.3389/fpls.2019.00157.

Cherr, C.M., Scholberg, J.M.S. and McSorley, R.M. (2006). Green manure approaches to crop production: A synthesis. Agron. J., 98: 302–31.

Clark, A. (2007). Managing Cover Crops Profitably, third ed., Handbook Series Book 9, Sustainable Agriculture Network.

Cotrufo, M.F., Wallenstein, M.D., Boot, C.M., Denef, K. and Paul, E. (2013). The microbial efficiency-matrix stabilisation (MEMS) framework integrates plant litter decomposition with soil organic matter stabilisation: Do labile plant inputs form stable soil organic matter? Global Change Biol., 19: 988–995.

Decker, A.M., Clark, A.J., Meisinger, J.J., Mulford, F.R. and McIntosh, M.S. (1994). Legume cover crop contributions to no-tillage corn production. Agron. J., 86: 126–135.

Delgado, J.A., Groffman, P.M., Nearing, M.A., Goddard, T., Reicosky, D., Lal, R., Kitchen, N.R., Rice, C.W., Towery, D. and Salon, P. (2011). Conservation practices to mitigate and adapt to climate change. J. Soil Water Conserv., 66: 118–129.

Drinkwater, L.E. and Snapp, S.S. (2007). Nutrients in agro-ecosystems: Rethinking the management paradigm. Advances in Agron., 92: 163–186.

FAO. (2010). Climate-smart agriculture—Policies, practices and financing for food security. Adaptation and Mitigation. Rome: FAO.

Fayzalla, E.A., El-Barougy, E. and El-Rayes, M.M. (2009). Control of soil-borne pathogenic fungi of soybean by biofumigation with mustard seed meal. J. Applied Sci., 9: 2272–2279.

Frazier, I., Noellemeyer, E., Figuerola, E., Erijman, L., Permingeat, H. and Quiroga, A. (2016). High quality residues from cover crops favor changes in microbial community and enhance C and N sequestration. Global Ecol. Cons., 6: 242–256.

Gladstones, J.S. (1976). The Mediterranean white lupin. J. Agric. West. Aust., 17: 70–74.

Hamzaev, A.X., Astanakulov, T.E., Ganiev, I.M., Ibragimov, G.A., Oripov, M.A. and Islam, K.R. (2007). Cover crops impact on irrigated soil quality and potato production in Uzbekistan. *In*: Lal, R., Sulaimenov, M., Stewart, B.A., Hansen, D.O. and Doraiswamy, P. (eds.). Climate Change and Terrestrial Carbon Sequestration in Central Asia. Taylor and Francis Group, London, UK.

Hargrove, W.L. (1991). Cover crops for clean water. Soil Water Conserv. Soc., Ankeny, IA, 198.

Hartwig, N.L. and Ammon, H.U. 2002. Cover crops and living mulches. Weed Sci., 50: 688–699.

Hessen, D.O., Göran, I., Agren, T.R., Anderson, J.J., Elser, P.C. de Ruiter. (2004). Carbon sequestration in ecosystems: The role of stoichiometry. 85: 1179–1192.

Hoorman, J.J., Islam, R. and Sundermeier, A. (2009). Sustainable Crop Rotations with Cover Crops, Ohio State University Extension Factsheet. https://ohioline.osu.edu/factsheet/SAG-9.

Hoorman, J.J., Sundermeier, A.P. and Islam, R. (2014). Growing and Managing Forage Cover Crops, Ohio State University Extension Factsheet SAG-XX-15. http://mccc.msu.edu/wp-content/uploads/2016/10/OH_2015_Using-Cover-Crops-as-Forage-051815.pdf.

Hoorman, J.J. and Sundermeier, A.P. (2017). Using Cover Crops to Improve Soil and Water Quality. https://ohioline.osu.edu/factsheet/anr-57#.

Hu, H.W., Chen, D. and Ji-Zheng He. (2015). Microbial regulation of terrestrial nitrous oxide formation: Understanding the biological pathways for prediction of emission rates. FEMS Microbiology Reviews, 39(21): 729–749.

IPCC. (2007). Climate Change 2007 – Synthesis Report, Contributions of working groups I, II, and III to the 4th assessment report of the intergovernmental panel on climate change, Geneva.

Islam, K.R. (2010). Cover Crops Impact on Soil Quality and Crop Yield, ASA-CSA-SSA International Meetings, October 31–November 4, Long Beach, Ca.

Islam, K.R. and Didenko, N.O. (2019). Impact of sustainable agricultural management practices on soil quality and crop productivity. Final Report (FSA3-18-63886-0), CRDF-Global, Washington D.C., USA.

Islam, K.R., Roth, G., Rahman, M.A., Didenko, N.O. and Reeder, R.C. (2020). Cover crop complements flue gas desulfurised gypsum to improve no-till soil quality? Communications in Soil Science and Plant Analysis (in press).

Jokela, W.E., Grabber, J.H., Karlen, D.L., Balser, T.C. and Palmquist, D.E. (2009). Cover crop and liquid manure effects on soil quality indicators in a corn silage system. Agron. J., 101: 727–737.

Karlen, D.L., Wollenhaupt, N.C., Erbach, D.C., Berry, E.C., Swan, J.B., Eash, N.S. and Jordahl, J.L. (1994). Crop residue effects on soil quality following 10-years of no-till corn. Soil Tillage Res., 31: 149–167.

Landers, J.N. (2001). How and why the Brazilian zero tillage explosion occurred? pp. 29–39. *In*: Stott, D.E., Mohtar, R.H. and Steinhardt, G.C. (eds.). Sustaining the Global Farm, Selected Papers from the 10th International Soil Conservation Organisation, Purdue University, West Lafayette.

Lazzeri, L. and Manici, L.M. (2001). Allelopathic effect of glucosinolate-containing plant green manure on Pythium sp. and total fungal population in soil. Hort. Science, 36: 1283–1289.

Lichtenberg, E., Hanson, J.C., Decker, A.M. and Clark, A.J. (1994). Profitability of legume cover crops in the Mid-Atlantic region. J. Soil and Water Conservation, 49: 562–565.

Long, T., Blok, V. and Coninx, I. (2016). Barriers to the adoption and diffusion of technological innovations for climate-smart agriculture in Europe: Evidence from the Netherlands, France, Switzerland, and Italy. Journal of Cleaner Production, 112: 9–21. DOI: 10.1016/j.jclepro.2015.06.044.

Lu, Y.C., Watkins, K.B., Teasdale, J.R. and Abdul-Baki, A.A. 2000. Cover crops in sustainable food production. Food Reviews International, 16: 121–157.

Magdoff, F. and van Es, H. (2010). Building soils for better crops. Sustainable Soil Management, third ed., Handbook Series Book 10, Sustainable Agriculture Research and Education (SARE) programme/National Institute of Food and Agriculture, U.S. Department of Agriculture.

Mitchell, D.C., Castellano, M.J., Sawyer, J.E. and Pantoja, J. (2013). Cover crop effects on nitrous oxide emissions: Role of mineralisable carbon. Soil Sci. Soc. Am. J., 77: 1765–1773.

Murdoch, W.W., Chesson, J. and Chesson, P.L. (1985). Biological control in theory and practice. Amer. Nat., 125: 344–266.

Murrell, E.G., Lemmon, M.E., Luthe, D.S. and Jason, P.K. (2019). Cover crop species affect mycorrhizae-mediated nutrient uptake and pest resistance in maize. Renew Agric. Food Syst. https://doi.org/10.1017/s1742170519000061.

Ngouajio, M., McGiffen, M.E. and Hutchinson, C.M. (2003). Effect of cover crop and management system on weed populations in lettuce. Crop Prot., 22: 57–64.

Phatak, S.C., Dozier, J.R., Bateman, A.G., Brunson, K.E. and Martini, N.L. (2002). Cover crops and conservation tillage in sustainable vegetable production. *In*: Edzard van Santen (ed.). Making Conservation Tillage Conventional: Building a Future on 25 Years of Research. Proc. 25th Annual Southern Conservation Tillage Conference for Sustainable Agriculture, Alabama Agric. Experiment Station, Auburn University, AL 36849, USA.

Pearson, L.C. (1967). Principles of Agronomy, Reinhold Publishing Corporation, New York, N.Y.

Pieters, A.J. (1917). Green manuring: A review of the American Experiment Station literature – 1. Agron. J., 9: 62–82.

Pieters, A.J. (1927). Green Manuring Principles and Practices. John Wiley & Sons, Inc., New York, N.Y.

Pimentel, D., Harvey, C., Resosudarmo, P., Sinclair, K., Kurz, D., McNair, M., Crist, S., Shpritz, L., Fitton, L., Saffouri, R. and Blair, R. (1995). Environmental and economic costs of soil erosion and conservation benefits. Science, 267: 1117–1123.

Plaza-Bonilla, D., Nolot, J.M., Passot, S., Raffaillac, D. and Justes, E. (2016). Grain legume-based rotations managed under conventional tillage need cover crops to mitigate soil organic matter losses. Soil Tillage Res., 156: 33–43.

Potter, M.J., Davies, K. and Rathjen, A.J. (1998). Suppressive impact of glucosinolates in brassica vegetative tissues on root lesion nematode *Pratylenchus neglectus*. J. Chem. Ecol., 24: 67–80.

Rosenthal, J.L. (1990). The fruits of revolution: Property rights, litigation and French agriculture, 1700–1860. The J. Econ. History, 50: 438–44.

SARE/CTIC. (2016). Cover Crop Survey Analysis.

Scholberg, J.M.S., Dogliotti, S., Leoni, C., Cherr, C.M., Walter, L.Z. and Rossing, A.H. (2010). Cover crops for sustainable agro-ecosystems in the Americas. pp. 23–58. *In*: Eric Lichtfouse (ed.). Genetic Engineering, Biofertilisation, Soil Quality and Organic Farming. ISBN 978-90-481-8740-9 e-ISBN 978-90-481-8741-6; DOI 10.1007/978-90-481-8741-6. Springer Dordrecht, Heidelberg, London, New York.

Shcherbak, L., Millar, N. and Robertson, G.P. (2014). Global meta-analysis of the nonlinear response of soil nitrous oxide (N_2O) emissions to fertiliser nitrogen. PNAS, 111: 9199–204.

Shennan, C. (2008). Biotic interactions, ecological knowledge and agriculture. Phil. Trans. R. Soc., B, 363: 717–739.

Smil, V. (2001). Enriching the Earth: Fritz Haber, Carl Bosch and the Transformation of World Agriculture, Cambridge, MA.

Snapp, S.S., Swinton, S.M., Labarta, R., Mutch, D., Black, J.R., Leep, R., Nyiraneza, J. and O'Neil, K. (2005). Evaluating cover crops for benefits, costs and performance within cropping system niches. Agronomy J., 97: 322–332.

Snapp, S., Date, K.U., Kirk, W., O'Neil, K., Kremen, A. and Bird, G. (2007). Root, shoot tissues of *Brassica juncea* and Cereal Secale promote potato health. Plant Soil, 294: 55–72.

Sujan, D., Oliver, J.B., O'Neal, P. and Addesso, K.M. (2019). Management of flat-headed apple tree borer (*Chrysobothris femorata* Olivier) in woody ornamental nursery production with a winter cover crop. Pest Management Science, 75: 1971–1978.

Sullivan, P. (2003). Overview of cover crops and green manures. Appropriate Technology Transfer for Rural Areas. Available at Web site http://www.attra.ncat.org.

Sultani, M.I., Gill, M.A., Anwar, M.M. and Athar, M. (2007). Evaluation of soil physical properties as influenced by various green manuring legumes and phosphorus fertilization under rain fed conditions. Int. J. Environ. Sci. Tech., 4: 109–118.

Teasdale, J.R. (1998). Cover crops, smother plants and weed management. pp. 247–270. *In*: Hatfield, J.L., Buhler, D.D. and Stewart, B.A. (eds.). Integrated Weed and Soil Management. Ann Arbor Press, Chelsea, MI.

Tonitto, C., David, M.B. and Drinkwater, L.E. 2006. Replacing bare fallows with cover crops in fertiliser-intensive cropping systems: A meta-analysis of crop yield and N dynamics. Agriculture, Ecosystems and Environment, 112: 58–72.

van der Ploeg, J.D. (2008). The New Peasantries: Struggles for Autonomy and Sustainability in an Era of Empire and Globalisation, London, Earth Scan.

Wen, L., Lee-Marzano, S., Ortiz-Ribbing, L.M., Gruver, J., Hartman, G.L. and Eastburn, D.M. (2017). Nematode management suppression of soil-borne diseases of soybean with cover crops. Plant Disease, 101: 1918–1928.

Zhu, X., Burger, M., Doane, T.A. and Horwath, W.R. (2013). Ammonia oxidation pathways and nitrifier denitrification are significant sources of N_2O and NO under low oxygen availability. P Natl. Acad. Sci., USA 110: 6328–6333.

Zwane, E.M. (2019). Capacity development for scaling up climate-smart agriculture innovations. DOI: http://dx.doi.org/10.5772/intechopen.84405.

2

Benefits of Cover Crops on Agronomic Crop Yield

Kateryna Chorna

1. Introduction

Cover cropping is one of the integral components of sustainable agriculture. Cover crops are plants that are grown between agronomic crops to provide labile organic matter, recycle nutrients, cover and mulch of soil rather than being harvested from the field for grains or forage (Giller, 2001). They are effective in controlling soil erosion; improving soil quality; conserving and recharging soil moisture; suppressing weeds, pests, and diseases; and enhancing biodiversity and wildlife in an agro-ecosystem—an ecologically-balanced system managed and shaped by humans. As cover crops protect the soil and improve soil fertility to support crop production, they are also called green manure crops (Fageria et al., 2005).

Cover crops are various grains, legumes and other herbaceous crops that alternate with agronomic crops to improve soil conditions and provide significant benefits to the succeeding main crops. They are grown both individually and in blends in agro-ecosystems. The use of a blend of several types of cover crops is a very promising and holistic agricultural method because it is expected to supply a variety of biochemical compounds to the soil and provide unique characteristics necessary for a stable high yield of crops. Cover crops are planted in early autumn and provide a 'living mulch' throughout the growing season (Scholberg et al., 2010; Hartwig and Ammon, 2002). They provide protection and make improvements in the quality of water resources during the winter months. Soil-plant-water ecosystems are strongly affected by environmental factors and it is the cover crops that help minimise problems, such as erosion and leaching of nutrients during the growing season. Cover crops provide a favourable soil microclimate and are involved in reducing soil degradation (Anderson et al., 2001; Sarrantonio and Gallandt, 2003). The advantages of using cover crops include fast growth, high biomass and convenience when applied into the soil. In a very short period, they can cover the surface of the soil and prevent weed growth by competing for space, water, light and nutrients.

The concept of cover cropping is not a new agricultural technology. The use of cover crops as green manure to close nutrient cycles dates backs to the 19th century. The practice has been justified by increasing global populations and the search for affordable and economically-beneficial fertilisers necessary to support food security. In the 1870s, *Mucuna pruriens* became widely used as a cover crop culture, by producers due to its accessibility, with the aim of improving and/or maintaining soil fertility. Mucuna was being used first in the American territories of California and Guatemala (Buckles et al., 1998).

Institute of Water Problems and Land Reclamation, National Academy of Agricultural Sciences of Ukraine, Kyiv, Ukraine.

In the United States, legumes have been used to increase agronomic crop yields (Russell, 1913). As shown in the teachings by McNeill and Winiwarter (2004), Asian countries used lentils and peas as cover crops, thereby ensuring stable cereal production. North American agriculture, at the beginning of the last century, was closely linked to the use of annual winter crops. However, due to the greater availability, economics and convenience of synthetic and chemical fertilisers, cover crop use was ignored and abandoned by the middle of the 20th century (Tonitto et al., 2006). In Uruguay, vetch and oats were grown as cover crops during the same period. However, the vast majority of agricultural producers continued to use the chemical fertilisers; they were only interested in mass production of monocultures and profitable crops using industrial farming techniques. The Dust Bowl, a period of intense duststorms in the 1930s of United States began to change agricultural policies and priorities for sustainable management of soil-plant-water ecosystems. The severe droughts and wind erosion caused widespread damage to agriculture and the environment as a result of conventional plowing and other management practices. Due to the extensive use of agricultural machinery, native grasses, that were in the upper layers of the soil and retained both soil and moisture, were replaced with annual crops (Ganzel, 2003). Brazil has also been affected by erosion and soil degradation caused by monocropping, excessive use of chemically reactive fertilisers and widespread use of deep plowing. Therefore, an increasing interest appeared in maintaining soil productivity to increase crop production and limit negative impacts on the environment with the integration of cover crops in sustainable agriculture worldwide (Drinkwater and Snapp, 2007; Calegari, 2003; Sullivan, 2003).

2. Role of Cover Crops on Ecosystems

Ecosystem services are a combination of benefits that agriculture acquires through certain conditions. These ecosystem services ensure sustainable development of the agricultural sector, as well as create optimal conditions for protecting the environment. Ecosystem services include regulating, provisioning and supporting functions (Blanco-Canqui et al., 2015; Daily et al., 1997).

Sustainable management of soil can provide many ecosystem services related to the regulation of water and air quality, climatic conditions, food supply, etc. Cover cropping, as one integral component of sustainable agriculture, plays a significant role in ecosystem services as it affects the soil properties and processes. The usage of cover crops allows for controlling water and wind erosion, increasing soil organic matter (SOM) content, recycling nutrients, inhibiting weed and pest growth, increasing crop yields and even affecting greenhouse gas emissions (Bastian et al., 2013; Schipanski et al., 2014; Blanco-Canqui et al., 2015).

Many crops can be used as cover crops. Legumes and grass crops, as well as other non-legumes, are mainly used for this purpose. However, in recent years, there has been more interest in growing mustard, phacelia, rapeseed, buckwheat and oilseed radish. The choice of crops to be used as cover crops depends on when the species can be planted, the cost and the purpose of use. Very often legumes are used as cover crops, justified by the fact that they can capture nitrogen (N) from the atmosphere via biological N-fixation and add it to the soil. Legumes also tend to attract beneficial insects. It is a proven tool in the fight against soil erosion and a means of increasing the SOM content in the soil profile. There are summer annual legumes, winter annual legumes and biennial and perennial legumes.

In pursuance of sustainable agriculture, a cover crop chart was developed by the scientists of the USDA-Agricultural Research Service to assist producers in making decisions regarding the use of cover crops (Fig. 1). To develop this chart, a periodic table of chemical elements was used as the basis. This cover crop chart accumulates knowledge about 66 types of crops that can be used, either separately or as part of a mixture of crops. Using this chart, information can be obtained regarding the plant growth period and quality, the water demand of crops, plant architecture, nutrient cycling and pollination characteristics, and required planting depths (Fig. 1). Using this chart (Fig. 1), it is possible to compare the characteristics and potential of different cover crops, evaluate their ability to reduce erosion, accumulation of SOM and the redistribution and recycling of nutrients in the soil.

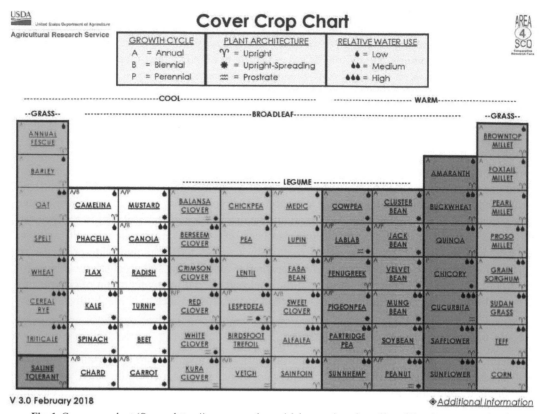

Fig. 1 Cover crop chart (*Source*: https://www.ars.usda.gov/plains-area/mandan-nd/ngprl/docs/cover-crop-chart/).

Also, one can determine the timing, norms and depth of planting of cover crops; their relationship with soil salinity and density and/or compaction; and their effect on weed suppression and microbial diversity and more.

The use of cover crops in agriculture has many diverse advantages. First, they are the number one tool in the fight against soil erosion. Also, with the regular use of cover crops, there is a steady improvement in the soil's physical and biological properties and the provision of the necessary nutrients to the culture that will be grown henceforth. Cover crops provide a covered soil surface and over time, improve both the surface and groundwater quality. Using these herbaceous plants as cover crops can significantly suppress the establishment and growth of weeds.

2.1 Cover Crop Biomass Production and Nutrient Recycling for Agronomic Crops

Of all the essential nutrients that plants require for normal growth, the main focus is always on three primary elements: nitrogen (N), potassium (K), and phosphorus (P) contents in soil (Magdoff and van Es, 2009). Lack of these nutrients in the soil is most obvious and reflected on the crops. Other than these nutrients, there is a limitation of calcium, magnesium, sulfur, iron, zinc, copper, magnesium, molybdenum, boron and manganese. Nitrogen, P and K, as chemical fertilisers, play key roles in soil fertility and crop productivity; however, excessive and unbalanced use of reactive fertilisers is associated with poor crop yield and adverse environmental consequences.

In recent years, the influence of cover crops on soil-nutrient cycling and plant growth has been studied by scientists (Cherr et al., 2006; Magdoff and van Es, 2009; Miller et al., 1992; Dabney et al., 2010; Blanco-Canqui et al., 2011, 2015; Wendling et al., 2015). The dynamics of nutrients in cover crops depend on a decrease in leaching, soil erosion and the fixation of atmospheric N_2. An increase in the content of nutrients, their storage and balance depends on the proper selection of the cover crops and the purpose of their use. It is known that legumes can fix and recycle atmospheric N_2 to provide an available form of N (NO_3 or NH_4) to the succeeding crops to supplement chemical fertilisation. This is especially important when growing crops on low productive and degraded soils (Tanimu et al., 2007).

Generally, legume cover-crop biomass decomposes much faster than grass-cover crop biomass, contributing to the N mineralisation in soil. This is explained by a rather low ratio of carbon to nitrogen (C:N < 20), while the ratio for grassy cover crop biomass C:N is much higher. When cover crop residues are plowed under or incorporated into the soil, N is more rapidly mineralised. There is a negative aspect to this practice—a rapid mineralisation of biomass N accumulates a reactive form of N as highly soluble NO_3, which is prone to leaching and/or surface runoff losses (Dabney et al., 2010; Pantoja et al., 2016).

Based on studies by Cherr et al. (2006), it was reported that the concentration of N is 1.9–3.6 per cent for moderate legumes, 2.6–4.8 per cent for tropical legumes, and 0.7–2.5 per cent for non-legumes. When converted to C:N, these values are 8–15, 11–21 and 16–57, respectively. During the growing season of the integumentary culture, the biomass N content changes its value. The closer the end of the cropping cycling is, the less the N that remains in the cover crop biomass. With increasing planting density, the canopy closes earlier and the biomass begins to accumulate N actively. However, planting too tightly can lead to reduced plant growth (Scholberg et al., 2010).

Cover crop residues with a lower C:N decompose and release N faster. Studies have shown that if hairy vetch and crimson clover (legumes) are used as cover crops, the biomass decomposition and release of N is much faster if grown independently than when grown with high C:N rye as a grass cover crop (Ranells and Wagger, 1996). The stems and roots of the plants, in general, contain a higher C:N than the leaves. Based on several studies, it is indicated that leaves decompose and release N much faster than stems and roots. So, for example, legume cover-crop biomass decomposes and releases N five times faster than its stems. In moderate conditions, leguminous plant residues in the soil decompose for about 15 weeks. For tropical conditions, the time of complete decomposition and growth of N is shortened to only two to six weeks. However, it is difficult to determine the exact amount of N released from the roots due to their rapid turnover. Griffin et al. (2000) indicated that 56 per cent of the total biomass and 32 per cent of the total accumulated N are from roots when growing alfalfa; 46 per cent and 28 per cent, respectively, for rye; and 38 per cent and 19 per cent, respectively, when growing hairy vetch and rye. In contrast, the advantage of using radish as a cover crop is that it produces a large amount of biomass. It prevents leaching of nutrients in the soil, as it can absorb them in a short time. The oilseed radish biomass reaches as much as 9.8 ton/ac (Fig. 2, Sundermeier, 2010). Below 50°F, radish started to die and upon decomposition, it recycled nutrients and provided favourable conditions for the succeeding crops (Table 1).

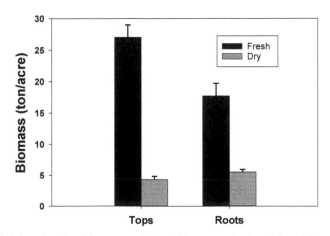

Fig. 2 Oilseed radish below- (root) and aboveground (shoot) biomass production (Adopted from Sundermeier, 2010).

Table 1 Nutrient recycling by oilseed radish (Based on Dry Biomass) for agronomic crops (adopted from Sundermeier, 2010).

Nutrients	Concentration (%)		Amounts of Nutrients (lbs/acre)		
	Tops	Roots	Tops	Roots	Total
Nitrogen	3.8	2.5	331	272	603
Phosphorus	0.7	0.6	61	65	126
Potassium	5.8	5.1	505	555	1,060

Table 2 Nutrient recycling by oilseed radish with Austrian Winter pea for agronomic crops (Hoorman and Islam, 2012, Personal Communications).

Biomass Nutrients	Concentration (%)	Range (lbs/acre)
Nitrogen	1.89	125–180
Phosphorus	0.97	65–90
Potassium	3.27	200–300
Sulfur	0.81	50–75
Calcium	2.17	150–200
Magnesium	0.26	15–25

Hoorman and Islam (2007, 2010) conducted several field research studies related to the recycling of nutrients using cover crops to support growing agronomic crops. Table 2 shows the concentration of nutrients in the biomass of Austrian winter pea and oilseed radish and their nutrient recycling for succeeding crops.

Sainju et al. (2005, 2009) have reported on the aboveground (shoot) and belowground (root) biomass of various cover crops, as well as C and N contents, over the years. Their studies showed that an increase in the aboveground biomass yield and the content of C and N changed when using different types of cover crops.

As seen in Table 3, the use of a mixture of hairy vetch and cereal rye led to a larger amount of biomass and a higher concentration of C and N (Sainju et al., 2005). For example, N was transferred from the hairy vetch to cereal rye, which increased the N concentration and the value of biomass to recycle nutrients. Also, when growing bi-culture or blends, increased seeding rates were used. As there is no competition between cereal rye and hairy vetch, when mixed together, they can complement each other and do not interfere with the growth of the plants. When growing bi-culture, a higher concentration of C is observed than when growing monocultures. The C:N in bi-culture (hairy vetch + rye) ranged from 10 to 32 and the biomass production ranged from 5.7 to 8.2 ton/acre.

Root biomass plays an important functional role in a terrestrial ecosystem. By increasing the content of C and N and their stoichiometric relationship, it can help to accumulate SOM and improve soil quality, which, consequently, has a positive effect on producing high-crop yields. Kuo et al. (1997) reported that cover crop roots can contribute from 7–43 per cent of the total plant biomass and the amount of organic carbon introduced in the soil during the growing season varies from 400–1,460 kg C/ha.

Carbon in SOM accumulated from roots tends to be more persistent than that of the SOM formed from aboveground biomass (shoot) and the soil aggregates formed by the action of roots and their metabolites are more stable (Sainju et al., 2005, 2009). Based on data presented in Table 4, it can be concluded that the root biomass develops in the same way as the shoot biomass.

Sainju et al. (2005, 2009) also reported that the C and N content of biomass is higher in a bi-culture (hairy vetch + rye) system, as they produce a higher biomass due to the increased seeding rates in bi-cultures. When compared, the C:Ns between the roots and shoots, a wide difference

Table 3 The effect of using cover crops on aboveground biomass and the content of carbon and nitrogen in a three-year study (adopted by Sainju et al., 2005).

Cover Crop*	Biomass Yield	Concentration		Content		C:N
		C	N	C	N	
	Mg ha^{-1}	g kg^{-1}		kg ha^{-1}		
2000						
Weeds	1.65d**	370b	15b	587d	25d	24b
Rye	6.07b	430a	15b	2670b	68c	29a
Vetch	5.10c	394ab	33a	2006c	165b	12c
Vetch/rye	8.18a	366b	38a	3512a	310a	10c
2001						
Weeds	0.75d	391b	20b	277d	15b	20c
Rye	3.81b	448a	8d	1729b	32b	57a
Vetch	2.44c	398b	32a	964c	76a	12c
Vetch/rye	5.98a	434a	14c	2693a	84a	32b
2002						
Weeds	1.25c	375b	15b	476c	23b	21b
Rye	2.28b	434a	11b	986b	25d	40a
Vetch	5.16a	361b	36a	2094a	167a	10c
Vetch/rye	5.72a	381b	33a	2260a	186a	11c

*   Cover crops are cereal rye (rye), hairy vetch (vetch), hairy vetch/rye bi-culture (vetch/rye), and winter weeds (weeds).
** Number followed by the different letter within a column of a year were significantly different at P ≤ 0.05 by the Least Square Means test.

Table 4 The effect of using cover crops on belowground biomass, the content of Carbon (C) and Nitrogen (N) and root/shoot in a three-year study (adopted by Sainju et al., 2005).

Cover crop*	Biomass yield	Nutrient content		C/N	Root/shoot
		C	N		
	kg ha^{-1}	kg ha^{-1}			
2000					
Weeds	280b**	73b	3.0b	24b	0.13a
Rye	174b	60b	1.3c	45a	0.03b
Vetch	147c	59b	4.0b	14c	0.03b
Vetch/rye	421a	154a	8.1a	19c	0.05b
2001					
Weeds	423c	130c	5.0b	25b	0.56a
Rye	772ab	250ab	6.9b	33a	0.20b
Vetch	656b	280b	10.6a	20c	0.27b
Vetch/rye	880a	269a	10.6a	25b	0.15b
2002					
Weeds	175b	57c	1.7b	33a	0.14ab
Rye	395a	137a	3.6a	38a	0.17a
Vetch	236b	78bc	4.2a	19b	0.05c
Vetch/rye	372a	130ab	4.0a	32a	0.10bc

*   Cover crops are cereal rye (rye), hairy vetch (vetch), hairy vetch and rye biculture (vetch/rye), and winter weeds (weeds).
** Number followed by the different letter within a column of a year are significantly different at P ≤ 0.05 by the Least Square Means test.

is observed in their values to regulate decomposition over time. This may be one of the factors responsible for the time difference in the decomposition of the cover-crop biomass when killed or incorporated. The crop residues of the belowground part (root) of the culture decompose over an extended period of time and therefore, release N more slowly. While the root system accumulates SOM and improves soil quality, the shoot biomass provides the subsequent culture with the necessary N after mineralisation (Sainju et al., 2009). The ratio of root and shoot depends on the type of integumentary culture and the duration of growth of the plants.

2.2 Cover Crops and Agronomic Crop Yields

It is expected that cover crops will have no direct effect on crop yield but rather a composite, indirect effect to influence the crop yield. The composite effect is via soil quality improvement; increased water and nutrient availability; weeds, disease and pest control; and a compatible microclimatic environment.

To better understand the impact of cover crops on agronomic crop yield, widespread data were collected from farmers by the Sustainable Agricultural Research and Extension (SARE)/ Conservation Tillage Information Centre (CTIC) and National Cover Crop Survey during the 2012–2016 growing years (Myers et al., 2019). Farmers, who planted cover crops on some fields but not on others, and who otherwise managed those fields, similarly, were asked to report on respective crop yields (Table 5). Though not all farmers had comparable fields with and without cover crops to report on, there were still several hundred farmers who provided crop-yield data each year. The significantly largest crop-yield differences were reported after the drought year of 2012, with average reported crop yield increases as much as 9.6 per cent in corn and 11.6 per cent in soybeans. Based on the high corn and soybean prices following the 2012 drought year, cover crops provided a substantial profit boost for farmers that year.

While several hundred farmers reporting data represents a good-sized data set, these were, however, self-reported numbers on crop yields. Moreover, it was clear that crop yields from field to field varied, with few fields having crop yield losses after introducing cover crops and with some fields showing no difference at all. Many farmers reported a crop-yield increase on their farms, but individual experiences varied. While the SARE/CTIC survey data set by far is the largest set available on the impact of cover crops on agronomic crop yields, it is worthwhile to mention that other cover-crop studies have reported a range of crop-yield impacts, from minor losses to minor increases in corn yields. For soybeans, several studies have shown that crop yields are unchanged with cover crops, while others have shown a modest improvement in crop yields.

Data presented in Table 6, based on the SARE/CTIC survey (using 2015 and 2016), show how yields change in response to duration of cover cropping in a field. Results showed that both corn and soybean yield increased with years of cover cropping. The corn yield increase was more than 0.5–3 per cent over a period of one to three years. Likewise, soybean yield increased by more than 2 per cent in one year and about 5 per cent after five years of cover cropping. These results showed a consistent increase in crop yield in response to temporal effects of cover cropping.

Table 5 Per cent increase in yield for corn and soybeans following cover crops versus comparably managed fields with no cover crops.[1]

Year	Corn	Soybeans
2012	9.6%	11.6%
2013	3.1%	4.3%
2014	2.1%	4.2%
2015	1.9%	2.8%
2016	1.3%	3.8%

[1] Data is from the SARE/CTIC National Cover Crop Surveys conducted annually for crop years 2012–2016.

Table 6 Per cent increase in corn and soybean yields after one, three and five years of consecutive cover crop use on a field, based on a regression analysis of data for crop years 2015–2016.[1]

Crop	One Year	Three Years	Five Years
Corn	0.52%	1.76%	3%
Soybeans	2.12%	3.54%	4.96%

[1] Figures shown are an average of yields from the 2015–2016 growing seasons, with yield data obtained from about 500 farmers each year through the SARE/CTIC National Cover Crop Survey.

The use of cover crops as green manure has a significant effect on increasing the yield of annual crops. Studies conducted by Ladha et al. (1996) showed an increase in agronomic crop-grain yield when using green manure crops compared to conventional fertilisation (grain yield increases of 0.5–1.4 Mg/ha, with green manure N ranging between 81–162 kg/ha). This can be explained by the N mineralisation from biomass residues of green manure crops via microbial cycling of carbon and nitrogen. Depending on the type of green manure crops, the amount of N assimilated by the subsequent main crop ranges between 4–30 per cent (Jackson, 2000; Fageria, 2007). Vyn et al. (1999) noted in their research that corn yields were consistently highest after red clover, and lowest after one-year when rye grass was used as a cover crop. Legume-cover crops, due to their higher ability for biological N-fixation with rhizobium bacteria, in comparison with non-leguminous cover crops, are more promising for obtaining high crop yields. However, different types of legume-cover crops have different rates of biological N-fixing capacities.

George et al. (1998) studied the effect of cover crops on rice yield. Their results on the use of green manure vs. conventional chemical fertilisers showed a yield increase between 1–2 Mg/ha, which corresponds to about 20–40 kg N/ha additional N uptake by the rice crop (George et al., 1998). Likewise, Becker et al. (1995) showed the benefits of using *Sesbania rostrata* as a cover crop in rice production. The cover cropping was able to replace 35–90 kg of introduced urea under different field conditions. Ladha et al. (1992) also concluded that green manure increased lowland rice yields between 45–130 per cent. Several studies reported the positive influence of *Sesbania* on crop yields in Thailand, India, Senegal and Nigeria. Fageria (2007) reported that the inclusion of *Sesbania* in crop rotation increased rice yield between 0.8–2.8 Mg/ha in the Philippines.

Becker et al. (1988) noted that when cover crops were included for 45–60 days, a significant increase in rice yield was observed and quantitatively replaced between 50–100 kg N/ha of chemical fertilisers. It is reported that leguminous green manure crops can accumulate about 2.6 kg N/ha/day. Experimental studies performed by Ladha et al. (1988) indicated that the amount of N supplied from raspberry clover and hairy vetch ranged between 72–149 kg N/ha. When managing the soil with green manure, the SOM content increases, improving the soil's physical and chemical properties and resulting in greater total root surface and subsequent higher absorption of nutrients and water by plants (Boparai et al., 1992). Mandal et al. (2003) reported that green manure increased both rice and wheat crop yields (Table 7).

The use of green manure, such as *Sesbania rostrata* increases SOM content, recycles nutrients, improves soil aggregation and reduces soil compaction—all reasons for the increased crop yields. Moreover, when combining green manure crops with supplemental N fertilisation, the rice yields were higher than the yield obtained with only an equivalent amount of N fertiliser. Studies by Fageria and Baligar (1996) also showed increased yields of beans and rice when green manure was used instead of N fertiliser (Table 8). When green manure is used before the high rainfall season, there is a moderate residual impact on the subsequent rice harvest during the dry season (Morris et al., 1989). Studies in Taiwan and the Philippines reported that tomato yields directly benefitted from the use of green manure during the rainy season (between 38–120 kg N/ha). In this study, when soybean was used as a green manure, it completely replaced chemical N fertilisation, providing high yields even in poor soil conditions (Thonnissen et al., 2000). The effect of cover crops on tomatoes and the subsequent maize crop was similar to the effect of applying 120 kg N/ha, yielding positive

Table 7 Effect of green manure and nitrogen treatments on grain yield of rice and wheat (Source: Mandal et al., 2003).

Treatment	Rice Yield (Mg/ha)				Wheat Yield (Mg/ha)			
	N (kg/ha)				N (kg/ha)			
	N0	N60	N120	Mean	N0	N60	N120	Mean
Fallow	3.15	4.61	5.19	4.32	2.75	3.89	4.07	3.57
S. rostrata	4.25	5.51	5.78	5.18	3.86	4.41	4.51	4.26
S. aculeata	4.19	5.49	5.73	5.14	3.74	4.36	4.49	4.20
Green gram incorporation	4.17	5.47	5.72	5.12	3.68	4.34	4.48	4.17
Mean	3.94	5.27	5.61		3.51	4.25	4.39	
	LSD (0.05)				LSD (0.05)			
Treatment (T)	0.495				0.287			
Nitrogen (N)	0.632				0.41			
T x N	1.109				NS*			

NS* – non-significant at P < 0.05

Table 8 Response to fertilisation of rice and common bean grown in rotation in Cerrado and Varzea acid soils (adapted from Fageria, 2007; Fageria and Baligar, 1996).

Fertility Level	Rice Yield (Mg/ha)[1]	Common Bean Yield (Mg/ha)[1]
Oxisol of Cerrado[2]		
Low	1.7b	1.2c
Medium	2.1a	1.8b
High	2.1a	2.2a
Medium + green manure	2.4a	1.5a
F-test	*	**
Inceptisol of Varzea[3]		
Low	4.3b	2.9b
Medium	5.5a	6.6a
High	5.5a	8.5a
Medium + green manure	6.3a	8.2a
F-test	**	**

[1] Values are averages of three crops grown in rice-bean rotation.
*, ** Significant at the 0.05 and 0.01 probability levels, respectively. Within the same column, means followed by the same letter do not differ significantly at the 05 probability levels by Tukey's test.
[2] Cerrado soil fertility levels for rice were low (without addition of fertilisers); medium (50 kg N ha⁻¹, 26 kg P ha⁻¹, 33 kg K ha⁻¹, 30 kg ha⁻¹ fritted glass material as a source of micronutrients); high (all the nutrients were applied at the double the medium level). *Cajanus cajan* L. was used as a green manure at the rate of 25.6 Mg ha⁻¹ green matter. For common bean the fertility levels were low (without addition of fertilisers); medium (35 kg N ha⁻¹, 44 kg P ha⁻¹, 42 kg K ha⁻¹, 30 kg ha⁻¹ fritted glass material as a source of micronutrients) and high (all the nutrients were applied at the double the medium level).
[3] Varzea soil fertility levels for rice were low (without addition of fertilizers); medium (100 kg N ha⁻¹, 44 kg P ha⁻¹, 50kg K ha⁻¹, 40 kg ha⁻¹ fritted glass material as a source of micronutrients); and high (all the nutrients were applied at the double the medium level). *Cajanus cajan* L. was used as a green manure at the rate of 28 Mg ha⁻¹ green matter. For common bean, the fertility levels were low (without addition of fertilisers); medium (35 kg N ha⁻¹, 52 kg P ha⁻¹, 50 kg K ha⁻¹, 40 kg fritted glass material as a source of micronutrients) and high (all the nutrients were applied at the double the medium level).

results. Appropriate use of green manure crops, in combination with chemical fertilisers, is expected to ensure sustainable crop production.

Tonitto et al. (2006) indicated that the crop yield, when applying the recommended doses of chemical fertilisers after non-legume cover crops, did not significantly differ from the crop yield after a fallow period. Several studies have shown that corn yields, when using legume cover crops, were 12 per cent lower as compared to conventional farming practices; however, the use of N from legumes did not significantly affect the sorghum yield. Analysing the data from research studies suggest that the uptake of N by the main crop from preceding legume cover crops, in general, was 28 per cent less than the chemical N fertilisations. However, the use of the required amount of N in conventionally fertilised systems showed an increase in the crop yields by only 10 per cent as compared to green manure (Fig. 3). The ultimate influence of the legume cover crops on the main crop yield can be seen in Fig. 4 (Tonitto et al., 2006). The release of N from legumes exceeding 180 kg N/ha showed a positive effect of crop yields (the average yield increase was 5 per cent as compared to conventional farming with N fertilisers). With the release of N from legumes in an amount of 110–180 kg N/ha, no significant difference in yield was observed among agronomic systems.

There are a few negative aspects of green manuring or cover crops if they are not properly used in agricultural production systems. These include N immobilisation. If green manure is not applied in advance before planting a cash crop, N deficiency is observed, since there was not enough time for green manure to decompose and recycle N for main crops. This leads to a decrease in crop yields. Based on several studies, the most optimal time for terminating the cover crop by plowing down or incorporating into the soil is two to four weeks before planting the main crop. As a result, the time will be sufficient for the decomposition of plant residues and the subsequent release of available N to the crop, leading to stable yields (Dabney et al., 1996). It is important that the use of certain types of cover crops must be appropriate for the cropping systems, based on the local climatic and soil

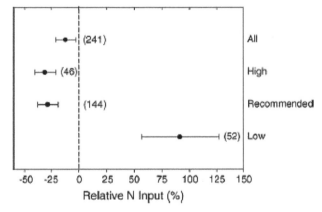

Fig. 3 Relative N added to cash crop in treatments (unfertilised and legume cover crop) compared to control (fertilised and winter bare fallow) systems grouped by fertiliser additions to conventional systems. Mean values and 95 per cent confidence intervals of the back-transformed response ratios are shown (number of comparisons in parentheses). *Source*: Tonitto et al., 2006.

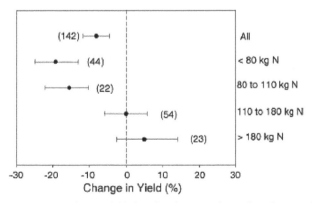

Fig. 4 Effect of legume cover crops on cash crop yields in units of percent change from the control (recommended fertiliser rate following winter bare fallow) grouped by legume N input rate. Mean values and 95 per cent confidence intervals of the back-transformed response ratios are shown (number of comparisons in parentheses). *Source*: Tonitto et al. (2006).

conditions. For example, the use of green manure in dry arid regions can adversely affect the yield of subsequent crops, as it creates a soil moisture deficit under such conditions conducive for optimum crop growth. In contrast, under cool climatic conditions, a dense layer of cover crop biomass inhibits soil warming and affects field workability. This negatively affects the germination of seeds and establishment of good stands of crops, thus reducing the grain yield. An important factor in reducing main crop yields is the excessive accumulation of N in the soil from rapid mineralisation of legume cover crop biomass. As a result, early lodging of main crops may occur. Therefore, it is necessary to use appropriate cover crops, N fertilisation and the total N content of soils.

References

Anderson, S., Gundel, S., Pound, B. and Thriomphe, B. (2001). Cover Crops in Small-holder Agriculture: Lessons from Latin America. ITDG Publishing, London, UK, 136 pp.

Bastian, O., Lupp, G., Syrbe, R. and Steinhauser, R. (2013). Ecosystem services and energy crops—spatial differentiation of risks. Ekologia, 32: 13–29.

Becker, M., Ladha, J.K., Watanable, I. and Ottow, J.C.G. (1988). Stem-nodulating legumes as green manure for lowland rice. Philippines J. of Crop Sci., 13: 121–127.

Becker, M., Ali, M., Ladha, J.K. and Ottow, J.C.G. (1995). Agronomic and economic evaluation of *Sesbania rostrata* green manure establishment in irrigated rice. Field Crops Res., 40: 135–141.

Blanco-Canqui, H., Mikha, M.M., Presley, D.R. and Claassen, M.M. (2011). Addition of cover crops enhance no-till potential for improving soil physical properties. Soil Sci. Soc. Am. J., 75: 1471–1482.

Blanco-Canqui, H., Shaver, T.M., Lindquist, J.L., Shapiro, C.A., Elmore, R.W., Francis, C.A. and Hergert, G.W. (2015). Cover crops and ecosystem services: insights from studies in temperate soils. Agron. J., 107: 449–2474.

Boparai, B.S., Singh, Y. and Sharma, B.D. (1992). Effect of green manure (*Sesbania aculeata*) on physical properties of soil and growth of rice-wheat and maize-wheat cropping system. Int. Agrophys, 6: 95–101.

Buckles, D., Triomphe, B. and Sain, G. (1998). Cover crops in hillside agriculture. IDRC, Ottawa, Canada, pp. 218.

Calegari, A. (2003). Cover Crop Management, Conservation Agriculture: Environment, Farmers – Experiences, Innovations, Socio-economy, Policy. Kluwer, Dordrecht, 191–199.

Cherr, C.M., Scholberg and McSorley, R. (2006). Green manure approaches to crop production: A synthesis J.M.S.is. Agron. J., 98: 302–319.

Dabney, S.M., Schreiber, J.D., Rothrock, C.S. and Johnson. J. (1996). Cover crops affect sorghum seedling growth. Agron. J., 88: 961–970.

Dabney, S.M., Delgado, J.A., Meisinger, J.J., Schomberg, H.H., Liebig, M.A., Kaspar, T., Mitchell, J. and Reeves, W. (2010). Using cover crops and cropping systems for nitrogen management. Advances in Nitrogen Management, Chapter 9: 230–281.

Dabney, S.M., Delgado, J.A., Meisinger, J.J., Schomberg, H.H., Liebig, M.A. and Kaspar, T. (2010). Using cover crops and cropping systems for nitrogen management. Advances in Nitrogen Management for Water Quality, Soil Water Conserv. Soc., Ankeny, IA, 231–282.

Daily, G.C., Matson, P.A. and Vitousek, P.M. (1997). Ecosystem services supplied by the soil. Nature's Services: Societal Dependence on Natural Ecosystems, Island Press, Washington, DC, 113–132.

Drinkwater, L.E. and Snapp, S.S. (2007). Nutrients in agro-ecosystems: Rethinking the management paradigm. Adv. Agron., 92: 163–186.

Fageria, N.K. and Baligar, V.C. (1996). Response of lowland rice and common bean grown in rotation to soil fertility levels on a Varzea soil. Fert. Res., 13: 13–20.

Fageria, N.K., Baligar, V.C. and Bailey, B.A. (2005). Role of cover crops in improving soil and row crop productivity. Commun. Soil Sci. Plant Analy., 36: 2733–2757.

Fageria, N.K. (2007). Green manuring in crop production. J. of Plant Nutr., 30(5): 691–719.

Ganzel, B. (2003). The Plow that Broke the Plains, Farming in the 1930s, Wessels Living History Far, York, Nebraska. https://livinghistoryfarm.org/farminginthe30s/machines_05.html.

George, T., Buresh, R.J., Ladha, J.K. and Punzalan, G. (1998). Recycling *in situ* of legume-fixed and soil nitrogen in tropical lowland rice. Agron. J., 90: 429–437.

Giller, K.E. (2001). Nitrogen Fixation in Tropical Cropping Systems, second edition, CAB International, Wallingford, UK, pp. 423.

Griffin, T., Liebman, M. and Jemison, J. (2000). Cover crops for sweet corn production in a short-season environment. Agron. J., 92: 144–151.

Hartwig, N.L. and Ammon, H.U. (2002). Cover crops and living mulches. Weed Sci., 50: 688–699.

Hoorman, J.J. and Islam, R. (2007). Nutrient Recycling with Manure and Cover Crops, Cover Crops and Soil Organic, Indiana CCA Conference Proceedings.

Hoorman, J.J., Islam, R., Sundermeier, A. and Reeder, R. (2009). Using cover crops to convert to No-till Crop and Soil, Agronomy for Practicing Professionals, 42.

Hoorman, J.J. and Islam, R. (2010). Understanding Soil Microbes and Nutrient Recycling, SAG-16, Agriculture and Natural Resources, Ohio State University.

Jackson, L.E. (2000). Fates and losses of nitrogen from a nitrogen-15 labelled cover crop in an intensively-managed vegetable system. Soil Sci. Soc. Amer. J., 64: 1404–1412.

Kuo, S., Sainju, U.M. and Jellum, E.J. (1997). Winter cover crop effects on soil organic carbon and carbohydrate. Soil Sci. Soc. Am. J., 61: 145–152.

Ladha, J.K., Pareek, R.P. and Becker, M. (1992). Stem-nodulating legume-rhizobium symbiosis and its agronomic use in lowland rice. Adv. Soil Sci., 20: 147–192.

Ladha, J.K., Kundu, D.K., Angelo-Van, C.M.G., Peoples, M.B., Carangal, V.R. and Dart, P.J. (1996). Legume productivity and soil nitrogen dynamics in lowland rice-based cropping systems. Soil Sci. Soc. Am. J., 60: 183–191.

Magdoff, F. and van Es, H. (2009). Building Soils for Better Crops: Sustainable Soil Management. ISBN 978-1-888626-13-1, 3.

Mandal, U.K., Singh, G., Victor, U.S. and Sharma, K.L. (2003). Green manuring: Its effect on soil properties and crop growth under rice-wheat cropping system. Eur. J. Agron., 19: 225–237.

McNeill, J.R. and Winiwarter, V. (2004). Breaking the sod: Humankind, history and soil. Science, 304: 1627–1629.

Miller, M.H., Beauchamp, E.G., Vyn, T.J., Stewart, G.A., Lauzon, J.D. and Rudra, R. (1992). The use of cover crops for nutrient conservation, Final Report on Sweepted Project, DSS No. XSE 89-0082-(302). Ser. No. 01686-9-0142/01-xse, 176 pp.

Morris, R.A., Furoc, R.E., Rajbhandari, N.K., Marqueses, E.P. and Dizon, M. (1989). Rice response to water-tolerant green manures. Agron. J., 81: 803–809.

Myers, R., Weber, A. and Tellatin, S. (2019). Cover Crop Economics Opportunities to Improve Your Bottom Line in Row Crops, Ag Innovative Series Technical Bulletin, pp. 1–9. www.sare.org/cover-cropeconomics.

Pantoja, J.L., Woli, K.P., Sawyer, J.E. and Barker, D.W. (2016). Winter rye cover crop biomass production, degradation and nitrogen recycling, soil fertility and crop nutrition. Agron. J., 108.

Ranells, N.N. and Wagger, M.G. (1996). Nitrogen release from grass and legume cover crop monocultures and bicultures. Agron. J., 88: 777–782.

Russell, E.J. (1913). The Fertility of the Soil, first ed., Cambridge University Press, London, 126 pp.

Sainju, U.M., Whitehead, W.F. and Singh, B.P. (2005). Bi-culture legume – cereal cover crops for enhanced biomass yield and carbon and nitrogen. Agron. J., 97: 1403–1412.

Sainju, U.M. (2009). Cover Crops and Crop Yields, Chapter: Bicultural Legume-Cereal Cover Crops for Sustaining Crop Yields and Improving Soil and Environmental Quality. Nova Science Publishers, New York, NY, 113–147.

Sarrantonio, M. and Gallandt, E. (2003). The role of cover crops in North American cropping systems. J. Crop Prod., 8: 53–74.

Schipanski, M.E., Barbercheck, M., Douglas, M.R., Finney, D.M., Haider, K., Kaye, J.P., Kemanian, A.R., Mortensen, D.A., Ryan, M.R., Tooker, J. and White, C. (2014). A framework for evaluating ecosystem services provided by cover crops in agro-ecosystems. Agric. Sys., 125: 12–22.

Scholberg, J., Dogliotti, S., Leoni, C., Cherr, C., Zotarelli, L. and Rossing, W. (2010). Cover Crops for Sustainable Agrosystems in the Americas. Springer Science + Business Media B.V. ISBN 978-90-481-8740-9.

Sullivan, P. (2003). Overview of cover crops and green manures, ATTRA Publication No. IP024, National Centre for Appropriate Technology, Butte, MT, 1–16.

Sundermeier, A.P. (2010). Nutrient management with cover crops. J. NACAA, ISSN 2158-9429, 3.

Tanimu, J., Iwuafor, E.N.O., Odunze, A.C. and Tian, G. (2007). Effect of incorporation of leguminous cover crops on yield and yield components of maize. World J. of Agric. Sci., 3: 243–249.

Thonnissen, C., Midmore, D.J., Ladha, J.J., Holmer, R.J. and Schmidhalter, U. (2000). Tomato crop response to short-duration legume green manures in tropical vegetable systems. Agron. J., 92: 245–253.

Tonitto, C., David, M.B. and Drinkwater, L.E. (2006). Replacing bare fallows with cover crops in fertilizer-intensive cropping systems: A meta-analysis of crop yield and N dynamics. Agric. Ecosyst. Environ., 112: 58–72.

Vyn, T.J., Janovicek, K.J., Miller, M.H. and Beauchamp, E.G. (1999). Spring soil nitrate accumulation and corn response to preceding small-grain N fertilisation and cover crops. Agron. J., 91: 17–24.

Wendling, M., Büchi, L., Amosse, C., Sinaj, S., Walter, A. and Charles, R. (2015). Nutrient accumulation by cover crops with different root systems. Aspects Appl. Biol., 129: 91–96.

3

Potential and Challenges of Growing Cover Crops in Organic Production Systems

Sutie Xu,[1] *Sindhu Jagadamma,*[1,*] *Renata Nave Oakes,*[2] *Song Cui,*[3] *Erin Byers*[4] and *Zhou Li*[5]

1. Introduction

Organic farming has long been recognised nationally and internationally as a method for improving agricultural sustainability and resilience. Consumers have also become increasingly selective regarding their food choices, associated health and environmental consequences. As a result, the demand for organic products from the public market sector has increased drastically in the past few decades. However, without the use of synthetic inputs (e.g., chemical fertilisers, pesticides and herbicides), organic systems are typically less productive than conventional systems in terms of cash-crop biomass production (De Ponti et al., 2012; Sacco et al., 2015; Taheri et al., 2017) due to the challenges in maintaining soil fertility and controlling weeds, pests and diseases (Tu et al., 2006; Dawson et al., 2008; Francis and Van Wart, 2009; Liebman et al., 2009; Casagrande et al., 2016; Shennan et al., 2017). Common nutrient sources used in organic systems, including cattle and poultry manure, along with compost, are effective in maintaining soil-nutrient status, but also have many limitations and drawbacks compared to inorganic sources, such as the presence of pathogens and contaminants, imbalanced nutrient content, slow release of nutrient content and high transportation cost. Incorporating cover crops into organic systems could help address those concerns to a certain extent and eventually improve the holistic sustainability. However, successfully maintaining and improving soil health using cover cropping requires a systematic approach integrated with strategic species selection and a diligent management plan (Parr et al., 2011; Spargo et al., 2016; Wittwer et al., 2017).

It has been shown that the environmental footprint of production systems can be reduced substantially by reducing the frequency and intensity of tillage (Blevins et al., 1983; Golabi et al., 1995). However, intensive tillage is often used as a mechanical method for weed control in organic systems (Cavigelli et al., 2008; Vakali et al., 2011; Larsen et al., 2014), which brings criticism

[1] Department of Biosystems Engineering and Soil Science, University of Tennessee, Knoxville, TN, USA.

[2] Department of Plant Sciences, University of Tennessee, Spring Hill, TN, USA.

[3] School of Agribusiness and Agriscience, Middle Tennessee State University, Murfreesboro, TN, USA.

[4] Faculty of Environmental Sciences and Natural Resource Management, Norwegian University of Life Sciences, Ås, Norway.

[5] College of Animal Science, Guizhou University, Guiyang, Guizhou, China.

* Corresponding author: sjagada1@utk.edu

regarding the environmental stewardship of organic production. Cover crops can suppress weeds through physical hindrance of weed seed germination and seedling growth (Mohler and Teasdale, 1993; Creamer et al., 1996). Thick cover crop stands can block the transmittance of sunlight, diminish the quantity and quality of light reaching the soil surface, and reduce weed-seedling emergence and vigour (Teasdale and Mohler, 1993). Furthermore, literature also indicates that certain species used as cover crops can also suppress weed growth via allelopathic effects (Jabran et al., 2015). Therefore, cover cropping-based weed control strategies incorporated with reduced/conservation tillage have become popular across different organic systems recently (Mirsky et al., 2012; Mirsky et al., 2013).

In addition to weed control, cover cropping also plays a critical role in building and maintaining soil fertility, accumulating soil organic matter, improving soil structure (Villamil et al., 2006), controlling erosion (Qi and Helmers, 2010) and enhancing economical profitability (Wayman et al., 2017). To optimise agronomic, economic and ecological outcomes from cover cropping, appropriate species selection and effective termination approaches are critically important (Keene et al., 2017; Wallace et al., 2017; Bavougian et al., 2019). This chapter will cover the challenges and opportunities of cover cropping in organic systems. We will discuss general species selection strategies for nutrient availability and weed control needs in organic grain, vegetable and crop-livestock integrated production systems. In addition, cover crop termination challenges and strategies applicable to organic systems will also be discussed.

2. Cover Crop Species Selection in Organic Systems

In general, most cover-crop species are either annuals or short-lived perennials (including biennials) with no long-lived roots competing with subsequent cash crops for soil nutrients. Cool-season cover crops are very popular in organic systems because the majority of agronomic cash crops are warm-season annuals (e.g., corn [*Zeamays* L.], soybean [*Glycine max* L. Merr.], sorghum [*Sorghum bicolor* L.]) with production curves peaking during the summer, posing little competition for resource demands with cool-season cover crops. Popular cool-season cover-crop species include clover (*Trifolium* spp.), alfalfa (*Medicago sativa* L.), vetch (*Vicia* spp.), yellow sweet clover (*Melilotus officinalis* L.), Austrian winter pea (*Pisum sativum* L.), cereal rye (*Secale cereale* L.), wheat (*Triticum aestivum* L.), triticale (× *Triticosecale* Wittmack), barley (*Hordeum vulgare* L.), oat (*Avena sativa* L.), annual ryegrass (*Lolium multiflorum* L.) and *Brassica* spp.

In the southeast and southwest regions of the United States with longer and warmer growing seasons, warm-season cover crops can be integrated into fall vegetable production systems (Creamer and Baldwin, 2000; Wang et al., 2008). A survey conducted in the southeast showed that 79 per cent of the producers, who adopt sustainable management practices, reported using cover crops and 91 per cent of them used cool-season and 55 per cent used warm-season species of cover crops (O'Connell et al., 2014). Popular warm-season cover crops include Sunn hemp (*Crotalaria juncea* L.), cowpea (*Vigna unguiculata* [L.] Walp.), sorghum, sudangrass (*Sorghum bicolor* L.), sorghum-sudangrass hybrid, foxtail millet (*Setaria italica* [L.] P. Beauv.), pearl millet (*Pennisetum glaucum* [L.] R. Br.), buckwheat (*Fagopyrum esculentum* Moench) and many warm-season beans, such as lablab (*Lablab purpureus* [L.] Sweet), velvet (*Mucuna pruriens* [L.] D.C.), jack (*Canavalia ensiformis* [L.] D.C.), and scarlet runner beans (*Phaseolus coccineus* L.).

In many organic systems, non-leguminous cover crops are included because they can scavenge excess plant-available nitrogen (N) from the soil solution after the main crop growing season, reducing the potential leaching loss of N. O'Connell et al. (2015) showed that sorghum-sudangrass was better in moderating soil N loss than cowpea, foxtail millet and cowpea-foxtail millet mix. These trapped nutrients in cover-crop biomass can later be released to the subsequent cash crops after termination. Though non-leguminous cover crops can benefit the soil fertility on a long-term basis, they usually cannot provide sufficient available N quickly, as their residues decompose slowly and may cause microbial immobilisation of N due to the high C:N ratio. Leguminous cover crops, on the other hand, can supply a greater amount of total, as well as plant-available N to a cropping system (Cui et al., 2014; Inwood et al., 2015). It is estimated that leguminous winter annual cover crops can add 100–200 kg N ha⁻¹ year⁻¹ depending on the species (Ott and Hargrove, 1989; Fageria et al., 2005; Drinkwater et al., 2008). Besides grasses and legumes, certain forb species, such as buckwheat and oilseed rape (*Brassica napus* L.) are highly efficient at utilising mineral phosphorus and thus can make more soluble phosphorus available to subsequent cash crops (Hinsinger, 2001; Raghothama, 2005). Cover crops that belong to the *Brassica* family are characterised by large taproot systems and excellent cold tolerance. They can reduce soil compaction and can be effectively incorporated into livestock-integrated organic systems for winter grazing. Overall, information related to the use of warm-season cover crops in organic systems is extremely limited on a national-to-international level.

2.1 Cover Crop Species Selection for Organic Grain Systems

Hairy vetch (*Vicia villosa* Roth) and cereal rye (Fig. 1) are the most commonly used cover crops in grain systems because of their great cold tolerance (Teasdale, 1996), outstanding weed suppression capacity and soil fertility improvement. It was found that hairy vetch can effectively reduce more than half of weed density (Mischler et al., 2010) and increase corn yield when grown in combination with the application of organic amendments (Spargo et al., 2016). A study that investigated the effect of 16 different winter annual cover crops showed that hairy vetch was the optimal choice in terms of maintaining corn yield (Parr et al., 2011). Although cereal cover crops can lead to N immobilisation, the decreased soil N availability can suppress weeds, which is beneficial for the succeeding leguminous cash crops (Reberg-Horton et al., 2012; Wells et al., 2013). Moreover, cereal rye has been shown to provide great weed suppression through the production of allelochemicals (Schulz et al., 2013) in addition to its N-scavenging capacity (Weston, 1996).

Some other leguminous cover crops, such as crimson clover (*Trifolium incarnatum* L.), balansa clover (*Trifolium michelianum* Savi.) and subterranean clover (*Trifolium subterraneum* L.) contain higher total N, but fail to maintain stable organic corn yield (Parr et al., 2011). In general, leguminous cover crops are more effective in providing soil inorganic N in humid temperate regions, while cereal rye and annual rye grass are better options for building soil health because they can accumulate more soil organic carbon and N due to higher biomass inputs compared to Austrian winter pea, hairy vetch and canola (*Brassica napus* L.) (Kuo et al., 1997a, 1997b).

Cover crops are also used in rotational organic grain production systems. In a corn-soybean rotation system in Maryland, long-term organic management using crimson clover before corn and cereal rye before soybean, in tandem with organic manure, accumulated more soil organic carbon and N than conventional systems (Teasdale et al., 2007). The overall soil fertility of an organic system with a four-year rotation of tomato (*Lycopersicon esculentum* Mill.) safflower (*Carthamus tinctorius* L.)-corn-bean (*Phaseolus vulgaris* L.) with manure application benefited from cover cropping using bi-culture of oat and vetch before main crop planting (Clark et al., 1998). Similarly, cereal rye, as a winter cover crop in an organic corn–soybean–oat–alfalfa system with composted

Fig. 1 Cover crop mixtures in organic corn production systems: (a) a mixture of crimson clover, cereal rye, and hairy vetch; (b) a mixture of black oats and crimson clover (adapted from Treadwell, 2009).

manure application, produced higher corn and soybean yields than conventional corn–soybean rotation systems (Delate and Cambardella, 2004).

Single-species cover crops are easy to incorporate into cropping systems; however, they can only provide specific functions. Mixtures of legumes and non-legumes (Fig. 1), on the other hand, can offer a multitude of benefits simultaneously (Kramberger et al., 2014; Smith et al., 2014; Tosti et al., 2014; Chu et al., 2017). Therefore, information abounds on use of different mixtures of cool-season cover crops in continuous or rotational cropping systems. Double species mixtures of legume and grain cover crops, such as hairy vetch/cereal rye or winter pea/cereal rye, showed promise in suppressing weeds in organic corn production, especially when starter nutrient sources, such as poultry litter was applied (Wells et al., 2016; Vann et al., 2017). A combination of cereal rye/hairy vetch and timothy grass (*Phleum pretense* L.)/red clover (*Trifolium pretense* L.) were both economical to integrate into an organically-managed corn-soybean rotation system with the rye/vetch combination showing more promise in increasing corn/soybean yields in the early years (Smith et al., 2011a). Other studies found that a combination of cereal rye and crimson clover suppressed weeds and conserved soil moisture in organic cotton (*Gossypium hirsutum* L.) systems (Vann et al., 2018), and a combination of hairy vetch and barley controlled weeds and sustained yield in organic wheat systems (Halde et al., 2014). In contrast, there is also evidence of decreased benefits from mixed cover crops compared to single-species cover crops, such as cereal rye (Blesh et al., 2019). Regardless, mixtures of different cover crop species can potentially provide multi-functional benefits if appropriate species selection and management techniques are followed.

2.2 Cover Crops Species Selection for Organic Vegetable Systems

Many studies have found that using cover crops can be successful in organic vegetable systems, including lettuce (*Lactuca sativa* L.) (Ngouajio et al., 2003), pepper (*Capsicum annuum* L.) (Hutchinson and McGiffen, 2000; Isik et al., 2009), broccoli (*Brassica oleracea* L. [Italica group]) (Wyland et al., 1996), zucchinis (*Cucurbita pepo* L.) (Canali et al., 2013), tomato (Wang et al., 2009), etc. Because of the specific functionalities of cover cropping, a proper selection of cover crops can offer similar benefits in organic vegetable production systems, as in the case of organic grain systems (Gaskell and Smith, 2007; Price and Norsworthy, 2013). Generally, cover crops such as cereal rye, crimson clover, cowpea, vetch species, radish (*Raphanus* L.), alfalfa, oats, buckwheat, sudangrass and sorghum × sudangrass hybrids are often used in organic vegetable production systems (Dorais, 2007; Price and Norsworthy, 2013).

Many studies of cover crops in vegetable systems focused on hairy vetch and cereal rye because of their outstanding effects on nutrient retention and weed suppression, respectively (Leavitt et al., 2011). Many of these studies showed that hairy vetch can promote vegetable yield, but its effect lacks consistency. Additionally, both yield and soil health benefits are usually enhanced if a combination of grass and legume cover crops are used (Rogers et al., 2015). For example, a mixture of rye-vetch incorporated into organic tomato production was shown to stimulate soil respiration and microbial biomass, suggesting greater microbial activity and enhanced nutrient cycling processes (Nair and Ngouajio, 2012). Similarly, incorporation of hairy vetch and rye was shown to offer similar pepper yield as when compost amendment was used (Delate et al., 2003).

The dynamics of legume-to-grass ratio and the overall dry matter density of cover crops are complex, with seeding rate and variation in growing degree days being the most dominating variables and less impacts from N fixation on nutrient availability benefits (Brennan and Boyd, 2012a). High seeding rate mixtures of bell bean (*Vicia faba* L.), wooly pod vetch (*Vicia dasycarpa* Ten.), purple vetch (*Vicia benghalensis* L.), Austrian winter pea and oat, for example, offered great weed suppression capacity and economic return (Brennan et al., 2009). A 90/10 mixing ratio of various legumes (bell bean, vetches and winter pea) to rye provided the greatest balance of dry matter biomass (Brennan et al., 2011) and high-quality residue that could be decomposed rapidly (Brennan et al., 2013) after termination. Furthermore, a high seeding rate of legume-rye mixture was critical for increasing early-season N accumulation in aboveground biomass and decreasing the risk of N leaching in organic vegetable production systems (Brennan and Boyd, 2012b).

Planting time of cover crops is critical for maximising agronomic outcomes. Some studies reported that cover crop species and planting time interactively influence to offer the best agronomic outcomes. For instance, a study on organic vegetable systems in Tennessee showed that when cover crop planting time is September, single species grasses, such as rye, triticale, or wheat successfully suppressed weeds (Rogers et al., 2015). Additionally, grain/legume mixture and single species legume (e.g., hairy vetch, winter pea, crimson clover) increased both nutrient supply and weed suppression as compared to no-cover crop control. When the planting date was delayed to November, grass/legume mixture outperformed grass monoculture and a mixture of rye or wheat with vetch were better options.

Although incorporating cover crops into organic vegetable systems can successfully suppress weeds, sometimes they cannot sustain vegetable yield due to insufficient N (Leavitt et al., 2011). The organic vegetable systems may also be negatively affected by the lowering of soil temperature during planting seasons caused by cover cropping, especially in areas with short growing season and cooler climate (Hoyt et al., 1994). Hence, region-specific studies are necessary for providing recommendations on cover-crop management in organic vegetable production.

2.3 Cover Crops Species Selection for Crop-Livestock Integrated Organic Systems

Cover crops by definition are not intended to be grazed or harvested; however, interest is growing in the potential benefits of utilising cover crops in integrated crop-livestock systems, especially when forage supply is limited (Franzluebbers and Stuedemann, 2014). The increased demand for feedstuff and variable climatic conditions throughout the United States are driving the interest in integrating cover crops with livestock production. Cover crops may provide high amounts of feed with high nutritional value to enhance livestock performance at times when pasture availability and quality are low (Blanco-Canqui et al., 2015). The USDA-National Organic Program requires that 30 per cent of a ruminant's dry matter intake must come from pastures during the grazing season, revealing an opportunity to increase adoption of forage cover crops in organic farming, along with increased consumer demand (Inwood et al., 2015). These systems should be managed carefully to optimise animal nutrition content at the time of grazing, avoid erosion from animal traffic and increase rather than decrease soil organic matter (Franzluebbers and Stuedemann, 2015).

Traditional organic farming used to mix crops and livestock, with livestock adding economical values while providing justification for growing the forages needed to support the land by replenishing and rejuvenating the soil quality and provide other benefits, such as N recycling (Clark, 2009). In organic systems, forage-based feeds, especially pastures, are intended to do more than support healthy digestive processes, but also enhance animal health and welfare. The economic benefits generated from forage cover crops can be enhanced through increased animal production and reduced supplemental feeding costs while enhancing soil quality and increasing long-term environmental benefits; therefore adding both short- and long-term economic value within operations (Faé et al., 2009).

In most organic agriculture systems, annual forage crops are selected to be used as cover crops because they have the potential to add economic value and provide increased flexibility of management decisions in organic crop rotations (Inwood et al., 2015). Annual forages may be especially useful when a farm is transitioning to organic systems due to rapid establishment and highly competitive growth habits and high yields. For example, sorghum-sudangrass hybrid incorporated into an annual-forage rotation system (wheat/crimson clover followed by sorghum-sudangrass) can provide great weed suppression capacity and forage nutritive value based on an organic transitional study (Inwood et al., 2015).

Perennial forage yields are usually lower, while integrating them as cover crops in organic crop rotations can potentially increase soil quality and fertility (Cavigelli et al., 2008; Inwood et al., 2015), reduce weed pressure (Cavigelli et al., 2013), and reduce pests and insects (Katsvairo et al., 2007) for subsequent crops, compared to crop rotations with only annual crops. Perennial species commonly used in forage systems that could be integrated into organic forage production systems include legumes, such as alfalfa, red clover, white clover (*Trifolium repens* L.), alsike clover (*Trifolium hybridum* L.), or birds foot trefoil (*Lotus corniculatus* [L.]), or perennial grasses, such as orchardgrass (*Dactylisglomerata* L.), Kentucky bluegrass (*Poa pratensis* L.), reed canary grass (*Phalaris arundinacea* L.), timothy, perennial ryegrass (*Lolium perenne* L.), Bahiagrass (*Paspalum notatum* Flugge), or bermudagrass (*Cynodon dactylon* [L.] Pers.) (Ball et al., 2012; Inwood et al., 2015). Among these species, alfalfa, red clover and orchardgrass is most commonly used in southeast United States and adapted to regional climate and soil type. Alfalfa provides high forage quality feed combined with a stable forage yield throughout the growing season. Red clover tolerates acidic and poorly-drained soils better than alfalfa and provides great yield when placed under rotational stocking. Nonetheless, both alfalfa and red clover are known for their high biological N fixation rate, making them useful in organic systems that are often limited by N availability. Alternatively, orchardgrass can be used to add functional diversity to legume monocultures, such as alfalfa and red clover, which can improve productivity, reduce risks of bloat and improve soil organic matter due to differing root structure (Inwood et al., 2015). In practice, the forage biomass can be promoted if legumes and grasses are planted together (Inwood et al., 2015) due to transfer of N, particularly between legumes and grasses (Dhamala et al., 2017) and stimulation of additional N-fixation by legumes, in some cases increasing N yields of the mixture versus pure legumes (Nyfeler et al., 2011).

Cover crops have been identified as important components of diversified crop-and-forage rotations (Snapp et al., 2005), which can reduce soil erosion and nutrient losses through runoff from these systems. One of the main benefits from increased plant diversity is the overall productivity of the system (Soder et al., 2007), resulting in increased soil quality and health. Most of the time, higher yields can be observed with higher plant diversity due to the ability of mixed systems to use resources more efficiently than a monoculture cropland (Hector et al., 1999). In addition, these cover crops can make forage available during the cooler months, providing an additional economic benefit while increasing diversity in cropping systems.

Grazing animals play a key role in modifying plant diversity in grazing lands. Livestock can be very selective when grazing, selecting either plant parts or plant species (Soder et al., 2007). In addition, over-grazing can result in changes in plant morphology, such as decreased leaf-stem ratio, or in the case of under-grazing, the canopy starts producing reproductive stems, decreasing the overall nutritive value. Therefore, it is extremely important to understand the interrelationship between plant and animal systems in a mixed and diverse grazing land system. Integrating grazing into cropping systems requires crops that complement these livestock production systems (Şentürklü et al., 2018). Within an integrated crop-livestock system, crop (e.g., corn, field pea [*Pisum sativum* L.]/barley mix) residues can also be grazed by livestock, increasing sustainability and efficiency of these operations.

Overall, forage crops are essential for the success of organic systems without the reliance on many common, but prohibited, inputs (Clark, 2009). The diverse mixtures optimise weed suppression, N-fixation and animal nutrition, which fit well within the traditional principles of organic farming.

3. Cover Crop Termination in Organic Systems

3.1 Termination Method

One of the main drawbacks of short-duration cover crops is that some of them tend to reseed often (e.g., yellow sweet clover, buckwheat), posing managerial challenges to the producers. Hence, a good strategy for cover-crop termination is essential for the successful establishment of the subsequent cash crops. This is more important in organic cropping systems, in which the use of herbicides is prohibited and the cover crops need to be terminated mechanically. Organic production systems rely mostly on heavy tillage for cover crop termination; however, producers are becoming increasingly aware that the ecological benefits obtained from the chemical-free management of organic farms may be negated by energy-intensive and soil-eroding tillage operations. Innovative cover-crop termination options that enable decreased frequency and intensity of tillage with no compromise on farm income have evolved in recent years.

Termination of cover crop mixtures with the sweep plow under-cutter can successfully suppress weeds and increase organic corn and soybean yields, while termination with the field disk can decrease the soybean yield compared to no cover crop control (Wortman et al., 2013). A novel means to mechanically terminate cover crops is to use a roller crimper. The roller crimper was originally developed by the U.S. Department of Agriculture – Agricultural Research Service (USDA-ARS), National Soil Dynamics Laboratory in Auburn, Alabama (Fig. 2a, 2b) based on the no-till techniques followed in Brazil (Ashford and Reeves, 2003) (Fig. 2c). It was later modified to have metal slats welded perpendicular to the cylinder, arranged in a chevron pattern, which can crimp the cover crops with decreased vibration, thus resulting in no disturbance to the soil and roots (Mirsky et al., 2012; Parr et al., 2014).

Demonstrated by the Rodale Institute, the lush stand of rye and hairy vetch can be turned into 5-inch thick, weed-suppressing mulch with only one pass of the roller crimper (Fig. 3a), and the weeds were successfully suppressed in the following soybean growing season (Fig. 3b). It has been found that using roller crimpers to kill hairy vetch can sustain the same organic corn production as disk-tillage systems (Teasdale et al., 2012). Compared to using flail mower, another mechanical termination method for cover crops, roller crimper can terminate cereal rye more effectively to control weed biomass and sustain organic soybean yields as long as the rye popularised by the Rodale Institute (Kutztown, PA, USA) (Fig. 2d). The steel cylinder of the biomass is sufficient in quantity before termination (Smith et al., 2011b). Roller crimping is preferred to mowing due to less fuel-and-labour demand, effective weed suppression by distributing the cover crop residues more evenly and creating highly persistent mulch with a lower decomposition rate (Mirsky et al., 2013; Wayman et al., 2015). To avoid residue build up, cover crops should be roll-crimped parallel to the direction of cash crop planting (Reberg-Horton et al., 2012). Nonetheless, mechanical termination strategies should be fine-tuned for each organic farm because cover crop growth stages and biomass

Fig. 2 Roller crimper types: (a) & (b) developed by the USDA-ARS National Soil Dynamics Laboratory; (c) original Brazilian type; (d) modified type designed by the Rodale Institute (adapted from Kornecki et al., 2006; Zinati et al., 2019).

Fig. 3 The effectiveness of roller crimper: (a) rye and hairy vetch rolled into a 12.5-cm-thick weed-suppressing mulch; (b) successful weed suppression after rolling in the cover crops (adapted from Sayre, 2003).

yields vary, depending on different cover crop and main crop species, soil nutrient levels and climatic conditions (Smith et al., 2011b).

3.2 Termination Time

Besides using the most appropriate mechanical termination tool, the termination time must be chosen wisely, depending on the species planted and their growth stages, for the successful termination of cover crops. Before being terminated, cover crops should attain sufficient biomass to suppress weeds and be at the correct phenological stage for a high killing efficiency (Wayman et al., 2015). When a roller crimper is used, a complete termination can only be achieved when rolling is delayed to coincide with full plant maturity, which also depends on species (e.g., rolling between half-way anthesis and early milk stage for cereal rye, between late flowing and early pod set for hairy vetch and at late flowering for crimson clover) (Mirsky et al., 2012; Reberg-Horton et al., 2012; Wayman et al., 2015; Keene et al., 2017). A study in North Carolina showed that crimson clover can attain peak biomass growth earlier than other species, such as hairy vetch, common vetch (*Vicia sativa* L.), cereal rye, berseem clover (*Trifolium alexandrinum* L.), Austrian winter pea and lupin (*Lupinus angustifolius* L.) (Parr et al., 2011). In the same study, only late-terminated hairy vetch produced equal or more corn than N-fertilised conventional production systems, while early termination of hairy vetch or the use of other cover crops decreased corn production (Parr et al., 2011). However, termination should not be over-delayed. Very late termination may delay the planting of main cash crops and suppress their production due to the excessive growth of weeds before termination. In addition, over-growth of cover crops can lead to seed setting and re-growth during the main crop growing season, so that they become weeds (Mischler et al., 2010; Carr et al., 2012; Wayman et al., 2015). Furthermore, rolling at maturation may not be suitable for all cover crops, for example, barley, which grows quickly after the anthesis stage, making it a challenge to determine the right termination time (Wayman et al., 2015).

Besides the termination time, multiple factors including seeding time of both cover crops and cash crops, seeding rates and the local climate should be considered for obtaining the maximum benefits from cover crops to organic production systems (Mirsky et al., 2012). In drier environments with limited moisture, termination needs to be done several days or weeks prior to planting the main crops for the soils to rebuild moisture (Reberg-Horton et al., 2012; Wells et al., 2016) while termination can be conducted on the same day right before planting if the soil moisture content is not limited (Mirsky et al., 2013).

4. Future Research Needs

There is plenty of information regarding the establishment and management of cover crops in organic grain, vegetable and forage systems. However, region-specific information, especially on species selection, is limited. In addition, little is known about the potential of cover cropping in organic fruit production systems, which are often confined to marginal lands. As far as cover-crop species are concerned, the focus so far has been on cool-season cover crops with little information on the potential and challenges of growing warm-season cover crops in organic systems. In the future, more emphasis should be given to determine the suitability, economic management and ecosystem benefits from locally adapted, diversified cover-crop mixtures that can be easily integrated into the simplified organic grain and vegetable systems as well as into the organic crop-livestock integrated systems. Finally, although reduced tillage-based cover-crop termination strategies have emerged in recent years (e.g., use of roller crimper, sweep plow under-cutter), more innovation in this area is warranted to avoid the heavy reliance on intensive tillage in organic systems.

References

Ashford, D.L. and Reeves, D.W. (2003). Use of a mechanical roller-crimper as an alternative kill method for cover crops. Amer J. Altern. Agric., 18: 37–45.

Ball, D.M., Hoveland, C.S. and Lacefield, G.D. (2002). Southern Forages, third ed., Potash & Phosphorus Inst. and the Foundation for Agronomic Res., Norcross, GA.

Bavougian, C.M., Sarno, E., Knezevic, S. and Shapiro, C.A. (2019). Cover crop species and termination method effects on organic maize and soybean. Biol. Agric. Hort., 35: 1–20.

Blanco-Canqui, H., Shaver, T.M., Lindquist, J.L., Shapiro, C.A., Elmore, R.W., Francis, C.A. and Hergert, G.W. (2015). Cover crops and ecosystem services: Insights from studies in temperate soils. Agron. J., 107: 2449–2474.

Blesh, J., VanDusen, B.M. and Brainard, D.C. (2019). Managing ecosystem services with cover crop mixtures on organic farms. Agron. J., 111: 826–840.

Blevins, R.L., Smith, M.S., Thomas, G.W. and Frye, W.W. (1983). Influence of conservation tillage on soil properties. J. Soil Water Conser., 38: 301–305.

Brennan, E.B., Boyd, N.S., Smith, R.F. and Foster, P. (2009). Seeding rate and planting arrangement effects on growth and weed suppression of a legume-oat cover crop for organic vegetable systems. Agron. J., 101: 979–988.

Brennan, E.B., Boyd, N.S., Smith, R.F. and Foster, P. (2011). Comparison of rye and legume–rye cover crop mixtures for vegetable production in California. Agron. J., 103: 449–463.

Brennan, E.B. and Boyd, N.S. (2012a). Winter cover crop seeding rate and variety affects during eight years of organic vegetables: I. Cover crop biomass production. Agron. J., 104: 684–698.

Brennan, E.B. and Boyd, N.S. (2012b). Winter cover crop seeding rate and variety affects during eight years of organic vegetables: II. Cover crop nitrogen accumulation. Agron. J., 104: 799–806.

Brennan, E.B., Boyd, N.S. and Smith, R.F. (2013). Winter cover crop seeding rate and variety effects during eight years of organic vegetables: III. Cover crop residue quality and nitrogen mineralisation. Agron. J., 105: 171–182.

Canali, S., Campanelli, G., Ciaccia, C., Leteo, F., Testani, E. and Montemurro, F. (2013). Conservation tillage strategy based on the roller crimper technology for weed control in Mediterranean vegetable organic cropping systems. Eur. J. of Agron., 50: 11–18.

Carr, P.M., Anderson, R.L., Lawley, Y.E., Miller, P.R. and Zwinger, S.F. (2012). Organic zero-till in the northern US Great Plains Region: Opportunities and obstacles. Renewable Agric. Food Sys., 27: 12–20.

Casagrande, M., Peigné, J., Payet, V., Mäder, P., Sans, F.X., Blanco-Moreno, J.M. and Cooper, J. (2016). Organic farmers' motivations and challenges for adopting conservation agriculture in Europe. Organic Agric., 6: 281–295.

Cavigelli, M.A., Teasdale, J.R. and Conklin, A.E. (2008). Long-term agronomic performance of organic and conventional field crops in the mid-Atlantic region. Agron. J., 100: 785–794.

Cavigelli, M.A., Mirsky, S.B., Teasdale, J.R., Spargo, J.T. and Doran, J. (2013). Organic grain cropping systems to enhance ecosystem services. Renewable Agric. Food Sys., 28: 145–159.

Chu, M., Jagadamma, S., Walker, F.R., Eash, N.S., Buschermohle, M.J. and Duncan, L.A. (2017). Effect of multispecies cover crop mixture on soil properties and crop yield. Agric. Environ. Letters, 2: 170030.

Clark, A.E. (2009). Forages in organic crop-livestock systems. pp. 85–112. *In*: Francis, C. (ed.). Organic Farming: The Ecological System, Agronomy Monographs, 54. ASA, CSSA, SSSA, Madison, WI.

Clark, M.S., Horwath, W.R., Shennan, C. and Scow, K.M. (1998). Changes in soil chemical properties resulting from organic and low-input farming practices. Agron. J., 90: 662–671.

Creamer, N.G., Bennett, M.A., Stinner, B.R., Cardina, J. and Regnier, E.E. (1996). Mechanisms of weed suppression in cover crop-based production systems. Hort. Sci., 31: 410–413.

Creamer, N.G. and Baldwin, K.R. (2000). An evaluation of summer cover crops for use in vegetable production systems in North Carolina. Hort. Sci., 35: 600–603.

Cui, S., Zilverberg, C.J., Allen, V.G., Brown, C.P., Moore-Kucera, J., Wester, D.B. and Phillips, N. (2014). Carbon and nitrogen responses of three old world bluestems to nitrogen fertilisation or inclusion of a legume. Field Crops Res., 164: 45–53.

Dawson, J.C., Huggins, D.R. and Jones, S.S. (2008). Characterising nitrogen use efficiency in natural and agricultural ecosystems to improve the performance of cereal crops in low-input and organic agricultural systems. Field Crops Res., 107: 89–101.

Delate, K., Duffy, M., Chase, C., Holste, A., Friedrich, H. and Wantate, N. (2003). An economic comparison of organic and conventional grain crops in a long-term agro-ecological research (LTAR) site in Iowa. Amer. J. Altern. Agric., 18: 59–69.

Delate, K. and Cambardella, C.A. (2004). Agro-ecosystem performance during transition to certified organic grain production. Agron. J., 96: 1288–1298.

De Ponti, T., Rijk, B. and Van Ittersum, M.K. (2012). The crop yield gap between organic and conventional agriculture. Agric. Sys., 108: 1–9.

Dhamala, N.R., Rasmussen, J., Carlsson, G., Søegaard, K. and Eriksen, J. (2017). N transfer in three-species grass-clover mixtures with chicory, ribwort plantain or caraway. Plant Soil, 413: 217–230.

Dorais, M. (2007). Organic production of vegetables: State of the art and challenges. Can. J. Plant Sci., 87: 1055–1066.

Drinkwater, L.E., Shipanski, M., Snapp, S.S. and Jackson, L.E. (2008). Ecologically based nutrient management. pp. 159–208. *In*: Agricultural Systems: Agro-ecology and Rural Innovation for Development, Amsterdam, the Netherlands: Elsevier.

Faé, G.S., Sulc, R.M., Barker, D.J., Dick, R.P., Eastridge, M.L. and Lorenz, N. (2009). Integrating winter annual forages into a no-till corn silage system. Agron. J., 101: 1286–1296.

Fageria, N.K., Baligar, V.C. and Bailey, B.A. (2005). Role of cover crops in improving soil and row crop productivity. Comm. Soil Sci. Plant Analy., 36: 2733–2757.

Francis, C. and Van Wart, J. (2009). History of organic farming and certification. Organic Farming: The Ecological System (organic farming), 3–17.

Franzluebbers, A.J. and Stuedemann, J.A. (2014). Temporal dynamics of total and particulate organic carbon and nitrogen in cover crop grazed cropping systems. Soil Sci. Soc. Amer. J., 78: 1404–1413.

Franzluebbers, A.J. and Stuedemann, J.A. (2015). Does grazing of cover crops impact biologically active soil carbon and nitrogen fractions under inversion or no tillage management? J. Soil Water Conser., 70: 365–373.

Gaskell, M. and Smith, R. (2007). Nitrogen sources for organic vegetable crops. Hort. Tech., 17: 431–441.

Golabi, M.H., Radcliffe, D.E., Hargrove, W.L. and Tollner, E.W. (1995). Macropore effects in conventional tillage and no-tillage soils. J. Soil Water Conser., 50: 205–210.

Halde, C., Gulden, R.H. and Entz, M.H. (2014). Selecting cover crop mulches for organic rotational no-till systems in Manitoba, Canada. Agron. J., 106: 1193–1204.

Hector, A., Schmid, B., Beierkuhnlein, C., Caldeira, M.C., Diemer, M., Dimitrakopoulos, P.G. and Harris, R. (1999). Plant diversity and productivity experiments in European grasslands. Science, 286(5442): 1123–1127.

Hinsinger, P. (2001). Bioavailability of soil inorganic P in the rhizosphere as affected by root-induced chemical changes: A review. Plant Soil, 237: 173–195.

Hoyt, G.D., Monks, D.W. and Monaco, T.J. (1994). Conservation tillage for vegetable production. Hort. Tech., 4: 129–135.

Hutchinson, C.M. and McGiffen, M.E. (2000). Cowpea cover crop mulch for weed control in desert pepper production. Hort. Sci., 35: 196–198.

Inwood, S.E.E., Bates, G.E. and Butler, D.M. (2015). Forage performance and soil quality in forage systems under organic management in the southeastern United States. Agron. J., 107: 1641–1652.

Isik, D., Kaya, E., Ngouajio, M. and Mennan, H. (2009). Weed suppression in organic pepper (*Capsicum annuum* L.) with winter cover crops. Crop Protection, 28: 356–363.

Jabran, K., Mahajan, G., Sardana, V. and Chauhan, B.S. (2015). Allelopathy for weed control in agricultural systems. Crop Protection, 72: 57–65.

Katsvairo, T.W., Wright, D.L., Marois, J.J. and Rich, J.R. (2007). Transition from conventional farming to organic farming using Bahiagrass. J. Sci. Food Agric., 87: 2751–2756.

Keene, C.L., Curran, W.S., Wallace, J.M., Ryan, M.R., Mirsky, S.B., Van Gessel, M.J. and Barbercheck, M.E. (2017). Cover crop termination timing is critical in organic rotational no-till systems. Agron. J., 109: 272–282.

Kornecki, T.S., Price, A.J. and Raper, R.L. (2006). Performance of different roller designs in terminating rye cover crop and reducing vibration. Applied Eng. Agric., 22: 633–641.

Kramberger, B., Gselman, A., Kristl, J., Lešnik, M., Šuštar, V., Muršec, M. and Podvršnik, M. (2014). Winter cover crop: the effects of grass–clover mixture proportion and biomass management on maize and the apparent residual N in the soil. Eur. J. Agron., 55: 63–71.

Kuo, S., Sainju, U.M. and Jellum, E.J. (1997a). Winter cover crop effects on soil organic carbon and carbohydrate in soil. Soil Sci. Soc. Amer. J., 61: 145–152.

Kuo, S., Sainju, U.M. and Jellum, E.J. (1997b). Winter cover cropping influence on nitrogen in soil. Soil Sci. Soc. Amer. J., 61: 1392–1399.

Larsen, E., Grossman, J., Edgell, J., Hoyt, G., Osmond, D. and Hu, S. (2014). Soil biological properties, soil losses and corn yield in long-term organic and conventional farming systems. Soil Tillage Res., 139: 37–45.

Leavitt, M.J., Sheaffer, C.C., Wyse, D.L. and Allan, D.L. (2011). Rolled winter rye and hairy vetch cover crops lower weed density but reduce vegetable yields in no-tillage organic production. Hort. Sci., 46: 387–395.

Liebman, M., Davis, A.S. and Francis, C. (2009). Managing weeds in organic farming systems: an ecological approach. Organic farming: the ecological system. Agron. Monog., 54: 173–195.

Mirsky, S.B., Ryan, M.R., Curran, W.S., Teasdale, J.R., Maul, J., Spargo, J.T., Moyer, J., Grantham, A.M., Weber, D., Way, T.R. and Camargo, G.G. (2012). Conservation tillage issues: Cover crop-based organic rotational no-till grain production in the mid-Atlantic region, USA. Renewable Agric. Food Sys., 27: 31–40.

Mirsky, S.B., Ryan, M.R., Teasdale, J.R., Curran, W.S., Reberg-Horton, C.S., Spargo, J.T. and Moyer, J.W. (2013). Overcoming weed management challenges in cover crop-based organic rotational no-till soybean production in the eastern United States. Weed Tech., 27: 193–203.

Mischler, R., Duiker, S.W., Curran, W.S. and Wilson, D. (2010). Hairy vetch management for no-till organic corn production. Agron. J., 102: 355–362.

Mohler, C.L. and Teasdale, J.R. (1993). Response of weed emergence to rate of *Vicia villosa* Roth and *Secale cereale* L. residue. Weed Res., 33: 487–499.

Nair, A. and Ngouajio, M. (2012). Soil microbial biomass, functional microbial diversity, and nematode community structure as affected by cover crops and compost in an organic vegetable production system. Applied Soil Ecol., 58: 45–55.

Ngouajio, M., McGiffen Jr, M.E. and Hutchinson, C.M. (2003). Effect of cover crop and management system on weed populations in lettuce. Crop Protection, 22: 57–64.

Nyfeler, D., Huguenin-Elie, O., Suter, M., Frossard, E. and Lüscher, A. (2011). Grass-legume mixtures can yield more nitrogen than legume pure stands due to mutual stimulation of nitrogen uptake from symbiotic and non-symbiotic sources. Agric. Ecosys. Environ., 140: 155–163.

O'Connell, S., Grossman, J.M., Hoyt, G.D., Shi, W., Bowen, S., Marticorena, D.C. and Creamer, N.G. (2014). A survey of cover crop practices and perceptions of sustainable farmers in North Carolina and the surrounding region. Renewable Agric. Food Sys., 30: 550–562.

O'Connell, S., Shi, W., Grossman, J.M., Hoyt, G.D., Fager, K.L. and Creamer, N.G. (2015). Short-term nitrogen mineralization from warm-season cover crops in organic farming systems. Plant Soil, 396: 353–367.

Ott, S.L. and Hargrove, W.L. (1989). Profits and risks of using crimson clover and hairy vetch cover crops in no-till corn production. Amer. J. Alter. Agric., 4: 65–70.

Parr, M., Grossman, J.M., Reberg-Horton, S.C., Brinton, C. and Crozier, C. (2011). Nitrogen delivery from legume cover crops in no-till organic corn production. Agron. J., 103: 1578–1590.

Parr, M., Grossman, J.M., Reberg-Horton, S.C., Brinton, C. and Crozier, C. (2014). Roller-crimper termination for legume cover crops in North Carolina: Impacts on nutrient availability to a succeeding corn crop. Commun. Soil Sci. Plant Analy., 45: 1106–1119.

Price, A.J. and Norsworthy, J.K. (2013). Cover crops for weed management in southern reduced-tillage vegetable cropping systems. Weed Tech., 27: 212–217.

Qi, Z. and Helmers, M.J. (2010). Soil water dynamics under winter rye cover crop in central Iowa. Vadose Zone J., 9: 53–60.

Raghothama, K.G. (2005). Phosphorus and plant nutrition: an overview. pp. 355–378. *In*: Sims, J.T. and Sharpley, A.N (eds.). Phosphorus: Agriculture and the Environment, Agron. Monogr. 46. ASA, CSSA, and SSSA, Madison, WI.

Reberg-Horton, S.C., Grossman, J.M., Kornecki, T.S., Meijer, A.D., Price, A.J., Place, G.T. and Webster, T.M. (2012). Utilizing cover crop mulches to reduce tillage in organic systems in the southeastern USA. Renewable Agric. Food Sys., 27: 41–48.

Rogers, M.A., Wszelaki, A.L., Butler, D.M., Inwood, S.E. and Moore, J.L.C. (2015). Fall Cover Crop Selection and Planting Dates in Tennessee, UT Extension, Institute of Agriculture, The University of Tennessee. https://extension.tennessee.edu/publications/Documents/W235-I.pdf (Accessed on June 10, 2019).

Sacco, D., Moretti, B., Monaco, S. and Grignani, C. (2015). Six-year transition from conventional to organic farming: effects on crop production and soil quality. Eur. J. Agron., 69: 10–20.

Sayre, L. (2003). Introducing a Cover Crop Roller without All the Drawbacks of a Stalk Chopper Rodale Institute. https://betuco.be/CA/Conservation%20Agriculture%20-%20%20New%20Farm%20Research%20Cover%20crop%20roller%20.pdf (Accessed on July 26, 2019).

Schulz, M., Marocco, A., Tabaglio, V., Macias, F.A. and Molinillo, J.M. (2013). Benzoxazinoids in rye allelopathy-from discovery to application in sustainable weed control and organic farming. J. Chem. Ecol., 39: 154–174.

Şentürklü, S., Landblom, D.G., Maddock, R., Petry, T., Wachenheim, C.J. and Paisley, S.I. (2018). Effect of yearling steer sequence grazing of perennial and annual forages in an integrated crop and livestock system on grazing performance, delayed feedlot entry, finishing performance, carcass measurements, and systems economics. J. Animal Sci., 96: 2204–2218.

Shennan, C., Krupnik, T.J., Baird, G., Cohen, H., Forbush, K., Lovell, R.J. and Olimpi, E.M. (2017). Organic and conventional agriculture: a useful framing? Annual Rev. Environ. Resour., 42: 317–346.

Smith, R.G., Barbercheck, M.E., Mortensen, D.A., Hyde, J. and Hulting, A.G. (2011a). Yield and net returns during the transition to organic feed grain production. Agron. J., 103: 51–59.

Smith, A.N., Reberg-Horton, S.C., Place, G.T., Meijer, A.D., Arellano, C. and Mueller, J.P. (2011b). Rolled rye mulch for weed suppression in organic no-tillage soybeans. Weed Sci., 59: 224–231.

Smith, R.G., Atwood, L.W. and Warren, N.D. (2014). Increased productivity of a cover crop mixture is not associated with enhanced agro-ecosystem services. PLoS ONE, 9: e97351.

Snapp, S.S., Swinton, S.M., Labarta, R., Mutch, D., Black, J.R., Leep, R. and O'Neil, K. (2005). Evaluating cover crops for benefits, costs and performance within cropping system niches. Agron. J., 97: 322–332.

Soder, K.J., Rook, A.J., Sanderson, M.A. and Goslee, S.C. (2007). Interaction of plant species diversity on grazing behaviour and performance of livestock grazing temperate region pastures. Crop Sci., 47: 416–425.

Spargo, J.T., Cavigelli, M.A., Mirsky, S.B., Meisinger, J.J. and Ackroyd, V.J. (2016). Organic supplemental nitrogen sources for field corn production after a hairy vetch cover crop. Agron. J., 108: 1992–2002.

Taheri, F., Azadi, H. and D'Haese, M. (2017). A world without hunger: organic or GM crops? Sustainability, 9: 580.

Teasdale, J.R. and Mohler, C.L. (1993). Light transmittance, soil temperature, and soil moisture under residue of hairy vetch and rye. Agron. J., 85: 673–680.

Teasdale, J.R. (1996). Contribution of cover crops to weed management in sustainable agricultural systems. J. Production Agric., 9: 475–479.

Teasdale, J.R., Coffman, C.B. and Mangum, R.W. (2007). Potential long-term benefits of no-tillage and organic cropping systems for grain production and soil improvement. Agron. J., 99: 1297–1305.

Teasdale, J.R., Mirsky, S.B., Spargo, J.T., Cavigelli, M.A. and Maul, J.E. (2012). Reduced-tillage organic corn production in a hairy vetch cover crop. Agron. J., 104: 621–628.

Tosti, G., Benincasa, P., Farneselli, M., Tei, F. and Guiducci, M. (2014). Barley-hairy vetch mixture as cover crop for green manuring and the mitigation of N leaching risk. Eur. J. Agron., 54: 34–39.

Treadwell, D. (2009). Introduction to Cover Cropping in Organic Farming Systems. eXtension. https://articles. extension.org/pages/18637/introduction-to-cover-cropping-in-organic-farming-systems (Accessed on July 27, 2019).

Tu, C., Louws, F.J., Creamer, N.G., Mueller, J.P., Brownie, C., Fager, K., Bell, M. and Hu, S. (2006). Responses of soil microbial biomass and N availability to transition strategies from conventional to organic farming systems. Agric. Ecosys. Environ., 113: 206–215.

Vakali, C., Zaller, J.G. and Köpke, U. (2011). Reduced tillage effects on soil properties and growth of cereals and associated weeds under organic farming. Soil Tillage Res., 111: 133–141.

Vann, R.A., Reberg-Horton, S.C., Poffenbarger, H.J., Zinati, G.M., Moyer, J.B. and Mirsky, S.B. (2017). Starter fertiliser for managing cover crop-based organic corn. Agron. J., 109: 2214–2222.

Vann, R.A., Reberg-Horton, S.C., Edmisten, K.L. and York, A.C. (2018). Implications of cereal rye/crimson clover management for conventional and organic cotton producers. Agron. J., 110: 621–631.

Villamil, M.B., Bollero, G.A., Darmody, R.G., Simmons, F.W. and Bullock, D.G. (2006). No-till corn/soybean systems including winter cover crops. Soil Sci. Soc. Amer. J., 70: 1936–1944.

Wallace, J.M., Williams, A., Liebert, J.A., Ackroyd, V.J., Vann, R.A., Curran, W.S., Keene, C.L., VanGessel, M.J., Ryan, M.R. and Mirsky, S.B. (2017). Cover crop-based, organic rotational no-till corn and soybean production systems in the mid-Atlantic United States. Agric., 7: 34.

Wang, Q., Klassen, W., Li, Y. and Codallo, M. (2009). Cover crops and organic mulch to improve tomato yields and soil fertility. Agron. J., 101: 345–351.

Wang, G., Ngouajio, M., McGiffen, M.E. and Hutchinson, C.M. (2008). Summer cover crop and management system affect lettuce and cantaloupe production system. Agron. J., 100: 1587–1593.

Wayman, S., Cogger, C., Benedict, C., Burke, I., Collins, D. and Bary, A. (2015). The influence of cover crop variety, termination timing and termination method on mulch, weed cover and soil nitrate in reduced-tillage organic systems. Renewable Agric. Food Sys., 30: 450–460.

Wayman, S., Kucek, L.K., Mirsky, S.B., Ackroyd, V., Cordeau, S. and Ryan, M.R. (2017). Organic and conventional farmers differ in their perspectives on cover crop use and breeding. Renewable Agric. Food Sys., 32: 376–385.

Wells, M.S., Reberg-Horton, S.C., Smith, A.N. and Grossman, J.M. (2013). The reduction of plant-available nitrogen by cover crop mulches and subsequent effects on soybean performance and weed interference. Agron. J., 105: 539–545.

Wells, M.S., Brinton, C.M. and Reberg-Horton, S.C. (2016). Weed suppression and soybean yield in a no-till cover-crop mulched system as influenced by six rye cultivars. Renewable Agric. Food Sys., 31: 429–440.

Weston, L.A. (1996). Utilisation of allelopathy for weed management in agro-ecosystems. Agron. J., 88: 860–866.

Wittwer, R.A., Dorn, B., Jossi, W. and Van Der Heijden, M.G. (2017). Cover crops support ecological intensification of arable cropping systems. Scientific Reports, 7: 41911.

Wortman, S.E., Francis, C.A., Bernards, M.A., Blankenship, E.E. and Lindquist, J.L. (2013). Mechanical termination of diverse cover crop mixtures for improved weed suppression in organic cropping systems. Weed Science, 61(1): 162–170.

Wyland, L.J., Jackson, L.E., Chaney, W.E., Klonsky, K., Koike, S.T. and Kimple, B. (1996). Winter cover crops in a vegetable cropping system: Impacts on nitrate leaching, soil water, crop yield, pests and management costs. Agric. Ecosys. Environ., 59: 1–17.

Zinati, G., Reddivari, L. and Kemper, D. (2019). Reduced Tillage Increases Nutrient Concentrations in Stored Winter Squash, Rodale Institute. https://rodaleinstitute.org/science/articles/reduced-tillage-increases-nutrient-concentrations-in-stored-winter-squash (Accessed on July 26, 2019).

4

Cover Crops in Vegetable Production and Urban Farming in Sub-Saharan Countries

Michael Kwabena Osei,[1,]* *Mavis Akom,*[1] *Joseph Adjebeng-Danquah,*[2]
Kenneth Fafa Egbadzor,[3] *Samuel Oppong Abebrese,*[2] *Kwabena Asare Bediako*[1]
and *Richard Agyare*[2]

1. Introduction

Global food production has not kept pace with the increasing population growth. There is a decline in soil fertility and accelerated desertification on marginal soils worldwide, including Ghana. Decades of cropping have resulted in unbalanced soil fertility, reduced levels of soil organic matter (OM) and abundance of marginal and degraded soils. Efforts to revive global agricultural productivity, including in Africa, must deal with degraded soils in many parts of the region. In Ghana, the total land area is 23,853,900 ha and 57.1 per cent (13,628,179 ha) of it is suitable for agriculture. However, most of the soils are of low inherent fertility. The coarse nature of the soils has an impact on their physical properties, and water stress is common during the growing season. Extensive areas of the land, particularly the interior savannah zone, have suffered from severe soil erosion and land degradation in various forms. The soil nutrient depletion rates in Ghana are projected as 35 kg N, 4 kg P, and 20 kg K ha^{-1} (Bationo, 2015). The extent of nutrient depletion is widespread in all the agro-ecological zones with nitrogen and phosphorus being the most deficient nutrients. Nutrients removed from the soils by crop harvesting have not been replaced through the use of corresponding amounts of plant nutrients in the form of organic and inorganic fertilisers. There is, therefore, a steady decline in crop-yield levels and increased food production is due mostly to extension of the area under cultivation. The average yields of most of the crops are 20–60 per cent below their achievable yields, indicating that there is a significant potential for improvement.

While Ghana has one of the highest soil nutrient depletion rates in Sub-Saharan Africa (SSA), it has one of the lowest rates of annual inorganic fertiliser application—only 8 kg per hectare on an average, in contrast to the Abuja Declaration target of 50 kg per hectare (Bationo, 2015). The average fertility status of soils in the different agro-ecological zones is presented in Table 1. The major processes of soil degradation in Ghana are biological (loss of organic matter), chemical (depletion of nutrients, salinity and acidification), and physical (erosion, compaction, crusting and iron pan formation).

[1] CSIR-Crops Research Institute, P.O. Box 3785, Kumasi, Ghana.
[2] CSIR-Savannah Agricultural Research Institute, P.O. BOX 52, Tamale-Nyanpala, Ghana.
[3] Ho Technical University, Faculty of Applied Sciences and Technology, Dept. of Agro Enterprise Development, Ho, Ghana.
* Corresponding author: oranigh@hotmail.com

Table 1 Soil fertility status of the various agro-ecological zones.

Agro-ecological zones	Soil pH	Organic C	Total N	Available P	Available K
	(%)	(mg/kg)			
High Rain-forest	3.8–5.5	1.52–4.24	0.12–0.38	0.12–5.4	63.6–150.4
Forest-Transition	5.1–6.4	0.59–0.99	0.04–0.16	0.30–4.75	8.3–72.5
Semi-Deciduous Forest	5.5–6.2	1.59–4.8	0.15–0.42	0.36–5.2	62–84.8
Coastal Savanna	5.6–6.4	0.61–1.24	0.05–1.16	0.28–4.1	48–58.7
Guinea Savanna	6.2–6.6	0.51–0.99	0.05–0.12	0.18–3.6	46.2–55.3
Sudan Savanna	6.4–6.7	0.48–0.98	0.06–0.14	0.06–1.8	37–44.5

Source: Bationo, 2015

Most efforts to increase fertiliser use in SSA over the past decade have focused on subsidies and targeted credit programmes. Where external inputs are expensive, cover crops grown on site can help to maximise the benefits of external inputs. A cover crop is a crop planted primarily to manage soil erosion, soil fertility, soil quality, water, weeds, pests, diseases, biodiversity and wildlife in an agro-ecosystem, which is an ecological system managed and largely shaped by humans across a range of intensities to produce food, feed, or fibre. Cover crops provide efficient, low-cost sources of N. They improve soil structure, increase the soil's biological activity and help to control pests. A major advantage of cover crops is their capacity to control noxious weeds, such as *Imperata cylindrical*, that are choking out crops in many regions. The biomass generated by cover crops can also be used as feed for animals. Evidence from West Africa presented in this chapter shows that cover crops help revive degraded land and sustain intensive agricultural practices.

One of the main uses of cover crops is to improve soil fertility and these types of cover crops are referred to as 'green manure'. They are used to manage a range of soil macronutrients and micronutrients. Of the various nutrients, the impact that cover crops have on nitrogen management has received utmost attention from researchers and farmers because nitrogen is often the most limiting nutrient in crop production. These cover crops are often grown for a specific period and then plowed under, before reaching full maturity in order to improve soil fertility and quality. Cover crops can also improve soil quality by increasing SOM levels through the input of cover crop biomass over time. Increased SOM content enhances soil structure as well as the water and nutrient-holding and buffering capacity of soil. It can also lead to increased soil carbon sequestration, which has been promoted as a strategy to help offset the rise in atmospheric carbon dioxide levels.

Current vegetable production systems require an intensive amount of work and inputs, and if not properly managed, could have detrimental effects on soil and the environment. Practices, such as intensive tillage, increased herbicide use and reduced organic matter inputs add additional stress on the sustainability of vegetable production systems. Growers, therefore, need the best practices to make production systems sustainable without compromising farm productivity and profitability. Cover crops serve as a valuable production tool in preserving the environmental sustainability of vegetable cropping systems and render numerous benefits to soil, vegetable crops and the grower. They are not harvested, but rather are planted to improve soil quality and provide other benefits for crop production and the environment. Before planting the next vegetable crop, most cover crops need to be cut down. The shoots can be chopped (or mowed) and left as mulch on the soil surface or incorporated into the soil.

In recent times, there has been an increased interest in cover crops grown as improved short fallow in agricultural research. Leguminous cover crops grown as improved short fallow have shown high agronomic potential (Peoples et al., 1995). Cover crops used in cropping systems, such as in a rotation, enhance biological mechanisms and serve as a potential substitute for chemical inputs. These can be incorporated into the soil before planting cash crops and left on the soil surface as mulch. The high cost of inputs, such as chemical fertilisers and other agro-chemicals, has created an urgent need for an alternative to the use of agro-chemicals for vegetable production in Ghana (ISSER, 2005). Due to the high cost of inputs (i.e., fertilisers, insecticides, fungicides, tools, etc.), farmers resort to the use of cover crops as a sustainable farming practice in order to break even.

Using cover corps will not only produce higher quality produce but will also help restore a healthy ecosystem in the soil. This chapter focuses on the use of cover crops in vegetable production and urban farming with an emphasis on Ghana. It discusses vegetable production and urban farming in Ghana, characteristics and examples of commonly used cover crops in vegetable production, types of cover crops used in vegetable production and the prospects and challenges of using cover crops in vegetable production.

2. Vegetable Production and Urban Farming in Ghana

The proportion of Ghana's population living in urban communities continues to increase. As indicated in the 2010 population and housing Census report, urbanisation increased from 9.4 per cent in 1931 to 50.9 per cent in 2010 (Ghana Statistical Service, 2014). Considering Ghana's standard of urbanisation and its trend, one can confidently say that Ghana's urban dwellers are currently more prolific than rural dwellers. Urban residency, however, varies from region to region within the country. The Greater Accra region has a higher urban population than rural, while the contrary is true for the Upper east and west regions, for instance.

Farming, and for that matter vegetable cultivation, is conventionally a trade for rural dwellers. More than 80 per cent of the population of some administrative districts in Ghana today are farmers. The Ho west district in the Volta region is an example with 88.7 per cent farmers (Ghana Statistical Service, 2014a). Despite this, many people in urban areas also engage in agriculture for their livelihood. At least 22 per cent of urban households in Ghana today are involved in agriculture production. This average, however, varies from 4.4 per cent in the Greater Accra region to 51.6 per cent in Brong Ahafo (Ghana Statistical Service, 2014b). This shows the significance of agriculture in terms of employment, even to city dwellers. Urban farmers continue to increase as urbanisation itself increases (Fig. 1). This trend has resulted in a corresponding increase in urban vegetable cultivation, which is demand-driven.

Vegetables are normally classified as indigenous or exotic. Some of the indigenous vegetables are wild but serve an important role in the diet of many people, especially the rural poor (Maundu, 2013). Two examples of such vegetables are the wild amaranth (*Amaranthus viridis*) and lettuce (*Launaea taraxacifolia*). Despite their importance, attention is not given to the cultivation of wild vegetables; therefore, their existence is threatened due to some farming practices, including the application of herbicides (Boutin et al., 2013). Some indigenous vegetables, like *Solanum microcapon* L. and the naturalised *Corchorus olitorius* with high market demand are cultivated alongside exotic vegetables

Fig. 1 Urban vegetable farming in Accra, Ghana (*Source*: Allen, 2010).

on commercial scales in cities of Ghana. Some introduced vegetables currently grown throughout Ghana are tomato (*Solanum lycopersicum* L.) and eggplant (*Solanum melongena*), which originated in Peru/Ecuador and India, respectively. The list of vegetables cultivated in urban areas in Ghana is long, with some more or less concentrated to particular localities, while others are cultivated widely. Indigenous and naturalised vegetables are relatively easier to cultivate as compared to the exotic types that originate from different climates, such as the temperate and Mediterranean regions.

It is more difficult to cultivate exotic vegetables than indigenous ones, with the former having higher market demand; therefore, their cultivation tends to be more lucrative than the latter. Notable exotic vegetables cultivated in commercial quantities in Ghana include lettuce (*Lactuca sativa* L.), cabbage (*Brassica oleracea* L. var. *capitata* L.), spring onion (*Allium fistulosum* L.), sweet pepper (*Capsicum annum* L.), cucumber (*Cucumis sativus* L.) and carrot (*Daucus carota*). The challenges associated with the cultivation of these vegetables, however, create a good business opportunity for interested farmers, especially those in urban areas.

Urban vegetable producers are usually migrants from the countryside with experience in farming hitherto their relocation. This is especially true for farmers in Accra and Kumasi, the two largest cities in Ghana that also attract the highest number of migrants. Most of these farmers do not have the requisite education and training for white collar jobs in the cities. Hence, they go into farming when jobs for which they came to the cities become illusive. These farmers take advantage of the available land and sometimes even the water resources as well as market opportunities to enter into farming. Some of the farmers also do other menial jobs in addition to vegetable cultivation.

Undeveloped lands belonging to institutions, either government or private, are usually the available spaces used for urban vegetable cultivation. Similar to the Council for Scientific and Industrial Research (CSIR) open space vegetable cultivation in Accra are Korle-Bu, Tema – Accra motor way, Dzorwulu, La and Roman Down in Ashaiman. Common characteristics of all these farming sites are the types of vegetables, which are mainly exotic, and the irrigation water that they obtain from filthy drains. The source of water for urban vegetable cultivation is thus a health concern in most cases (Keraita et al., 2014). Most farmers use a watering can for irrigation; however, some farmers are now using water-pumping machines.

Crop rotation is another common practice for these farmers. Farmers in Ashaiman and Tema sometimes rotate their vegetables with maize, which they sell as fresh corn. CSIR-area farmers, however, usually rotate one vegetable with another. The demand for vegetables in Kumasi makes their cultivation lucrative, similar to what goes on in Accra. Many farmers in Kumasi cultivate vegetables along the drains. Land at portions of these drains is not suitable for construction. This becomes an advantage for farmers, who use the land to cultivate vegetables and use the water in drains for irrigation. Notable vegetable growing areas in Kumasi are Kwame Nkrumah University of Science and Technology (KNUST), Asokore, Agriculture College Farm and Dayname Hotel. In less upstart urban areas, like Ho, cultivation of local vegetables like okra (*Abelmoscus esculetus*) and hot pepper (*Capsicum frutescens* L.) dominates exotic vegetables. However, most of the exotic vegetables, such as cabbage and spring onion, cultivated in Accra and Kumasi, are found in the smaller cities, especially during the rainy season. These are produced for local consumption and have a ready market. There is, however, a high demand for the exotic vegetables in these smaller cities, as a considerable amount of such exotic vegetables from different localities are imported and sold in their local markets.

In addition to vegetables, varieties of other crops, such as maize, cocoyam, plantain and cassava are also found in backyard gardens of the cities. These crops are produced on small scale, mainly for home consumption. It is common to find one or a few stands of plantains or cassava, or mounds of yam in gardens in the cities. Few vegetables, either planted as mono or mixed crops for home consumption, are also found in many homes in the cities of Ghana. These are different from the commercial vegetable farms in the cities. Crops of interest to city gardeners vary from home-to-home and from city-to-city. In general, vegetables are produced on a larger scale than non-vegetables, like maize and yams, in the cities. However, some amount of commercial farming involving non-vegetables are also found in the cities. At the University of Ghana, for instance, undeveloped lands are usually cultivated with a variety of commercial scale crops including vegetables, legumes, cereals, and roots and tubers.

Vegetables play an important role in the diet of the individual and their cultivation also contributes to the economic growth of the country. Urban vegetable cultivation will continue to be a lucrative business for several reasons. First, increased urbanisation has a corresponding increase in demand for not only food, but also quality food. Vegetables cultivated near consumers, therefore, will get to a ready market, making it profitable. Secondly, many people who make their way into the cities but do not have good jobs, will use any available space to cultivate different crops for their livelihood. In areas where it is affordable, advantage will be taken of advancements in technology to obtain borehole water for irrigation. Currently, shallow wells are dug to supply water for vegetable cultivation in some coastal towns, like the Keta and Denu municipalities in the Volta Region. Vegetable cultivation will continue to be a good business in the urban areas of Ghana. To make the best out of urban vegetable cultivation, however, employment of modern technology is the key. In addition, the best effort must be made to reduce contamination, such as fecal material from wastewater, microbial contamination and agro-chemical residue (Gonzalez et al., 2016). Last but not least, advantage can be taken of the government's planting for food and job policy, which has vegetables as the target crops, as a means of creating jobs for youth (Ghana-Agriculture Sector Policy Note, 2017).

3. Uses and Importance of Cover Crops in Vegetable Production and Urban Farming

Weed Suppression: Many species of weed flourish throughout the crop production period with consequent management practices that increase production costs. Cover crops play a pivotal role in vegetable production systems by reducing weed germination and establishment and herbicide requirements (Fageria et al., 2005). Weed suppression by cover crops has been attributed to competition for resources (light, nutrients and water), niche disruption and phytotoxic and allelopathic effects (Dabney et al., 2001; Hutchinson and McGiffen, 2000). Cover crops influence weeds while they grow (living mulches) and as residue (dead mulch) when killed (Moyer et al., 2000). Weed control is usually best achieved in dense cover-crop plantings and when cover crops are maintained for the longest time possible (Smeda and Putnam, 1988). Hutchinson and McGiffen (2000), for example, as observed in season-long weed control with the use of cowpea mulch in a pepper field.

Erosion Control: Soil becomes exposed to erosion in the absence of adequate vegetation ground cover or crop residues on its surface, particularly under intensive tillage in vegetable production systems. Consequently, the loss of topsoil by wind and water erosion occurs, resulting in deterioration of physical, chemical and biological properties of the soil and further decline in productivity of most croplands (Dabney et al., 2001). Cover crops are of paramount importance in reducing water runoff and soil erosion, thus contributing to improved soil productivity. Cover crop provides vegetation cover to reduce the effect of raindrops that otherwise would detach soil particles and culminate in erosion. Furthermore, cover-crop root systems help to hold soil in place against raindrop impact and water and wind erosion by anchoring soil aggregates, thereby reducing aggregate breakdown and erosion and increasing SOM content, which improves soil water infiltration and holding capacity (Fageria et al., 2005).

Organic Matter Addition: Decomposition of residues from cover crops furnishes the soil with organic matter. In most agricultural soils, average organic matter content is generally low. Hence, an increase in organic matter through the decomposition of cover crops will be of tremendous benefit to these soils. Organic matter from cover crops releases available nutrients to plants and improves nutrient cycling by increasing the population and activities of soil fauna, such as earthworms and millipedes, which help create air pore spaces in the soil. In addition, organic matter improves crop yields by improving soil physical conditions, such as structure, aggregate stability, water holding and buffering capacities and porosity. This results in increased soil permeability and aeration, which facilitate crop emergence and promotes root growth (Carter and Stewart, 1996; Fageria et al., 2005).

Disease and Insect Control: Cover crops can be used in rotation to control diseases and insects by breaking pest cycles. Cover crops have been reported to create a favourable environment for beneficial predators and parasitoid insects, which serve as prey for harmful insects and vectors of disease-causing organisms that cause significant damage to vegetable crops. In addition, cover crops have been reported to control many soil-borne pathogenic fungal diseases and nematodes in succeeding cash crops (Fageria et al., 2005). Again, cover crops used as mulches may release chemicals that influence pest dispersal and reproduction on vegetable crops (Rice, 1984).

Fixing of Atmospheric Nitrogen: Leguminous cover crops have the ability to fix nitrogen from the atmosphere for their growth and the growth of succeeding crops. Nitrogen fixation occurs through a symbiotic association between leguminous crops and nitrogen-fixing bacteria present in the nodules of plant roots. The plant provides food and shelter to the bacteria, as the bacteria fixes nitrogen for plant growth. Furthermore, stem, leaf and root residues left in the field from leguminous cover crops contain high nitrogen content. The decomposition of residues from legumes commonly releases nitrogen into the soil for use by succeeding vegetable crops and, consequently, reduce grower's dependence on nitrogen fertiliser (Hartwig and Ammon, 2002; Singh et al., 2004). Greater soil nitrogen availability arising from the use of a leguminous cover crop increases the amount of marketable fruits, biomass yields and relative fitness of vegetables (Campiglia et al., 2014).

Nutrient Scavenging and Recycling: Cover crops can increase the nutrient-use efficiency of vegetable farming systems. Cover-crop roots penetrate deep into soil layers to scavenge and bring up nutrients, which may have otherwise leached during the growing season to the upper soil layers. Following decomposition of such cover crops, nutrients are released and used by succeeding vegetable crops (Dabney et al., 2001; Hartwig and Ammon, 2002).

3.1 Characteristics and Examples of Commonly Used Cover Crops in Vegetable Production

Many types of plants are used as cover crops. The common ones include legumes, grasses and brassicas (Baldwin and Creamer, 2006; Marr et al., 1998). The choice of a cover crop depends on its purpose within the cropping system. Among the purposes is included whether it has to add available nitrogen to the soil or scavenge nutrients (Lenzi et al., 2009). For instance, legumes add nitrogen to the system whilst others, mostly the grasses, take up available soil nitrogen (Maughan and Drost, 2016). The nitrogen-fixing ability of legumes improves when it is inoculated with the right strain of inoculant (Maughan and Drost, 2016). Cover crops with high and easily degraded biomass, like brassicas, are desirable if the objective is to provide large amounts of organic residue (Marr et al., 1998).

When the objective is to use the cover crop as surface mulch or to incorporate it into the soil, the ones with large broad leaves and creeping habit are preferred. Such cover crops are also good for preventing erosion and weed control (Marr et al., 1998; Baldwin and Creamer, 2006). Other cover crops with deep and larger root systems, like radish, are good for soils with compaction problems (Maughan and Drost, 2016). The growth duration and prevailing weather conditions (summer or winter) will also affect the choice of a cover crop. Some cover crops require a degree of coldness, while others do not tolerate cold. The growth duration of the cover crop should also not interfere with the intended main crop. This makes early maturing legumes and grasses a popular choice (Baldwin and Creamer, 2006; Maughan and Drost, 2016; Danso et al., 2014). In addition, rotation of the cover crop with the main crop is important when the objective is to control diseases and pests, such as nematodes. Some cover crops are known to have an allelopathic effect on other crops, so it is important to be mindful of which crops to follow in a rotation (Lenzi et al., 2009). The common legumes and non-legumes suitable for specific uses, including for use in Ghana, are covered in the following sections.

Legumes: The following legume cover crops: *Crotalaria retusa, C. jucea*, cowpea (*Vigna unguiculata*), *Mucuna pruriens*, and pigeon pea (*Cajanus cajan*) are capable of rapidly improving soil fertility by fixing nitrogen. These cover crops are often included in crop rotation plans for effective management of soil fertility. Legume cover crops are, however, not as effective for weed control because they are slow to establish, do not produce a significant amount of biomass and tend to breakdown more quickly than grasses, reducing their ability to suppress late-emerging annual weeds.

Non-legume Cover Crops: The non-legume cover crops are mostly small-seeded annual grasses, such as brachiaria (*Brachiaria ruziziensis*), finger millet (*Eleusine coracana*) and sorghum (*Sorghum bicolor*). These non-legume cover crops produce a significant amount of biomass, thereby offering better mulches for weed control, moisture conservation and mitigation of erosion. They are characteristically warm-season annual grasses that grow well in hot conditions and produce a large amount of biomass. Planting is done by drilling at 9–18 kg per acre seed rate. To get the most growth from these non-legume cover crops, nitrogen fertiliser (18–36 kg/ac.) must be applied. If incorporated at a young stage, the nitrogen will be re-released for the following crop. Some of the grass cover crops (e.g., *Sorghum bicolor*) are very effective in suppressing weeds and have been shown to have allelopathic and biofumigant properties that improve vegetable productivity (Danso et al., 2014; Baldwin and Creamer, 2006).

4. Types of Cover Crops Used in Vegetable Production in Ghana

Cover crops used in vegetable production in Ghana can be grouped into legumes and non-legumes, based on their uses. The legumes are grown primarily for their nitrogen-fixing potential and high nitrogen content in their biomass. The non-legumes are grown for their abilities to prevent erosion and capture and recycle nutrients in the soil. Some common cover crops used in vegetable production in Ghana are as follows:

Source: CSIR – Crops Research Institute

Cowpeas are important grain legumes in Ghana's cropping system. They are grown mainly for grains, green manure and animal fodder. The variety used are mostly semi-erect to erect in stature with a maturing period ranging between 55–77 days after sowing (MoFA, 2015). The crop is adapted to a wide range of soils and grown in all the agro-ecology zones in the country. Evenly distributed rainfall of at least 500 mm is required throughout the growing season (Adu-Dapaah et al., 2005). Some varieties have yield and biomass potential of 2.9 and 3.5 ton ha⁻¹, respectively (MoFA, 2015). Other varieties are moderately resistant to insect pests, especially thrips and *Striga hermonthica* disease.

Source: CSIR – Crops Research Institute

Groundnut is an important food and cash legume in Ghana. The crop is grown throughout the country with a majority of the production taking place in the northern Guinea and Sudan savannah zones of the country (Tanzubil et al., 2017). The varieties used are creeping, semi erect and erect in stature. Groundnuts are grown for their grains, green manure, soil-fertility improvement via nitrogen fixation and as fodder for animals. The days to maturity for the varieties range from 90–120 days with a potential kernel yield between 2.0–2.7 ton ha⁻¹.

Source: CSIR – Crops Research Institute

Soybean is relatively a new crop in Ghana as compared to other legumes. The crop is cultivated on a small scale by farmers in the country for its grains, green manure, animal feeds and weed control (*Striga hermonthica*). Soybean is well adapted to a wide range of climatic and soil conditions, with an annual rainfall of not less than 700 mms, well distributed throughout the growing season (Asafo-Adjei et al., 2005). The crop grows best in the Guinea savannah and forest-savannah transitional zones. The maturity period for the varieties in Ghana range from 85–130 days after planting with a grain yield potential of 1.5–2.8 ton ha^{-1} (MoFA, 2015). Soil cultivars serve as trap crops for *Striga hermonthica* disease.

***Mucuna* sp.** is an important plant for green manure, fallow and forage. The crop improves the fertility of the soil by fixing nitrogen and helps in the control of weeds, such as *Imperata cylindrical* (Buckles, 1995). *Mucuna* is occasionally used for human consumption in the preparation of soups and stew. The plant grows well in most parts of the country and thrives on most soil types. The crop has a high dry matter content and a yield range from 5–12 tha^{-1} depending on the rainfall pattern (Cook et al., 2005). The most common species grown in the country is *Mucuna pruriens*.

Sorghum is an important cereal crop, especially in the savannah agro-ecological zone of the country. The crop is used in foods, beverages and livestock feed. It is mostly cultivated in the savannah agro-ecological zones of the country, but it can adapt to different environmental conditions. The crop grows well with an annual average rainfall range of 500–800 mm (Sani et al., 2013).

Source: gardja.org

Pearl Millet is mostly grown in the Guinea and Sudan savannah agro-ecological zone of the country. The crop is mainly grown as a food crop, fodder and building material (roofing and fencing). The varieties used in the country have a maturing period range of 70–75 days after sowing and a potential yield range of 1.9–2.1 ton ha^{-1} (MoFA, 2015).

Source: ghananewsagency.org

5. Growth and Management of Cover Crops in Vegetable Production and Urban Farming

Cover crops in Urban Vegetable Systems: Integrating cover crops into vegetable production systems is important due to the numerous advantages they provide. However, a proper analysis of the production system needs to be evaluated before its implementation (Maughan and Drost, 2016). Growers need to consider the type of cover crop to use, the season to cultivate it, cultural practices required for its growth, when and how to incorporate it into the soil and crop rotations in cover-crop vegetable-production systems (Marr et al., 1998). Cover crops should be integrated into the production system in such a way that the growth of the cover crop does not overlap with the time for growing the vegetable crops (Maughan and Drost, 2016). A careful integration of cover crops into urban vegetable production systems increases the chances of maximising the successes of vegetable production systems (Maughan and Drost, 2016).

Methods of Planting Cover Crops: Land preparation for growing cover crops usually follows the practice of making seed-beds or ridges using farm implements or manpower (manually).

However, in urban vegetable farming, land preparation is mainly done using a hand hoe and cutlass (manpower) due to the small landholdings of urban vegetable farmers, which makes it impossible to use tractor-mounted farm implements (Marr et al., 1998). Animal-drawn implements, such as a mold-board plow, are sometimes used. The field should be levelled to ensure optimum planting depth and uniform crop establishment (Maughan and Drost, 2016). Cover crops can be planted using mechanical or manual methods. However, the method of planting depends on the size of the grower's field and farm implements available. In Ghana, cover crops are mostly planted manually by either broadcasting or row planting. Broadcasting seeds of cover crops may be easy, fast and an inexpensive method compared to row planting. However, it requires more seeds and leads to poor field establishment and weed infestations (SARE, 2012). Row planting, on the other hand, utilises optimum plant spacing and efficient use of land, involves optimum seed requirement and an early closure of the canopy results in reduced soil compaction and weed infestation. Plant cover crops when there is adequate moisture in the soil to support germination and emergence.

***Field Establishment of Cover Crops*:** The benefits derived from growing cover crops depend significantly on the management practices employed in the field. The time of planting influences the benefits obtained from growing cover crops and should be manipulated so that plant establishment and maximum biomass production will be attained before they are incorporated in the soil (Maughan and Drost, 2016). Under rain-fed urban vegetable farming in southern Ghana, cover crops can be planted towards the end of the major season (June) to take advantage of the residual moisture that provides plant cover during the dry season (July–August). They can also be replanted at the end of the minor season in December to provide plant cover before the beginning of the rainy season in February (personal communication). However, in northern Ghana, cover crops can be planted at the end of the short rainy season in October to provide plant cover for the long dry season, which lasts until May. The cover crops should be protected from bush fires and free-grazing animals. Additional irrigation may be required during the season to ensure maximum growth and development. High seed rate is required to ensure a denser plant population and early ground cover (Maughan and Drost, 2016). Some cover crop species, especially cereals, are fast-growing, establish rapidly and are able to provide enough ground cover within a short period (Marr et al., 1998). Such cover crops are able to close the canopy quickly and minimise weed competition and soil dryness. However, broad-leaf cover crops establish slowly at the seedling stage but grow at a faster rate and improve the nutrient status of the soil through biological N-fixation before termination (Marr et al., 1998).

***Termination (Killing) of Cover Crops*:** Termination refers to killing or incorporating cover crops into the soil (Maughan and Drost, 2016). Killing can be done using tillage or with non-selective herbicide. If herbicides are used, ensure that the safety period has elapsed before planting the main crop to avoid herbicide effects on the main crop (SARE, 2012). Incorporating cover crops under the soil is the most common method of working cover crops into the soil (Hendrickson, 2009). The time to terminate the growth of a cover crop may be influenced by the type of cover crop grown (SARE, 2012). For example, leguminous and non-leguminous crops may vary in their rate of decomposition and, therefore, should be terminated at different times. The stage of growth of cover crops also influences the rate of decomposition and release of nutrients (Maughan and Drost, 2016).

Cover crops killed at the vegetative stage or just before flowering decompose faster and release nutrients for early crop establishment. However, cover crops terminated after flowering become woody and decompose at a slower rate (Peet, 2001). Cereal cover crops should be terminated before full vegetative growth to avoid nutrient lock-up in plant tissue (SARE, 2012). Cover crops allowed to set seeds before termination may reseed and become weeds in the vegetable crop (Maughan and Drost, 2016). In addition, the time of terminating cover crops also depends on the method of cultivating the vegetable crop (SARE, 2012). For instance, growing vegetables under plastic mulch will require that the cover crop be incorporated ahead of time to allow for decomposition before laying the plastic. Also, the time of killing the cover crop depends on the nature of the cover-crop species. Some cover crops exhibit allelopathic properties that can affect the germination of the vegetable seeds (Hendrickson, 2009). Such cover crops should be killed and incorporated into the soil before planting the vegetable crop. In general, cover crops should be incorporated into the soil two-to-four weeks before planting the desired vegetable crop (Maughan and Drost, 2016). To aid early decomposition and easy land preparation for the vegetable crop, cover crops should be slashed (shred) before being incorporating into the soil (Maughan and Drost, 2016). Ensure that the cover crop is completely killed to avoid rejuvenation in the succeeding vegetable crop. Under no-till, cover crops can be slashed to provide a layer of mulch on the soil surface before planting the main cash crop (Walters et al., 2005).

Rotation in Cover Crop Vegetable Farming: Harnessing the benefits of cover crops in urban vegetable farming requires proper planning and effective management. Although inclusion of cover crops into urban vegetable rotation systems in Ghana is challenging and uncommon due to small farm sizes and high demand of vegetables, the role of cover crops in soil health and quality overrides the challenges. The challenges of cover-crop vegetable rotations vary from location to location. In addition, cover crops vary in their ability to improve the soil and fit well into a rotation system; therefore, you should experiment with different cover crops in the locality and finally select what is most beneficial to your condition (Hendrickson, 2009). Experiment with the rotation system on smaller, manageable plots before expanding to cover the entire field.

The compatibility of different cover crops and the vegetable crop in a rotation sequence is essential in improving the physico-chemical and biological activities of the soil, as well as increasing vegetable production (Iowa State University). Carefully plan and integrate cover crops into a rotation system to avoid failure of the system, which may be detrimental to vegetable cultivation (SARE, 2012). The rotation system should be such that you can adjust and adapt to changes in weather conditions. In order to build the health and quality of the soil and improve organic matter content and break the build-up of pest associated with cover crops, growers have to use cover crops in a rotation system. Rotate leguminous and non-leguminous cover crops to enhance the health of the soil and reduce the use of inorganic fertilisers (Iowa State University). Cover crops can also be integrated into vegetable rotation systems as living mulches by planting them in between rows and as boarder plants (Marr et al., 1998). This system can be used to control pests and diseases of the main vegetable crop by attracting the insects on to the cover crop. Vegetable crops that have pesticidal effects on soil-borne pathogens should be included in the rotation system. For example, amaranth could be included in the cover-crop vegetable-rotation system to control nematodes (personal experience). Cover crops used as living mulches should be managed well to prevent them from competing with the vegetable for scarce resources (SARE, 2012).

6. Prospects and Challenges of Using Cover Crops in Vegetable Production

The use of cover crops in urban vegetable production provides a wide range of opportunities to exploit the full potential of the chosen crops, either in isolation or in combination. Cover crops can be grown either simultaneously, sequentially, or in an alternate manner with the vegetable crops.

Prospects of Cover Crops: The opportunities range from improving soil fertility to maintaining good ground cover during fallow periods. Cover crops generally face the prospect of decreasing soil erosion by reducing the impact of rain drops, resulting in improved soil structure and quality for growing crops. The crop residue left on the soil after mechanical killing or using herbicides also provides crop residues to increase soil organic matter. This also enhances the infiltration, thereby increasing the amount of water available for use by the crops. Cover crops also have the potential of breaking the disease cycle through the replacement of host plants with non-hosts during the off-season. In some cases, cover crops with deep rooting systems can scavenge soil nutrients from deeper soil levels and make them available on the soil surface through leaf drop. In the era of climate change, cover crops are mostly utilised to provide ground cover to conserve soil moisture, maintain optimum soil temperature and help in the buildup of soil microorganisms. The choice of cover crop should, however, depend on compatibility with the vegetable crop, season and the type of the farming system adopted.

Improving Soil Structure and Fertility: Cover crops have the potential to improve the soil fertility in intensive vegetable crop production. This stems from the fact that most vegetable growers practice intensive mono-cropping over several years, leading to mining of soil nutrients, especially when there is limited application of soil amendments. The use of carefully selected cover crops can help replenish soil nutrients. Leguminous cover crops can be selected to provide ground cover and capture atmospheric N through biological N-fixation, as well as leaf drop (Blanco-Canqui et al., 2015).

Cover crops with deep and extensive root systems have the potential to scavenge nutrients leached into deeper layers of the soil (Dabney et al., 2001; Hartwig and Ammon, 2002). These crops can rapidly develop their extensive root systems after planting, to maximise the ability to scavenge from levels that are beyond the reach of the desired crops (Hunter et al., 2014). Cover crops, particularly leguminous crops, have the potential to encourage the buildup of populations of beneficial microorganisms, such as bacteria and fungi, which form mycelia that help to bind the aggregates of the soil particles together. This association, called mycorrhizae, enables the production of water-insoluble protein, which plays a key role in the binding of organic matter with soil particles (Wright and Upadhaya, 1998).

Creation of Soil Microclimate: The use of cover crops has the potential to increase the biological activity of most soils through the provision of raw materials for soil organic matter. Cover crops often create favourable environments for soil microorganisms through moisture retention, regulation of temperature and a source of food for them. In addition, the availability of organic matter improves the diversity and mass of soil microorganisms (Drinkwater et al., 1995). Maximum benefit can be obtained when high biomass-producing cover crops are used. Non-legumes that have the potential as cover crops include sudangrass and ryegrass and can be used to provide soil cover and enhance the biological activities of the soil (Snapp et al., 2005; Smith et al., 2011). Alternatively, leguminous crops, such as clover, can be used in combination with cereals to provide SOM and N-fixation. The growth rate of the crops should be taken into consideration since fast-growing crops are needed for rapid ground cover for the intended benefit. When cover crops are grown in association with vegetables, they tend to create a microclimate that ensures optimum temperature and humidity for optimum growth and yield of the vegetables (Blanco-Canqui et al., 2015).

Soil and Water Conservation: Cover crops also have the potential of being planted to provide a high percentage of ground cover for oil conservation. In some cases, non-leguminous crops with very dense rooting systems are planted in alleys and along the banks of rivers and dams that are used in dry season of vegetable production. The provision of soil cover by these cover crops reduces soil crusting and, subsequently, prevents surface water runoff, thereby conserving soil. The benefits of cover crops are mainly derived through their ability to reduce excessive water runoff and soil erosion, improve water infiltration and enhance soil moisture retention (Mallory et al., 1998; Sainju and Singh, 1997). Cover crops also prevent moisture loss by preventing excessive evaporation. Since less water is lost through evaporation, more will be available for crop use (Corak, 1991). The use of mixtures of grasses and legumes provides the added advantage of soil conservation and addition of nitrogen to the soil for the following crop. Soil-moisture retention under mulched fields has been found to be higher than fields with no cover, apart from the main crop in a no-till field versus conventionally tilled fields (Blevins et al., 1971). This was attributed to decreased evaporation and, consequently, increased moisture storage under the no-till, mulch-allowed plots.

Cover Crop as Vegetative Mulch: Living mulches are cover crops that are planted to co-exist either before or with a main crop and are usually maintained to provide a live ground cover throughout the growing season (Hartwig and Ammon, 2002). When cover crops are maintained as living mulch, they do not need to be replanted every year; the crops are able to regenerate themselves or maintain their vegetation during the off-season (Balkcom et al., 2012). Living mulches continue to grow after the main crops have been harvested, thereby suppressing weeds that might have otherwise colonised the field. The crop chosen as living mulch can be an annual or perennial plant. When legumes are used as living mulch, they have the potential to fix N, thereby complementing the N made available through SOM.

Pests and Disease Management: Carefully selected and managed cover crops can be used as a pest management strategy in intensive vegetable crop production. Certain cover crops, such as velvet bean, have the potential to produce allelo-chemicals that interfere with the growth and development of pests and disease pathogens (Szabo and Tebbet, 2002). When used in combination with no-till farming practices, cover crops help to maintain good ground cover for several seasons without any disturbance to the soil structure, thereby allowing development of the microbial population. Cover crops also interfere with the disease cycle through the replacement of host plants with non-hosts during the off-season. In some cases, the crop mixtures result in a reduced spread of disease pathogens, thereby slowing down outbreaks in vegetable farms (Entz et al., 2002). The cover crops also tend to create a favourable environment for beneficial organisms that will lead to rich soils capable of supporting the plants to combat pest and disease incidences. The cover crops provide food that is needed to support the population of these beneficial organisms for enhanced activity on the fields. According to Phatak and Diaz-Perez (2012), growing vegetables in association with certain cover crops results in less damage compared to crops grown alone. When some crops are attacked, they send chemical signals that attract the natural enemies of the pests to control their population (van Nouhuys and Kaartinen, 2007; Tumlinson et al., 1993).

Cover Crops in Weed Management in Vegetable Farms: Cover crops have the potential of being used to control weeds in vegetable farms, either as living mulch or as crop residue (Teasdale et al., 2007). These crops need to have the ability to out-compete and smother weeds through rapid growth and shading (Teasdale, 1996). Some of these crops can physically suppress the weeds, or through the production of natural chemicals, prevent the growth of the weeds through allelopathy (Reddy, 2001). The purpose of planting off-season cover crops is to produce plant residue that impedes the germination and establishment of weed populations on the field. Besides serving as mulch, the cover crop residues also provide partial weed control during early crop growth before complete development of the crop canopy (Teasdale, 1996). Cereals, such as rye, have been used to reduce the biomass and density of weeds in soybean fields for enhanced performance (Liebl et al., 1992; Moore et al., 1994). Integration of cereal cover crops, like rye, into a no-till, conservation-oriented farming system aid in early-season weed management and significantly reduce the use of herbicides (Norsworthy et al., 2011).

6.1 Cover Crop Challenges Associated with Urban Vegetable Production

In the era of population explosion and challenges with attaining food security and poverty alleviation, there is the need to develop a sustainable agricultural production system that will cause less harm to the environment. The march towards food sufficiency has been stalled by diminishing natural resources, intensive cultivation practices without addition of the necessary soil amendments, climate variability and declining soil fertility—all of which has led to low productivity. The use of conservation tillage practices, such as cover cropping, have also faced several challenges that are preventing their holistic adoption in most farming systems. Some of these challenges include:

Technical knowhow: In most cover-crop-based farming, particularly no-till conservation agriculture, special techniques are needed to manage the crop residue, either as live mulch or incorporated into the soil. The timing, incorporation technique and subsequent planting of the target crops are very important for maximum benefit. However, these techniques, as well as equipment, are not readily available to most vegetable farmers in developing countries. The introduction of user-friendly, low-cost techniques for the utilisation of cover crops in vegetable cultivation is very helpful for most vegetable farmers to produce healthy organic vegetables for consumption.

Technological challenges: One of the major challenges facing the utilisation of cover crops in vegetable cultivation is the lack of appropriate technologies that will enable their incorporation into the soil. Unlike arable crops, like cereals and legumes, that have various equipment and machinery for slashing and planting the seeds (Bhan and Behera, 2014), most vegetable crops require well-prepared seed-beds, which in most cases is not possible in vegetable production. The adoption of no-tillage systems in vegetable production can be hampered by scarcity of suitable equipment, like no-till planters. This often results in establishment and stand problems in vegetable farms (Morse, 1998).

Choice of cover crop: The choice of cover crops should be influenced by several factors. The farmer should consider the goal of planting the cover crop, the type of vegetable to be grown, the type of management practices to be adopted and the length of the growing period. For instance, the farmer should choose a cover crop or crop mixture that is easy to maintain, can provide adequate ground cover to reduce erosion, can produce enough organic matter and must not be in competition for nutrients, water and space with the main crop. The cover crop must also not attract pests and disease pathogens unto the field (Sarrantonio, 2012). Cover crops intended to be planted at the end of the season to provide ground cover should be tolerant to extreme temperature, which is very pronounced, especially in savannah ecologies. The chosen crop should be adaptable to dry areas and can maintain optimum growth and development in order to retain its biomass for the intended benefit. In temperate climates, cover crops that are planted to go through the winter are expected to be cold-tolerant with the potential to regrow during the new season. Care should be taken to prevent bush fires that may destroy the biomass and expose the soil to wind erosion.

Water availability during offseason: For cover crops to maintain their vegetation during the offseason, there is the need to provide adequate moisture. In savannah regions, the availability of water for growth of offseason cover crops can be a challenge due to the scarcity of water. Cover crops require soil water for their initial growth and this can be problematic during the onset of the dry season (Balkcom et al., 2012). Competition from cover crops can negatively affect crop yields when they are not well managed. They need to be killed in time to prevent them from competing with the main crops (Munawar et al., 1990). After killing, maximum benefit can be derived from the soil water conservation role played by the crop residue (Unger and Vigil, 1998).

Cover crop-vegetable crop compatibility: Cover crops, like vegetable crops, also require essential nutrients for their growth and development. For the maximum benefits to be derived, careful selection should be made in such a way that the performance of the vegetables will not be compromised. Certain cover crops have been found to exhibit allelopathic and inhibitory effects, which have the potential to negatively affect the desired crops, including vegetables (Adler and Chase, 2007; Zasada et al., 2006; Udensi et al., 1999). When the wrong crops are chosen as cover crops, they could compete with the vegetables for nutrients, moisture and space (Sarrantonio, 2012). Grasses have a slower decomposition rate and most often microorganisms tend to immobilise the available nitrogen during the decomposition process. In that case, the nutrients remain tied to the microorganisms, rendering them unavailable for utilisation by the vegetables. If slow-decomposing crops are to be used as cover crops, they should be incorporated far ahead of time to enable complete decomposition.

Cover crops as alternative hosts for pests and diseases: Though cover crops serve several beneficial purposes in vegetable crop production, they may serve as alternative hosts for pests if they attract insects and other pathogens other than beneficial organisms (Lu et al., 2000). Although some cover crops have the potential to prevent the growth of pests and disease pathogens, some others serve as suitable habitats for the growth of other pathogens, particularly nematodes. For instance, some leguminous crops, such as cowpeas, are susceptible to the reniform nematode, *Rotylenchulus reniformis* (Robinson, 2014) and variable levels of susceptibility or resistance to *Meloidogyne incognita*. Care should be taken in the choice of compatible cover crops for intensive vegetable crop cultivation.

Timeframe for biomass decomposition: The benefits of cover crops are derived following long-term practice for the crop residue to be decomposed, build up the microbial population for the soil structure to be maintained and nutrients to be released slowly for use by the crops. However, vegetables are mostly produced under intensive cultivation without ample time for the regeneration of the soil. There is the need to understand the interaction between the physical, chemical and biological processes involved in the crop residue management and nutrient release processes (Abrol and Sangar, 2006).

Threats of annual bush fires: Intensive utilisation of cover crops in a farming system requires that crop residues are left on the soil after the season, or in some cases, the residual moisture is used to seed leguminous cover crops to provide live mulch cover during the off-season to prevent wind erosion. To ensure that these crops or crop residues remain intact, the areas are often fenced to prevent grazing by animals. Burning is avoided in cover crop-based systems to ensure that there is gradual decomposition of the crop residue and improvement of the soil organic-matter content. However, most crop production areas are prone to annual bush fires that tend to deprive the soil of the needed crop residue buildup that will improve the organic matter content of the soil. Prevention of bushfires through the creation of fire belts, along with protection from grazing by animals, would be very crucial in deriving the full benefits of cover crops in vegetable production.

Management of cover crops: For a holistic and better understanding of the expected benefits of cover crops in vegetable crop cultivation, there is the need to select the right cover crop that will be compatible with the vegetable crop of choice. Efforts should be made to select crops that will ensure better utilisation of nutrients, moisture and space. Carefully selected crops ensure maximum benefit for both the farmer and the soil life for subsequent crops.

References

Abrol, I.P. and Sangar, S. (2006). Sustaining Indian agriculture-conservation: Agriculture the way forward. Current Sci., 91: 1020–2015.

Adler, M.J. and Chase, C.A. (2007). A comparative analysis of the allelopathic potential of leguminous summer cover crops: cowpea, sunn hemp and velvet bean. Hort. Sci., 42: 289–293.

Adu-Dapaah, H., Afun, J.V.K., Asumadu, H., Gyasi-Boakye, S., Oti-Boateng, C. and Padi, H. (2005). Cowpea Production Guide, Food Crop Development Project, Ministry of Food and Agriculture, CSIR-Crops Research Institute and CSIR-Savanna Agricultural Research Institute, Print Clemana Ventures, Kumasi, Ghana.

Allen, W. (2010). Urban Agriculture in Accra, Ghana. In: Gardening the Community. https://gardeningthecommunityblog.wordpress.com/2010/07/06/urban-agriculture-in-accra-ghana-2. July 6, 2010.

Baldwin, K.R. and Creamer, N.G. (2006). Cover Crops for Organic Farms, North Carolina, USA: North Carilina Cooperative Extension Service.

Balkcom, K., Schomberg, H., Reeves, W., Clark, A., Baumhardt, L., Collins, H., Delgado, J., Duiker, S., Kaspar, T. and Mitchell, J. (2012). Managing cover crops in conservation tillage systems. pp. 44–61. In: Clark, A. (ed.). Managing Cover Crops Profitably, third ed., Sustainable Agriculture Research and Education (SARE) Handbook Series Book 9, United Book Press, Inc.

Bationo, A., Lompo, F. and Koala, S. (1998). Research on nutrient flows and balances in West Africa: State-of-the-art. Agric. Ecosys. Environ., 71: 19–35.

Bhan, S. and Behera, U.K. (2014). Conservation agriculture in India—Problems, prospects and policy issues. Intern. Soil Water Conser. Res., 2: 1–12.

Blanco-Canqui, H., Shaver, T.M., Lindquist, J.L., Shapiro, C.A., Elmore, R.W., Francis, C.A. and Hergert, G.W. (2015). Cover crops and ecosystem services insights from studies in temperate soils. Agron. J., 107: 2449–2474.

Blevins, R.C., Cook, D. and Phillips, S.H. (1971). Influence of no-tillage on soil moisture. Agron. J., 63: 593–596.

Boutin, C., Strandberg, B., Carpenter, D., Mathiassen, S.K. and Thomas, P.J. (2013). Herbicide impact on non-target plant production: What are the toxicological and ecological implications. Environ. Pollut., 185: 295–306.

Brown, M.W. and Glenn, D.M. (1999). Ground cover plants and selective insecticides as pest management tools in apple orchards. J. Econ. Entom., 92: 899–905.

Buckles, D. (1995). Velvet bean: A new plant with a history. Econ. Bot., 49: 13–25.

Campiglia, E., Radicetti, E., Brunetti, P. and Mancinelli, R. (2014). Do cover crops species and residue management play a leading role in pepper productivity? Scientia Horticulturae, 166: 97–104.

Carter, M.R. and Stewart, B. (1996). Structure and Organic Matter Storage in Agricultural Soils. CRC Press, Boca Raton, Florida.

Cook, B.G., Pengelly, B.C., Brown, S.D., Donnelly, J.L., Eagles, D.A., Franco, M.A., Hanson, J., Mullen, B.F., Partridge, I.J., Peters, M. and Schultze-Kraft, R. (2005). Tropical Forages: An Interactive Selection Tool [CD-ROM], CSIRO, DPI&F (Qld), CIAT and ILRI, Brisbane, Australia.

Corak, S.J., Frye, W.W. and Smith, M.S. (1991). Legume mulch and nitrogen fertiliser effects on soil water and corn production. Soil Sci. Soc. Amer. J., 55: 1395–1400.

CSIR/Savanna Agricultural Research Institute. (2007). Sorghum Production Guide, Food Crops, Development Project, Ministry of Food and Agriculture, p. 1.

Dabney, S.M., Delgado, J.A. and Reeves, D.W. (2001). Using winter crops to improve soil and water quality. Commun. Soil Sci. Plant Analy., 32: 1221–1250.

Danso, G., Dechsel, P., Obuobie, E. and Forkuor, G. (2014). Urban vegetable farming sites, crops and cropping practices. pp. 7–27. *In*: Pay Drechsel and Keraita, B. (eds.). Irrigated Urban Vegetable Production in Ghana: Characteristics, Benefits and Risk Mitigation. Colombo, Sri Lanka: International Water Management Institute (IWMI).

Drinkwater, L.E., Letourneau, D.K., Workneh, F., van Bruggen, A.C.H. and Shennan, C. (1995). Fundamental differences between conventional and organic agro ecosystems in California. Ecol. Appl., 5: 1098–1112.

Entz, M.H., Baron, V.S., Carr, P.M., Meyer, D.W., Smith Jr., S.R. and McCaughey, W.P. (2002). Potential of forages to diversify cropping systems in the Northern Great Plains. Agron. J., 94: 240–250.

Fageria, N.K., Baligar, V.C. and Bailey, B.A. (2005). Role of cover crops in improving soil and row crop productivity. Commun. Soil Sci. Plant Analy., 36: 2733–2757.

Ghana Statistical Service (2014a). 2010 Population and Housing Census, District Analytical Report, Ho West District.

Ghana Statistical Service (2014b). 2010 Population and Housing Census Report, Urbanisation.

Gonzalez, Y.S., Dijkxhoorn, Y., Koomen, I., Maden, E., Herms, S., Joosten, F. and Mensah, S.A. (2016). Vegetable business opportunities in Ghana. Ghana Vegetable Sector Reports, Wageningen, The Netherlands.

Hartwig, N.L. and Ammon, H.U. (2002). Cover crops and living mulches. Weed Sci., 50: 688–699.

Hendrickson, J. (2009). Cover crops on the intensive market farm. Report of Center for Integrated Agricultural Systems (CIAS), a research center for sustainable agriculture in the College of Agricultural and Life Sciences, University of Wisconsin-Madison.

Hutchinson, C.M. and McGiffen Jr., M.E. (2000). Cowpea cover crop mulch for weed control in desert pepper production. Hort. Sci., 35: 196–198.

Institute of Statistical, Social and Economic Research (ISSER). (2005). The state of Ghanaian economy in 2004, Accra.

Iowa State University Extension and Outreach website: Cover Crops in Vegetable Production Systems. Accessed February 23, 2019. www.extension.iastate.edu/cover-crops-vegetable-production-systems.

Keraita, B., Silverman, A., Amoah, P. and Asem-Hiablie, S. (2014). Quality of irrigation water used for urban vegetable production. Drechsel, P. and Keraita, B. (eds.). Irrigated Urban Vegetable Production in Ghana, second ed., IWMI 2014.

Lenzi, A., Antichi, D., Bigongiali, F., Mazzoncini, M. and Migliorini, P. (2009). Effect of different cover crops on organic tomato production. Renew. Agric. Food Sys., 24: 92–101.

Liebl, R., Simmons, F.W., Wax, L.M. and Stoller, E.W. (1992). Effect of rye (*Secale cereale*) mulch on weed control and soil moisture in soybean (*Glycine max*). Weed Tech., 6: 838–846.

Lu, Y.-C., Watkins, K.B., Teasdale, J.R. and Abdul-Baki, A.A. (2000). Cover crops in sustainable food production. Food Rev. Intern., 16: 121–157.

Mallory, E.B., Posner, J.L. and Baldock, J.O. (1998). Performance, economics and adoption of cover crops in Wisconsin cash grain rotations: On-farm trials. Amer. J. Altern. Agric., 13: 2–11.

Marr, C.W., Janke, R. and Conway, P. (1998). Cover Crops for Vegetable Growers. Kansas, USA: Kansas State University.

Maundu, P. (2013). African leafy vegetables come out of the shade. *In*: Hendrickson, J. (ed.). Cover Crops on the Intensive Market Farm, University of Wisconsin Centre for Integrated Agricultural Systems, College of Agricultural and Life Sciences, Madison, Wisconsin.

Maughan, T. and Drost, D. (2016). Introduction to Cover Crops for Vegetable Production in Utah. Utah State University Extension website. www.extension.usu.edu.Horticulture/CoverCrops/2016-01.

Ministry of Food and Agriculture (MOFA). (2015). Agricultural Extension Policy (abridged version). Accra, Ghana, Ministry of Food and Agriculture.

Ministry of Food and Agriculture. (2015). Catalogue of Crops Varieties Released and Registered in Ghana, vol. 1.

Moore, M.J., Gillespie, T.J. and Swanton, C.J. (1994). Effect of cover crop mulches on weed emergence, weed biomass and soybean (*Glycine max*) development. Weed Tech., 8: 512–518.

Morse, R. (1998). Keys to successful production of transplanted crops in high-residue, no-till farming systems. Proceedings of the 21st Annual Southern Conservation Tillage Conference for Sustainable Agriculture, July 1998.

Moyer, J.R., Blackshaw, R.E., Smith, E.G. and McGinn, S.M. (2000). Cereal cover crops for weed suppression in a summer fallow-wheat cropping sequence. Can. J. Plant Sci., 80: 441–449.

Munawar, A., Blevins, R.L., Frye, W.W. and Saul, M.R. (1990). Tillage and cover crop management for soil water conservation. Agron. J., 82: 773–777.

Norsworthy, J.K., McClelland, M., Griffith, G. and Bangarwa, S.K. (2011). Evaluation of cereal and brassicaceae cover crops in conservation-tillage, enhanced, glyphosate-resistant cotton. Weed Tech., 25: 6–13.

Phatak, S.C. and Diaz-Perez, J.C. (2012). Managing pests with cover crops. pp. 25–33. *In*: Clark, A. (ed.). Managing Cover Crops Profitably, third ed., Sustainable Agriculture Research and Education (SARE) Handbook Series Book 9. United Book Press, Inc.

Peet, M. (2001). Cover Crops and Living Mulches, North Carolina State University. Accessed from www.ncsu.edu/sustainable/cover/cover.html.

Peoples, M.B., Herridge, D.F. and Ladha, J.K. (1995). Biological nitrogen fixation and efficient source of nitrogen for sustainable agricultural production. Plant Soil, 174: 3–28.

Reddy, K.N. (2001). Effects of cereal and legume cover crop residues on weeds, yield, and net return in soybean. Weed Technology, 15: 660–668.

Rice, E.L. (1984). Allelopathy, Physiological Ecology: A Series of Monographs, Texts, and Treatises, second ed., Academic Press, New York.

Robinson, G. (2014). Geographies of Agriculture: Globalization, Restructuring and Sustainability. 1–331. 10.4324/9781315839509.

Sani, R.M., Haruna, R. and Sirajo, S. (2013). Economics of Sorghum (*Sorghum bicolor* (L.) Moench) production in Bauchi local government area of Bauchi State, Nigeria. *In*: 2013 AAAE Fourth International Conference, September 22–25, 2013, Hammamet, Tunisia (No. 161644), African Association of Agricultural Economists (AAAE).

SARE. (2012). Managing Cover Crops Profitably, third ed., Sustainable Agriculture Research and Education (SARE) Handbook Series Book 9.

Sainju, U.M. and Singh, B.P. (1997). Winter cover crops for sustainable agricultural systems: Influence on soil properties, water quality and crop yields. Hort. Sci., 32: 21–28.

Sarrantonio, M. (2012). Selecting the best cover crops for your farm. pp. 12–15. *In*: Clark, A. (ed.). Managing Cover Crops Profitably, third ed., Sustainable Agriculture Research and Education (SARE) Handbook Series Book 9. United Book Press, Inc.

Singh, Y., Singh, B., Ladha, J.K., Khind, C.S., Gupta, R.K., Meelu, O.P. and Pasuquin, E. (2004). Long-term effects of organic inputs on yield and soil fertility in the rice-wheat rotation. Soil Sci. Soc. Amer. J., 63: 1350–1358.

Smeda, R.J. and Putnam, A.R. (1998). Cover crop suppression of weeds and influence on strawberry yields. Hort. Sci., 23: 132–134.

Smith, R.G., Gareau, T.P., Mortensen, D.A., Curran, W.S. and Barbercheck, M.E. (2011). Assessing and visualising agricultural management practices: A multivariable hands-on approach for education and extension. Weed Tech., 25: 680–687.

Snapp, S.S., Swinton, S.M., Labarta, R., Mutch, D., Black, J.R., Leep, R., Nyiraneza, J. and O'Neil, K. (2005). Evaluating cover crops for benefits, costs and performance within cropping system niches. Agron. J., 97: 322–332.

Szabo, N.J. and Tebbett, J.R. (2002). The chemistry and toxicity of Mucuna species. pp. 120–141. *In*: Flores, M., Eilittä, M., Myhrman, R., Carew, L. and Carsky, R. (eds.). Mucuna as a Food and Feed: Current Uses and the Way Forward. Workshop held April 26–29, 2000 in Tegucigalpa, Honduras. CIDICCO, Honduras.

Tanzubil, P.B., Buah, S.S.J., Iddrisu, A., Wih, K., Anyeembey, J.B. and Walier, P.A. (2017). Guide to Profitable, Sustainable Groundnut Production in the Northern Ghana, ICRISAT, p. 20.

Teasdale, J.R. (1996). Contribution of cover crops to weed management in sustainable agriculture systems. J. Prod. Agric., 9: 475–479.

Teasdale, J.R., Brandsaeter, L.O., Calegari, A. and Neto. F.S. (2007). Cover crops and weed management. pp. 49–64. *In*: Upadhyaya, M.K. and Blackshaw, R.E. (eds.). Non-chemical Weed Management, CAB Int., Chichester, UK.

Tumlinson, J.H., Lewis, W.J. and Vet, L.E.M. (1993). How parasitic wasps find their hosts. Sci. Amer., 26: 145–154.

Udensi, U.E., Akobundu, I., Ayeni, A.O. and Chikoye, D. (1999). Management of Congo grass (*Imperata cylindrical*) with velvet bean (*Mucuna pruriens* var. *utilis*) and herbicides. Weed Tech., 13: 201–208.

Unger, P.W. and Vigil, M.F. (1998). Cover crops effects on soil water relationships. J. Soil Water Conser., 53: 241–244.

van Nouhuys, S. and Riikka Kaartinen, R. (2007). A parasitoid wasp uses landmarks while monitoring potential resources. Proceedings of the Royal Society B: Biol. Sci., 275: 377–385.

Walters, S.A., Young, B.G. and Nolte, S.A. (2007). Cover crop and pre-emergence herbicide combinations in no-tillage fresh market cucumber production. J. Sustain. Agric. 30: 5–19.

Wright, S.F. and Upadhaya, A. (1998). A survey of soils for aggregate stability and glomalin, a glycoprotein produced by hyphae of arbuscular mycorrhizal fungi. Plant Soil, 198: 97–107.

Zasada, I.A., Klassen, W., Meyer, S.L.F., Codallo, M. and Abdul-Baki, A.A. (2006). Velvet bean (*Mucuna pruriens*) extracts: Impact on *Meloidogyne incognita*, *Lycopersicon esculentum* and *Latuca sativa* survival. Pest Manag. Sci., 62: 1122–1127.

5

Algorithms to Optimise Cropping Diversity with Cover Crops

Romashchenko, M,[1,*] *Matiash, T,*[1] *Bohaienko, V,*[2] *Kovalchuk, V,*[1]
Lukashuk, V[1] and *Saydak, R*[1]

1. Introduction

Anthropogenic activities, such as mechanised agriculture, have transformed the earth, and today, about 12 per cent of the world's land is used for agriculture and another 22 per cent is used for pastures and rangelands. In other words, today, agronomic croplands occupy nearly 18 M km^2, pastures take up another 34 M km^2 and urban areas use roughly 2.5 M km^2 (Turner et al., 1993; Ramankutty and Foley, 1998; Goldewijk, 2001). Altogether, these three ecosystems currently occupy over a third of the global land area.

Agronomic crops, especially their diversification, can provide several benefits to the terrestrial ecosystems. With living mulch of cover crops, cropping diversity reduces agro-ecosystem disservices. Other than diverse food production, crop diversity breaks down pest- and-disease cycles, including plant diseases, insects and weed infestations, and can serve as extended and living mulches, provide organic matter, recycle organically-bound nutrients, reduce chemical fertilisation and other reactive chemicals use, reduce surface runoff and soil erosion, improve soil structures, conserve soil moisture, improve soil quality, increase crop yields, create new markets, and strengthen rural communities by creating green jobs.

Of the over half a million plant species on the earth, global populations currently rely on just four major agronomic crops (wheat, rice, maize and soybean) for more than three-fourths of our food security (Fig. 1). These 'major' crops are grown in a few exporting countries, usually as monocultures, and are highly dependent on chemical inputs (such as fertilisers, agrochemicals) and irrigation. About 7.4 billion people depend on the productivity of these major crops, not just for their direct food needs, but increasingly as raw materials for livestock, aquaculture feeds and bioenergy systems.

Planning of agricultural production is one of the strategic factors of economic activities associated with the sustainable development of agriculture, which enables considerable long-term impacts on ecological and environmental processes and/or properties. The basis for such planning is the concept of cropping diversity (crop rotation with and without cover crops), which has immense biological, ecologic and economic impacts (Turner et al., 1993; Ramankutty and Foley, 1998; Goldewijk, 2001; Kaminsky and Boyko, 2014a,b; Kovalenko, 2012; Gadzalo et al., 2015). Crop

---

[1] Institute of Water Problems and Land Reclamation, National Academy of Agrarian Sciences of Ukraine, Kyiv, Ukraine.

[2] VM Glushkov Institute of Cybernetics, National Academy of Sciences of Ukraine, Kyiv, Ukraine.

* Corresponding author: mi.romashchenko@gmail.com

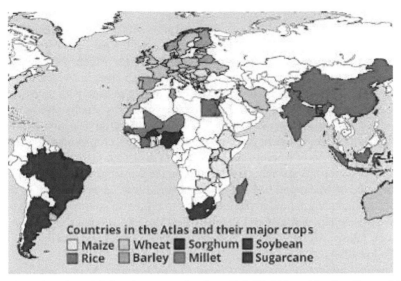

Fig. 1 Countries in the Atlas and their major crops (taken from *Global Yield Crop Atlas* (http://www.yieldgap.org)).

rotation prevents loss of biodiversity, improves of soil fertility, reduces weed infestation and the spread of pests, and has an impact on soil moisture dynamics in agro-ecosystems.

Mathematical modelling is commonly used in agricultural research and is one of the most effective methods to optimise crop rotation for economically viable and environmentally compatible agro-ecosystems (Verhunova, 2000; Kovalenko, 2007; Schönhart et al., 2009; Nuppenau, 2011; Yurkevych et al., 2011; Gadzalo et al., 2015; Osman et al., 2015). The important principles often considered when constructing models for crop rotation optimisation are the influence of crop type and a sequence of predecessors on economic yields of specific crops with all possible variations (Verhunova, 2000; Gadzalo et al., 2015). Criterion of optimality can be stated here as an achievement of maximal value of net profit or gross product with environmental restrictions imposed by a need to preserve soil quality (Verhunova, 2000; Yurkevych et al., 2011). Development of mathematical tools for routine selection of crop rotations with the widest consideration of various economic, environmental and ecological factors is important to develop effective decision-making support systems in agriculture.

2. Problem Definitions for the Conditions of Ukraine

Historically, Ukraine is one of the most productive agrarian countries in the world with an abundance of high-quality soil and a suitable weather to grow both winter and spring crops. Ukrainian climate is close to that of Kansas in the United States: slightly drier and cooler during the summer months and colder and wetter during the winter months, but close enough for comparison. Average annual precipitation in Ukraine is approximately 600 mm (24 inches), including roughly 350 mm during the growing season (April to October). Rainfall amounts are typically higher in western and central Ukraine and lower in the south and east.

Of Ukraine's total land area of 60 M ha, about 54 per cent is classified as agricultural land, 9 per cent under pasture and 4 per cent under hay. Agriculture includes cultivated land (grains, oilseed crops, forages, potatoes and vegetables and fallow), gardens, orchards, vineyards and permanent meadows and pastures. Winter wheat, spring barley and corn are the country's main agronomic crops. Sunflowers and sugar beets are the main industrial crops. However, agricultural land use has shifted abruptly since Ukraine declared independence from the Soviet Union, in 1991. The State and collective farms were officially dismantled in 2000 and farm properties were divided among the farm workers in the form of land shares. However, most new shareholders have leased their land back to newly-formed private agricultural associations. The sudden loss of State agricultural subsidies had an enormous effect on economic, political and social aspects of Ukrainian agriculture.

Between 1991 and 2000, crop-planted area dropped by about 5 per cent, from 32 M hectares to 30.4 M, and the area decreased for almost every category of crops, except for sunflowers. Forage-crop area plunged by nearly 40 per cent, concurrent with a steep slide in livestock inventories and feed demand. Chemical fertiliser use fell by 85 per cent over a decade with an associated 50 per cent decrease in grain production. The emergence from the Soviet-style collective economy enabled farmers to make increasingly market-based decisions regarding crop selection and management, which contributed to increased efficiency in both the livestock and crop-production sectors.

Ukrainian farmers currently employ a variety of crop diversity, including four or more crops, or sometimes only two crops. A six-year crop rotation in the winter grain region will often include two consecutive years of wheat and one season of 'clean fallow', during which no crop is sown. The main reason for including fallow in the rotation is to replenish soil-moisture reserves and it is more widely used in southern and eastern Ukraine, where drought is not uncommon. A typical crop sequence might be fallow, winter wheat, winter wheat, sunflowers, spring barley, and corn. Wheat almost always follows fallow.

Several crop rotations include consecutive years of a forage crop. An example of such a rotation would be fallow, two years of winter wheat and four years of perennial forage. The perennial forage is usually alfalfa; farmers will get three-to-four cuttings per year, or five if the crop is irrigated. In southern Ukraine, clean fallow is frequently omitted and crop rotation will likely include sugar beets and/or sunflower—the region's main industrial crops. A typical seven-year rotation might include winter wheat, winter barley, sugar beets, winter wheat, winter barley, sunflowers and corn. Most field crops, including grains, sunflowers and sugar beets are not irrigated. Traditionally, irrigation is used only on forage crops and vegetables. About 5 per cent of grains and 10 per cent of potatoes, vegetables, and forage crops are irrigated (http://wdc.org.ua/en/node/29).

Finding a solution to crop rotation optimisation problems, in terms of search for their sequential structure, is one of trends in the development of sustainable agriculture. In recent years, climate change has exerted an adverse impact on zonal placement of crops in Ukraine. Within Ukraine, there are four physical and geographical zones that are distinguished according to the type of landscapes: steppe, forest-steppe, zone of mixed forests (Polissya region) and Ukrainian Carpathians (Marynych et al., 1985). They are formed in accordance with specific balance ratios of heat and humidity, which affect plant growth conditions, soil formation and agricultural usage of landscapes. These ratios directionally vary from north to south. Due to global climate-change effects, the balance of heat and humidity in Ukraine has slightly changed over time and this is demonstrated in Fig. 3 (Romashchenko et al., 2015).

Growing of crops, like soybeans and corn, has become problematic in the steppe zone of Ukraine due to climate changes. In the zones of forest-steppe and Polissya, hydrothermal conditions favourable for growing grain and leguminous crops were formed.

Moreover, in Ukraine, a decrease in the number of crops used in crop rotations was observed and the practice of growing highly profitable crops without taking crop rotations into account was

Fig. 2 A typical six-year cropping diversity sequence in Ukraine.

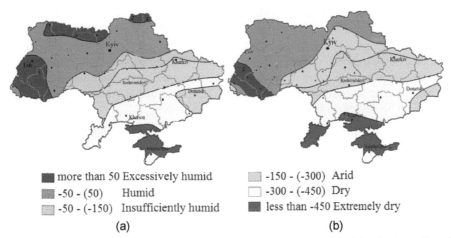

more than 50 Excessively humid   -150 - (-300) Arid
-50 - (50)   Humid   -300 - (-450) Dry
-50 - (-150) Insufficiently humid less than -450 Extremely dry

(a)   (b)

Fig. 3 Zoning of the territory of Ukraine according to the annual climatic water balance (a) by the data collected between 1960 and 1990; (b) between 1990 and 2015.

spreading (Kaminsky and Boyko, 2014a,b). This trend is also observed in other countries. In most cases, structure of crop area is ignored and its importance is determined by a market demand for a culture, the division of crops on profitable and non-profitable, high expenses rate and relatively low selling price for non-profitable (but agriculturally important) crops. At the same time, crop rotations are saturated with highly profitable hygrophilous crops (soybeans, corn and vegetables) with high demand for nutrients, water and other agrochemicals (Ramankutty and Foley, 1998). In response to the reduction of livestock sector, the integration of fodder crops in crop rotations, especially perennial legumes, decreased greatly. Therefore, farmers are inclined to use synthetic fertilisers with variable rates of their application instead of the use of organic or integrated ones. An economically viable and environmentally-compatible balanced agricultural production is possible only if the crop rotation principles are applied.

As the natural processes occurring during crop growth is complex, the main approach used to develop models of crop rotations is (Detlefsen, 2004; Schönhart et al., 2009) based on expert scoring (Garcia et al., 2005; Kovalenko, 2007; Dury et al., 2012) of predecessor influence on crop yield. It is worth noting that such scoring mainly considers the agronomic part of the process with very little attention paid to the economic viability and environmentally compatible factors. In its base, the formalised problem of crop rotation selection is a combinatorial optimisation problem. The simplest numerical method that solves it is the NP-complete brute force search method (Trevisan, 2011). Such approaches as the formulation of the problem as a linear programming problem (Detlefsen, 2004), use of branch and bounds (Alfandari et al., 2015; Santos et al., 2015), or heuristic methods (Pavón et al., 2009; Lee et al., 2015) are aimed at reducing the complexity of optimal crop rotation search. In our example, a recent approach was used to solve multi-objective problems that consider both economic viability and ecological compatibility criteria during the process of crop rotation development (Pavón et al., 2015).

3. Methods and Algorithms to Select for Optimal Crop Rotation

We considered two types of expert scoring on the basis of which the overall crop rotation score is calculated: • score $\varepsilon_{ij}^{(1)} \leq 100$ of the 'predecessor-crop' pair and score $\varepsilon_{ij}^{(2)} \leq 100$ of a crop depending on its year of return. The overall crop rotation score is the sum of minimums of these scores:

$$\varepsilon = \frac{1}{N} \sum_{i=1}^{N} \min\left(\varepsilon_{k(N+i-1)\%N,k_i}^{(1)}, \varepsilon_{k_i,y_i}^{(2)} \right), \tag{1}$$

where k_i is an i-th crop in crop rotation, y_i is its year of return, N is the length of crop rotation.

Optimal cropping diversity [the one with the highest score (1)] is proposed to be built for three or less priority crops. The length of crop rotation is defined as the maximum from the optimal years of return for priority crops, which in turn are calculated as min y: $\varepsilon_{ky}^{(2)} = 100$. Such a single-objective issue with optimal crop-rotation selection is the problem of combinatorial optimisation. To solve it effectively, we considered two alternatives to the brute force method – the modifications of branch-and-bounds and genetic algorithms. The recursive branch and bounds algorithm that performs left-to-right search can be described as follows:

- Fix the first M crop in the proposed crop rotation and consider K branches where K is the number of crops that can be included in the rotation. For the branch i set $k_{M+1} = C_i$, where C_i, $i = 1, \ldots, K$ is an i-th crop from the full list of considered crops.

- For each branch $\varepsilon_b \leq \dfrac{1}{N}\left(\sum_{i=1}^{M+1} \min\left(\varepsilon_{k(N+i-1)\%N,k_i}^{(1)}, \varepsilon_{k_i,y_i}^{(2)}\right) + (N - M - 1)\cdot 100\right)$, we assumed that scores for the last $N - M - 1$ crop in the rotation are maximal. If for some branch $\varepsilon_b > \varepsilon_{\max}$ where $\varepsilon_{\max}$ is the maximal score from previously evaluated variants of crop rotation, recursively do a search in it with $M + 1$ fixed crops. In the opposite case, recursive search in the branch is not carried out. Recursively find a solution of the problem starting from the arbitrary initial variant and $M = 0$.

Heuristic genetic algorithm (Goldberg, 1989) allows for obtaining approximate solutions for the considered problem. It can be stated in the following way:

- Consider an initial population of potential solutions (crop rotations) of a given size; crossover operation is defined as an operation of copying a 'predecessor-crop' pair from the first operation (crop rotation) to the second. Potential solutions to which crossover operation is applied were chosen, weighted randomly with scores (1) as weight coefficients. The pair to be copied was determined and weighted randomly with $\min\left(\varepsilon_{k(N+i-1)\%N,k_i}^{(1)}, \varepsilon_{k_i,y_i}^{(2)}\right)$ scores as weight coefficients.

- Mutation operation is defined as a change of a randomly selected crop in a selected crop rotation; an iteration of the algorithm consists in execution of a crossover operation, after which a mutation operation is executed with a given probability. If the score (1) of the crop rotation generated by a crossover operation is greater than the lowest score among crop rotations in the population, the new crop rotation replaces the earlier crop rotation with the lowest score; iterative process finishes when the difference between maximal and minimal crop rotation scores in the population becomes less than a given value.

While the genetic algorithm is an algorithm of random search, it resulted in crop rotation close by the value of goal function, but different by the structure. To obtain a fixed solution, if genetic algorithm converges to solutions about a single local maximum, after the completion of its work, we proposed to additionally perform an algorithm to a certain number of greedy search iterations.

3.1 Selection of Crop Rotation According to Economic and Ecologic Factors

The problem of optimal crop-rotation selection does not always consider economic and ecologic factors; instead restricts itself within the agronomic ones. An algorithm that integrates all three components is proposed, considering that:

- an agronomic expert scoring of crop-growing effectiveness, depending on the predecessor, is the percentage of maximal yield of this crop that can be obtained in current situation;

- considering the cost of other conditions (such as chemical fertiliser) as optimal, variable part of the total expenses is the cost of fertilisers;

- negative environmental effect from fertilisers usage is proportional to the number of nitrates applied in the soil.

According to abovementioned assumptions, we defined crop rotation score as a sum of aggregated economic and ecologic scores:

$$\varepsilon_1 = \sum_{i=1}^{N} \left(P_{k_i} Y_{k_i} \min\left(\varepsilon^{(1)}_{k(N+i-1)\%N,k_i}, \varepsilon^{(2)}_{k_i,y_i} \right) - F_{k_i} - EN_{k_i} \right) \tag{2}$$

where P_{k_i} is the price of the crop, Y_{k_i} is the optimal crop yield, F_{k_i} is the cost of applied fertilisers, E is the coefficient of negative environmental impact, N_{k_i} is the amount of applied fertiliser especially nitrogen.

As we can only predict crop prices in future periods, we considered them as fixed factors or, in the presence of retrospective data, predicted them using methods of time series analysis—the method of linear regression. Ecologically and economically optimal yield is determined for each crop, using the balance method (Lazer and Mikheiev, 2006) to calculate the rate of nutrient applications depending on targeted crop yields:

$$H = \frac{Y \cdot B - M \cdot Kg - D_0 P_0 K_0}{K_d} \tag{3}$$

where H is the rate of specific nutrient application, kg/ha; Y is the targeted yield for the selected crop, t/ha; B is the rate of nutrients removal from the soil by the crop, kg/t; M is the soil nutrient contents, kg/ha; K_g and K_d are the coefficients of nutrients usage by plants from the soil and applied fertilisers, %; P_0 is the rate of organic fertilizers application, t/ha; D_0 is the nutrient content in organic fertilisers, kg/t; K_0 is the coefficient of nutrient usage from organic fertilisers, %.

To determine an optimal crop yield, we maximised the following goal function by Y_{k_i} variable with constraints on maximal and minimal crop yields by the gradient method:

$$f(Y_{k_i}) = P_{k_i} Y_{k_i} \min\left(\varepsilon^{(1)}_{k(N+i-1)\%N,k_i}, \varepsilon^{(2)}_{k_i,y_i} \right) - F_{k_i} - EN_{k_i} \to \max.$$

At the same time, for each Y_{k_i} value, we calculated the required amount N_{k_i}, P_{k_i}, K_{k_i} of nutrients to be applied, taking into account the amount of fertilisers applied in the previous two years. After that, we did a search for such set and the amount of fertilisers from the existing database that would minimise the expenses F_{k_i}.

We proposed to solve this problem of convex optimisation with constraints by the gradient method with projection on acceptable solutions set using pseudo-inversion operator. Thus, we performed a selection of optimal crop rotation considering economic and ecologic factors by solving three embedded optimisation problems.

At the highest level, combinatorial problem of optimal crop rotation selection is solved by the above-described method with the goal function (2). The calculation of goal function requires the finding of optimal yield level regarding the price for a crop, the cost of fertilisers that must be applied to ensure the required yield under optimal conditions and the adverse environmental effects of nitrate-based fertiliser application. This problem is solved by the gradient method with numerical calculation of goal function derivative. We obtained the goal function value, solving the problem of optimal selection of a set and amount of fertilisers that provide the required amount of nutrients. As the calculation of goal function (2) is computationally complex, we used genetic algorithm only to obtain solutions to the combinatorial problem.

3.2 Application of Algorithms to Assess Crop Rotation Used in Ukraine

Brute force, branch-and-bounds and genetic algorithms were implemented as a module of a client-server decision support system, which is being developed at the Institute of Water Problems and Land Reclamation, National Academy of Sciences, Kyiv, Ukraine. The algorithms were executed on the server side and implemented in C++ language as an extension of php-interpreter and interact with the client part via php-script. We used *trader* php-module for prediction of crop prices.

The performance of the proposed algorithms of optimal crop rotation selection with goal function (1) were tested for the case of two priority crops with a sufficiently large optimal return period. For this purpose, we selected peas and soybeans grown in central and northern subzones of the steppe zone of Ukraine.

Solution of the problem by the brute force algorithm took 27.44s. The resulting optimal crop rotation has a 95.6 point score and consisted of the following crops:

- green corn
- soybeans
- spring barley
- green corn
- peas
- winter barley

The branch-and-bounds algorithm took 22.81s to get the same result, giving a 16.8 per cent speedup. Genetic algorithm was tested with the following values of parameters:

- population size – 20
- maximal number of iterations – 50000
- mutation probability – 0.01
- iterative process finishes when the difference between maximal and minimal values of crop rotation scores in the population is less than 2. Genetic algorithm's working time was equal to 1.30s, which is significantly less than in the cases of complete or restricted search.

The proposed methods were used to assess the effectiveness of crop rotations specific for the different growing conditions of Ukraine. Scoring results are presented in Table 1. We assessed the cases of monoculture and short-crop rotations that consist of the most profitable crops in Ukraine: soybeans, corn, winter wheat and sunflower. Effectiveness of such crop rotations was modelled and the results show that only corn or corn with soybeans may be used in short crop rotations or as a monoculture.

In the next series of computational experiments, we modelled the effectiveness of crop rotation described in Lukashuk (2015) and appropriately estimated it having maximal efficiency. The crop rotation consists of the following crops for the conditions of Polissya region:

- alfalfa
- potato

Table 1 Effectiveness scores for short-crop rotations.

Crop Rotation	Climatic Zones	Crops Scores	Average Score of Crop Rotation
Soybeans-soybeans	Steppe, forest steppe	Soybeans 50	50
Corn-corn	Polissya region	Corn 50	50
Corn-corn	Forest steppe	Corn 89	89
Soybeans-corn-corn	Forest steppe	Soybeans 25 Corn 96 Corn 89	75
Winter wheat-sunflower	Forest steppe	Winter wheat 20 Sunflower 10	15
Peas-winter wheat-sunflower	Forest steppe	Peas 25 Winter wheat 75 Sunflower 25	41.6
Soybeans-winter wheat -sunflower	Steppe	Soybeans 30 Winter wheat 80 Sunflower 15	41.6

- corn
- peas
- sugar beet
- spring barley
- winter wheat

Efficiency score for crop rotation obtained using the developed software was equal to 87 points, which corresponds with known expert scores.

Selection of optimal crop rotation for three fixed crops (sugar beet, winter wheat, alfalfa) resulted in the following crop rotation:

- alfalfa (100 points efficiency)
- winter wheat (100 points efficiency)
- sugar beet (100 points efficiency)
- oats (90 points efficiency)
- winter wheat (70 points efficiency)
- green corn (100 points efficiency)

The mean crop rotation score was equal to 93.3 with a standard error of a mean of 5 points (93.3 ± 5). There are several existing alternative crop rotations with efficiency close to optimal; however, the fact that our crop rotation (by its type and structure) is similar to the initial one, was a demonstration of correctly-used database of expert scores and selection algorithms suitable for the Ukraine conditions. We also obtained similar results for two crop rotations described in (Kovalenko, 2005), which, from the expert point of view, are optimal for the conditions of the sufficiently wet subzone of the forest-steppe zone in Ukraine. While the first crop rotation with a 94 score consists of clover, winter wheat, sugar beet, corn and spring barley, the second one having 80 per cent grain crops consists of clover, winter wheat, buckwheat, corn and barley, respectively. Both have an estimated efficiency score equal to 93 points, which corresponds with known expert assessments for these crop rotations. To compare these crop rotations with the proposed one by our recently developed software, optimal crop rotation was found for two fixed crops—sugar beet and spring barley. The result consisted of barley (100 points efficiency), clover (89 points), sugar beet (97 points) and green corn (98 points). The proposed crop rotation is shorter than the original one described in (Kovalenko, 2005) but coincides with it in other aspects.

An algorithm for selecting optimal crop rotation based on economic and ecologic factors was tested for the case of one fixed crop—corn. Growth conditions were taken as follows:

- soils – non-degraded sandy sod-podzolic on the eluvium of massive-crystalline rocks at the depth of 50 to 100 cm;
- climatic subzone – forest steppe with variable wetting at the level of 480 to 570 mm;
- region of Ukraine – Kyiv region;
- irrigation is absent;
- depth of plowing layer – 50 cm; and
- manure application rate – 5 ton/ha.

Weight coefficient that modelled negative environmental impact in equation (2) was assumed to be equal to 2, which corresponds to the absence of a significant effect from the application of nitrate-based fertilisers to the soil. Economic efficiency of fertilisation depends on the prices of chemical fertilisers and cultivated crops. When fertilisation is highly profitable, the additional potential benefit from the use of long crop rotations is not significant enough, due to the increase of economic risks from changes in cultivated crops. In this case, it is economically viable to select short-term crop rotations or a monoculture. With low fertilisation efficiency, long-term crop rotations become more economically viable.

In Ukraine, short-term crop rotations are used on large farms with only two or three highly economically-profitable crops included in crop rotation. A scenario was conducted in the case of two different levels of prices for corn, in order to experimentally verify how the selected or proposed models reflect and/or validate these assumptions. The results containing a list of crops in three-year crop rotation and the required amount of fertilisers to support their production, are presented in Tables 2 and 3.

Results of computational experiments show that at high prices for corn, maximal profit can be obtained by growing it as a monoculture with high fertilisation rates. When price decreases, the use of a short crop rotation with reduced (compared with the first case) fertilisation rates become more profitable.

Table 2 Results of crop rotation and fertilisers used in the case of high prices for corn.

Years 1 to 3: Corn	Yield - 7 ton/ha Price - 4500 UAH/ton [equivalent to $ 173)			
	N (kg/ha)	P (kg/ha)	K (kg/ha)	Price (UAH/ha)
	0.00	59.7	20.7	1121.2 ($43.1)

Table 3 Results of crop rotation and fertilisers selection in the case of low prices for corn.

Year 1: Potatoes	Yield – 34 ton/ha Price – 1000 UAH/ton (equivalent to $ 38.5)			
	N (kg/ha)	P (kg/ha)	K (kg/ha)	Price (UAH/ha)
	0.00	0.00	313.1	5740.1 ($ 220.8)
Year 2: Winter wheat	Yield – 5.55 ton/ha Price – 5300 UAH/ton (equivalent to $ 204)			
	N (kg/ha)	P (kg/ha)	K (kg/ha)	Price (UAH/ha)
	0.00	0.00	179.1	1433.1 ($55.1)
Year 3: Corn	Yield – 7 ton/ha Price – 3500 UAH/ton (equivalent to $ 134.6)			
	N (kg/ha)	P (kg/ha)	K (kg/ha)	Price (UAH/ha)
	0.00	20.4	6.7	379.8 ($ 14.6)

4. Conclusion

Three algorithms for solving the single-objective optimisation problem of finding optimal crop rotation (with and without cover crops on the base of the model with expert scores) have been determined, especially for conditions like those found in Ukraine. The branch-and-bounds algorithm resulted in obtaining a solution ~ 16 per cent faster compared with that of the complete search. However, working time of both algorithms was significant when selecting crop rotations longer than six years. The genetic algorithm that gave approximate solutions was significantly faster and allowed for obtaining approximations in the controlled time, regardless of the duration of crop rotation. The conducted analysis of short-term crop rotation used have shown the adequacy of the recommendations derived from simulation results to be applicable for developing sustainable agricultural practices. We also proposed a statement and solution algorithm for the multi-objective optimisation problem of a suitable crop rotation selection by taking economic and ecologic criteria into consideration. The costs of chemical fertilisation under other optimal conditions were considered as expenses. Negative impact of agricultural activity on the environment was estimated by the amount of nitrate-based fertilisation to the soil; however, the problem consists of three embedded optimisation problems.

Due to its complexity, we solve it with a heuristic genetic algorithm. By solving the multi-objective problem, we simulated the efficiency of growing corn as a monoculture, depending on the level of market prices for it. The simulation showed a decrease in profit from such practices when the market price fluctuates, especially at low prices. We plan to further develop the proposed mathematical models and algorithms by integrating additional ecologic criteria, crop yield-influencing factors and economic and climatic uncertainties associated with them.

References

Alfandari, L., Plateau, A. and Schepler, X. (2015). A branch-and-price-and-cut approach for sustainable crop rotation planning. Eur. J. Oper. Res., 241: 872–879.

Detlefsen, N. (2004). Crop rotation modelling. Proceedings of the EWDA-04 European Workshop for Decision Problems in Agriculture and Natural Resources, pp. 5–14.

Dury, J., Schaller, N., Garcia, F., Reynaud, A. and Bergez, J.E. (2012). Models to support cropping plan and crop rotation decisions: A review. Agron. Sust. Dev., 32: 567–580.

Gadzalo, Ya.M., Kaminsky, V.F. and Saiko, V.F. (2015). Crop rotations in agriculture of Ukraine. Collection of Scientific Papers, Agriculture, 1: 3–6.

Garcia, F., Guerrin, F. and Martin-Clouaire, R. (2005). The human side of agricultural production management: the missing focus in simulation approaches. pp. 203–209. *In*: Proc. MODSIM 2005 Conference, Melbourne, Australia.

Goldberg, D.E. (1989). Genetic Algorithms in Search, Optimisation and Machine Learning, Reading. MA: Addison-Wesley.

Goldewijk, K.K. (2001). Estimating global land use change over the past 300 years: The HYDE Database. Global Biogeochem. Cycles, 15: 417–433.

Kaminsky, V.F. and Boyko, P.I. (2014a). Strategy of development and implementation of crop rotations in Ukraine (Part 1). Collection of Scientific Papers, NSC Institute of Agriculture of NAAS, 3: 3–9.

Kaminsky, V.F. and Boyko, P.I. (2014b). Strategy of development and implementation of crop rotations in Ukraine (Part 2). Collection of Scientific Papers, NSC Institute of Agriculture of NAAS, 4: 3–11.

Kovalenko, N.P. (2005). History of optimisation of soil-protective crop rotations in the second half of the XX - early XXI century on the basis of economic and mathematical modelling (in Ukrainian). *In*: Adamen, F.F., Vergunov, V.A. and Vergunova, I.N. (eds.). Fundamentals of Mathematical Modelling of Agro-biocenoses, Kyiv, Ukraine, Nora-Print.

Kovalenko, N.P. (2007). Optimisation of the structure of cultivated areas and specialised crop rotations by economic-mathematical modelling method (in Ukrainian). Collection of Scientific Papers of the Institute of Sugar Beet of NAAS, 9: 245–251.

Kovalenko, N.P. (2012). Development and improvement of crop rotations for the conditions of insufficient humidification in Ukraine: Historical retrospective (in Ukrainian). Bull. Poltava State Agrarian Aca., 4: 27–32.

Lazer, P.N. and Mikheievle, K. (2006). Tools and technologies of information organisation in agriculture, Teaching Guide for Agriculture Students (in Ukrainian). Kherson, KhDU Publishing House.

Lee, G., Bao. C., Langrene, N. and Zhu, Z. (2015). Choosing crop rotations under uncertainty: A multi-period dynamic portfolio optimisation approach. 21st International Congress on Modelling and Simulation, Gold Coast, Australia, pp. 1084–1090.

Lukashuk, V.P. (2014). Productivity of grain-fodder crop rotation depending on the system of fertilization and basic tillage on meadow drained soils of the Left-Bank Forest-Steppe (in Ukrainian). International Economic Relations, The Caucasus. Economic and Social Analysis Journal of Southern Caucasus, No. 02/03/2014, pp. 26–29.

Marynych, A.M., Pashchenko, V.M. and Shyshchenko, P.H. (1985). The Nature of the Ukrainian SSR. Landscapes and Physical-geographical Zoning (in Ukrainian). Kyiv, Naukova Dumka.

Nuppenau, E.A. (2011). Linking Crop Rotation and Fertility Management by a Transition Matrix: Spatial and Dynamic Aspects in Programming of Ecosystem Service, paper Presented at the EAAE 2011 Congress Change and Uncertainty Challenges for Agriculture, Food and Natural Resources, August 30th–September 2, 2011, ETH Zurich, Zurich, Switzerland.

Osman, Ju., Inglada, Jo. and Dejoux, J.F. (2015). Assessment of a Markov logic model of crop rotations for early crop mapping. Computers Electronics Agric., 113: 234–243.

Pavón, R., Brunelli, R. and von Lücken, C. (2009). Determining optimal crop rotations by using multiobjective evolutionary algorithms, KES 2009, Part I, LNAI 5711, pp. 147–154.

Ramankutty, N. and Foley, J. (1998). Characterising patterns of global land use: An analysis of global croplands data. Global Biogeochem. Cycles, 12: 667–685.

Romashchenko, M., Tarariko, Yu., Shatkovskyi, A., Saydak, R. and Soroka, Yu. (2015). Scientific principles of the development of farming agriculture systems in the zone of Ukrainian Steppe (in Ukrainian). Bulletin of Agrarian Sci., 10: 5–9.

Santos, L.M.R., Munari, P., Costa, A.M. and Santos, R.H.S. (2015). A branch-price-and-cut method for the vegetable crop rotation scheduling problem with minimal plot size. Eur. J. Oper. Res., 245: 581–590.

Schönhart, M., Schmid, E. and Schneider, U.A. (2009). Crop Rota—A Model to Generate Optimal Crop Rotations from Observed Land Use, Working Papers 452009, Institute for Sustainable Economic Development, Department of Economics and Social Sciences, University of Natural Resources and Life Sciences, Vienna.

Trevisan, L. (2011). Combinatorial Optimisation: Exact and Approximate Algorithms, Stanford University, CA, USA.

Turner, I.I.B.L., Moss, R.H. and Skole, D.L. (1993). Relating Land Use and Global Land Cover Change: A Proposal for an IGBP-HDP Core Project, Rep. 24, Int. Geosphere-Biosphere Programme, Stockholm.

Verhunova, I.M. (2000). Fundamentals of Mathematical Modelling for the Analysis and Prognosis of Agronomic Processes (in Ukrainian). Kyiv, Nora-print.

Yurkevych, Ye.O., Kovalenko, N.P. and Bakuma, A.V. (2011). Agro-biological Basis of Crop Rotations of Ukrainian Steppe (in Ukrainian). Odessa, VMV.

6

Sustainable Suppression of Weeds through Ecological Use of Cover Crops

Shawn T Lucas

1. Introduction

The management of agro-ecosystems involves extensive and frequent ecosystem disturbances. Events like clearing native vegetation, tillage and sudden inputs of readily available soluble nutrients contrast with natural ecosystems. Weeds have often been described as indicators of ecosystem disturbance because they are essentially early successional plants that take advantage of the pervasive disturbances encountered in agro-ecosystems (Gaba et al., 2014). Because they compete with crops and can cause significant yield losses (Chandler et al., 1984; Knezevic et al., 2002), producers have struggled with controlling weeds in crop production systems since the earliest days of agriculture. Weed-management practices have ranged from simple hand-pulling and physical smothering to more complex tillage, mechanical weeding and stale seed-bed techniques (Zimdahl, 2013).

Cover crops are an important management component within sustainable agro-ecosystems. Benefits of well-managed cover crops include reduced erosion (Reeves, 1994; De Baets et al., 2011), improved soil organic-matter content (Kuo et al., 1997; Puget and Drinkwater, 2011; Poeplau and Don, 2015), promotion of good soil structure (Hermawan and Bomke, 1997; Liu et al., 2005; Blanco-Canqui et al., 2011), healthy microbial communities and improved soil function (Schutter and Dick, 2002; Nair and Ngouajio, 2012; Lucas et al., 2014). Cover crops also scavenge and translocate nutrients (Ranells and Wagger, 1997; Tonitto et al., 2006; Komainda et al., 2018) and provide a source of N input (Hargrove, 1986; Blevins et al., 1990). Another often-reported benefit of cover crops is suppression of weed populations (Creamer et al., 1996; Liebman and Davis, 2000; Teasdale et al., 2007).

Several factors have given rise to new interest in using cover crops for weed suppression in recent years. Increasing resistance among weeds to glyphosate and other herbicides has become problematic (Beckie, 2006; Powles, 2008; Peterson et al., 2018). Strategic use of cover crops is seen as a having a role in combating this problem (Lamichhane et al., 2015; Wiggins et al., 2016; Marochi et al., 2018). Similarly, for producers managing organic systems or low-input systems, weed pressure has long been recognised as a major challenge. In certified organic systems in particular, weeds are a significant impediment to overcome for producers considering a transition from non-organic production systems (Stonehouse et al., 1996; Bond and Grundy, 2001). Cover crops are a critical component of holisticweed management strategies in organic and low-input systems (Barberi, 2002; Leibman et al., 2009).

College of Agriculture, Community, and the Sciences, Kentucky State University, 400 East Main Street, Frankfort, KY 40601.
Work conducted by Dr. Shawn Lucas at Kentucky State University is supported by USDA-NIFA Evans-Allen funding.

Understanding how to use cover crops as a weed management tool is important for those who aim to manage agro-ecosystems to maintain or improve productivity, while minimizing environmental impacts and expenses. To use cover crops successfully for weed control, producers need to have knowledge about the mechanisms by which these crops can reduce weed pressure. Understanding which crops are optimal for a region, particular weed problems, and crop production schemes is of critical importance. Information on optimal management of specific cover crops for maximum benefit, as well as challenges that can arise in cover crop systems, is also important. This chapter attempts to provide general scientific information on the practical use of cover crops for suppression of annual weeds in modern agro-ecosystems.

2. Mechanisms by which Cover Crops can Suppress Weeds

Actively growing cover crops, as well as the residues from terminated cover crops, can inhibit weed establishment, growth and productivity via several mechanisms. Living cover crops affect weed populations through direct competition for ecosystem resources, allelopathic interactions and interference with weed-seed germination and seedling emergence (Lemessa and Wakjira, 2015; Teasdale and Daughtry, 1993; Osipitan et al., 2018). The mulch or incorporated residues associated with terminated cover crops can impact weeds by physically or chemically inhibiting germination of weed seeds, releasing weed-inhibiting allelopathic compounds during decomposition, preventing weed emergence and immobilising nitrogen during decomposition (Teasdale et al., 2007; Teasdale and Mohler, 1993; Creamer et al., 1996). There is also evidence that cover crops can affect weeds via interactions with the soil microbial community by increasing the activity of fungi that infect weeds (Jordan et al., 2000; Lou et al., 2016; Mohler et al., 2012).

2.1 Allelopathy

Nature is filled with examples of complex biochemical interactions between organisms. A skunk releases a spray of mercaptans to ward off a predator. Streptomycetes release aminoglycosides, having antibiotic properties, to suppress competing bacteria in the harsh soil ecosystem. It should be no surprise that members of the kingdom Plantae also emit active biochemical compounds, known as allelochemicals, which can influence their neighbours or give them competitive advantages. Allelopathy is a biochemical interaction between plants in an ecosystem. This interaction can be stimulatory or inhibitory, but in agro-ecosystems research, there has been significant focus on exploiting inhibitory interaction for the purpose of weed suppression (Putnam et al., 1983; Khanh et al., 2005; Kelton et al., 2012; Jabran et al., 2015). Despite first being described and named in the early twentieth century by Austrian botanist, Hans Molisch, our understanding of allelopathic interactions is still developing, but these interactions are being increasingly recognised as a potential tool in integrated, sustainable farming operations (Molisch, 1937; Jabran et al., 2015).

Many cover crops have been reported to have allelopathic activity (Table 1). As seen in Table 1, many different types of compounds having different modes of action can be involved in alleopathic interactions. These compounds have numerous possible processes for being released into the soil (Kelton et al., 2012). Some are released from living plants as root exudates or volatiles while others are released from decomposing plant tissues (Table 1). Producers using cover crops having allelopathic properties need to understand the cover crop itself, the effects of the cover crop on target weeds and potential unintended effects, such as impacts on subsequent crops or non-target impacts of the crop in the agro-ecosystem (Lemessa and Wakjira, 2015).

When using allelopathic cover crops as part of weed control strategy, producers need to consider multiple factors that can influence the effectiveness of the management practice. Different cultivars of a crop may have greater or lesser allelopathic potential. Numerous studies have found cultivar to be a determinant in the allelopathic activity of cover crops, such as rye (Reberg-Horton et al., 2005), wheat (Baghestani et al., 1999), sunflower (Leather, 1983), sorghum (Alsaadawi et al., 1986), oats (Fay and Duke, 1977) and others (Bhowmik, 2003). One should also understand how allelopathic

Table 1 Selected cover crops with allelopathic properties. Compound types and potential release mechanisms are included.

Cover Crop	Scientific Name	Allelopathic Compounds	Compound Type	Compound Release Mechanism	References
Brassicas					
Rapeseed	*Brassica naupus* L.	Allyl isothiocyanate Methyl isothiocyanate	Glucosinolates Isothiocyanates	Root exudates, Decomposition products	Boydston and Hang, 1995; Haramoto and Gallandt, 2004
Mustards	*Brassica juncea* L. *Brassica nigra* L. *Sinapsis alba* L.	Various	Glucosinolates Isothiocyanates	Root exudates, Decomposition products	Haramoto and Gallandt, 2004; Narwal and Haouala, 2001
Radish	*Raphanus* spp.	Various	Glucosinolates Isothiocyanates	Root exudates, Decomposition products	Kunz et al., 2016; Malik et al., 2008
Turnip	*Brassica rapa* L.	Various	Glucosinolates Isothiocyanates	Root exudates, Decomposition products	Petersen et al., 2001; Uremis et al., 2009
Grains					
Barley	*Hordeum vulgare* L.	Gramine Hordenine Others	Alkaloids Others	Leaf and root decomposition, root exudates	Kremer and Ben-Hammouda, 2009
Oats	*Avena sativa* L.	Scopoletin	Coumarin	Root exudates	Fay and Duke, 1977
Rye	*Secale cerale* L.	Benzoxazinoids Phenolic acids	Hydroxamic acid Phenolic acids	Root exudates, Decomposition products	Barnes and Putnam, 1987; Chou and Patrick, 1976
Triticale	x *Triticosecale*	Benzoxazinoids	Hydroxamic acid	Root exudates, Decomposition products	Reiss et al., 2018
Wheat	*Triticum aestivum* L.	Benzoxazinoids Phenolic acids	Hydroxamic acid Phenolic acids	Root exudates, Decomposition products	Reiss et al., 2018; Wu et al., 2001
Legumes					
Alfalfa	*Medicago sativa* L.	Coumarin, *trans*-cinnamic acid *o*-coumaric acid	Phenolic acids	Leaf decomposition products	Chon and Kim, 2002
Berseem clover	*Trifolium alexandrinum* L.	Volatile Aldehydes	Aldehydes	Decomposition products	Bradow and Conncik, 1990
Red clover	*Trifolium pratense* L.	Phenolic acids	Phenolic acids	Decomposition products	Ohno et al., 2000
Sweet clover	*Melilotus alba* Desr *Melilotus officinalis* L.	Coumarins	Coumarins	Decomposition products	Blackshaw et al., 2001; Moyer et al., 2007; Wu et al., 2016
White clover	*Trifolium repens* L.	Phenolic acids	Phenolic acids	Decomposition products	Macfarlane et al., 1982a; Macfarlane et al., 1982b
Hairy vetch	*Vicia villosa* Roth	Cyanamide	Cyanamide	Decomposition products	Fuji, 2003; Ercoli et al., 2007; Hill et al., 2007; Teasdale et al., 2007a

Table 1 Contd. ...

...Table 1 Contd.

Cover Crop	Scientific Name	Allelopathic Compounds	Compound Type	Compound Release Mechanism	References
Field pea	*Pisum sativum* L.	Pisatin	Phytoalexin	Decomposition products	Kato-Noguchi, 2003
Cow Pea	*Vigna unguiculata* L.	Unknown	Unknown	Decomposition products	Adler and Chase, 2007; Hill et al., 2007;
Sunn Hemp	*Crotalaria juncea* L.	Unknown, possibly hydroxynorleucine	Amino acid	Decomposition products	Adler and Chase, 2007; Javaid et al., 2015
Others					
Buckwheat	*Fagopyrum esculentum* Moench.	Gallic acid	Phenolic	Root exudates	Iqbal et al., 2003 Kalinova et al., 2007
		(+)-Catechin	Phenolic	Root exudates	Iqbal et al., 2003
Sunflower	*Helianthus annuus* L.	various	Sesquiterpenes	Decomposition products Root exudates	Leather, 1983; Macías et al., 2006
Sorghum	*Sorghum bicolor* (L.) Moench	Dhurrin Sorgoleone	Cyanogenic glycoside Hydroquinone	Decomposition products Root exudates	Weston et al., 2013
Sorghum-Sudangrass	*Sorghum bicolor* (L.) Moench × *Sorghum Sudanese* (P.)	Dhurrin Sorgoleone	Cyanogenic glycoside Hydroquinone	Decomposition products Root exudates	Weston et al., 1989; Czarnota et al., 2003

plants interact with the environment. Lemerle et al. (2001) found that most of the yield variability (81 per cent) observed in a study conducted on wheat could be explained by the interaction of the cultivar with the environment. Einhellig (1987) found that environmental stresses, such as elevated temperatures or moisture stress could enhance allopathic inhibition of other plants. Others have found that allelopathic interactions are dependent on soil texture and soil organic matter content (Blum and Shafer, 1988).

A challenge for those working with allelopathic cover crops is that the considerations described above are generally not well-understood (Weston, 1996; Kelton et al., 2012; Lemessa and Wakjira, 2015). Weston (1996) points out that attributing cover crop impacts on weeds to allelopathy can be challenging when cover crops can also affect the weed population through other means, such as smothering, shading, or other competitive processes. Environmental effects can be subtle, varying from season to season. Those working with allelopathic plants should be aware that allelopathic interactions often involve specific donor plants, affecting specific target plants (Weston and Duke, 2003). In addition, when target plants have optimal growth conditions, allelopathic impacts from cover crops or cover crop residues are often minimised (Einhellig, 1987). When choosing cover crops for weed control purposes, given the complexity of allelopathic interactions and the need for more understanding of allelopathy in general, producers should consider all weed control aspects of the cover crop rather than rely solely on allelopathic potential.

3. Interference with Weed Germination and Emergence

Cover crops can inhibit weed-seed germination through several mechanisms. As discussed in the previous section, some cover crops release allelopathic compounds and some of these compounds can inhibit seed germination (Putnam et al., 1983; Przepiorkowski and Gorski, 1994). Much of the information on allelopathic inhibition of weed seed germination comes from studies on small grain cover crops, such as annual rye. Putnam et al. (1983) found 43–100 per cent inhibition of germination and emergence in selected weeds. They attributed this to release of allelochemicals on the soil surface as rye residues decomposed. Similarly, Przepiorkowski and Gorski (1994) found that aqueous extracts of rye inhibited up to 50 per cent of seed germination in horseweed (*Conyza canadensis* L.) and willow herb (*Epilobiumciliatum* Rafin). Creamer et al. (1996) used leached or non-leached residues of rye, barley, crimson clover, hairy vetch, or a mixture of all four to test inhibition of germination in nightshade (*Solanum ptycanthum* Dun.) and yellow foxtail (*Setaria glauca* L.). All individual cover crops, as well as the mixture, showed inhibition of nightshade seed germination, while rye and barley also inhibited yellow foxtail seed germination. In addition to the inhibitory effects of hairy vetch and crimson clover observed by Creamer et al. (1996), others have also found that leguminous plants can inhibit seed germination. White et al. (1989) also found hairy vetch and crimson clover to inhibit germination of wild mustard (*Sinapsis arvensis* L.) and morning glory (*Ipomoea lacunosa* L.) and some crops including corn and cotton. As discussed previously, it is often difficult to characterise the impacts of allelopathy and differentiate those effects from other inhibitory mechanisms.

Teasdale and Mohler (1993) attempted to demonstrate that other cover-crop factors aside from allelopathy could inhibit weed-seed germination. They hypothesised that cover-crop residues on the soil surface would interfere with light transmittance, soil moisture and soil temperature – parameters essential for triggering seed germination. They were interested in phytochrome dynamics in seeds because red light activates phytochrome, which in turn stimulates the germination process, while far-red light deactivates phytochrome and inhibits germination. Their work showed that both rye and hairy vetch residues reduced red light transmittance and reduced emergence of weeds. They also demonstrated that cover-crop residues maintained lower soil temperatures, which can also reduce germination. Teasdale and Daughtry (1993) saw similar results in a study that focused on hairy vetch. Kruk et al. (2006) further confirmed the importance of cover-crop influence on red and far-red light transmittance. They observed that wheat suppressed germination and emergence of *Galinsoga parviflora* by reducing the red light to far-red light ratio. They further showed that the effect was reversible when light coming through the canopy was forced through far-red light filters. Lawley et al. (2012) also attributed the weed-suppressive effect of a fall forage radish cover crop to interference with red light transmittance. These light transmittance effects are specific for weed seeds that reside near the soil surface.

In addition to reducing germination by blocking light and keeping the soil cooler, the dense mulches associated with terminated cover crops, if managed appropriately, act as a physical barrier to weed emergence. Mohler and Teasdale (1993) found that this mulch effect only happened at very high residue densities and that, at the time of their study, achieving these densities was not common in the fields of northeastern United States. The mulch effect of a cover crop is dependent on even residue coverage on the soil surface. This coverage is not achieved when cover crops are terminated by mowing, stalk chopping, or herbicide spray (Moore et al., 1994; Teasdale and Rosecrance, 2003). Generation of a high cover-crop biomass and termination of cover crops, particularly small grains, using a crimper-roller, greatly improves the mulch effect of cover crops on weed emergence (Mirsky et al., 2011; Reberg-Horton et al., 2012; Wayman et al., 2015). Reberg-Horton et al. (2012) highlighted the importance of biomass production for the mulch-based suppression of weeds and found that about 9,000 kg ha⁻¹ is suitable for suppressing weed emergence. Mirsky et al. (2011) saw evidence of weed suppression, compared to control plots that received no cover crop residue, at slightly lower biomass levels and found suppression to be a function of both planting and termination date (Fig. 1), which are determining factors in biomass production. Wayman et al. (2015) also note the importance of cover-crop species and varieties in determining ground coverage. In their study, the grains they used (rye and barley) generated higher biomass than the vetch varieties studied and the Aroostook rye variety had significantly higher biomass (9,000 kg ha⁻¹) than all other cover crops.

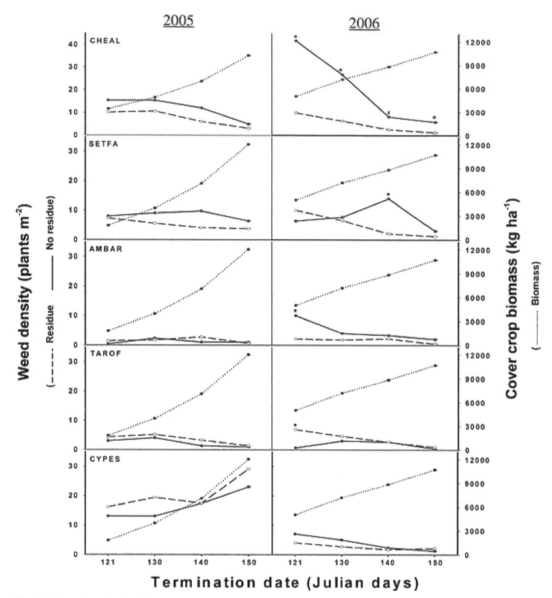

Fig. 1 Effects of termination date on cover crop biomass and weed ground coverage in 2005 and 2006 in a study by Mirsky et al. (2011). They compared plots that received cover crop residues to those receiving no residues at various covercrop termination dates. They conducted ANOVA and means comparisons (Tukey-Kramer Method) on selected weed species in their trial (CHEAL, common lamb's quarters; SETFA, giant foxtail; AMBEL, common ragweed; TAROF, dandelion; CYPES, yellow nutsedge). For each termination date, an asterisk (*) indicates a significant difference in weed density between no-residue and residue plots at α = 0.05. Figure from Mirsky et al. (2011).

Much like allelopathic interactions, germination and emergence inhibition by cover crops has only recently begun to be understood. Cover-crop management technology that influences cover crop effects on germination and emergence, such as crimper-rollers, is also relatively new and still being improved upon (Mirsky et al., 2011; Kornecki, 2015). Effects on germination and emergence need to be considered as part of the potential weed-suppressive traits any particular cover crop can have and should not be the sole basis for choosing a specific cover crop.

4. Weed-suppressive Cover Crop Interactions with the Soil Microbial Community

It is becoming increasingly apparent that the ways in which producers manage their soils can have an impact on the microbial communities in those soils. Researchers have observed numerous ways by which inputs or soil disruption have altered the soil microbial community and some have suggested that the soil microbial community can be managed to facilitate positive agro-ecological outcomes (Chaparro et al., 2012; Lucas et al., 2014). Cover crops are an option in managing the soil microbial community. Kumar et al. (2004) found that tomatoes grown after a hairy vetch fall cover crop had improved disease resistance and delayed leaf senescence. They associated this improved health with increased expression of certain nitrogen uptake and disease response genes. Mattoo and Abdul-Baki (2006) expanded on the work by Kumar et al. (2004) and suggested that microbes influenced by the hairy vetch restudies may have elicited the gene responses seen in the tomato plants. Using fatty acid methyl ester biomarkers extracted from soils, Lucas et al. (2014) found that hairy vetch residues promoted increased fungal biomass and that this fungal biomass was correlated with soil aggregate formation. These studies and others suggest that agro-ecosystem-management strategies can involve manipulation of the microbial community.

While actively managing the soil microbial community is a developing area of research, there is evidence that cover crops can influence weed populations via interactions with the soil microbial community. Jordan et al. (2000) noted that studies by Zobel et al. (1997) and Van Der Heijden et al. (1998) had shown that arbuscular mycorrhizal fungi in soils could have effects on the productivity, diversity and community dynamics of plants in an agro-ecosystem. Jordan et al. (2000) also noted that Johnson and Pfleger (1992) had observed that cover crops increased the abundance and diversity of arbuscular mycorrhizal fungi in soils. These findings led Jordan et al. (2000) to speculate that cover crops could be used to influence soil arbuscular mycorrhizal fungi for the purpose of affecting the weed populations in agro-ecosystems. There has been recent evidence to support this idea as a viable approach. Mohler et al. (2012) saw reduced emergence of several annual weeds, including velvet leaf (*Abutilon theophrasti*), lamb's quarters (*Chenopodium album*), Powell's amaranth (*Amaranthus powellii*), giant foxtail (*Setariafaberi*) and cockspur (*Echinochloa crus-galli*), following the incorporation of green manures that consisted primarily of peas (*Pisum sativum* L.). They isolated the pathogens *Fusarium oxysporum* and *Fusarium chlamydosporum* from weed-seeds collected from soils that had received green manures. They inferred that the reduced weed emergence they observed may have been due to increased pathogen levels that were influenced by green manure incorporation. Lou et al. (2016) also saw weed suppression after a red clover cover crop. They concluded that the suppression was due to a combination of allelopathy and microbial infection of weed seeds and seedlings. They suggest a phased weed inhibition where, for a period after cover crop residue deposition, the allelopathic interactions are most important. However as some soil microbes break down allelochemicals, other members of the soil microbial community become more important and inhibit weeds through seed and/or seedling infection.

Like other aspects of cover crop suppression of weeds previously discussed, the concepts involved in managing soil microbes to control weeds are relatively new and developing. More research needs to be conducted on specific interactions between cover crops and microbial communities and related effects on weed populations. Potential impacts from these microbial interactions on subsequent crops need to be researched as well.

5. Cover Crop Competition with Weeds

Perhaps the most direct way by which cover crops impact weed populations is through direct competition. Since weeds are plants that can fill a niche in a disturbed ecosystem, the goal in using cover crops as competitors is to preempt weeds from filling that niche or to displace them if they have been established. Cover crops that successfully do this are those that take up the light, nutrients, water, space and other resources that the weeds would otherwise garner. Some of the mechanisms already discussed are inherently competitive processes. Allelopathy, stemming from living cover crops, is a type of natural 'chemical warfare' by which the cover crop is inhibiting a potential competitor, thereby creating a competitive advantage for its presence in the agro-ecosystem (Rice, 1984). The previously described studies by Teasdale and Doughtry (1993), Teasdale and Mohler (1993), Kruk et al. (2006), and Lawley et al. (2012) are all examples of cover-crops suppressing weeds through competition for light. It has been established that certain cover crops, particularly small grains, are adept at scavenging N and other nutrients (Ranells and Wagger, 1997; Tonitto et al., 2006; Komainda et al., 2018). Establishment of winter cover crops early in fall leads to greater stand densities of these nutrient-scavenging plants (Mirsky et al., 2011), which would leave less available N for early emerging weeds in spring. Further, termination of high biomass gramineous crops can also cause immobilisation of N as residues are decomposed by the soil microbial community (Wells et al., 2013). Wells et al. (2013) suggest that this strategy may be most suited for cover crops that precede a nitrogen-fixing legume crop, such as soybean. Incorporating a legume, such as hairy vetch into the cover crop, would alleviate the nitrogen stress on weeds associated with immobilisation as the legume residues decompose and release N. Others have seen that cover crops have interfered with weed productivity through competition for water (Mayer and Hartwig, 1986). Biomass density helps determine how well cover crops compete with weeds for space. The work of McLenaghen et al. (1996) indicated that weed ground-cover density decreased as cover crop coverage increased. They observed 52 per cent ground cover with weeds when no cover crops were grown, while there was only 4 per cent weed coverage in their highest density cover crop treatment (white mustard), which saw 94 per cent ground coverage with the cover crop. The work by Mirsky et al. (2011) described earlier also shows that, in the five annual weeds they studied, in general, as cover crop biomass increases, weed ground coverage decreases.

Cover crop competition, microbial interactions, effects on germination and emergence and allelopathy are different facets of weed control via cover crops. They are not mutually exclusive and a general understanding of all these mechanisms will help producers, researchers and extension agents make cover-crop-management decisions to improve weed control in agro-ecosystems. When this understanding is coupled with good ecological management considerations, weed control from cover crops can be maximised.

6. Agroecological Considerations for Weed Control with Cover Crops

In sustainably managed agro-ecosystems, the goal for weed control is generally not the total elimination of weeds, but rather suppressing weed populations, such that weed populations do not have significant negative impacts, particularly negative yield impacts, on the desired crops (Buhler, 2003). To achieve this goal, producers, the researchers and extension agents who work with them, need to have an understanding of whole cropping-system designs, of which cover crops are an important component. A whole cropping system design includes integrated approaches that recognise that every management practice can potentially have an impact on multiple aspects of the agro-ecosystem, including soil health, crop productivity and weed population dynamics. Weed populations will respond to many management practices, including tillage, fertility treatments, crop rotations and cover crops (Buhler, 2003; Sanyal et al., 2008). The objective for those wishing to manage weeds in an agro-ecosystem is to integrate these different practices so that deleterious impacts of weeds are minimised. To do this, producers need to take an integrated ecological approach to the problem. According to Sanyal et al. (2008), there are four basic components of integrated weed control: (1) suppress weed growth; (2) suppress or prevent weed seed production; (3) deplete the weed seed-bank (the population of weed seeds in the soil); and (4) control spread of weeds. Cover crops, through the mechanisms described previously, can be useful tools in addressing all of these components.

Using an ecological approach to weed control, producers should have an understanding of which weeds are most problematic in their fields and additionally, they should understand the lifecycles of these weeds as well as their crops (Buhler, 2003; Sanyal et al., 2008; Liebman et al., 2009). They need to know the critical period of a particular crop with respect to the problem of weed's lifecycle (Knezevic et al., 2002). The critical period is defined as that period in a crop's lifecycle where it is most vulnerable to impacts of weeds (Zimdahl, 1988; Knezevic et al., 2002). During this period, weeds must be managed to minimise their potential for crop losses. This often occurs early in the crop's lifecycle when it faces competition from weeds that germinated earlier than, or concurrently with, the crop (Knezevic et al., 1995; Martin et al., 2001).

Fall or overwinter cover crops that suppress weed germination, emergence, or growth, the following spring can potentially affect weeds during the critical period. To maximise this effect, Kruidhof et al. (2011) suggest that cover crop termination and residue placement be synchronised with the period at which a given species of problematic weed is most susceptible to the impacts of the suppressive mechanisms of cover crop (Fig. 2). It should be noted that some weed species may escape, depending on emergence time, but synchronising, such that the most problematic weeds are affected, could give the desired crop a competitive advantage during its critical period. Summer cover crops, such as buckwheat or sorghum sudangrass, can be strategically used as smother crops during a crop rotation to reduce weed groundcover during the reproductive phases of the weed lifecycle, thus reducing weed-seed production and deposition into the seed-bank. Some evidence of this type of seed-bank control is seen in the work of Mirsky et al. (2010), where they observed that crop rotation containing summer buckwheat meant a reduction in the weed-seed bank for three weed species: *Chenopodium album*, *Abutilon theophrasti* and *Setaria* spp. This rotation also contained other cover crops (yellow mustard and winter rape) that could affect weed populations and thus attributing the seed-bank reduction solely to buckwheat, was not possible. More research confirming seed-bank reduction due to summer cover crops is needed. Cultural practices, such as stale seed bed (as described in Caldwell and Mohler, 2001) or cover crops that stimulate weed germination, only to be followed by a smother crop or a competitive cash crop in the rotation, can also deplete the weed seed-bank. Seed-degrading pathogenic microbes that may be stimulated by cover crop interactions, as described previously, could also reduce viable weed seeds in the agro-ecosystem. Cover crops, used with integrated approaches through multiple growing seasons in a carefully planned cropping system, could reduce weed pressure and reduce the spread of weeds in that agro-ecosystem.

Managing the cover crop aspects of an integrated weed control system takes care and planning (Sanyal et al., 2008). Regionally and seasonally appropriate cover crops that are complimentary to the rest of the cropping system design should be chosen. Cool season covers, such as rye, wheat

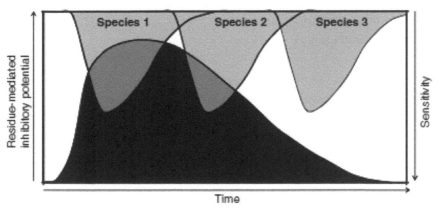

Fig. 2 A schematic diagram of the synchrony between cover crop residue weed suppression effectiveness and three weed species in an inhibition hypothesis proposed by Kruidhof et al. (2011). The area in black represents cover crop residue weed suppression potential, while the three areas in grey represent the potential vulnerability to cover crop effects of three weed species that emerge at different times. The areas where black and grey areas overlap represent the time period with the potential for the most weed suppression due to cover crop effects. Note that weed species 3 will not be affected by cover crop suppression mechanisms. Figure from Kruidhof et al. (2011).

and hairy vetch should be planted as early in the fall as possible, within the constraints of the overall cropping system, to allow for acclimation to colder temperatures and sufficient biomass generation (Mirsky et al., 2011). Termination timing for fall-planted covers varies for cover crops. Some cover crops, such as hairy vetch, should be terminated before they set seed (when about 60 per cent of plants are flowering is a good rule of thumb for termination of vetch) to prevent vetch from becoming part of the weed-seed bank (Sarrantonio, 1994). For grains, the anthesis stage seems to be ideal when using a crimper roller (Mirsky et al., 2009). Fast-growing summer cover crops, such as buckwheat or sorghum sudangrass, should be planted when temperatures are warm enough to support their rapid growth. There is some flexibility with these crops due to their growth habits and evidence suggests that even if planted after weeds have emerged, they can overcome the weeds and have an impact on the weed-seed bank (Gibson et al., 2011). Like vetch, buckwheat should be terminated before seed set when it is used as a cover crop or there will be a large number of 'weedy' late season or spring buckwheat volunteers (Bjorkman et al., 2008). Cover-crop species that grow well in a region should be chosen since weed control is linked to cover-crop biomass. Even within a species, producers should make certain that they have cultivars that will thrive. For example, Teasdale et al. (2004) found that hairy vetch cultivars developed in Alabama at Auburn University did not overwinter in New York fields. This chapter has primarily discussed weed control strategies with cover crops that are best suited for controlling annual weeds. Perennial weeds can be more challenging to control with cover crops because they are better competitors than annual weeds, often having structures, like rhizomes or tubers, providing nutrient storage below the soil surface, in addition to having vigorous seed production (Teasdale et al., 2007). Other intensive cultural practices need to be employed in addition to cover crops to combat perennial weeds. Teasdale et al. (2007) provide a brief review of controlling perennial weeds and obtaining additional resources from the literature for that purpose.

Specific information on regionally appropriate cover crops is available through Land Grant Extension services in the United States. Producers should consult these services when making cover crop decisions. Several excellent, producer-oriented, online briefs on using cover crops for weed control are available through the eXtension.org platform (Schonbeck, 2011; Schonbeck, 2015). Additionally, there are several, relatively new, regional cover-crop councils that promote cover crop use and provide decision tools and resource guides for their specific regions. These organisations include the Midwest Cover Crops Council (mccc.msu.edu), the Southern Cover Crops Council (southerncovercrops.org), the Northeast Cover Crops Council (northeastcovercrops.com) and the Western Cover Crops Council, which at the time of this writing, is being formed. Additional information on using a crimper-roller (including implement design plans and suggested vendors) and for using cover crops for weed control (and more) in organic systems is available through the Rodale Institute (rodaleinstitute.org).

7. Conclusion

Cover crops can be an important part of an integrated weed control programme in agro-ecosystems. Different cover crops can have varying and multiple effects that suppress weeds. Mechanisms involved in weed suppression by cover crops include competition, allelopathy, interference with weed germination and emergence and promotion of microbial community members that can inhibit or attack weed-seeds or seedlings. Research on all of these mechanisms and their interactions with weeds and crops in specific regions is needed to better optimise suppression of weeds with cover crops. Other specific areas where research is needed include the development of highly allelopathic cover-crop cultivars, further development of regionally suitable cover crop varieties and refining of the timing of cover-crop planting and termination for maximum weed suppression during critical periods for crops. Use of cover crops for weed control does not happen in a vacuum, but rather is an important component of a whole cropping system design that is managed ecologically. Other considerations in these systems include specific knowledge of weed life-cycles, tillage practices, stale seed bedding, crop rotation schemes and selection of competitive crop cultivars. Producer-oriented information on cover crops is available through Land Grant Extension services, online resources and regional cover-crops councils. As herbicide costs and resistance continue to increase and producers and consumers continue to have growing interest in sustainably-managed food crops, including certified organic and regenerative cropping systems, interest in cover crops for weed control will continue to grow. This interest will sharpen the need for more research in this area to facilitate effective weed control with cover crops in ecologically-managed agro-ecosystems.

References

Adler, M.J. and Chase, C.A. (2007). Comparison of the allelopathic potential of leguminous summer cover crops: Cowpea, Sunn hemp and velvet bean. Hort. Sci., 42: 289–293.

Alsaadawi, I.S., Al-Uqaili, J.K., Alrubeaa, A.J. and Al-Hadithy, S.M. (1986). Allelopathic suppression of weed and nitrification by selected cultivars of *Sorghum bicolor* (L.) Moench. J. Chem. Ecol., 12: 209–219.

Baghestani, A., Lemieux, C., Leroux, G.D., Baziramakenga, R. and Simard, R.R. (1999). Determination of allelochemicals in spring cereal cultivars of different competitiveness. Weed Sci., 47: 498–504.

Bàrberi, P.A.O.L.O. (2002). Weed management in organic agriculture: Are we addressing the right issues? Weed Res., 42: 177–193.

Barnes, J.P. and Putnam, A,R. (1987). Role of benzoxazinones in allelopathy by rye (*Secale cereale* L.). J. Chem. Ecol., 13: 889–906.

Beckie, H.J. (2006). Herbicide-resistant weeds: Management tactics and practices. Weed Tech., 20: 793–814.

Bhowmik, P.C. (2003). Challenges and opportunities in implementing allelopathy for natural weed management. Crop Protection, 22: 661–671.

Björkman, T., Bellinder, R., Hahn, R. and Shail Jr., J. (2008). Buckwheat Cover Crop Handbook: A Precise Tool for Weed Management on Northeastern Farms. Cornell University.

Blackshaw, R.E., Moyer, J.R., Doram, R.C. and Boswell, A.L. (2001). Yellow sweet clover, green manure, and its residues effectively suppress weeds during fallow. Weed Sci., 49: 406–413.

Blanco-Canqui, H., Mikha, M.M., Presley, D.R. and Claassen, M.M. (2011). Addition of cover crops enhances no-till potential for improving soil physical properties. Soil Sci. Soc. Amer. J., 75: 1471–1482.

Blevins, R.L., Herbek, J.H. and Frye, W.W. (1990). Legume cover crops as a nitrogen source for no-till corn and grain sorghum. Agron. J., 82: 769–772.

Blum, U. and Shafer, S.R. (1988). Microbial populations and phenolic acids in soil. Soil Bio. Biochem., 20: 793–800.

Bond, W. and Grundy, A.C. (2001). Non-chemical weed management in organic farming systems. Weed Res., 41: 383–405.

Boydston, R.A. and Hang, A. (1995). Rapeseed (*Brassica napus*) green manure crop suppresses weeds in potato (*Solanum tuberosum*). Weed Tech., 9: 669–675.

Bradow, J.M. and Connick, W.J. (1990). Volatile seed germination inhibitors from plant residues. J. Chem. Ecol., 16: 645–666.

Buhler, D.D. (2003). Weed biology, cropping systems and weed management. J. Crop Prod., 8: 245–270.

Caldwell, B. and Mohler, C.L. (2001). Stale seed-bed practices for vegetable production. Hort. Sci., 36: 703–705.

Chandler, J.M., Hamill, A.S. and Thomas, A.G. (1984). Crop Losses Due to Weeds in Canada and the United States: Special Report of the Losses Due to Weeds Committee, Weed Science Society of America, Champaign, IL.

Chaparro, J.M., Sheflin, A.M., Manter, D.K. and Vivanco, J.M. (2012). Manipulating the soil microbiome to increase soil health and plant fertility. Biol. Fert. Soils, 48: 489–499.

Chon, S.U. and Kim, J.D. (2002). Biological activity and quantification of suspected allelochemicals from alfalfa plant parts. J. Agron. Crop Sci., 188: 281–285.

Chou, C.H. and Patrick, Z.A. (1976). Identification and phytotoxic activity of compounds produced during decomposition of corn and rye residues in soil. J. Chem. Ecol., 2: 369–387.

Creamer, N.G., Bennett, M.A., Stinner, B.R., Cardina, J. and Regnier, E.E. (1996). Mechanisms of weed suppression in cover crop-based production systems. Hort. Sci., 31: 410–413.

Czarnota, M.A., Rimando, A.M. and Weston, L.A. (2003). Evaluation of root exudates of seven sorghum accessions. J. Chem. Ecol., 29: 2073–2083.

De Baets, S., Poesen, J., Meersmans, J. and Serlet, L. (2011). Cover crops and their erosion-reducing effects during concentrated flow erosion. Catena, 237–244.

Einhellig, F.A. (1987). Interactions among allelochemicals and other stress factors of the plant environment. pp. 343–357. *In*: Waller, G.R. (ed.). Allelochemicals: Role in Agriculture and Forestry, Amer. Chem. Soc., Washington, D.C.

Ercoli, L., Masoni, A., Pampana, S. and Arduini, I. (2007). Allelopathic effects of rye, brown mustard and hairy vetch on redroot pigweed, common lambs quarter and knotweed. Allelopathy J., 19: 249–256.

Fay, P.K. and Duke, W.B. (1977). An assessment of allelopathic potential in Avena germ plasm. Weed Sci., 25: 224–228.

Fujii, Y. (2003). Allelopathy in the natural and agricultural ecosystems and isolation of potent allelochemicals from Velvet bean (*Mucuna pruriens*) and Hairy vetch (*Vicia villosa*). Biol. Sci. in Space, 17: 6–13.

Gaba, S., Fried, G., Kazakou, E., Chauvel, B. and Navas, M.L. (2014). Agro-ecological weed control using a functional approach: A review of cropping systems diversity. Agron. Sustainable Develop., 34: 103–119.

Gibson, K.D., McMillan, J., Hallett, S.G., Jordan, T. and Weller, S.C. (2011). Effect of a living mulch on weed seed banks in tomato. Weed Tech., 25: 245–251.

Haramoto, E.R. and Gallandt, E.R. (2004). *Brassica* cover cropping for weed management: review. Renew. Agric. Food Sys., 19: 187–198.

Hargrove, W.L. (1986). Winter legumes as a nitrogen source for no-till grain sorghum. Agron. J., 78: 70–74.

Hermawan, B. and Bomke, A.A. (1997). Effects of winter cover crops and successive spring tillage on soil aggregation. Soil Tillage Res., 44: 109–120.

Hill, E.C., Ngouajio, M. and Nair, M.G. (2007). Allelopathic potential of hairy vetch (*Vicia villosa*) and cowpea (*Vigna unguiculata*) methanol and ethyl acetate extracts on weeds and vegetables. Weed Tech., 21: 437–444.

Iqbal, Z., Hiradate, S., Noda, A., Isojima, S.I. and Fujii, Y. (2003). Allelopathic activity of buckwheat: isolation and characterization of phenolics. Weed Sci., 51: 657–662.

Jabran, K., Mahajan, G., Sardana, V. and Chauhan, B.S. (2015). Allelopathy for weed control in agricultural systems. Crop Protection, 72: 57–65.

Javaid, M.M., Bhan, M., Johnson, J.V., Rathinasabapathi, B. and Chase, C.A. (2015). Biological and chemical characterisations of allelopathic potential of diverse accessions of the cover crop Sunn Hemp. J. Amer. Soc. Hort. Sci., 140: 532–541.

Johnson, N.C. and Pfleger, F.L. (1992). Vesicular-arbuscular mycorrhizae and cultural stresses. Mycorrhizae in Sustainable Agric., 54: 71–99.

Jordan, N.R., Zhang, J. and Huerd, S. (2000). Arbuscular-mycorrhizal fungi: Potential roles in weed management. Weed Res. (Oxford), 40: 397–410.

Kalinova, J., Vrchotova, N. and Triska, J. (2007). Exudation of allelopathic substances in buckwheat (*Fagopyrum esculentum* Moench). J. Agric. Food Chem., 55: 6453–6459.

Kato-Noguchi, H. (2003). Isolation and identification of an allelopathic substance in *Pisum sativum*. Phytochem., 62: 1141–1144.

Kelton, J., Price, A.J. and Mosjidis, J. (2012). Allelopathic weed suppression through the use of cover crops. *In*: Weed Control, InTech.

Khanh, T.D., Chung, M.I., Xuan, T.D. and Tawata, S. (2005). The exploitation of crop allelopathy in sustainable agricultural production. J. Agron. Crop Sci., 191: 172–184.

Knezevic, S.Z., Weise, S.F. and Swanton, C.J. (1995). Comparison of empirical models depicting density of *Amaranthus retroflexus* L. and relative leaf area as predictors of yield loss in maize (*Zea mays* L.). Weed Res., 35: 207–214.

Knezevic, S.Z., Evans, S.P., Blankenship, E.E., Van Acker, R.C. and Lindquist, J.L. (2002). Critical period for weed control: The concept and data analysis. Weed Sci., 50: 773–786.

Komainda, M., Taube, F., Kluß, C. and Herrmann, A. (2018). Effects of catch crops on silage maize (*Zea mays* L.): Yield, nitrogen uptake efficiency and losses. Nutrient Cycl. Agroecosyst., 110: 51–69.

Kornecki, T.S., Price, A.J. and Balkcom, K.S. (2015). Cotton population and yield following different cover crops and termination practices in an alabama no-till system. J. Cotton Sci., 19: 375–386.

Kremer, R.J. and Ben-Hammouda, M. (2009). Allelopathic Plants. 19. Barley (*Hordeum vulgare* L.). Allelopathy J., 24: 225–242.

Kruidhof, H.M., Gallandt, E.R., Haramoto, E.R. and Bastiaans, L. (2011). Selective weed suppression by cover crop residues: effects of seed mass and timing of species' sensitivity. Weed Res., 51: 177–186.

Kruk, B., Insausti, P., Razul, A. and Benech-Arnold, R. (2006). Light and thermal environments as modified by a wheat crop: Effects on weed seed germination. J. Appl. Ecol., 43: 227–236.

Kumar, V., Mills, D.J., Anderson, J.D. and Mattoo, A.K. (2004). An alternative agriculture system is defined by a distinct expression profile of select gene transcripts and proteins. Proc. National Acad. Sci., 101: 10535–10540.

Kunz, C., Sturm, D.J., Varnholt, D., Walker, F. and Gerhards, R. (2016). Allelopathic effects and weed suppressive ability of cover crops. Plant, Soil Environ., 62: 60–66.

Kuo, S., Sainju, U.M. and Jellum, E.J. (1997). Winter cover crop effects on soil organic carbon and carbohydrate in soil. Soil Sci. Society Amer. J., 61: 145–152.

Lamichhane, J.R., Devos, Y., Beckie, H.J., Owen, M.D., Tillie, P., Messéan, A. and Kudsk, P. (2017). Integrated weed management systems with herbicide-tolerant crops in the European Union: lessons learnt from home and abroad. Crit. Rev. Biotech., 37: 459–475.

Lawley, Y.E., Teasdale, J.R. and Weil, R.R. (2012). The mechanism for weed suppression by a forage radish cover crop. Agron. J., 104: 205–214.

Leather, G.R. (1983). Sunflowers (*Helianthus annuus*) are allelopathic to weeds. Weed Sci., 31: 37–42.

Lemerle, D., Verbeek, B. and Orchard, B. (2001). Ranking the ability of wheat varieties to compete with *Lolium rigidum*. Weed Res., 41: 197–209.

Lemessa, F. and Wakjira, M. (2015). Cover crops as a means of ecological weed management in agro-ecosystems. J. Crop Sci. Biotech., 18: 123–135.

Liebman, M. and Davis, A.S. (2000). Integration of soil, crop and weed management in low-external-input farming systems. Weed Res. (Oxford), 40: 27–48.

Liebman, M., Davis, A.S. and Francis, C. (2009). Managing weeds in organic farming systems: an ecological approach. Organic farming: the ecological system. Agron. Monog., 54: 173–195.

Liu, A., Ma, B.L. and Bomke, A.A. (2005). Effects of cover crops on soil aggregate stability, total organic carbon and polysaccharides. Soil Sci. Soc. Amer. J., 69: 2041–2048.

Lou, Y., Davis, A.S. and Yannarell, A.C. (2016). Interactions between allelochemicals and the microbial community affect weed suppression following cover crop residue incorporation into soil. Plant Soil, 399: 357–371.

Lucas, S.T., D'Angelo, E.M. and Williams, M.A. (2014). Improving soil structure by promoting fungal abundance with organic soil amendments. Applied Soil Ecol., 75: 13–23.

Macías, F.A., Fernández, A., Varela, R.M., Molinillo, J.M., Torres, A. and Alves, P.L. (2006). *Sesquiterpene lactones* as allelochemicals. J. Nat. Products, 69: 795–800.

Macfarlane, M.J., Scott, D. and Jarvis, P. (1982a). Allelopathic effects of white clover 1. Germination and chemical bioassay. NZ J. Agric. Res., 25: 503–510.

Macfarlane, M.J., Scott, D. and Jarvis, P. (1982b). Allelopathic effects of white clover 2. Field investigations in tussock grasslands. NZ J. Agric. Res., 25: 511–518.

Malik, M., Norsworthy, J., Culpepper, A., Riley, M. and Bridges, W. (2008). Use of wild radish (*Raphanusraphanistrum*) and rye cover crops for weed suppression in sweet corn. Weed Sci., 56: 588–595.

Marochi, A., Ferreira, A., Takano, H.K., Oliveira Junior, R.S. and Ovejero, R.F.L. (2018). Managing glyphosate-resistant weeds with cover crop associated with herbicide rotation and mixture. Ciência e Agrotecnologia, 42: 381–394.

Martin, S.G., Van Acker, R.C. and Friesen, L.F. (2001). Critical period of weed control in spring canola. Weed Sci., 49: 326–333.

Mattoo, A.K. and Abdul-Baki, A. (2006). Crop genetic responses to management: evidence of root-to-leaf communication. pp. 221–230. *In*: Uphoff, N., Ball, A.S., Fernandes, E.H., Husson, O., Laing, M., Palm, C. and Thies, J. (eds.). Biological Approaches to Sustainable Soil Systems. CRC Taylor & Francis: Boca Raton, FL.

Mayer, J.B. and Hartwig, N.L. (1986). Corn yields in crown vetch relative to dead mulches. Proc. Northeast Weed Sci. Soc., 40: 34–35.

McLenaghen, R.D., Cameron, K.C., Lampkin, N.H., Daly, M.L. and Deo, B. (1996). Nitrate leaching from ploughed pasture and the effectiveness of winter catch crops in reducing leaching losses. NZ J. Agric. Res., 39: 413–420.

Mirsky, S.B., Curran, W.S., Mortensen, D.A., Ryan, M.R. and Shumway, D.L. (2009). Control of cereal rye with a roller/crimper as influenced by cover crop phenology. Agron. J., 101: 1589–1596.

Mirsky, S.B., Gallandt, E.R., Mortensen, D.A., Curran, W.S. and Shumway, D.L. (2010). Reducing the germinable weed seed-bank with soil disturbance and cover crops. Weed Res., 50: 341–352.

Mirsky, S.B., Curran, W.S., Mortensen, D.A., Ryan, M.R. and Shumway, D.L. (2011). Timing of cover-crop management effects on weed suppression in no-till planted soybean using a roller-crimper. Weed Sci., 59: 380–389.

Mohler, C.L. and Teasdale, J.R. (1993). Response of weed emergence to rate of *Vicia villosa* Roth and *Secale cereale* L. residue. Weed Res., 33: 487–499.

Mohler, C.L., Dykeman, C., Nelson, E.B. and Ditommaso, A. (2012). Reduction in weed seedling emergence by pathogens following the incorporation of green crop residue. Weed Res., 52: 467–477.

Molisch, H. (1937). Der EinflusseinerPflanze auf die andere-Allelopathie, Fischer, Jena.

Moore, M.J., Gillespie, T.J. and Swanton, C.J. (1994). Effect of cover crop mulches on weed emergence, weed biomass and soybean (*Glycine max*) development. Weed Tech., 8: 512–518.

Moyer, J.R., Blackshaw, R.E. and Huang, H.C. (2007). Effect of sweet clover cultivars and management practices on following weed infestations and wheat yield. Can. J. Plant Sci., 874: 973–983.

Narwal, S.S. and Haouala, R. (2013). Role of allelopathy in weed management for sustainable agriculture. pp. 217–249. *In*: Allelopathy, Springer, Berlin, Heidelberg.

Nair, A. and Ngouajio, M. (2012). Soil microbial biomass, functional microbial diversity, and nematode community structure as affected by cover crops and compost in an organic vegetable production system. Applied Soil Ecol., 58: 45–55.

Ohno, T., Doolan, K., Zibilske, L.M., Liebman, M., Gallandt, E.R. and Berube, C. (2000). Phytotoxic effects of red clover amended soils on wild mustard seedling growth. Agric. Ecosys. Environ., 78: 187–192.

Osipitan, O.A., Dille, J.A., Assefa, Y. and Knezevic, S.Z. (2018). Cover crop for early season weed suppression in crops: Systematic review and meta-analysis. Agron. J., 110: 2211–2221.

Petersen, J., Belz, R., Walker, F. and Hurle, K. (2001). Weed suppression by release of isothiocyanates from turnip-rape mulch. Agron. J., 93: 37–43.

Przepiorkowski, T. and Gorski, S.F. (1994). Influence of rye (*Secale cereale*) plant residues on germination and growth of three triazine-resistant and susceptible weeds. Weed Tech., 8: 744–747.

Poeplau, C. and Don, A. (2015). Carbon sequestration in agricultural soils via cultivation of cover crops—A meta-analysis. Agric. Ecosys. Environ., 200: 33–41.

Powles, S.B. 2008. Evolved glyphosate-resistant weeds around the world: Lessons to be learnt. Pest Manage. Sci., 64: 360–365.

Puget, P. and Drinkwater, L.E. (2001). Short-term dynamics of root and shoot derived carbon from a leguminous green manure. Soil Sci. Soc. Amer. J., 65: 771–779.

Putnam, A.R., DeFrank, J. and Barnes, J.P. (1983). Exploitation of allelopathy for weed control in annual and perennial cropping systems. J. Chem. Ecol., 9: 1001–1010.

Ranells, N.N. and Wagger, M.G. (1997). Nitrogen-15 recovery and release by rye and crimson clover cover crops. Soil Sci. Soc. Amer. J., 61: 943–948.

Reberg-Horton, S.C., Burton, J.D., Danehower, D.A., Ma, G., Monks, D.W., Murphy, J.P. and Creamer, N.G. (2005). Changes over time in the allelochemical content of ten cultivars of rye (*Secale cereale* L.). J. Chem. Ecol., 31: 179–193.

Reberg-Horton, S.C., Grossman, J.M., Kornecki, T.S., Meijer, A.D., Price, A.J., Place, G.T. and Webster, T.M. (2012). Utilising cover crop mulches to reduce tillage in organic systems in the southeastern USA. Renewable Agric. Food Sys., 27: 41–48.

Reeves, D.W. (1994). Cover crops and rotations. pp. 125–172. *In*: Hatfield, J.L. and Stewart, B.A. (eds.). Advances in Soil Science—Crop Residue Management. Lewis Publishers, CRC Press Boca Raton, FL, USA.

Reiss, A., Fomsgaard, I.S., Mathiassen, S.K. and Kudsk, P. (2018). Weed suppressive traits of winter cereals: Allelopathy and competition. Biochem. Sys. Ecol., 76: 35–41.

Rice, E.L. (1984). Manipulated Ecosystems: Roles of Allelopathy in Agriculture, Allelopathy, Academic Press, New York.

Sanyal, D., Bhowmik, P.C., Anderson, R.L. and Shrestha, A. (2008). Revisiting the perspective and progress of integrated weed management. Weed Science, 56: 161–167.

Sarrantonio, M. (1994). Northeast Cover Crop Handbook, Soil Health Series, Rodale Institute, Kutztown, PA.

Schonbeck, M. (2011). Plant and Manage Cover Crops for Maximum Weed Suppression. eXtension [Online]. Retrieved from https://articles.extension.org/pages/18525/plant-and-manage-cover-crops-for-maximum-weed-suppression.

Schonbeck, M. (2015). How Cover Crops Suppress Weeds. eXtension [Online]. Retrieved from https://articles.extension.org/pages/18524/how-cover-crops-suppress-weeds.

Schutter, M.E. and Dick, R.P. (2002). Microbial community profiles and activities among aggregates of winter fallow and cover-cropped soil. Soil Sci. Soc. Amer. J., 66: 142–153.

Stonehouse, D.P., Weise, S.F., Sheardown, T., Gill, R.S. and Swanton, C.J. (1996). A case study approach to comparing weed management strategies under alternative farming systems in Ontario. Can. J. Agric. Economics/Revue Canadienne d'agroeconomie, 44: 81–99.

Teasdale, J.R. and Daughtry, C.S. (1993). Weed suppression by live and desiccated hairy vetch (*Vicia villosa*). Weed Sci., 41: 207–212.

Teasdale, J.R. and Mohler, C.L. (1993). Light transmittance, soil temperature, and soil moisture under residue of hairy vetch and rye. Agron. J., 85: 673–680.

Teasdale, J.R. and Rosecrance, R.C. (2003). Mechanical versus herbicidal strategies for killing a hairy vetch cover crop and controlling weeds in minimum-tillage corn production. Amer. J. Altern. Agric., 18: 95–102.

Teasdale, J.R., Devine, T.E., Mosjidis, J.A., Bellinder, R.R. and Beste, C.E. (2004). Growth and development of hairy vetch cultivars in the northeastern United States as influenced by planting and harvesting date. Agron. J., 96: 1266–1271.

Teasdale, J.R., Brandsaeter, L.O., Calegari, A., Neto, F.S., Upadhyaya, M.K. and Blackshaw, R.E. (2007). Cover crops and weed management. pp. 49–64. *In*: Updadhyaya, M.K. and Blackshaw, R.E. (eds.). Non-chemical Weed Management: Principles, Concepts and Technology. CABI Publishing, Wallingford, UK.

Tonitto, C., David, M.B. and Drinkwater, L.E. (2006). Replacing bare fallows with cover crops in fertiliser-intensive cropping systems: A meta-analysis of crop yield and N dynamics. Agric. Ecosys. Environ., 112: 58–72.

Uremis, I., Arslan, M., Uludag, A. and Sangun, M. (2009). Allelopathic potentials of residues of 6 *Brassica* species on Johnsongrass [*Sorghum halepense* (L.) Pers.]. Afr. J. Biotech., 8: 3497–3501.

Van Der Heijden, M.G., Klironomos, J.N., Ursic, M., Moutoglis, P., Streitwolf-Engel, R., Boller, T. and Sanders, I.R. (1998). Mycorrhizal fungal diversity determines plant biodiversity, ecosystem variability and productivity. Nature, 396: 69.

Wayman, S., Cogger, C., Benedict, C., Burke, I., Collins, D. and Bary, A. (2015). The influence of cover crop variety, termination timing and termination method on mulch, weed cover and soil nitrate in reduced-tillage organic systems. Renewable Agric. Food Sys., 30: 450–460.

Wells, M.S., Reberg-Horton, S.C., Smith, A.N. and Grossman, J.M. (2013). The reduction of plant-available nitrogen by cover crop mulches and subsequent effects on soybean performance and weed interference. Agron. J., 105: 539–545.

Weston, L.A., Harmon, R. and Mueller, S. (1989). Allelopathic potential of sorghum-Sudangrass hybrid (sudex). J. Chem. Ecol., 15: 1855–1865.

Weston, L.A. (1996). Utilisation of allelopathy for weed management in agro-ecosystems. Agron. J., 88: 860–866.

Weston, L.A. and Duke, S.O. (2003). Weed and crop allelopathy. Crit. Rev. Plant Sci., 22: 367–389.

Weston, L.A., Alsaadawi, I.S. and Baerson, S.R. (2013). Sorghum allelopathy—From ecosystem to molecule. J. Chem. Ecol., 39: 142–153.

White, R.H., Worsham, A.D. and Blum, U. (1989). Allelopathic potential of legume debris and aqueous extracts. Weed Sci., 37: 674–679.

Wiggins, M.S., Hayes, R.M. and Steckel, L.E. (2016). Evaluating cover crops and herbicides for glyphosate-resistant Palmer amaranth (*Amaranthus palmeri*) control in cotton. Weed Tech., 30: 415–422.

Wu, C.X., Zhao, G.Q., Liu, D.L., Liu, S.J., Gun, X.X. and Tang, Q. (2016). Discovery and weed inhibition effects of coumarin as the predominant allelochemical of yellow sweet clover (*Melilotus officinalis*). Int. J. Agric. Biol., 18: 168–175.

Zimdahl, R.L. (1988). The concept and application of the critical weed- free period. pp. 145–155. *In*: Altieri, A.L.A. and Liebman, M. (eds.). Weed Management in Agro-ecosystems: Ecological Approaches. CRC Press, Boca Raton, FL.

Zimdahl, R.L. (2013). Fundamentals of Weed Science, fourth ed., Academic Press, San Diego, CA, USA.

Zobel, M., Moora, M. and Haukioja, E. (1997). Plant co-existence in the interactive environment: Arbuscular mycorrhiza should not be out of mind. Oikos, 78: 202–208.

7

Cover Crops for Pests and Soil-borne Disease Control and Insect Diversity

Nataliia Didenko,[1,]* *Vira Konovalova,*[2] *Somayyeh Razzaghi,*[3] *Alimata Bandaogo,*[4]
Sougata Bardhan[5] and *Alan Sundermeier*[6]

1. Introduction

Cover crop is a biological primer to support agricultural sustainability with enhanced ecosystem services. Cover crops, when incorporated alone or as blends in agronomic, horticultural and other systems to support ecological biodiversity, provide habitats and food for beneficial predators and act as non-host crops for nematodes and other pests (Lu et al., 2000; Wen et al., 2017; LaRose and Myers, 2019).

Cover crops are generally grown without any intention of harvesting their grain or biomass, either partly or completely, during their growing season or at the end of their cropping season. Both the below- and above-ground biomass of cover crops are often incorporated into the soil by plowing or left in the soil after rolling or killed by herbicides or winter-killed at the maximum vegetative growth. The aim is to provide organic mulch to suppress weeds, minimise soil erosion, conserve moisture, return organically-bound labile nutrients (e.g., biologically-fixed nitrogen or recycling of accumulated nutrients), or improve biodiversity and release useful secondary metabolites (e.g., glucosinolates) to the soil to control pests, nematodes and other soil-borne diseases (Campiglia et al., 2009; LaRose and Myers, 2019).

2. Methods of Biocontrol by Cover Crops

Natural biocontrol services provide for sustainable management of pest populations and rely on complex communities, natural enemies and the integration of cultural practices including the use of mulches, trap crops, cover crops and other mechanisms that modify insect and pest behaviour to reduce the effect of insect herbivores and soil-borne diseases on food and feed crops (Shelton and Badenes-Perez, 2006; Liburd et al., 2008). There are two important approaches to biocontrol services based on the characteristics and placement of the cover crops or trap plants: (1) the trap cropping approach and (2) the biochemical approach (Table 1).

[1] Institute of Water Problems and Land Reclamation, Kyiv, Ukraine.
[2] Askaniyska State Agricultural Experimental Station, Institute of Irrigated Agriculture, National Academy of Agrarian Sciences of Ukraine, Khersons'ka, Ukraine.
[3] Dept. of Soil Science and Plant Nutrition, University of Çukurova, Adana, Turkey.
[4] Institute of Environment and Agricultural Research (INERA),Bobo-Dioulasso, Burkina Faso.
[5] University of Missouri, Columbia, MO, USA.
[6] The Ohio State University Extension, Bowling Green, OH 43210, USA.
* Corresponding author: 9449308nd@gmail.com

Table 1 Potential cover crops and their roles or effects on insects, pestd, nematodes and ssoil-borne disease control.

Cover Crops	Potential Benefits to Control Pests and Diseases
Grasses	
Wheat	Control diseases and suppress nematodes.
Annual ryegrass	Nematode suppression and soil-borne pathogens control.
Cereal rye	Nematode reduction. *R. solani/F. virguliforme*. Do not host root-knot nematodes/soil-borne diseases.
Sudan-sorghum	Soil-borne pathogens (*Verticillium wilt*).
Sorghum	Manage cotton bollworm with increasing parasitism by *Trichogramma chilonis*.
Brassicas	
Radish	Nematode suppression and reduction.
Rapeseed/canola	Nematode reduction and control soil-borne pathogens (*Verticillium wilt*).
Giant mustard/Turnip weed	Control soil-borne fungal diseases (*Pythium* spp.). Control nematodes.
Yellow rocket	Control diamond-back moth.
Indian Mustard	Control *Sclerotium rolfsii, F. oxysporum, R. solani, M. phaseolina*, nematodes, and diamond-back moth.
Legumes	
Alfalfa	As a trap crop, it controls lygus bugs in cotton.
Cahaba white vetch	Increases soil-borne diseases yet suppresses root-knot nematodes.
Hairy indigo	Markedly reduces the numbers of several root nematodes of the genus Meloidogyne.
Showy crotalaria	Control and suppress nematodes.
Sunn hemp	Control nematodes and bean pod borer.
Velvet bean	Nematode reduction.
Others/blends	
Marigold	Control pollen beetle in cauliflowers. Trap nematodes.
Cabbage, rapes, and sunflower	Control of pollen beetle in cauliflowers.
Castor, millet, and soybean	Control groundnut leaf miner.
Corn and potato plants	Control wireworms in sweet potato fields.
Napier, Sudan, and molasses grasses	Enhances stem borer parasitoid abundance, thereby improving stem borer control.

Source: Djarwaningsih (1997); Potter et al. (1998); Lazzeri and Manici (2001); Wen et al. (2017); Sujan et al. (2019).

2.1 Trap Cropping Approach

Prior to the industrial revolution and the development of reactive agrochemicals (pesticides, herbicides, insecticides, fertilisers, etc.), cover crops, as green manures or trap crops, were used for nitrogen fertility and pest control in diverse cropping systems worldwide (Talekar and Shelton, 1993; Wen et al., 2017). Recently, renewed interest has emerged in cover crops as one of the critical components of integrated pest management (IPM) tools, in response to concerns for reportedly adverse effects of pesticides on beneficial insects, public health and the air-soil-water-plant ecosystems, chemical resistance and economics of agricultural sustainability. Cover crops, as trap crops, have been defined as 'plant stands that are, *per se* or via manipulation, deployed to attract, divert, intercept, and/or retain targeted insects or the pathogens they vector in order to reduce damage to the main crop' (Shelton and Badenes-Perez, 2006).

In a *conventional trap crop approach*, cover crops planted adjacent to high-value vegetable, fruit or agronomic crops are naturally more attractive to pests as either a food source or oviposition site than the main crop, thus preventing or making the pests less likely to attack the main crop (Javaid and Joshi, 1995). One of the most common examples of successful conventional trap cropping, which served as a major contributor to the IPM development in the central valley of California during the 1960s, is the use of alfalfa as a legume trap crop for controlling lygus bugs in cotton (Godfrey and Leigh, 1994) and which is still used globally today. Moreover, highly attractive squash varieties are widely used to minimise squash bugs and cucumber beetle infestations in cucurbitaceous crops (Pair, 1997).

In a *dead-end approach*, plants that are highly attractive to insects, but on which they and their offspring cannot survive, serve as a sink for pests, preventing their movement from the trap crop (cover crops) to the high-value crops later in the growing season (Badenes-Perez et al., 2004; Shelton and Nault, 2004). For example, yellow rocket (also called bitter cress, herb barbara, rocket cress, yellow rocket cress, winter rocket and wound rocket) is a biennial herb belonging to the mustard family that serves as a dead-end approach trap crop for the control of diamondback moth (Lu et al., 2004; Shelton and Nault, 2004; Badenes-Perez et al., 2005a,b). Sunn hemp, a legume, has been suggested as a dead-end trap crop to control the bean pod borer. High ovipositional preference for host plants, on which larvae do not survive, has also been reported, especially among Lepidoptera (Thompson and Pellmyr, 1991). However, this approach needs to be properly followed so it can intercept insect pests (e.g., field borders) and reduce pest damage to the main crop.

(Source: https://www.growjourney.com/prevent-stop-squash-vine-borers)

Fig. 1 Cover crops as trap crops to control insects and pests.

2.1.1 Placement and Extent of Trap Cropping

There is a myriad of microhabitats that temporarily vary in their attractiveness, suitability and effectiveness to insect pests and/or their natural enemies in terrestrial ecosystems (Kennedy and Storer, 2000). The most relevant components of the landscape structural ecology are those that refer to the spatial and temporal distribution of vegetation, as insects and their host plants interact and influence by size, fragmentation and connectivity of host systems (Tscharntke and Brandl, 2004). When cover crops are used in trap cropping, the following placement/deployment components need to be considered to achieve sustainable pest control success:

1) *Perimeter trap cropping* – The perimeter trap-cropping approach is based on the use of crops or cover crops that are naturally attractive to pests and insects as trap crops planted around the border of the main crop (Boucher et al., 2003), and is commonly used for insect control in IPM programmes. For example, the borders of early-planted potatoes have been used as a trap crop for the Colorado potato beetle, which moves to potato fields from overwintering sites next to the crop, becoming concentrated in the outer rows where it can be treated with insecticides, cultural practices, or other means (Hoy et al., 2000). Similar success has been reportedly achieved in commercially grown agricultural or horticultural fields with perimeter trap cropping to control

pepper maggot in bell peppers by using a trap crop of hot cherry peppers (Boucher et al., 2003). However, perimeter trap cropping does not always provide the best spatial design or consistent results in the control of pests and diseases (Kumari and Pasalu, 2003).

2) *Sequential trap cropping* – This approach is based on using conventional crops and cover crops as trap crops that are planted earlier and/or later than the main crop to attract targeted insects and pests; for example, the use of an early-season trap crop of potatoes to manage Colorado potato beetles (Hoy et al., 2000). Another example is the use of Indian mustard as a trap crop for the control of diamondback moth; however, this requires planting mustard two or three times through the cole crop season because Indian mustard has a shorter crop cycle than cabbage and other cole crops (Srinivasan et al., 1991; Pawar and Lawande, 1995). Field studies have also shown that dusky wireworms could be managed in strawberry fields by planting wheat as a trap crop one week before planting strawberries (Vernon et al., 2000).

3) *Multiple trap cropping* – In a multiple trap-cropping system, several species or blends are planted simultaneously as trap crops for the purpose of either managing multiple insects and pests at the same time, or enhancing the control of one insect pest by combining plants with growth stages that will increase attractiveness to the pest at different times. Over the years, cover crop blends of Chinese cabbage, marigolds, rapes and sunflowers have been used as trap crops for the control of pollen beetle in cauliflower. Other examples of multiple trap cropping include the blends of castor, millet and soybean to control groundnut leaf miner and the use of corn and potato plants combined as trap crops to control wireworms in sweet potato fields (Muthiah, 2003).

4) *Push-pull trap cropping* – This strategy is based on a blend of trap crops (pull component) with a repellent intercrop (push component) in diverse agro-ecosystems (Miller and Cowles, 1990; Khan et al., 2001). The trap crop attracts insects and pests and, combined with the repellent intercrop, diverts them away from the main crop. For example, either Napier or sudangrass as a trap crop planted around the main crop, and either desmodium or molasses grass planted within the field as a repellent intercrop, has greatly suppressed corn stem borers in African countries (Khan et al., 2001). The use of molasses grass as a repellent intercrop increases stem borer parasitoid abundance, thereby controlling the stem borer (Khan et al., 2000).

5) *Biodiversity trap cropping* – In biodiversity-assisted trap cropping, two approaches are generally considered to control herbivore damage on main crops. The 'natural enemies' approach is based on more diversified food sources (nectar, pollen, prey host species), allowing greater establishment of higher densities of predators and parasites to control and regulate pest populations in diverse vegetative habitats. Secondly, the 'resource concentration' approach is based on the principle that herbivores will not remain in sparsely populated crop stands because of fewer available resources for support and proliferation (Liburd et al., 2008). For example, sorghum can be planted as a trap crop to manage cotton bollworm and increase rates of parasitism by *Trichogramma chilonis* (Virk et al., 2004). The increase in parasitism of stem borers by *Cotesia* spp., when using molasses grass as an intercrop, further enhances the effectiveness of push-pull trap cropping (Khan and Pickett, 2004).

6) *Semi-chemical trap cropping* – This approach is based on principles underlying the effects of trap cropping on insect behaviour that are like those behind semio-chemicals and other behaviour-based methods for pest management (Foster and Harris, 1997). In conventional trap cropping, insects and pests attracted to the crops may be due to semio-chemicals (chemical substances) naturally produced by the trap crop and that attractiveness is enhanced by the application of semio-chemicals or regular crops that can act as trap crops after the application of semio-chemicals. One of the most successful examples of this strategy is the use of pheromone-baited plants that attract bark beetles to facilitate their control (Borden and Greenwood, 2000). Pheromone-baited fly traps hung on perimeter plants acting as trap crops have been suggested for fruit fly management in papaya orchards. The use of semi-chemical toxic baits may also enhance the effectiveness of trap crops (Vernon et al., 2002).

2.2 Biochemical Approach

In ecologically-balanced systems, pests- and disease-causing organisms are internally managed by natural enemies; best management practices, including the use of a suitable cover crop or blends of cover crops, favour beneficial organisms and promote disease-suppression mechanisms (van Bruggen and Semenov, 2000). Appropriate cover-crop-based systems are expected to reduce reactive chemical usage with an associated increase in biodiversity, disease supressiveness of soils, increased release of secondary metabolites as natural biocides (Tables 2 and 3), and thus, will minimise both the incidence and severity of soil-borne diseases (Cherr et al., 2006; Shennan, 2008).

2.2.1 Disease Control

Soil-borne plant diseases are caused by diverse and numerous pathogens that live in the soil and affect plant health by infecting the below-ground organs (roots, rhizomes, tubers, etc.) or, in some cases, above-ground organs (stem bases, plant crowns, or the vascular system). Another important point is the difference between purely biotrophic pathogens and those that can survive in soils as saprophytes (Michel et al., 2015). The most important groups of such organisms are the oomycetes, fungi and nematodes; meanwhile, soil-borne diseases caused by bacteria, viruses, or other microorganisms are less important. Soil-borne pathogens can be highly specialised and affect a broad range of plant hosts. The persistence of many soil-borne pathogens is often enhanced by long-term survival strategies and structures (such as chlamydospores, sclerotia, microsclerotia, oospores, or cysts). Populations of soil-borne pathogens increase and spread rapidly in the field through the planting or seeding of host plants.

The two major groups of plants in agro-ecosystems, monocotyledons and dicotyledons, are hosts to many different groups of soil-borne pathogens. Generally, monocotyledons (e.g., flowering plants) are considered non-host plants of *Verticillium* spp. (Pegg and Brady, 2002); Poaceae, the most important family of the monocotyledons, are non-host plants to *Phytophthora* spp. (Erwin and Ribeiro, 1996). Other important soil-borne pathogens, such as *Colletotrichum coccodes, Sclerotinia sclerotiorum,* or *Thielaviopsis basicola* do not, or rarely, infect monocotyledons. Pathogens that equally affect mono- and dicotyledons include *Rhizoctonia solani* or *Fusarium* species. In contrast, some soil-borne pathogens, such as *Gaeumannomyces graminis*, only infest monocotyledons, but not dicotyledons (Michel et al., 2015).

Cover crops suppress diseases by interfering with, and affecting, disease cycle phases such as dispersal, host infection, disease development, propagation, population buildup and survival of the pathogen. The presence of cover-crop mulches minimises pathogen dispersal via rainfall splashing, runoff and/or wind-borne processes (Everts, 2002; Cantonwine et al., 2007).

Cover crops, as non-host crop in cropping diversity, affect and control soil-borne pathogens in diverse ways. The cover crop roots exudate a range of functionally active organic compounds (e.g., organic acids, sugars, flavonoids, amino acids, etc.) in the rhizosphere during the growing period. These organic compounds can directly influence the soil microbial biomass and their community structures more than any other soil factors (Ladygina and Hedlund, 2010). Likewise, the incorporation of cover crop biomass added substantial amounts of readily useable carbon, essential nutrients and secondary metabolites to stimulate microbial activity and biodiversity and efficiency in the soil (Stark et al., 2008; Michel and Lazzeri, 2011). Incorporation of cover crop residues as sources of diverse organic chemicals greatly affects soil microbial populations with an associated increase in pathogen inhibitory activities, as was shown in *Phytopthor aroot rot* in alfalfa, *Verticillium wilt* in potato and *Rhizoctonia solani root rot* by changes in resident *Streptomyces* spp. community (Wiggins and Kinkel, 2005a,b). Such increases in soil microbial diversity and activity correlate with an effective decrease in the number of soil-borne pathogens, e.g., *Verticillium dahliae* (Michel and Lazzeri, 2011). Increased microbial activity and diversity, induced by a preceding blend of canola, rapeseed and barley as cover crops, has been shown to reduce several soil-borne potato diseases (Larkin et al., 2010). Certain groups of soil microbes (such as *Streptomyces* spp.) stimulated by the

Table 2 Type and concentration of glucosinolates detected in the shoot and root tissues of Brassicaceae Crops Before Incorporation in Soil in 2006 (Clemson, SC) and 2007 (Fayetteville, AR). [a Standard Error of Each Mean. b Abbreviation: nd, not detected].

Year	Indian – White mustard blend		Herb cress		Indian Mustard (Fumus F-L71)		Indian Mustard (Fumus F-E75)		Turnip		Indian ustard (Southern curled giant)		Oilseed rape	
Glucosinolates	Shoot	Root	Shoot	Root	Shoot	Root	Shoot	Root	Shoot	Root	Shoot	Root	Shoot	Root
2006							mmol g^{-1} tissue							
(2R)-2-hydroxybut-3-enyl	nd[b]	nd	nd	nd	nd	nd	nd	nd	2.8(0.3)[a]	1.2(01)	nd	nd	18.5(0.5)	9.4(2.6)
2-propenyl	72.3(11)	5.1(1.1)	nd	nd	5.8(0.1)	6.3(0.2)	12.8(0.1)	0.5(0.1)	nd	nd	38(1)	5.6 (0.1)	nd	nd
p-hydroxybenzyl	2.4(0.2)	nd	nd	nd	nd	nd	nd	nd	nd	nd	nd	nd	nd	nd
But-3-enyl	nd	nd	nd	nd	nd	nd	6.6(0.2)	nd	24.6(0.8)	nd	nd	nd	nd	nd
Benzyl	nd	nd	131.9(3.3)	50.6(1)	nd	nd	nd	nd	nd	nd	nd	nd	nd	nd
2-phenylethyl	4(0.2)	22.8(0.5)	nd	nd	0.7(0.1)	8.9(0.1)	1.4(0.1)	3.5(0.1)	2.2(0.6)	19.5(0.3)	4.3(0.5)	13.8(1.9)	2.7(0.9)	24.7(0.5)
Total	78.7(10.7)	27.9(0.4)	132(3.3)	50.6(1)	6.5(0.2)	15.2(0.2)	20.7(0.2)	4.0(0.1)	29.6(1.4)	20.8(0.3)	42.4(0.5)	19.3(1.9)	21.1(0.6)	34.1(2.1)
2007														
(2R)-2-hydroxybut-3-enyl	nd	nd	nd	nd	nd	nd	nd	nd	2(0.1)	2.7(0.1)	nd	nd	11.7(0.1)	14.4 (0.2)
2-propenyl	12.7(0.1)	nd	nd	nd	31.6(1.5)	11.2(0.1)	13.6(0.4)	nd	nd	nd	59.4(1.1)	nd	nd	nd
p-hydroxybenzyl	67.9(0.3)	nd	nd	nd	nd	nd	nd	nd	nd	nd	nd	nd	nd	nd
But-3-enyl	nd	nd	nd	nd	13.2(0.3)	nd	23(0.4)	nd	21.4(11.3)	4.7(0.2)	nd	nd	3.6(0.1)	2.9(0.1)
Benzyl	4.7(0.3)	nd	131.1(3)	12.7(0.3)	nd	nd	nd	nd	nd	nd	nd	nd	nd	nd
2-phenylethy	11.9(0.4)	7.1(0.5)	nd	nd	5.8(0.2)	60.4(2.5)	6.1(0.2)	3.4(0.5)	2.3(0.2)	35.3(1.9)	5.1(1.1)	0.5(0.1)	1.1(0.1)	51.6(2.6)
Total	87.1(0.4)	7.1(0.5)	131.2(2.9)	12.7(0.3)	50.7(1.5)	71.6(2.5)	42.7(1.1)	3.5(0.5)	25.7(11.3)	42.7(2.2)	64.5(2)	0.5(0.1)	16.4(0.1)	68.9(2.7)

Source: Bangarwa et al. (2011)

Table 3 Individual glucosinolate concentrations found in roots of crucifer cover crops. (Values correspond to the mean of the four experimental years for crops grown as sole crops and in mixtures. '0' means that glucosinolate concentrations were under the threshold of detection. Numbers in brackets represent standard errors).

Glucosinolate Names	Rape	White Mustard	Indian Mustard	Ethiopian Mustard	Turnip	Turnip Rape	Radish	Rocket
Dry Matter Production (ton/ha)								
Biomass	**0.54 (0.03)**	**0.38 (0.04)**	**0.36 (0.03)**	**0.37 (0.03)**	**0.67 (0.07)**	**0.63 (0.04)**	**0.84 (0.04)**	**0.07 (0.01)**
Glucosinolate conc (μmol g/ dry matter)								
Aliphatic								
Sinigrin	0.1 (0.03)	0.7 (0.14)	23.9 (2.15)	5.1 (0.51)	0	0.1 (0.04)	0.2 (0.07)	0.6 (0.24)
Glucoerucin	1 (0.09)	0.9 (0.17)	0.5 (0.19)	0.3 (0.12)	0.1 (0.07)	0.6 (0.07)	0.3 (0.04)	25.5 (2.39)
Glucoraphanin	0.2 (0.13)	0.1 (0.03)	0	0.1 (0.04)	0	1.7 (0.4)	0	1.8 (0.83)
Glucoraphanin	0	0	0	0	0	0	0	0
Glucoraphasatin	0	0	0	0	0	0	48.9 (2.56)	0
Gluconapin	0.1 (0.03)	0	0.1 (0.03)	0	0	1.8 (0.38)	0	0
Progoitrin	0.3 (0.06)	0.2 (0.08)	0.4 (0.29)	0.2 (0.09)	1 (0.25)	7.2 (0.85)	0	0.1 (0.08)
Glucobrassicanapin	0.5 (0.12)	0	0	0	0.3 (0.08)	1.3 (0.43)	0	0
Gluconapoleiferin	0.1 (0.02)	0	0	0	1.2 (0.11)	0.2 (0.06)	0	0
Glucoalyssin	0.1 (0.02)	0	0	0	0	0	0	0
aliphatic 6, 37	0.4 (0.1)	2.1 (0.2)	0.1 (0.05)	0.3 (0.07)	0.1 (0.04)	0.3 (0.06)	0	4.5 (1.13)
Aromatic								
Sinalbin	0	4.8 (0.29)	0.1 (0.04)	0.1 (0.03)	0 (0.02)	0	0	0.1 (0.05)
Gluconasturtiin	13.6 (0.74)	5 (0.35)	15.4 (1.09)	9 (0.67)	5.7 (0.87)	13.8 (0.92)	0.4 (0.08)	0.5 (0.18)
Glucotropaeolin	0	1.5 (0.18)	0	0	0	0	0	0
Indole								
4hydroxyglucobrassicin	0.1 (0.02)	0	0.1 (0.02)	0.1 (0.02)	0.7 (0.12)	0.5 (0.1)	0.1 (0.01)	0
Glucobrassicin	0.5 (0.06)	0.1 (0.02)	0.3 (0.04)	0.8 (0.21)	0.4 (0.07)	0.8 (0.07)	0.3 (0.05)	0
4methoxyglucobrassicin	0.2 (0.01)	0.1 (0.02)	0 (0.01)	0.1 (0.02)	0.2 (0.06)	0.3 (0.06)	0.2 (0.02)	0
Neoglucobrassicin	2 (0.13)	0.3 (0.03)	0.9 (0.19)	1.4 (0.26)	1.1 (0.18)	3.3 (0.32)	0.1 (0.02)	0.1 (0.06)
Indole 16.3 unknown	1 (0.69)	1.1 (0.28)	1.2 (0.44)	0.3 (0.11)	0.1 (0.07)	0.2 (0.1)	0	1.6 (0.86)
indole 18,503	0	0	0	0	0	0	0	1.3 (0.5)
indole 15.683	0	0	0	0	0	0	0.9 (0.15)	0

Source: Couëdel et al. (2018)

incorporation of cover crop biomass directly increased the growth and yield of alfalfa and potato (Wiggins and Kinkel, 2005a).

Another important way to suppress or reduce soil-borne pathogens through cover crops is the use of nematode catch crops, such as marigolds and mustards (Held et al., 2000). Everts (2002) reported that no-till pumpkin, grown on hairy vetch and blend of hairy vetch-rye cover crops, decreased *Plectosporium blight* incidence by 36 per cent and black rot by 50 per cent compared to those grown under conventional tillage without cover crops. Powdery mildew was less severe on a moderately-resistant variety of pumpkin than on a susceptible variety. However, this method is limited to a small number of cropxpathogen combinations (Michel et al., 2015).

A reduction of disease pressure with an increase in crop yield can be achieved with the use of cover crops as sources of green manures (Larkin et al., 2010; Wiggins and Kinkel, 2005b). Incorporating the shoot biomass (i.e., using a crop not only as cover crop, but also as green manure) can enhance the efficacy of this control method (Motisi et al., 2009). Cover crop residues, when fragmented and incorporated by plowing, can provide soil disinfestation via biochemical mechanisms (Gamliel et al., 2000). They also contain specific metabolites that may be toxic or exert biocidal effects on various pathogens (Tables 2 and 3). Biofumigation is based on the use of cruciferous plants with high contents of specific glucosinolates, which upon incorporation into soil are converted to toxic isothiocyanate compounds during and after decomposition (Kirkegaard, 2009). For example, *Brassica* residues (such as mustard crop) containing glucosinolates when decomposed resulted in the release and formation of bio-toxins, including isothiocyanates (Fig. 2), which provide control of diseases, weeds and parasitic nematodes (Weil and Kremen, 2007). *Brassica* species are quite effective in controlling Sclerotinia diseases in lettuce.

Other plant species widely used as green manures, belonging to the Poaceae family, also contain secondary chemical compounds or metabolites that are transformed into toxic substances during and after decomposition (Widmer and Abawi, 2002). Studies reported that *Verticillium wilt* incidence in potato was reduced when potato was grown after corn or sudangrass, compared to planting it after rape or winter peas. Plants that contain essential oils are another group of cover crop that can be used for the control of soil-borne pathogens (Gwinn et al., 2010).

As weeds are one of the opportunistic and host plants for pathogens, cover crops need to be established in a way to effectively suppress weeds for undermining their host effect on crops. Moreover, the incorporation of cover crop residue as a source of labile and fresh organic matter can lead to a temporary increase of certain organisms associated with soil-borne diseases (Hoitink

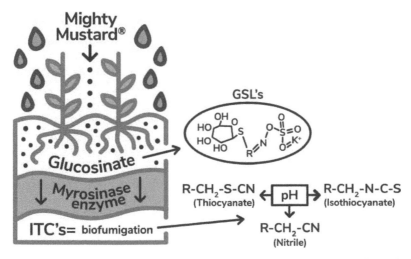

Fig. 2 Decomposition of mustard biomass and pathway of isothiocyanate formation for soil biofumigation. Credit (with permission): Mighty Mustard (mightymustard.com) and Pacific Northwest Farmers' Cooperative (pnw.coop).

and Boehm, 1999). The incorporating of higher amounts of readily decomposable organic matter increased competition for O_2 consumption in the soil, caused by the resulting intense macro- and microbial activity. In some cases, the cover crop can be a host for pathogens, but will not develop any disease symptoms itself. *Fusarium oxysporum* f. sp. *phaseoli* may prevail in leguminous cover crops when rotated with beans (Dhingra and Netto, 2001). When cover crops are slowly or not properly decomposed, populations of pathogens, such as *Pythium* spp. increase, causing severe epidemics (Manici et al., 2004). Cover crops may increase disease incidence for pathogens, such as *Sclerotium rolfsii*, with a wide host range (Widmer et al., 2002; Gilsanz et al., 2004). However, green manure crops have a low efficacy in controlling soil-borne diseases associated with main crops when compared to soil disinfestation with chemical substances (fumigation), heat, soil solarisation, or steaming.

2.2.2 Control of Nematodes

Nematodes are minute roundworms—an integral competent of soil ecosystems that interact directly and indirectly with growing crops (Scholberg et al., 2010; Wen et al., 2017). Several species of nematodes feed on young roots and weaker or stressed plants and introduce disease or vectors through feeding wounds. However, most nematodes are not plant parasitic in nature, but feed on, and interact with, many soil-borne microorganisms including fungi, bacteria and protozoa. Crop damage by the plant-parasitic nematodes results in a breakdown of plant tissue, such as lesions or yellow foliage; retarded growth of cells and stunted growth of shoots; or excessive growth, such as root galls, swollen root tips, or unnatural root branching (Wen et al., 2017). Once a nematode species is established in a field or greenhouse, it is usually extremely difficult to eliminate; however, if the community of nematodes contains diverse species, no single species will dominate the soil-crop ecosystems. In conventional cropping systems, plant-parasitic nematodes have abundant available food and a conducive soil environment for their survival and growth, often leading to rapid expansion and establishment of plant parasitic species, plant disease and crop yield loss.

Reduction of nematodes by cover crops and their metabolites is well-documented (Tables 2–4), as reported in numerous studies (Widmer et al., 2002; Wen et al., 2017). Crop rotation with cover crops that increased and supported biodiversity and efficiency prevent or control the onset of nematode infestation problems in healthy soil due to ecological balance and adjustment. Crop rotations, including such species, disrupt the life cycle of parasitic nematodes and reduce the risk of breakdown of inbred nematode resistance of commercial crops. Several leguminous and grassy cover crops, including Crotalaria, Mucuna, and Tagetes species, have shown to be non-hosts or suppressors of selected parasitic nematodes (Wang et al., 2007). Nematode suppression is also related to the beneficial effects of cover-crop residues on predatory nematodes and nematode-trapping fungi (Cherr et al., 2006). To control nematode populations and their adverse effects on potato crops, growers often planted two years of radish to improve potato production and lower pest control costs.

It is reported that small grain and *Brassica* cover crops are reportedly abating fungal and nematode problems in various agronomic and horticultural cropping systems (Ji et al., 2012; Mowlick et al., 2013; Wen et al., 2017). Cereal rye, as a hardy winter cover crop, effectively controls soybean cyst nematodes (Creech et al., 2008). *Brassicaceae* cover crops, such as brown mustard (*Brassica juncea* L.), winter canola (*B. napus* L.) and winter rapeseed (*B. napus* L.) are known to produce and contain high levels of glucosinolates (Tables 2 and 3), which are enzymatically converted into allelochemicals, such as isothiocyanates (ITCs) and related compounds (Fig. 2) when incorporated into the soil (Bangarwa et al., 2011; De Nicola et al., 2013). Allelopathic chemical compounds produced during the decomposition of plant residues share a similar mode of action for pathogen control as synthetic pesticides that are derivatives of ITCs (Table 4). These compounds act as bio-fumigants and are toxic to a wide spectrum of pathogens (Rudolph et al., 2015; Curto et al., 2016).

Several studies reported that radish, mustard and soybean nematode (SNB)-resistant sugar beets had similar levels of resistance to nematode populations (Smith et al., 2004; Wen et al., 2017).

Table 4 Soybean Cyst Nematode (SCN) egg counts (Number/100 g soil) Collected from cover crop treatments at Four Illinois Locations (2011 to 2013). *Source*: Wen et al. (2017).

Location	Cover Crop	2011	2012	2013
UIUC[x]	Canola	85a[w]	111	85
	Fallow	461b	153	80
	Mustard	305b	155	74
	Rapeseed	231ab	125	67
	Cereal rye	149ab	101	48
Ayres[y]				
	Fallow	885b	105	222b
	Rapeseed	219a	ND	ND
	Cereal rye	288ab	121	63a
WIU[z]	Canola	226ab	153ab	407ab
	Fallow	1836c	367b	2618b
	Mustard	679bc	73ab	2640b
	Rapeseed	459bc	9b	204a
	Cereal rye	205ab	101ab	2087ab
	Cereal rye no-till	58a	ND	ND
Hunt	Fallow	16	207b	0
	Rapeseed	19	48a	0
	Cereal rye	7	48a	0

[w] Means followed by the same letter are not significantly based on the multiple comparison test with Tukey's adjustment method at $\alpha = 0.05$. ND: no data were collected.

[x] Field trial at University of Illinois, Urbana-Champaign.

[y] Rapeseed did not overwinter in the Ayres field trials in 2012 and 2013, so soil was not collected from the rapeseed plots in those years.

[z] Field trial at western Illinois University Allison Farm. In 2012 and 2013, the cereal rye no-till treatment was not planted at the WIU location.

However, growth chamber studies indicated significant differences in the number of females, as well as in reproductive factors of nematodes under radish, mustard and sugar beets. Results suggested that the radish cultivars have a higher nematode reduction potential under optimal conditions than most of the mustard cultivars. It is also reported that some cover crops can enhance a resident parasitic nematode population if they are grown before or after another crop that hosts a plant-damaging nematode species. No cover crop is expected to function as a universal non-host for all parasitic nematodes, while in several cases, cover crops were shown to favour the growth of parasitic nematodes as well (Cherr et al., 2006; Sanchez et al., 2007).

2.2.3 Insects Diversity and Problems

Insect pests that have been associated with cover crops include green clover worm, Japanese beetle, bean leaf beetle, stinkbugs, true armyworm, black cutworm, seed corn maggot and wireworms (McMechan, 2018). Cover crops suppress and reduce populations of insect pests including aphids, beetles, caterpillars, leafhoppers, moths and thrips (Sarrantonio and Gallandt, 2003). The suppression and reduction of pest populations by cover crops is often associated with improvement in soil health via soil biological diversity and efficiency, soil chemical buffering and balancing, soil physical stability and the formation of protective niches of beneficial organisms, release of allelochemicals and changes in soil ecology (Sarrantonio and Gallandt, 2003; Tillman et al., 2004).

Cover crops support and promote biodiversity by creating more favourable habitat conditions for free-living bactivores and fungivores and other predators (Table 5). Beneficial insects reported

Table 5 Beneficial and pest insects attracted to common cover crop species.

Cover Crop	Beneficial Insects	Pest Insects
Buckwheat	Extra floral nectaries attract parasitic wasps, ladybugs; tachinid and hover flies; and lacewings	Tarnished plant bugs and aphids (aphids can act as a food source for beneficial).
Clovers	Parasitic wasps, big-eyed bugs, minute pirate bugs; ladybugs; tachinid flies and aphid midges	Spider mites and flower thrips (flower thrips can prey on spider mite eggs and provide food for several predatory insects).
Hairy vetch	Minute pirate bugs; ladybugs; predatory and parasitic wasps	Tarnished plant bugs.
Cahaba white Cereals	It supports beneficial insects yet attracts and supports ladybugs	Tarnished plant bug. Aphids.
Mixed field border plantings to provide year-round food for beneficial organisms: carrot family (umbels), sunflower family (composites), legume, mint, Yarrow and wild carrot provide nectar and pollen for adult phases of parasites and predators of many insect pests.		

Source: Plotkin (2012)

in cover crop studies are ground beetles, lady beetles, hover flies and spiders (McMechan, 2018). Combined with reduced proliferation of pests, cover crops and their decomposition by-products affect and reduce dispersal of visual and olfactory clues emitted by host crops, thus, resulting in effective insect pest suppression in agro-ecosystems (Tillman et al., 2004). However, in other cases, cover crops provide a shelter for insect pests as well.

In balanced ecosystems, insects and pests are generally in equilibrium or held in balance by their natural enemies or predators (Wen et al., 2017). These natural biocontrol organisms—called beneficial organisms in agricultural systems – include predator and parasitoid insects and diseases. Predators kill and eat other insects; parasitoids spend their larval stage inside another insect, which then dies as the invader's larval stage ends. However, in conventional agricultural systems, synthetic and reactive agrochemicals that kill insect pests also typically kill the natural enemies of insects. Diversifying, conserving and encouraging beneficial organisms are a key to achieving sustainable pest management in this age of climate-smart agriculture. By including cover crops in crop rotations, without or reduced application of synthetic agrochemicals (such as insecticides), beneficial organisms often are already in place at planting of spring or summer crops. However, incorporation of cover crops into the soil affects or disperses most of the beneficial organisms in agro-ecosystems.

Cover crops left on the soil surface may be living, temporarily suppressed, dying or dead mulch, which protect beneficial organisms and their habitat. Researchers have found that generalist predators, which feed on many species, may be an important biological control. During periods when pests are scarce or absent, several important generalist predators can subsist on nectar, pollen and alternative prey afforded by cover crops. This suggests that biocontrol services can be effective and improved by planting cover crops as habitat or food for the beneficial organisms in diverse agro-ecosystems. This strategy is important to adapt in warm and humid areas where pest pressures are generally heavy. Research showed that populations of beneficial insects, such as insidious flower bugs, big-eyed bugs and various lady beetles can attain high densities in vetch, clover and cruciferous cover crops (Table 5). These predators subsisted and reproduced on nectar, pollen, thrips and aphids, were established before key pests arrived.

Research throughout Georgia, Alabama, and Mississippi in the southern United States showed that when summer vegetables were planted amid 'dying mulches' of cool season cover crops, selected beneficial organisms moved into attack crop pests by sending chemical signals that attracted other beneficial insects. Lady beetles in cover-cropped systems help to control aphids attacking many crops. Maximising natural predator-pest interaction is the primary goal of a biologically-based IPM strategy and cover crop is a critical component to achieve the goal. Properly selected and managed, cover crops can enhance the soil health and field environment to favour beneficial organisms (LaRose and Myers, 2019).

3. Efficacy and Limitations of Cover Crops

Appropriate use of cover crops has the potential to control soil-borne diseases and insects and pests of agronomic crops and vegetables. However, their immediate efficacy is relatively less when compared to more radical chemical soil disinfestation or heat treatments. Crops are attacked by a complex of insects and pests because their effectiveness is relatively species-specific, which makes them less practical compared to other alternative IPM strategies. In contrast, the use of broad-spectrum insecticides can control a complex of insects and pests. The cost of chemical control is often lower as compared to the various associated costs of cover crops, especially in vegetables and other high-value crop production. Agronomic and logistical considerations, associated with managing cover crops, such as different planting dates and fertiliser requirements of the cover crop and main crop, are also likely to limit the practical use of cover cropping. Most importantly, pest management practices need to show consistent results. The consistent success of some cover-cropping systems has been highly variable, increasing the risk of economic loss to the grower.

Cover cropping is a knowledge-intensive practice and requires information on the temporal and spatial attractiveness of potential cover crops to maximise their effectiveness. In some cases, cover cropping may even require cooperation among growers because pests move freely between property boundaries. There are some changes caused by cover crops that may inadvertently put the main crop at risk if the cover crop harbours certain harmful insects and pathogens. Cover cropping does not entail a 'product' that can be sold, such as an insecticide or pesticide; however, cover cropping can be attractive to find natural and eco-friendly alternatives to conventional control practices, which are environmentally incompatible to increase ecosystem disservices.

References

Badenes-Perez, F.R., Shelton, A.M. and Nault, B.A. (2004). Evaluating trap crops for diamondback Moth (L.), Plutellaxylostella (Lepidoptera: Plutellidae). J. Econ. Entomol., 97: 1365–72.

Badenes-Perez, F.R., Nault, B.A. and Shelton, A.M. (2005a). Manipulating the attractiveness and suitability of hosts for diamondback moth (Lepidoptera: Plutellidae). J. Econ. Entomol., 98: 836–44.

Badenes-Perez, F.R., Shelton, A.M. and Nault, B.A. (2005b). Using yellow rocket as a trap crop for the diamondback moth, *Plutellaxylostella* (L.) (Lepidoptera: Plutellidae). J. Econ. Entomol., 98: 884–90.

Bangarwa, S.K., Norsworthy, J.K., Mattice, J.D. and Gbur, E.E. (2011). Glucosinolate and isothiocyanate production from *Brassicaceae* cover crops in a plasticulture production system. Weed Sci., 59: 247–254.

Borden, J.H. and Greenwood, M.E. (2000). Cobaiting for spruce beetles, Dendroctonusrufipennis (Kirby), and western balsam bark beetles, Dryocoetesconfusus Swaine (Coleoptera: Scolytidae). Can. J. For. Res., 30: 50–58.

Boucher, T.J., Ashley, R., Durgy, R., Sciabarrasi, M. and Calderwood, W. (2003). Managing the pepper maggot (Diptera: Tephritidae) using perimeter trap cropping. J. Econ. Entomol., 96: 420–32.

Campiglia, E., Paolini, R., Colla, G. and Mancinelli, R. (2009). The effects of cover cropping on yield and weed control of potato in a transitional system. Field Crop Res., 112: 16–23.

Cantonwine, E.G., Culbreath, A.K. and Stevenson, K.L. (2007). Effects of cover crop residue and pre-plant herbicide on early leaf spot of peanut. Plant Dis., 91: 822–827.

Cherr, C.M., Scholberg, J.M.S. and McSorley, R.M. (2006). Green manure approaches to crop production: A synthesis. Agron. Journal, 98: 308–319.

Creech, J.E., Westphal, A., Ferris, V.R., Faghihi, J., Vyn, T.J., Santini, J.B. and Johnson, W.G. (2008). Influence of winter annual weed management and crop rotation on soybean cyst nematode (*Heterodera glycines*) and winter annual weeds. Weed Sci., 56: 103–111.

Curto, Dallavalle, E., Matteo, R. and Lazzeri, L. (2016). Biofumigant effect of new defatted seed meals against the southern root-knot nematode, Meloidogyne incognita. Ann. Appl. Biol., 169: 17–26.

DeNicola, G.R., Montaut, S., Rollin, P., Nyegue, M., Menut, C., Iori, R. and Tatibouët, A. (2013). Stability of benzylic-type isothiocyanates in hydrodistillation-mimicking conditions. J. Agric. Food Chem., 61: 137–142.

Dhingra, O.D. and Netto, R.A.C. (2001). Reservoir and non-reservoir hosts of bean wilt pathogen, *Fusarium oxysporum* f. sp. Phaseoli. J. Phytopathol., 149: 463–467.

Djarwaningsih, T. (1997). *Indigofera hirsuta* L. plant resources of South-East Asia (PROSEA) No. 11: Auxiliary plants: 159–161. http://proseanet.org/prosea/e-prosea_detail.php?frt&id=3018.

Erwin, D.C. and Ribeiro, O.K. (1996). Phytophthora Diseases Worldwide, APS Press, St. Paul, MN, USA.

Everts, K.L. (2002). Reduced fungicide applications and host resistance for managing three diseases in pumpkin grown on a no-till cover crop. Plant Dis., 86: 1134–1141.

Foster, S.P. and Harris, M.O. (1997). Behavioural manipulation methods for insect pest-management. Annu. Rev. Entomol., 42: 123–46.

Gamliel, A., Austerweil, M. and Kritzman, G. (2000). Non-chemical approach to soil-borne pest management—organic amendments. Crop Prot., 19: 847–853.

Gilsanz, J.C., Arboleya, J., Maeso, D., Paullier, J., Behayout, E., Lavandera, C., Sanders, D.C. and Hoyt, G.D. (2004). Evaluation of limited tillage and cover crop systems to reduce N use and disease population in small acreage vegetable farms mirror image projects in Uruguay and North Carolina, USA. *In*: Bertschinger, L. and Anderson, J.D. (eds.). Proc. of XXVI IHC, Sustainability of Horticultural Systems, Acta Hort., 638: 163–169.

Godfrey, L.D. and Leigh, T.F. (1994). Alfalfa harvest strategy effect on Lygus bug (Hemiptera: Miridae) and insect predator population density: Implications for use as trap crop in cotton. Environ. Entomol., 23: 1106–18.

Gwinn, K.D., Ownley, B.H., Greene, S.E., Clark, M.M., Taylor, C.L., Springfield, T.N., Trently, D.J., Green, J.F., Reed, A. and Hamilton, S.L. (2010). Role of essential oils in control of Rhizoctonia damping off in tomato with bioactive monarda herbage. Phytopathology, 100: 493–501.

Held, L.J., Jennings, J.W., Koch, D.W. and Gray, F.A. (2000). Economics of trap cropping for sugar beet nematode control. J. Sugar Beet Res., 37: 45–55.

Hoitink, H.A.J. and Boehm, M.J. (1999). Biocontrol within the context of soil microbial communities: A substrate-dependent phenomenon. Annual Rev. Phytopath., 37: 427–446.

Hoy, C.W., Vaughn, T.T. and East, D.A. (2000). Increasing the effectiveness of spring trap crops for *Leptinotarsa decemlineata*. Entomol. Exp. Appl., 96: 193–204.

Javaid, I. and Joshi, J. (1995). Trap cropping in insect pest management. J. Sustain. Agric., 5: 117–36.

Ji, P., Koné, D., Yin, J., Jackson, K.L. and Csinos, A.S. (2012). Soil amendments with *Brassica* cover crops for management of Phytophthora blight on squash. Pest Manage. Sci., 68: 639–644.

Kennedy, G.G. and Storer, N.P. (2000). Life systems of polyphagous arthropod pests in temporally unstable cropping systems. Annu. Rev. Entomol., 45: 467–93.

Khan, Z.R., Pickett, J.A., van den Berg, J., Wadhams, L.J. and Woodcock, C.M. (2000). Exploiting chemical ecology and species diversity: Stem borer and striga control for maize and sorghum in Africa. Pest Manag. Sci., 56: 957–62.

Khan, Z.R., Pickett, J.A., Wadhams, L. and Muyekho, F. (2001). Habitat management strategies for the control of cereal stemborers and striga in maize in Kenya. Insect Sci. Appl., 21: 375–80.

Khan, Z.R. and Pickett, J.A. (2004). The 'push-pull' strategy for stemborer management: A case study in exploiting biodiversity and chemical ecology. pp. 155–164. *In*: Gurr, G., Waratten, S.D. and Altieri, M.A. (eds.). Ecological Engineering for Pest Management: Advances in Habitat Manipulations for Arthropods. CSIRO and CABI Publishing.

Kirkegaard, J. (2009). Biofumigation for plant disease control—From the fundamentals to the farming system. pp. 172–195. *In*: Walters, D. (ed.). Disease Control in Crops: Biological and Environmentally Friendly Approaches. Wiley-Blackwell, Oxford, UK.

Kumari, A.P.P. and Pasalu, I.C. (2003). Influence of planting pattern of trap crops on yellow stem borer, Scirpophagaincertulas (Walker), damage in rice. Ind. J. Plant Prot., 31: 78–83.

Ladygina, N. and Hedlund, K. (2010). Plant species influence microbial diversity and carbon allocation in the rhizosphere. Soil Biol. Biochem., 42: 162–168.

Larkin, R.P., Griffin, T.S. and Honeycut, C.W. (2010). Rotation and cover crop effects on soil borne potato diseases, tuber yield and soil microbial community. Plant Disease, 94: 1491–1502.

LaRose, J. and Myers, R. (2019). Impact of cover crops on natural enemies and pests. Cover Crop Facts, Cover Crop Resource Series. https://www.sare.org/Learning-Center/Topic-Rooms/Cover-Crops/Ecosystem-Services-from-Cover-Crops/Impact-of-Cover-Crops-on-Natural-Enemies-and-Pests.

Lazzeri, L. and Mannici, L.M. (2001). Allelopathic effect of glucosinolate-containing plant green manure on *Pythium* sp. and total fungal population in soil. Hort. Sci., 36: 1283–1289.

Liburd, O.E., Nyoike, T.W. and Scott, C.A. (2008). Cover, border and trap crops for pests and disease management. pp. 1095–1100. *In*: Capinera, J.L. (ed.). Encyclopedia of Entomology. Second ed., vol. 1A-C. Dordrecht, the Netherlands: Springer.

Lu, Y.C., Watkins, K.B., Teasdale, J.R. and Abdul-Baki, A.A. (2000). Cover crops in sustainable food production. Food Reviews Intern., 16: 121–157.

Lu, J., Liu, Y.B. and Shelton, A.M. (2004). Laboratory evaluations of a wild crucifer *Barbarea vulgaris* as a management tool for diamondback moth. Bull. Entomol. Res., 94: 509–16.

Manici, L.M., Caputo, F., Nicoletti, F., Leteo. F. and Campanelli, G. (2004). The impact of legume and cereal cover crops on rhizosphere microbial communities of subsequent vegetable crops for contrasting crop decline. Biol. Control, 120: 17–25.

McMechan, J. (2018). Insects in Cover Crops, Crop Production Clinic Proc. https://cropwatch.unl.edu/2018/insects-cover-crops.

Michel, V. and Lazzeri, L. (2011). Green manures to control *Verticillium wilt* of strawberry. IOBC/wprs Bulletin, 70: 81–86.

Michel, V.V., Urba, K. and Clarkson, J. (2015). Green manures and cover crops to reduce the pressure of soil-borne diseases in annual crops. Environmental Science, published 2015, EIP-Agri Focus Group Soil-borne Diseases, Minipaper.

Miller, J.R. and Cowles, R.S. (1990). Stimulodeterrent diversion: A concept and its possible application to onion maggot control. J. Chem. Ecol., 16: 3197–3212.

Motisi, N., Montfort, F., Faloya, V., Lucas, P. and Doré, T. (2009). Growing *Brassica juncea* as a cover crop, then incorporating its residues provide complementary control of Rhizoctonia root rot of sugar beet. Field Crops Res., 113: 238–245.

Mowlick, S., Takehara, T., Kaku, N., Ueki, K. and Ueki, A. (2013). Proliferation of diversified clostridial species during biological soil disinfestation incorporated with plant biomass under various conditions. Appl. Microbiol. Biotechnol., 97: 8365–8379.

Muthiah, C. (2003). Integrated management of leafminer (*Aproaeremamodicella*) in groundnut (*Arachis hypogaea*). Ind. J. Agric. Sci., 73: 466–68.

Pawar, D.B. and Lawande, K.E. (1995). Effects of mustard as a trap crop for diamondback moth on cabbage. J. Maharashtra Agric. Univ., 20: 185–86.

Pair, S.D. (1997). Evaluation of systemically treated squash trap plants and attracticidal baits for early-season control of striped and spotted cucumber beetles (Coleoptera: Chrysomelidae) and squash bug (Hemiptera: Coreidae) in cucurbit crops. J. Econ. Entomol., 90: 1307–1314.

Pegg, G.F. and Brady, B.L. (2002). *Verticillium Wilts*. CABI Publishing, Wallingford, UK.

Plotkin, J. (2012). Insects Attracted to Common Cover Crop Species, University of Maine. Originally published in Proceedings - NEVBC 1999; reviewed by T. Jude Boucher, U Conn IPM.

Potter, M.J., Davies, K. and Rathjen, A.J. (1998). Suppressive impact of glucosinolates in *Brassica* vegetative tissues on root lesion nematode *Pratylenchus neglectus*. J. Chem. Ecol., 24: 67–80.

Rudolph, R.E., Sams, C., Steiner, R., Thomas, S.H., Walker, S. and Uchanski, M.E. (2015). Biofumigation performance of four *Brassica* crops in a green chili pepper (*Capsicum annuum*) rotation system in southern New Mexico. Hort. Science, 50: 247–253.

Sanchez, E.E. Sánchez, Giayetto, A., Cichón, L., Fernández, D., Aruani, M.C. and Curetti, M. (2007). Cover crops influence soil properties and tree performance in an organic apple (*Malus domestica Borkh*) orchard in northern Patagonia. Plant Soil, 292: 193–203.

Sarrantonio, M. and Gallandt, E. (2003). The role of cover crops in North American cropping systems. J. Crop Prod., 8: 53–74.

Scholberg, J.M.S., Dogliotti, S., Leoni, C., Cherr, C.M., Walter, L.Z. and Rossing, A.H. (2010). Cover crops for sustainable agrosystems in the Americas. pp. 23–58. *In*: Eric Lichtfouse (ed.). Genetic Engineering, Biofertilisation, Soil Quality and Organic Farming.

Shelton, A.M. and Nault, B.A. (2004). Dead-end trap cropping: A technique to improve management of the diamondback moth, *Plutellaxylostella* (Lepidoptera: Plutellidae). Crop Prot., 23: 497–503.

Shelton, A.M. and Badenes-Perez, F.R. (2006). Concepts and applications of trap cropping in pest management. Annu. Rev. Entomol., 51: 285–308.

Shennan, C. (2008). Biotic interactions, ecological knowledge and agriculture. Phil. Trans. R. Soc., B: 363: 717–739.

Smith, H.J., Gray, F.A. and Koch, D.W. (2004). Reproduction of *Heterodera schachtii* Schmidt on resistant mustard, radish, and sugar beet cultivars. J. Nematol., 36: 123–130.

Srinivasan, K. and Krishna Moorthy, P.N. (1991). Indian mustard as a trap crop for management of major lepidopterous pests on cabbage. Trop. Pest Manag., 37: 26–32.

Stark, C.H., Condron, L.M., O'Callaghan, M., Stewart, A. and Di, H.J. (2008). Differences in soil enzyme activities, microbial community structure and short-term nitrogen mineralisation resulting from farm management history and organic matter amendments. Soil Biol. Biochem., 40: 1352–1363.

Sujan, D., Oliver, J.B., O'Neal, P. and Addesso, K.M. (2019). Management of flat-headed apple tree borer (*Chrysobothris femorata* Olivier) in woody ornamental nursery production with a winter cover crop. Pest Manage. Sci., 75: 1971–1978.

Talekar, N.S. and Shelton, A.M. (1993). Biology, ecology and management of the diamondback moth. Annu. Rev. Entomol., 38: 275–301.

Thompson, J.N. and Pellmyr, O. (1991). Evolution of oviposition behaviour and host preference in Lepidoptera. Annu. Rev. Entomol., 36: 65–89.

Tillman, G., Schomberg, H., Phatak, S., Mullinix, B., Lachnicht, S., Timper, P. and Olson, D. (2004). Influence of cover crops on insect pests and predators in conservation tillage cotton. J. Econ. Entomol., 97: 1217–32.

Tscharntke, T. and Brandl, R. (2004). Plant-insect interactions in fragmented landscapes. Annu. Rev. Entomol., 49: 405–30.

van Bruggen, A.H.C. and Semenov, A.V. (2000). In search of biological indicators for soil health and disease suppression. Applied Soil Ecol., 15: 13–24.

Vernon, R.S., Kabaluk, J.T. and Behringer, A.M. (2000). Movement of *Agriotes obscurus* (Coleoptera: Elateridae) in strawberry (Rosaceae) plantings with wheat (Gramineae) as a trap crop. Can. Entomol., 132: 231–41.

Vernon, R.S., Kabaluk, J.T. and Behringer, A.M. (2002). Aggregation of *Agriotes obscurus* (Coleoptera: Elateridae) at cereal bait stations in the field. Can. Entomol., 135: 379–89.

Virk, J.S., Brar, K.S. and Sohi, A.S. (2004). Role of trap crops in increasing parasitation efficiency of *Trichogramma chilonis* Ishii in cotton. J. Biol. Control., 18: 61–64.

Wang, Q.R., Li, Y.C. and Klassen, W. (2007). Changes of soil microbial biomass carbon and nitrogen with cover crops and irrigation in a tomato field. J. Plant Nutr., 30: 623–639.

Wen, L., Lee-Marzano, S., Ortiz-Ribbing, L.M., Gruver, J., Hartman, G.L. and Eastburn, D.M. (2017). Suppression of soil-borne diseases of soybean with cover crops. Plant Dis., 101: 1918–1928.

Weil, R. and Kremen, A. (2007). Thinking across and beyond disciplines to make cover crops pay. J. Sci. of Food Agric., 87: 551–557.

Widmer, T.L. and Abawi, G.S. (2002). Relationship between levels of cyanide in sudangrass hybrids incorporated into soil and suppression of *Meloidogyne hapla*. J. Nematol., 34: 16–22.

Widmer, T.L., Mitkowski, N.A. and Abawi, G.S. (2002). Soil organic matter and management of plant-parasitic nematodes. J. Nematol., 34: 289–295.

Wiggins, B.E. and Kinkel, L.L. (2005a). Green manures and crop sequences influence alfalfa root rot and pathogen inhibitory activity among soil-borne *Streptomycetes*. Plant Soil, 268: 271–283.

Wiggins, B.E. and Kinkel, L.L. (2005b). Green manures and crop sequences influence potato diseases a pathogen inhibitory activity of indigenous *Streptomycetes*. Phytopathology, 95: 178–185.

8

Cover Crops for Forages and Livestock Grazing

Riti Chatterjee

1. Introduction

Cover crops represent an essential component of sustainable agro-ecosystems. In recent years, the use of cover crops for grazing has gained renewed attention among livestock producers in response to adverse impacts of conventional management practices on soil health and crop productivity. Cover crops tend to be common annual forages, which livestock producers have been planting in pastures for decades, as these crops contribute high quality forage for livestock while enhancing agro-ecosystem services.

Cover crops offer many potential benefits to crop production and their use is becoming increasingly popular. Advantages of cover crops include greater and diverse biomass production, improved biodiversity, soil balancing, nutrient recycling, carbon sequestration, soil physical stability, nutrient recycling and erosion control (McVay et al., 1989; Zhu et al., 1989; Hoorman et al., 2009). Cover crops can be expensive to plant; however, through grazing of these crops, farmers can recover some of the expenses incurred. Grazing also adds nutrients to the field, further enhancing the productive capacity of the soil (Farney et al., 2018). Studies conducted by USDA-NRCS (2013) reported that the four keys for improving soil health and increasing soil organic matter (SOM)

Fig. 1 Integration of livestock for grazing on diverse cover crop mixtures (*Source*: David Brandt, 2018; Personal communication).

Bidhan Chandra Krishi Vishwavidyalaya, West Bengal, India.

are: (1) plant diversification to form microbial diversity in soils; (2) improvement of soil health by reducing disturbance (no/reduce tillage); (3) keeping plants growing throughout the year to feed soil microbes; and (4) keeping the soil covered to reduce soil erosion. However, many soil health experts and associates view the integration of livestock as a fifth key for rapidly improving soil health.

In a case study by Williams et al. (2018) during the winter grazing season between 2016 and 2017, it was found a farm spent around $108/ac. per year to plant, graze and terminate a winter cover crop. However, those cover crops also produced forage worth $329/ac., and with the added fertility to their soil from both the cover crop and livestock, the impact was valued at around $54/ac. While there are many beneficial aspects to the use of cover crops, it has remained difficult to economically justify their use to the farmers (Allison and Ott, 1987).

Farmers can begin grazing of their livestock after cover crops grow to at least 6–8 inches tall for grasses, or above 10 inches for *Brassicas* and legumes. Ideally, livestock must be moved twice per week, but another good option is to strip graze cover crops. With strip grazing, a set amount of forage that meets the needs of the livestock is allocated at a time. One important aspect to keep in mind is when livestock graze cover crops, do not allow them to overgraze a single area and make sure to leave at least four inches of plant cover. Once animals have grazed an area to this height, cattle should be rotated to another area to optimise future re-growth in this area. A good management practice is to monitor the amount of ground cover frequently for overgrazing or under grazing and adjust the rotation of cattle accordingly. Application of grazing practices on cover crops can aid in the distribution of manure and increase organic matter in soil. To reduce animal-induced soil compaction, it is better to plan and utilise a 'sacrifice area' during wet conditions (Smith et al., 2017).

Cover crops offer excellent nutritional opportunities for grazing ruminant livestock, whose presence can also improve the pasture and the succeeding cash crops, if managed properly. Generally, non-leguminous cover crops as pastures provide excellent nutrition to breed and grow livestock. The leguminous cover crops can provide high-quality hay for grazing, although excessive amounts of certain legumes are responsible for bloat or other anti-nutritional factors. The nutritive value of consumed forages, in combination with the different feeding and grazing system per trial period, influenced milk quality and performance. The replacement of feed supplementation, with an increase of the contribution of grazing in the rangeland, could maintain the milk yield and quality at the late stage of lactation (Manousidis et al., 2018). Legumes are often effective in improving animal performance when grazed and persist well in rotational grazing systems (Hoveland et al., 1988), as legumes have higher concentrations of crude protein, total non-structural carbohydrates, digestible dry matter and diverse organic acids with a lower concentration of fibre compared to grasses and others (Johnson et al., 1982). Therefore, legumes can be effective in supplementing grasses when added to the diet in a balanced proportion or at a rate of 15–30 per cent. Brandt and Klopfenstein (1986) reported that the quality of the legumes influence the amount needed for this response (15 per cent high-quality alfalfa versus 30 per cent medium-quality alfalfa). If legumes will be used in a crop rotation, it will act as a source of N in the form of supplemental cattle feeding.

Planting and managing cover crops that apparently produce no immediate outcome is still a sceptical practice throughout the world. The successful, widespread implementation of cover crop adoption is not likely to occur unless it encompasses an immediate economic value for the farmer, or unless it is either mandated or subsidised. An incorporation of ruminant livestock into the crop production system may help to encourage greater socio-economical acceptance. In addition, soil erosion and fertiliser needs could be reduced following this integration of improved grazing and forage systems using livestock. Due to their unique ability to utilise forages and sustain with minimal management, these ruminants fit into such an integrated system. These ruminant animals do not compete with humans for food and they recycle organically-bound nutrients through the decomposition of manure wherever they graze or are fed forages, including cover crops (Boehncke, 1985). Therefore, the integration of livestock with cover crops as an available forage resource can result in an economically viable, environmentally compatible, biologically efficient and socially acceptable alternative for those farmers who choose to invest the time and management necessary to make them successful.

2. Cover Crops as Forage and Grazing

It is well known that agriculture is taking place throughout a world that encompasses diverse topography, soils, climate and plant and animal communities (Cox and Atkins, 1979). Inclusion of the animal component into the crop production system can alter the nutrient dynamics and simultaneously, the soil fertility status over time. This change occurs differently according to the arrangement adopted, depending on both the intensity and frequency of animal grazing, as well as the crop rotation system (Alves et al., 2019).

Cover crops with a diverse range of herbaceous plants often provide valuable resources as forages to livestock as feed, which can help farmers improve livestock production and diversify farm incomes. The fast-growing herbaceous plants are also a climate-smart option, as they have the potential to reduce the environmental footprint of agriculture through carbon sequestration, reduction of greenhouse gas emissions and restoration of degraded lands. Cover crops, such as grasses, legumes, or *Brassica*, along with other feeding sources, need to be promoted for widespread adoption among farmers, especially small and medium-sized holdings. A combination of grasses and legumes need to be identified for sustainable cropping diversity that will foster the livelihood and environmental benefits of agronomic crop forage-based, socio-ecological systems with greater ecosystem services (Fig. 2).

Cover crops as pasture forages for livestock can be categorised into one of six categories: warm-season perennials, warm-season annuals, cool-season perennials, cool-season annuals, Brassicas and legumes. Each of these forage types has certain characteristics to meet the food, energy and nutritional requirements of livestock when they are at their peak production; however, none are able to completely satisfy the nutritional and balanced needs of livestock without mixing with other types or blending with others.

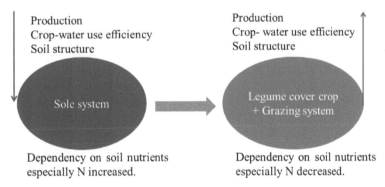

Fig. 2 Transformation of a sustainable farming system from sole cropping to inclusion of diverse cover crops with legumes + grazing animal (*Source*: Modified from Puckridge and French [1983] and Webber [1990]).

2.1 Warm-season Perennials

Warm-season perennial pastures tend to be the best grasses for livestock, as they have strong survivability and re-growth over time. Once established, these pastures continue to grow and produce over long periods. Warm-season perennial pastures, such as Bermuda grass, Bahia grass, Napier grass or Klein grass generally have a longer growing season than cool-season plants. As perennials, they re-grow from roots and/or rhizomes each year. Because they do not have to re-establish yearly, they maintain top forage production for longer periods. They also tend to be lower in digestibility and in protein content because of the fibre build-up during the warmer part of the growing season. Warm-season perennial grasses respond well to fertilisation and, with heavy fertilisation, can produce large amounts of hay or grazing per acre. If fertilised and managed properly, they work well in almost any livestock production programme.

Fig. 3 Typical warm-season annual and perennial grasses.

2.2 *Warm-season Annuals*

Warm-season annual grasses, such as the pearl millet, sudan-sorghum, crabgrass and forage sorghums play important roles in livestock production. As annuals, they are expensive to plant because land must be prepared and seeded annually. The annuals are the most expensive grasses for forage because they must be seeded annually on prepared land; the seed is expensive, there is a limited production season and they require high rates of fertility treatments. While warm-season annuals provide higher quality (digestibility) grazing than perennial warm-season grasses, their production period is shorter. Their primary role in forage production, however, is as high-quality hay.

2.3 *Cool-season Perennials and Annuals*

Cool season perennials, such as tall fescue, orchard grass, festulolium, brome grass and tall wheat grass are commonly used as forages; however, their growth and production depend on climate, soil and management practices. They generally do not offer high quality nutrition for maximum animal performance.

While the cool-season annuals, such as oats, wheat, rye, barley, triticale and ryegrass are expensive pastures due to their cost and availability to establish annually, they are high in nutritional value. Winter annuals are best adapted to stocker operations or to cow-calf combination programmes; however, due to their expense, annual pastures may not be the best types of pastures for dry pregnant cows, which can be maintained very well on less expensive forages, such as high-quality hay.

Fig. 4 Typical cool-season annual and perennial grasses.

Legumes

Legumes are one of the important components of forages to support livestock. Temperate legumes include clovers, medics, peas, vetch and alfalfa which can be over-seeded into permanent pastures or seeded with winter annual pastures. Legumes have the unique ability to fix their own nitrogen if properly inoculated (nitrogen-fixing bacteria is added to the legume seed before planting). They require high levels of P, K and in acidic soil, lime. Cool-season or temperate legumes produce most of their growth during the late winter-spring period, when they are very useful in beef-cattle operations.

Warm-season or tropical legumes, such as cowpea, soybean and peanut can provide high quality forage during the summer. However, they are used as a salvage crop in drought years when they do not 'yield' well as a row-crop (Bade and Dorsett, 2002). Generally, legumes produce higher quantities of protein than grasses. If properly inoculated, legumes have the capacity to fix atmospheric N, eliminating the need to apply N from commercial sources. Legumes also supply a considerable amount of nitrogen to the grass portion of the mixture. Legume forages might also be considered for a livestock operation.

Brassicas and Others

Brassica belongs to a plant family that includes oilseed radish, turnips, rapeseed and mustards. They are leafy and grow very rapidly. Thus, they help in weed suppression (because of production of glucosinolates), eliminate some soil-borne pathogens like fungi, nematodes and have the potential for lessening soil compaction.

Some of the *Brassicas* (e.g., turnips) form a large bulb just below the ground and break surface compaction, whereas radishes, with their longer tubers, help in lessening soil compaction in deeper soil layer. Rapeseed and mustards also do the same in deeper compaction. However, most *Brassica* species (except some rapeseeds) generally get winter-killed and decay rapidly due to their high-water content in the biomass. Thus, *Brassica* plants need to be planted in a mix with other species like grasses. However, they are increasingly planted as winter or rotational cover crops in vegetables and other crops that are also being used as livestock feed.

Annual Legumes and Pasture Legumes

Annual and short-lived perennial clovers are frequently used alone or in mixes with grasses as forage cover crops. While in mild environments, annual legumes can be sown in fall or spring, in cold-winter regions, annuals are planted in the spring after the danger of hard freeze has passed. Winter annuals (like subterranean clover) should be planted in autumn.

The cool season legume annual clover consists of mainly two types—arrow leaf and crimson clover. They are most productive in spring. Crimson matures earlier in spring than arrow leaf and provide less grazing in late spring. In the Piedmont region of the United States, arrow leaf may provide grazing until early June. These legumes are generally higher in protein and more digestible than cool-season annual grasses, particularly as the grasses mature in late spring. As a result, gains of 2.5 lbs/head/day and 260 lbs/acre can be expected during spring grazing when an annual clover is used. Apart from it, these legumes may contribute as much as 100 lbs of nitrogen (N)/acre via nitrogen fixation. Two forage legumes fit best with tall fescue, however, white clover and red clover.

White Clover

White clover is a low-growing legume that can tolerate close grazing. It mainly spreads by stolon. It helps with grazing in the fall, late winter and spring. Yields of white clover are usually not sufficient for it to be grown alone or as a hay crop, but it contributes a substantial amount of high-quality forage when produced with tall fescue. It grows best on moist soils and might die during hot, dry summers. However, some new varieties of white clover are more persistent and will either survive these conditions or return from seed.

Table 1 Characteristics of annual and perennial cover crops as sources of forage.

Species	Suitability	Duration (years)	Biomass (Mg/ha)	Strength/Quality	Cautions/Problems
Annuals					
Legumes					
Field peas	Pasture	1	2–2.5	Good quality N rich feed.	Annual planting, expensive.
Cowpea	Pasture cover crop	1	1.25–2.25	Good quality N. Very flexible. Prevent soil erosion.	Less adaptable to alkaline soil. Intolerant nematode resistant. Suppress weeds to waterlogged and saline conditions.
Austrian winter pea	Pasture cover crop Dought tolerant.	1	3–4	Good quality N and micronutrients. Rich feed. Chill-tolerant. Sensitive to soil pH.	Vulnerable to over-grazing. Low green manure.
Lentils	Pasture cover crop Stored feed	1	0.2–1.25	Good quality N. Controls weeds and plant diseases.	Intolerant to flooding, salinity and, frost.
Kidney beans	Pasture Stored feed	1	0.45–0.75	High quality N. Adaptable to wide range of soils.	Very sensitive to soil salinity and pH.
Soybeans	Pasture cover crop	1	2.5–7.15	High quality N. Flexible in harvest dates.	May cause bloat. Weed problem is quite high.
Berseem	Pasture cover crop	1	3.25	High quality N. Suppress weeds.	Susceptible to root-knot nematode. More tolerant to saline soils. Fast growing.
Hairy vetch	Pasture Cover crop	1	1.25–2.5	Good quality N. Tolerant to low fertile soil.	Low flood tolerant, wide pH range and, fluctuating winter.
Woolly pod vetch	Pasture Cover crop	1	2–4	Good quality N. Excellent allelopathic	Overgrazing can be toxic. Hard seed can be effect. Suppress weeds; Problematic.
Grasses					
Wheat	Pasture Stored feed	1	1.5–4	Good quality but N-poor feed.	Heavy N and H_2O user in spring. Intolerant to flood.
Triticale	Pasture Cover crop	1	2.25–4.25	Cross between wheat and rye. Control soil erosion. Less prone to lodging. Short growing season.	Lagging in grain quality. Susceptible to leaf rust.
Cereal rye	Pasture Stored feed	1	1.5–5	Good quality N. Prevent soil erosion	Can become a weed if tilled at wrong stage and weed problem.
Barley	Pasture Stored feed	1	1–5	Prevent soil erosion. Tolerates moderately alkaline conditions.	Poor performance in acid soil < pH 6.0.
Oats	Pasture Stored feed	1	1–5	Prevent weed infestation. Softens Soil and control erosion.	Prone to lodging in N-rich soil. Susceptible to nematodes.
Pearl millets	Pasture Stored feed	1	2–4	Does not contain prussic acid. Not susceptible to sugarcane aphid.	Sensitive to cold stress and nitrate poisoning.

Table 1 Contd. ...

...Table 1 Contd.

Species	Suitability	Duration (years)	Biomass (Mg/ha)	Strength/Quality	Cautions/Problems
Sorghum	Pasture Stored feed	1	15	Tolerant to drought. Regarded as emergency crop	Potential to prussic acid poisoning.
Sorghum-Sudan	Pasture	1	4–5	Prevent soil erosion. Resistant to weed.	Mature, frost-killed plants become quite woody.
Annual ryegrass	Pasture Stored feed	1	1–4.5	Prevent soil erosion. Moderate flood resistant.	Heavy N and H_2O user.
Brassicas					
Oilseed radish	Pasture	1	1.5	Highly digestible. Prevent soil compaction.	May become a weed if allowed to go to seed.
Turnip	Pasture	1	4.4	Reduce soil compaction. Good suppressor of weeds	May cause haemolytic anaemia, nitrate toxicity infertility, bloat, and goitre.
Rapeseed	Pasture Stored feed	1	1–2.5	Good suppressor of nematodes and weeds Great bio-fumigation potential.	Cannot withstand flood condition.
Kale (Biennial)	Pasture Feed	2	15.5–33.5	High protein, vitamins, and energy content.	Cannot withstand dry condition. Club root disease is a problem.
Cabbage	Pasture Feed	1	1.4	Reduce soil erosion. Good suppressor of weeds.	High in sugar. May lead cattle to acidosis. Can also cause haemolytic anaemia and goitre.
Broccoli	Pasture Feed	1	0.25–0.75	Good source of vitamins, minerals and fibre.	Excessive summer hot cause bloating.
Others					
Buckwheat	Pasture Stored feed	1	1–2	High source of protein. Easily digestible.	May cause rashes on light coloured cattle.
Phacelia	Pasture	1	4	Source of high-quality nectar and pollen. Drought tolerant.	May cause nitrate toxicity.
Perennials					
Legumes					
Alfalfa	Stored Feed	3–4		Excellent quality/ Excellent yield	May cause bloat. Poor persistence under grazing. Low tolerance to acidic or variably drained soil. Needs fall rest period.
Birds foot trefoil	Pasture	5+		High quality. No bloat hazards	Slow to establish Slow spring growth and re-growth.
	Stored feed			Good tolerance to acidic & variably drained soil.	Needs fall rest period. Unpalatable to horses.

Table 1 Contd. ...

...Table 1 Contd.

Species	Suitability	Duration (years)	Biomass (Mg/ha)	Strength/Quality	Cautions/Problems
Red clover	Pasture Cover Crop Stored feed	1–3		Excellent 1st-yearyield. Easy to establish. High quality. Good tolerance to acidic.	Difficult to dry for hay. May cause bloat. Stand thins rapidly. May cause temporary infertility in grazing sheep. Very competitive, drained soil especially with or variably other legumes.
White clover	Pasture	5+		Excellent quality and palatability. Good tolerance to close grazing.	May cause bloat. Low drought tolerance.
Kura clover	Pasture	5+		Persistent high quality.	Difficult to establish. May cause bloating.
Alsike clover	Pasture Stored feed	1–2		Very good tolerance to wet, acidic soils. Good quality.	Lower yield than red clover Regrowth yields low Stand thins rapidly May cause bloating.
Sweet clover	Cover crop Stored feed	2		Excellent soil builder. Opens up subsoil Excellent bee pasture.	Low palatability unless harvested early Coumarin varieties causes feeding difficulties; only 1 harvest.
Grasses					
Timothy	Stored feed	5+		Easy to establish Good tolerance to variable drainage Seed is inexpensive.	Poor summer production Poor persistence of late heading varieties under three-cut harvest system.
Smooth bromegrass	Pasture Stored feed	5+		Excellent spring/fall yield Good regrowth Better quality retention with maturity.	Large seed size may cause seeding challenges.
Meadow bromegrass	Pasture Stored feed	5+		Early spring growth Fast recovery after cutting or grazing Good winter-hardiness Good palatability.	Large seed size may cause seeding challenges Sensitive to flooding. Spreads less by rhizomes than smooth Brome grass.
Orchard grass	Pasture Stored feed	5		Very early pasture Excellent regrowth Very Good drought tolerance Good tolerance to close grazing. Very responsive to N.	Rapidly loses quality and palatability with maturity. Very competitive with other species Poor tolerance to variable drainage and icing.
Reed Canary grass	Pasture Stored feed	5+		Excellent yield on both variably drained and dry soils Good re-growth Very responsive to nitrogen.	Slow to establish First cut rapidly loses quality and palatability with maturity Poor tolerance to close grazing or frequent cutting.

Table 1 Contd. ...

...Table 1 Contd.

Species	Suitability	Duration (years)	Biomass (Mg/ha)	Strength/Quality	Cautions/Problems
Meadow fescue	Pasture Stored feed	5+		More suitable for managed gazing than as stored feed Grows in early spring and late fall. Tolerant to variably drained soil. More palatable than tall fescue. Prevents erosion in waterways.	Coated seed required Very competitive with other spp. Low drought tolerance Low quality with maturity Less persistent and lower yielding than tall fescue.
Tall fescue	Pasture Stored feed	5+		High yield Good summer growth grazing Good tolerance to acidic soil Good feed quality in fall for stockpile. Waterways.	Coarse leaves and low palatability. Need endophyte-free seed.
Perennial ryegrass	Pasture Stored feed	2–3		Excellent quality and palatability Establishes very quickly Good tolerance to close grazing.	Poor drought and heat tolerance Poor tolerance to variably drainedsoils Variable persistence.
Kentucky bluegrass	Pasture	5+		Good quality and palatability Good waterways.	Poor summer production. Very slow to establish. Low tolerance to close grazing seasonal yield.
Festulolium	Pasture	2–3	2.25–2.75	High disease resistant. Winter-hardy. High yield.	Cannot withstand poorly drained and less fertile soil.

Source: Hashemi, M. (1996). Forage Management; Perennial Forage Species for Pasture and Hay. https://ag.umass.edu/crops-dairy-livestock-equine/fact-sheets/perennial-legumes-grasses. This publication has been funded in part by Mass. Dept. of Agricultural Resources in a grant to the Massachusetts Farm Bureau Federation, Inc. and by Mass. Dept. of Environmental Protection, s319 Program.2. Managing Cover Crops Profitably, third edition. 2012.ISBN: 978-1-888626-12-4(pbk). Published by Sustainable Agriculture Research and Education (SARE).

There are mainly three types of white clover: large (e.g., Ladino clover, Patriot, Regal), intermediate (e.g., Durana, Osceola), and low growing (e.g., Dutch clover). Large or ladino types are higher yielding than other types and the intermediate types are well adapted to most sites and are prolific re-seeders. Intermediate white clovers are more tolerant of grazing and persist better than red clover (especially in some drought-prone and infertile sites). Consequently, white clover often fits better within tall fescue-based pastures that are continuously stocked or stocked in a way that leaves animals in the pasture while the clover is recovering from grazing.

Red Clover

Red clover is a short-lived perennial legume that is adaptable to a wide range of soils. Red clover is tall-growing, deeper rooted and higher yielding than white clover. It is more productive than white clover during periods of drought stress as well. However, even under the best of conditions, red clover stands start to thin in the second year. Moreover, red clover does not tolerate close grazing and will not produce or survive well in continuously stocked pastures. As a result, red clover will need to be re-planted every two to three years and must be used in well-managed, rotationally-grazed tall fescue pastures.

Pasture legumes are sown in mixtures with grasses and are sufficiently low growing enough to be grazed by livestock. They belong to the sub-family Papilionacae, which includes the following:

Pencil flower – Flowering plants of the legume family Fabaceae and contain numerous highly important pasture and forage species (e.g., Brazilian Lucerne, Caribbean stylo, etc.).

Tick-clover – Flowering plant of the family Fabaceae and sometimes called tick-trefoil, tick clover, hitch hikers, or beggar lice. There are dozens of species and the delimitation of the genus has shifted much over time. These are mostly inconspicuous legumes as few have bright or large flowers.

Butterfly peas – American vines of the legume family. Species include Centrosema angustifolium, Benth, Centrosema arenarium Benth, Centrosema Brazilian Benth, Centrosema dasyanthum Benth, Centrosema macranthum Hoehne, Centrosema macrocarpum, Centrosema plumeria Benth.

Lead tree – Small fast-growing mimosoid tree native to southern Mexico and northern Central America and is now naturalised throughout the tropics. Common names include white lead tree, jumbay, river tamarind, subabul and white popinac.

Phasey bean – Macroptilium atropurpureum, commonly referred to as purple bush bean or Siratro, is a perennial legume recognised by its climbing, dense green vines and deep purple flowers. The plant is indigenous to the tropical and subtropical regions of North, Central and South America, as far north as Texas in the United States and as far south as Peru and Brazil. It has been introduced for use as a food for stock to many tropical regions around the world. It has become an invasive pest plant in several areas, including the north-eastern coast of Australia (Queensland Department of Agriculture and Fisheries, 2016).

Table 2 Cover crop yield and forage quality.

Cover Crop	DM Yield (ton/ac)	Crude Protein (%)	NDF	Lignin (%)	TDN
Grasses					
Annual ryegrass	2.4	21.7	37.9	5.4	60.6
BMR sorghum-sudan	4.5	14.3	53.7	2.8	62.2
Forage oats	1.6	16.6	51.0	3.7	62.2
Grazing corn	6.4	13.4	32.7	3.3	48.4
Pearl millet	3.4	15.9	54.8	2.6	60.6
Sorghum-sudangrass	7.2	10.9	56.1	3.3	58.4
Teff	3.4	17.7	59.0	4.0	60.2
Legumes					
Berseem clover	1.1	22.4	38.5	6.6	60.9
Crimson clover	1.5	20.4	38.1	3.9	63.6
Forage peas	3.2	13.5	41.1	7.2	45.5
Lentils	0.6	14.8	49.8	4.8	52.2
Soybean	0.7	22.1	37.9	4.4	62.6
Sunn hemp	2.0	19.8	37.6	4.9	62.6
Others					
Buckwheat	1.7	13.6	42.4	7.3	58.0
Phacelia	0.4	21.4	34.2	4.2	63.7
Rox orange cane	10	12.7	51.3	3.0	63.2
Fodder beets	1.4	24.0	33.4	3.7	66.7
Sugar beet	3.1	21.7	29.3	3.3	68.6
Kale	1.4	23.2	39.0	4.5	65.2
Turnip	1.8	17.2	28.6	2.4	67.8

NDF = Neutral detergent fibre and TDN = Total digestible nutrients. *Source*: Brad Heins (2015).

Invasive Plant Siratro Macroptilium atropurpureum – Rich in protein, *M. atropurpureum* is commonly used for cattle pastures intercropped with grass, used in hay, or as a ground cover to prevent soil erosion and to improve soil quality (*Technical Report*, State of Queensland, Department of Agriculture and Fisheries).

Cowpea – Important in the diets of many societies. Many of the Vigna species are also valued as forage, cover and green manure crops in many parts of the world. Annual worldwide production of the various Vigna species is likely to approach 20 million ha and virtually all this production is in developing countries. The economic Vigna species exhibit several attributes that make them particularly valuable for inclusion in many types of cropping systems. They can be grown successfully in extreme environments (e.g., high temperatures, low rainfall and poor soils) with few economic inputs. Many of these species produce multiple edible products and these products provide subsistence farmers with a food supply throughout the growing season, as well as dry seeds that are easy to store and transport. For example, tender shoot tips and leaves of cowpeas can be consumed as soon as the plants reach the seeding stage and immature pods and immature seeds can be consumed during the fruiting stage. Harvested dry seed of all the Vigna crops can be consumed directly and seeds of several of the crops are commonly used to make flour or produce sprouts. Plant residues can be used as fodder for farm animals. Vigna food products exhibit many excellent nutritional attributes and these products provide a needed complement in diets comprised mainly of roots, tubers, or cereals.

2.4 Small Grains and Others

Rye, wheat and oats – They are mainly used in stocker programmes. However, rye and wheat are cold-tolerant than oats and can be widely grown across many different states. Rye gives an added advantage of producing more forage in late fall and late winter than wheat, but matures earlier in the spring. However, wheat will provide grazing about three weeks later in spring than rye. The growing season for oats and wheat is similar, but rye is the best choice for the land that will be ploughed in spring for a summer row crop because it matures in early spring. Wheat and oats are slightly more palatable than rye and cattle generally gain slightly faster than when grazing pure rye stands because rye can mature very rapidly. As a result, the rye forage quality can decrease very quickly. Triticale (a hybrid of rye-wheat) can be a good option, but it is not as grazing-tolerant and offers no substantive advantage over rye or wheat.

Buckwheat – When there is shortage of hay, buckwheat can be a good alternative forage crop. There is no question that cattle will happily eat buckwheat. In fact, buckwheat straw cannot be used as bedding for that reason. As for the forage value of buckwheat, if it is cut early in flowering (five to six weeks after sowing), the amount of protein is respectable (15–20 per cent) and digestibility is high. Longer periods to harvest decreases quality and does not increase yield substantially. The value of buckwheat as forage is comparable to the grain price if hay prices spike, so selling buckwheat as forage may be an option for buckwheat growers to consider in some years.

Buckwheat hay has good digestibility, but the protein content would need to be raised to make a balanced ration. For buckwheat growers, forage can be an option under certain conditions. First, if the crop has poor seed set evident in late August or early September, the forage value may be higher than the grain value. Low seed set at that time may happen if growth is especially lush, or if there have been prolonged high temperatures in early to mid-August; if hay prices are high because of local crop failures, using buckwheat for forage may be a good idea.

For dairy producers, buckwheat forage can substitute corn acreage that was not planted due to adverse weather. Sowing in the month of June, along with some nitrogen or manure application, can promote good vegetative growth. A double crop of annual forage is also possible. If buckwheat is planted in June for a late August harvest, then oats or triticale can be direct drilled for a fall harvest. However, in June there are plenty of attractive forage choices. A matter of concern regarding the use of buckwheat as cattle feed is skin rash, particularly on light-coloured cows, if they are fed a ration that is greater than 30 per cent buckwheat and they are in the sun (Bjorkman and Chase, 2019).

Annual ryegrass – Annual ryegrass is a highly productive cool season annual grass having excellent forage quality. This ryegrass is more productive on heavier soils (those with a high clay or loam content or moist low-lying soils) than on deep, well-drained sandy soils. In late spring, this forage is more productive and extends the spring grazing season. Ryegrass must be seeded in pure stands. It is necessary to mix ryegrass with rye and/or an annual clover so that high-quality forage can be maintained from late winter through spring (Table 3).

Table 3 The effect of a cool season annual mixture on stocker production.[†]

Average Daily Gain (lbs/hd/d)	Oats+[‡] Ryegrass	Ryegrass	Rye +Ryegrass	Triticale +Ryegrass	Wheat[§] +Ryegrass
Winter	1.2	0.7	1.4	1.1	1.2
Spring	2.5	2.6	2.4	2.1	2.4
Gain (lb/acre)	253	239	281	219	256
Cost of Gain ($/lb)	$0.29	$0.28	$0.25	$0.39	$0.28
Net Return ($/acre)	$110	$106	$144	$56	$115

† Adapted from Beck et al., 2007. J. Anim. Sci., 85: 536–544.

§ Stockers weighed between 500 and 575 lbs. Note that the stocking rate in this study began at 1.5 stockers/acre and additional calves were later added to maintain equal grazing pressure on each treatment (a research method called 'put-and-take'). In this study, grazing began in early winter (early January) and continued through early May in each system.

2.5 Warm-season Annual Forage Crops

Pearl millet – Pearl millet is a warm-season annual grass. Dwarf type of pearl millets, such as Tifleaf-3, have a higher percentage of leaves because of fewer stems and produce a good yield compared to the taller ones. It produces generally higher animal weight gains and is resistant to leaf spot diseases. Pearl millet is well adapted to sandy soils. Unlike sorghum-sudangrass hybrids and other members of the sorghum family, pearl millet does not cause prussic acid poisoning during periods of drought. Pearl millet yields quite well, even when subjected to drought or low soil pH. However, like all warm-season annuals, nitrate accumulation in drought-stressed crops may pose a significant risk to the health of ruminant animals that will graze them under such conditions. Pearl millet can be grazed or harvested at any growth stage. To optimise forage quality, however, grazing of pearl millet should start when plants accumulate 20–24 inches of growth and stockers should be removed when 6–12 inches of stubble remain. These rotational stocking methods also promote re-growth. Gains per acre greatly vary with growth conditions, grazing management techniques, condition of the animals, stocking rate and the number of days in the grazing period. In general, a stocking rate of 2–2.5 stockers (~ 600 lbs/stocker)/acre over an 80-to-100-day grazing period should be anticipated if rotational stocking is used.

Sorghum-sudangrass hybrids – Sorghum and sudangrass hybrids are high yielding and good in forage quality. Sometimes they also contain the brown-midrib trait. Varieties with the BMR trait have lower lignin levels, which can substantially increase the digestibility of their forage. However, none of the sorghum-sudangrass hybrids is as tolerant of high grazing pressure, low soil pH, or drought as pearl millet. This hybrid may pose a significant risk to stocker producers, since drought-stressed sorghum-sudan is not only at risk of toxic levels of nitrates, but it may also contain toxic concentrations of prussic acid (cyanogenic compounds). Prussic acid problems are also hazardous when the forage is under frost. In general, forage systems based on sorghum-sudangrass will provide slightly better ADGs than pearl millet-based forage systems. However, the maintenance of similar or higher gains/acre will require good growing conditions and excellent grazing management.

Crabgrass – Crabgrass is a warm-season annual forage. Though it is most widely known as a weed, it has excellent palatability and exceptionally high forage quality relative to other warm-season annuals and warm-season perennials. There is another distinct advantage of crabgrass—it readily re-seeds itself each year if it can produce a seed-head and mature. However, yields of crabgrass are quite variable, as they depend on the selection of a well-drained site, fertile soil and amount of rainfall. Forage yields for crabgrass generally range between 1–5 tons/acre, but one could expect yields to be 3–4 tons/acre. A stocking rate of ~ 1.5 stockers (~ 600 lbs/stocker)/acre could be expected from crabgrass. However, in trials from northern Florida, it found stockers grazing crabgrass gained 1.1 to 1.9 lbs/head/day. Research from other areas stated that one could expect ADGs of 1.5 to 1.8 lbs/head/day. The length of the grazing period for crabgrass ranges between 60–120 days. However, crabgrass appears to have the potential as a warm-season annual forage crop for stocker development, especially if rotationally stocked.

Others – Several other warm-season annual forages are forage sorghum, sudangrass, brown top millet and teff. Either because of poor yields, low quality, a pre-disposition to nitrate accumulation, or grazing management problems, these forage crops are generally not useful in stocker development programs and are not recommended.

2.6 Warm-Season Perennial Forage Crops

Hybrid Bermuda Grass – Hybrid Bermuda grass is a warm season perennial forage crop. Because of its high yield potential and, in some cases, increased digestibility, cattlemen grazing stockers should sow it. However, many years of selection, breeding and research have led to the release of several hybrid Bermuda grass varieties. However, few of these have shown to consistently provide high yields, increased digestibility and improved animal gains (Table 4).

From Table 2, we can see the best of the hybrid Bermuda grass varieties for stocker development is Tifton 85. Tifton 85 has been shown to produce the highest yield, digestibility, ADG, stocking rate, and gain-per-acre of any of the forage Bermuda grasses. Tifton 85 is clearly the best choice for new pastures for cattlemen in the Coastal Plain. Unfortunately, Tifton 85 lacks the cold tolerance of some other hybrid Bermuda grasses. Other hybrid Bermuda grasses such as Tifton 44, Russell, Tifton 78 and Coastal are used in cow-calf production systems and can also be used in stocker development systems. However, substantially more supplemental feeding will be necessary (relative to that on Tifton 85 pastures) to attain satisfactory ADG, stocking rate and gain-per-acre of stockers grazing these other varieties.

Table 4 Stocker performance on 'Pensacola' Bahia Grass and Preferred Bermuda Grass varieties in selected research trials in the coastal plain.

Species	State	ADG (lbs/hd/d)	Gain (lb/ac)	Stocking rate (hd/ac)	Grazing period (day)
Pensacola (Bahia)	GA[1]	1.0	222	1.5	131
Coastal	GA[1]	1.1	331	2.5	131
Coastal	TX[2]	1.0	279	3.0	92
Coastal	GA[3]	1.5	641	2.5	168
Tifton 44	GA[3]	1.6	681	2.5	168
Tifton 78	GA[4]	1.4	704	3.2	169
Tifton 85	GA[4]	1.5	1032	4.4	169
Tifton 85	TX[2]	1.7	465	3.0	92

[1] Utley et al., 1974. J. Anim. Sci., 38: 490–495.
[2] Rouquette et al., 2003. Beef Cattle Research in Tx. pp. 62–66.
[3] Utley et al., 1981. J. Anim. Sci., 52: 725–728.
[4] Hill et al., 1993. J. Anim. Sci., 71: 3219–3225.

Bahia grass – Bahia grass is well adapted to a wide range of soils in the Coastal Plain region and persists greatly under the hot, dry summers. However, it will not consistently support the stocking rate and live-weight gains per acre that hybrid Bermuda grasses can provide. Even improved Bahia grass hybrids produce lower yield, lesser digestibility, ADG, stocking rate and gain-per-acre than most Bermuda grass varieties. Besides, the rate of supplemental feed required to attain satisfactory ADG, stocking rate and gain-per-acre usually makes stocker development programmes on Bahia grass unprofitable. As a result, Bahia grass is not recommended as a forage system for stocker development (Hancock et al., 2014).

Forage turnips and oilseed radishes – Forage turnip is a member of the *Brassica* family that does best if planted in early fall. Although it does not provide as vigorous growth as some oilseed radish varieties, turnips are superior for grazing. The leafy top growth and the tuber are also good forage. Turnip top growth generally dies by late December. Turnips can be mixed with another forage species, such as cereal grass, for good results. Oilseed radishes are a popular cover crop. Radishes exhibit rapid fall growth, have a deep taproot, and provide good fall soil coverage (if planted early enough). They establish relatively easily when seed is broadcasted, which is good because radishes often need to be aerial seeded in late August or early September to achieve adequate fall growth. However, the residual herbicides applied to the preceding commodity crop must be reviewed to make sure none will affect radish establishment.

Oats and triticale – Oats are widely available and generally spring-planted for grain. Oats have the advantage of strong fall growth as a cover crop. Many first-time users of cover crops appreciate that oats will winter kill and require no special management in spring. However, the erosion control provided by oats in late winter and spring is less than that from overwintering cereals such as cereal rye, triticale, and wheat. Triticale is a cross between wheat and cereal rye, possessing some of the hardiness of rye and the shorter stature of wheat. Considered a good cover for erosion control and grazing, it also works well in combination with other cover crops, particularly legumes.

Annual ryegrass – Annual ryegrass is a completely different plant than cereal rye. It has fine blades and smaller stature, more like a turf grass. It is fast growing and easier to establish by broadcast method than other grass cover crops. Annual ryegrass needs to be planted earlier than cereal rye to achieve good fall cover. Ryegrass is conducive to inter-seeding applications in corn or soybean, as it is relatively shade tolerant. Annual ryegrass turns brown over the winter, but depending on planting date, location and winter conditions, will generally re-grow in the spring. Where it does overwinter, timely herbicide application in spring is needed to terminate it. Use of a cover crop-specific variety makes spring termination easier. Annual ryegrass makes good forage and, as a cover crop, it is known for rooting deeply in the soil and providing good erosion control.

3. Grazing Management Under Forage Crops

Grazing management can influence forage growth and utilisation, as well as animal performance. Limited grazing can begin in the fall as soon as the plants are well established and have 6–8 inches of accumulated growth. This ensures that root development is sufficient to prevent grazers from plucking the plant from the soil. Limited early grazing will improve tillering and increase stand density. However, it is critical that the pastures are not overgrazed during the early grazing period (i.e., maintain at least 2.5–3 inches of stubble height). This is also important in late winter when pastures start to recover from extreme cold. Allowing some re-growth to occur before putting significant grazing pressure on the pasture will significantly improve spring forage production. Therefore, achieving the proper balance between cattle stocking rate and forage growth rate is quite difficult. Forage growth varies during the growing season with changes in temperature and moisture conditions. The correct number of animals per acre in one week may be far too many the next. To best utilise the forage that is grown, plan to provide supplemental feed and/or conserved forage during periods of slow pasture growth so that pastures will not be overgrazed.

Another way to prevent damage to late-fall and winter pasture is to implement a rotational grazing programme. Rotational grazing systems (such as management-intensive grazing or MiG) allow the forage crop to better recover before being grazed again. Further, rotational grazing can substantially increase utilisation efficiency of forage (i.e., more of the forage that is produced ends up being consumed by the grazing animals) and this can increase the stocking rate that the forage system can sustain.

Another strategy to control grazing more tightly is a method called 'limit grazing'. Limit grazing is a system by which the animals are only allowed a brief opportunity to graze (usually one to two hours). Limit grazing works best when the cattle are allowed access at strategic times during the day. Cattle generally consume large quantities of forage in the morning (~ 6 to 8:30 a.m.) and mid-afternoon (3–5 p.m.) with a smaller bout around the time the sun sets. Timing a limit-grazing bout to align with one or more of these natural grazing behaviours during a day can allow the animals to obtain much of their diet from the available pasture, while minimising hoof traffic and other damage to the stand. Of course, this assumes that one has another pasture or lot and enough conserved forage and feed for the animals when they are not present in the limit-grazed pasture.

Regardless of the grazing system, it is important to measure how much forage is to offer, monitor the growth rate of the forage and manage how much forage is allocated to the herd. The forage can easily be measured using a grazing stick or rising plate meter. This data can then be entered into a spreadsheet that can display the total forage in each pasture or paddock and the growth rate. Thus, forage requirement and usage can be measured easily.

4. Livestock and Cover Crop Performance

Cover crop pastures and livestock are compliments to each other. A need has arisen for legume-based cover crops that maintain long-term soil fertility for succeeding crops because a periodic legume pasture serves to maintain an individual field in early stages of succession, lending stability to later-planted crops (Tothill, 1978).

Ley pasture also contribute to the availability of high-quality forage for livestock through either haying or grazing. However, haying must be considered a harvest because both the biomass and nutrients are removed from the field, where grazing is more complex. The grazing animal's removal and rapid cycling of the forage and all the effects that come with it, can exert additional beneficial effects if properly managed.

Wallace (1892) wrote nearly a century ago, "The Western farmer has now reached a point where, willing or not, he must elect to do one of three things: (1) continue his present robbery of the soil by continuously growing grain for sale in the world's markets and thus selling his land by piece-meal; (2) he may, by supplying nitrogen in the clovers and returning nothing in the form of manure, rob it more completely and reduce it to a more hopeless barrenness; or (3) he may draw on the winds of Heaven, by means of the miracle-working tubercle in the roots of clovers, and then by the judicious use of the manure made on the farm in various ways, restore the potash and phosphoric acid, trusting to the gradual disintegration of the rocks of which the soil is composed to keep up indefinitely their miracles aside." Investigators have since demonstrated the importance of P and K fertility for legumes around the globe. There is, perhaps, no better example of P and K fertilisers permitting the adoption of leguminous cover crops than in southern Australia (Webber, 1990) and New Zealand. Livestock manure remains another viable alternative to meeting the soil-fertility requirements of intensive use of cover crops. That ruminant manure contains the complete range of nutrients that plants require and roughly in the same proportions, should come as no surprise when both plant and animal are thought of as evolving from the same ecosystem.

As the ruminant consumes the cover crop, it also becomes included in the nutrient cycle of the field. In reviewing Australian pasture research, Hilder (1969) reported the nutrients retained by grazing ruminants are 25 per cent with cattle and 4 per cent with sheep. Researchers have found recently in North Dakota, of all the plant biomass ingested, beef cattle retained 28 per cent and sheep retained 15 per cent in confinement-reared animals where the forage was fed instead of grazed. These studies do not account for the mineral supplementation of the animals, which provides an additional supply of many minerals, including P, to the nutrient cycle and eventually the succeeding crops. While the nutrients retained are of importance to animal productivity, nearly 75 per cent or more of the nutrients consumed are returned to the field. This cycling of nutrients is of great importance in maintaining soil fertility and crop productivity. After passing through the animal, however, the stability and plant availability of the nutrients is changed.

Russell (1913) calculated half-life for humus from ryegrass at four years of age, N from farmyard manure at 25 years, soil nutrients in the prairie after being cropped at 10–45 years, and humus from un-manured field plots that are 600–1,700 years old. Such data emphasise the great differences in time among various nutrient cycles, especially when animals are involved. However, seemingly contradictory, a portion of the nutrient pool is also more quickly cycled on passing through ruminants. Nitrogen contained in the cover crop's leaves, for example, can be consumed and excreted in days. As such, N has again been made plant-available and greatly reduced the time necessary to be cycled. Solid and liquid animal wastes differ greatly in their elemental composition and immediate plant availability (Barrow, 1987). Solid wastes contain all the P, some stable forms of organic and inorganic N, and most of the minor nutrients. Urine contains mostly N, K and S (Hilder, 1969). We can readily observe evidence of such nutritional differences in pastures because legumes usually are stimulated near solid waste patches, while grasses are stimulated near urine. Therefore, nutrient cycling through manure can also be greatly influenced by the density of the grazing animals. Studies show that, at normal stocking rates of one animal unit/ac., pastures received little benefit from the animals because both liquid and solid waste was contained in small, dispersed patches (Peterson et al., 1956a,b). Hilder (1969) further reported that sheep may be of less use than cattle because they tended to congregate more and concentrate wastes in loafing areas. He also concluded that the fertility improvement of grazed, short-term pastures was caused more by the sheep's preference for the grasses, which would increase total legume growth and N function. The question of groundwater contamination from nutrients and the source of those nutrients, has been a source of controversy. Commoner (1970) expressed the first major concern over the contribution of fertiliser N to water quality problems. Nitrate concentrations in groundwater under forests, unfertilised pastures and grasslands generally are cited as less than 2 mg/l NO_3 and often less than 1 ppm. However, NO_3 concentration under fertilised crops and animal production areas are commonly more than 5 mg/l and have been reported as high as 100 mg/l (Pionke and Urban, 1985; Beck, 1985).

Grazing management, such as short duration grazing techniques, may hold promise for better manure distribution, and unlike permanent or semi-permanent pastures, short-term hay or cover crop pastures are rotated to cash crops. Shallow tillage operations may help distribute nutrient-rich patches for more efficient crop uptake and use. It has been found that inappropriate fertiliser applications are also sources of NO_3 contamination in pasture situations. During a five-year period, investigators monitored runoff from pastured watersheds on hillsides in eastern Ohio for water quality. In 7 per cent of the events, NO_3 concentrations exceeded 10 ppm, and 48 of the 64 events occurred within a three-day period following N application on the watershed. Owens et al. (1986) concluded that the closeness of the high NO_3 concentrations and fertilisation suggests that the fertiliser, and not animal manure, was the major contributor of NO_3 in this situation. Continuing the debate on the source of contaminants, however, may not be as useful as coming to a more thorough understanding of nutrient movement and leaching processes (Cameron and Haynes, 1986). Proper management, application timing and rates determine whether livestock manure is a soil amendment or a soil contaminant. If properly managed, cycling nutrients through livestock holds the potential to conserve cover crop nutrients for later crop uptake.

Grazing can also help manage a cover crop's water use, alleviating producers' concerns over competition for soil moisture between cover crops and cash crops. In the U.S. Great Plains, green-manure substitutes for fallow repeatedly have been discouraged due to high water use (Army and Hide, 1959), so there exist several approaches to limit water use to a tolerable level. Sims (1989) has developed in Montana a threshold level of water use, above which continued growth of the cover crop will be detrimental to wheat planted after the cover crop. While killing the cover crop with a tillage operation or chemicals is one possible solution, water use also can be limited by lessening leaf area periodically. Water use of cover crops in drought-prone areas can be managed to desirable levels by mowing or grazing, while retaining the presence and growth of the cover crop. Water use may not be as critical to succeeding crops in more humid areas. Under some situations, the cover crop's principal purpose may be to provide an actively transpiring surface to prevent the downward movement of mobile soil nutrients. Winter cereals, particularly rye, have been used most frequently in such situations and grazing management would have to reflect the needs of a rapidly transpiring cover crop. There is evidence that grazing can improve growth rate and thus maintain maximum water use, if producers achieve an optimal leaf area before grazing begins (Myers, 1969). The rate of grazing must then be managed to balance leaf removal with leaf regeneration capability. Under dense canopies, grazing may improve light penetration and increase overall interception of radiant energy by the transpiring leaves.

Though weed suppression is one of the many critical aspects of cover-crop management, teamed with grazing ruminants, weeds still can be managed selectively. Particularly where one or few plant species are present in the pasture, weeds are often preferred browsing. Forwood et al. (1989) found weed consumption to be greatest in cattle steers when same pastures of various grass-legume mixtures were least variable in composition. However, what motivates livestock to select certain plants is unknown, although researchers have stated that such selectivity is based upon the animal's need for a diverse diet (Arnold, 1982). Grazing animals also can introduce weeds to the pasture through seed dispersal. The proportion of seeds that pass through a ruminant is a function of the digestibility of seeds and their ability to remain viable (Watkins and Clements, 1978). These factors vary by plant species, but usually less than 10 per cent of ingested seeds remain viable. However, undesirable and foreign weeds, that are introduced by livestock, require attention and selective management.

Generally, cover crops are needed for surface cover, but sometimes too much surface residue also can be detrimental because it may prevent adequate seed placement in no-till planting operations or can reduce the soil temperature to the point of inhibiting early season crop growth (Wall and Stobbe, 1984). Integrating ruminant livestock into the system can offer an alternative management option to deal with each of these problems. Intensive grazing can help operations such as mowing, tilling, or using herbicides to control vegetation while making the transition to the cash crop.

5. Novel Grain–Forage–Livestock Systems

Here, we will discuss how novel forage-based livestock-cropping systems are successfully running in different parts of the world. This type of system has been used for decades in Australia (Grace et al., 1995), where self-regenerating subterranean clover and annual medic are growing in pasture-grain systems. There has been considerable interest in adapting these systems to the U.S. Northern Great Plains region (NPG). Sims and Slinkard (1991) concluded that black medichad potential for replacing summer fallow in a wheat-fallow cropping system in Montana. Long-term field trials demonstrated that 'George black' medic (Sims et al., 1985) success fully reseeded itself and boosted wheat yields by 1,300 kg ha^{-1} compared to wheat on summer fallow. In this system, black medic can be grazed during the fallow year. Self-regenerating annual medics can also be integrated into continuous grain production systems. Three annual medic species were established on a 40 ha farm in North Dakota in 1991. Thiessen-Martens and Entz (2001) determined that a large area of the NGP has enough heat and water resources for late-season growth, including seed production of several medic and sub-clover species.

6. Models of Crop Livestock Systems

Theoretically, crop and livestock-production systems seem to be mutually beneficial. Still, the separation of both into specialised production units has been the trend in most developed countries. Studying specific cases where the mutual benefits have been demonstrated may aid in the discovery of new cover crop-livestock systems that could be developed. As Puckridge and French (1983) and Webber (1990) describe, the situation of southern Australia in the late 1940s has a great resemblance to much of today's world, for example:

1. High demand of livestock product and high price of grains.
2. Falling cereal yields and increasing dependence upon N fertiliser.
3. Widespread soil erosion.
4. Farmers and the government are becoming concerned about environmental decline.

What followed was a switch from cropping systems without integrating livestock and rotation to systems in which rotation to leguminous cover crops depended economically upon the income from grazing sheep. This began another era in the evolution of Australian agriculture and these simultaneously raised the production of both crops and livestock. Furthermore, these new crop-livestock systems also resulted in increased crop water-use efficiency, better soil structure and greatly reduced dependency on mining soil nutrients, especially N.

Current limitations to a similar worldwide revolution may hinge on three critical factors: (1) an adequate breeding program specifically in search of grazing cover crops; (2) willingness and confidence in livestock management and markets; and (3) a coordinated, long-term government policy that may encourage the use of cover crops for fallow.

Here is an example of collaborative works of North Dakota State University, together with the Michael Fields Agricultural Institute, the University of Nebraska and the Kansas State University, which currently are evaluating cover crop systems that could substitute for fallow under the USDA's low-input sustainable agriculture programme. There, initial evaluations have concentrated on yellow blossom sweet clover, black medic and hairy vetch, although more than a dozen were introduced and native species have been tried. Nevertheless, no single legume seems to possess all the necessary traits for a broadly adapted Great Plains cover crop. Under these conditions, ease of stand establishment and seed cost are important features, along with low or easily managed water use and general pest resistance. In North Dakota and Nebraska trials, alternative legumes, such as black medic, have used less soil water than traditional legumes, like sweet clover, particularly at soil depths greater than 12 inches (Guldan et al., 1990). Both the sweet clovers and the medics seem to possess the diversity of germ plasm needed to breed new genotypes specifically as cover crops suited to regional conditions. However, no current programme exists. Developing a cover crop with the seed vigour and winter hardiness necessary for persistence, corn bed with low water use and a small enough seed size to remain economical is the current challenge.

Centralised feeding and processing have concentrated much on beef cattle, leaving most of the region largely deficient in weaned calves. Because the soils of this region could greatly benefit from retaining the cover crop biomass produced, haying does not seem a viable option here. However, many producers are reluctant to obtain either cattle or sheep to graze ley pasture because of a lack of experience and problems in time management during critical crop establishment periods. Recent comparisons of crops-only versus crop-beef cattle operations found that livestock can increase the total labour required on an average central North Dakota farm by 56 per cent, but only one-third of that additional time directly competed with crops during critical management periods. Net economic returns attributable to the added livestock increased whole farm income by 19 per cent. Despite the economic returns possible from adding livestock, given current markets, the required labour and management expertise needed with livestock still possess a barrier to broad-scale adoption. In most regions of the United States, producers must own the livestock for cover crop grazing and management. In contrast, contract grazing of ley pastures from nomadic sheepherders is a common practice in southern Australia. U.S. agricultural policy also must be considered when analysing the limitations to broad-scale use of cover crops for fallow in the plains. Mainly a wheat-producing region, plains farmers have been encouraged to grow wheat to keep global supplies adequate. It is unfortunate that the frequent encouragement of set-aside, or fallow, to help control supplies in the past few years was not coupled with encouraging the establishment of legitimate cover crops. Though such ley pastures have been established in some regions, they usually have had some restrictions for use as pasture or forage. With a limited demand for forages, making available government subsidised forage or pasture production has been perceived to be economically unfair to the unsubsidised forage producer. Thus, balancing the economic needs of individuals with the ecological needs of the landscape is going to be a global issue that must be resolved.

7. Holistic Crop Livestock Models

The importance of the link between plants and animals is at the core of both biodynamics and holistic resource management practices. According to Koepf (1981), this biodynamic movement might be considered one of the first organized attempts at reforming conventional 20 century agricultural production practices. The teachings were broad and all-encompassing, outlining ecological, economic, social and even spiritual changes that were suggested on the farm. However, the central theme was based upon the belief that an integrated crop-livestock production system is necessary for long-term soil fertility. The concepts of a well-planned crop rotation, occasional green manuring and the ability of composted livestock manure to replenish stable soil organic matter and nutrients sounds all too familiar given the current interest in 'alternative' agriculture (NRC, 1989).

Biodynamic methods emphasize the need for cultural practices that may promote net gains in the nutrients contained within the soil-plant system. While not excluding the use of synthetic fertilizers to do so, biodynamic practices encourage the use of ruminant livestock, particularly cattle, to transform the nutrients into forms that can be retained within this soil system. However, biodynamic principals have yet to be, and may never be, examined thoroughly by disinterested third party scientists, and the recognition and relative success of integrating crop and livestock production is noteworthy. Particularly where leguminous cover crops are warranted, the increased need and expense of P, K, and other nutrients should be considered. Careful and appropriate management of livestock manure, as demonstrated in decades-old biodynamic farms, could be at least part of the solution to these long-term fertility needs.

'Holistic resource management', a term coined and promoted by Savory (1988), is a more recent example of agricultural management strategies that link plant and animal performance. It is a goal-oriented system largely employed in permanent pasture and range situations that has challenged much of conventional range management thinking. It is an approach that views plants, animals, humanity, etc. as one single ecosystem functioning through four rudimentary processes (i.e., succession, the nutrient cycle, the water cycle and the flow of solar energy). Thus, the increasing interest in short-duration, rotational and multispecies grazing can be attributed, at least in part, to the concepts of holistic resource management. It has drawn attention to the connection between grazing animal activity, soil physical conditions and plant performance. While mostly seen as a system practiced on range plant communities, the concepts speak to agro-ecosystems as well. This management practice would suggest that crops in polyculture and integration with animals would be more productive and ecologically stable than monocultures without animals. Therefore, this type of theory will suggest a decreasing need for subsidies in the form of energy, fertilisers and pesticides. Here, biotic regulation would replace such subsidies within smaller field units of increasing biological and ecological complexity.

Holistic resource management theory would be able to contribute to the integration of cover crops and livestock in agriculture, but not before practical applications are well thought out. In most of farm situations, the time and cost of employing holistic resource-management concepts in using cover crops will be calculated using short-term economic efficiency and is often over-emphasised and the mutual benefits of managing the cover crop and the grazing ruminant in a same piece of land may not be fully realised. Thus, theory must be able to be used practically.

8. Adding Value to Beef and Dairy Products

Ultimately, we must find new ways to divert land out of grain production. Backgrounding beef on predominately forage rations (often pasture) before entry into the main feedlot system, by growing animals at relatively low rates of gain, is a means of providing a low-cost animal to the feedlot system. This part of the beef system is predicated on forage or pasture being a low-cost feed stuff (Mathison, 1993).

The immediate advantage of pasture finishing is the potential of relatively low cost of beef production, although this must be weighed against the cost of grain during times of low grain prices considering certain supply-demand relationships for beef markets. Research in Canada and the United States since the 1950s has reported that pasture-finished beef is feasible (Aalhus and Mandell, 1999). Problems with meat quality and consumer acceptance of pasture-finished beef, such as off-flavour (Mandell et al., 1998) and discoloured fat (Aalhus and Mandell, 1999) have occurred. However, some traits are confounded with respect to age of cattle and fat cover when forage vs. grain-finished beef are compared (Mandell et al., 1998). McCaughey and Cliplef (1995) reported that about 70 per cent of cattle finished on high quality pasture would meet the standards for finished beef on the Canadian grading system. This appears typical. The remaining 30 per cent finished after relatively short periods (30–75 days) on a high grain ration. Pasture-finished beef in this trial met standards for fat colour and was acceptable in taste panel studies for tenderness and flavour.

Other niche markets may develop for forage or pre-dominantly forage-finished beef based on enhanced human health. Forage-based rations are linked with relatively high concentrations of conjugated linoleic acid and omega-3 fatty acids in meat (Aalhus and Mandell, 1999) and dairy products (Jiang et al., 1996). Meat and dairy products may contain 1–2 per cent linoleic acid.

Health benefits derived from linoleic acid include anti-carcinogenic, anti-cachexic and anti-atherosclerotic properties (Aalhus and Mandell, 1999). Omega-3 fatty acids, also found in fishmeal (Mandell et al., 1997) as well as linolenic, eicosapentaeoic and docosahexaenoic acids, are enriched in meat from forage-finished rations (Aalhus and Mandell, 1999). While omega-3 fatty acids appear to have a role in the prevention of many age-related diseases, their role in mitigation of coronary heart disease has been most extensively studied and verified (Addis and Romans, 1989). In the future, forage-fed dairy and pasture-finished beef may have a role in niche markets for a health-conscious and aging society. More research is highly required to further elucidate these relationships.

9. Conservation Impact of the Forage

Animals feed mostly on crop residues in drier months of the year. Apart from the leguminous crop residues, such as groundnut haulms and cowpeas, other crop residues like sorghum, millet rice, wheat, maize and cotton are low in quality. Though groundnut and cowpea hays are used widely, they do not meet total requirements. There is a place in the farming system for legume crops grown specifically for hay and silage. Legumes that have shown some value for conservation are groundnuts, soybeans, cowpea, Mucuna or velvet bean and lablab. These legumes, if properly used in conjunction with improved native pasture, sown pasture and browse plants, will help reduce the heavy livestock live weight that are common in the dry season. So far, crop farmers are leading the way in forage legume feed conservation. Pasture agronomists have paid little attention to this area of feed conservation thus far. There is, therefore, an urgent need for research work to be carried out in fodder crop conservation as a source of feed for livestock (Nuru and Ryan, 1985).

The economic estimates above charge all the establishment costs of the cover crop to the cattle enterprise. However, the possibility of potential soil conservation benefits does not consider the stakeholders at all levels. The impact of grazing on soil conservation and the benefits of cover crops may depend on grazing management. Mismanagement in grazing, such as excessive rate biomass removal, high stocking densities and grazing during wet conditions can lead to decreased cover, may harm soil structure, and/or affect infiltration rates (Warren et al., 1986). On the other hand, with proper grazing management, these negative effects are diminished (Twerdoff et al., 1999; Agostini et al., 2012; Franchin et al., 2014; Blanco et al., 2017). Though haying reduces the amount of residue left on the soil surface, cover crops are still able to protect soil from erosion (Blanco-Canqui et al., 2013). Franzluebbers and Stuedemann (2008a,b) reported that in Georgia in the United States, cattle grazing around 90 per cent of the forage produced by cover crops for 2.5 years had minimal effects on soil physical properties in two cropping systems (i.e., corn or sorghum with a cereal rye cover crop and winter wheat with summer pearl millet cover crop under no-tillage and conventional tillage system). They also found that grazing of cereal rye and pearl millet did not affect soil bulk density of soil aggregates when compared with and without grazing the cover crop.

Grazing also has additional positive contributions on soil microbial biomass because soil microorganisms are essential to the sustainability of any cropping system. It also helps an agro-ecosystem through carbon recycling and nutrient management (Falkowski et al., 2008). Grazing can also affect microbial abundance (Franzluebbers and Studemann, 2015) and community interactions through increased manure inputs, which provide more easily decomposable organic matter than the plant material supplied in non-grazed systems. Besides, penetration resistance is used to understand soil compaction and surface crusting, which can potentially influence root growth and water movement within the soil. Therefore, previous studies suggest that grazing cattle on cropland in winter and early spring may increase penetration resistance of the soil (Clark et al., 2004; Franzluebbers and Stuedemann, 2008a; Radford et al., 2008; Fae et al., 2009).

In Ohio, Fae et al. (2009) noted that grazing of annual ryegrass and a mixture of winter rye and oats, managed under a no-tillage corn silage system, increased soil penetration resistance by 7–15 per cent the first year, but after one-year penetration resistance values decreased to levels similar to non-grazed. In addition, grazing of cover crops did not affect subsequent corn silage yields. A long-term study for nine years in North Dakota found that winter grazing of annual crop residues had no effect on water infiltration rates in the spring (Liebig et al., 2011). Much of the soil organic matter benefits of cover crops in no-till systems have been extracted from the retention of plant root. A study conducted in Canada (Mapfumo et al., 2002) examined grazing intensity on root growth in triticale cultivation. In this experiment, grazing intensity was based upon plant height. Data was taken at the start and end of grazing, with grazing being initiated at 11, 12 and 21 cm and ended at 3, 4, and 6 cm for light, moderate, and heavy grazing intensities. These researchers reported that root mass production did not differ because of grazing intensity. It has found that root mass production was almost 4,000 kg/ha in the top 30 cm of the soil profile with about 60 per cent of the root mass in the top 15 cm.

Kierkegaard et al. (2015) found in their experiment in Australia, that grazing had no substantial effect on rooting depth of root mass in wheat unless the wheat was grazed during early growth, when the primary root system was still developing. Thus, both studies suggest that properly managed grazing has little impact on root growth of small grains. Therefore, it is safe to assume that properly managed grazing of cover crops would not negatively affect the soil organic matter benefits of cover crops. Another study found that grazing of cover crops removes plant carbon, but recycling of carbon through manure enhances carbon accumulation in the soil (Soussana et al., 2004; Blanco-Canqui et al., 2015).

10. Potential Barriers

Cover crop as forage is becoming a good option for obtaining some crop yield, grazing of your livestock on the same field and as maintenance of soil health on the farm. However, this practice is not being widely adopted, so there must be some reasons behind this. Most farm operations are no longer integrated with individuals having knowledge and expertise in a field (specialization in either crop or cattle production). For these integrated crop-cattle systems to happen on a larger scale, partnership is highly needed. When considering the use of cover crops for forage, there is a need to consider management practices, including planting dates of both the cash crop and cover crop, harvest date, fertility management and herbicide use, as all of these will affect the whole system. Moreover, there are concerns over weed management in the cover crop. This has happened with both cereal rye and annual ryegrass. Recently, it was reported that cover crops also serve as hosts for pests, which is serious matter of concern (McMechan et al., 2017).

Lastly, grazing cover crops can be used to cut production costs and, in many cases, result in economic returns exceeding the costs of establishment. The limited literature available suggests that grazing cover crops will not hamper the soil benefits of adding cover crops into a cropping system. In fact, grazing may have additional positive contributions on soil microbial biomass through manure addition. Therefore, there are potential opportunities for both economic incentives and soil benefits for crop producers to incorporate grazing of cover crops into their agro-ecosystems. However, incorporating forage production into current cropping systems greatly increases the need for timely management since the window of opportunity for forage production is quite narrow (Drewnoski et al., 2018).

11. Conclusion

Forage crops and pastures will provide the bedrock for a sustainable agriculture system. The edible parts of plants, other than separated grain, can provide feed for grazing animals or can be harvested for feeding (Allen et al., 2011). Therefore, forages play an important role in the cattle industry while enhancing crop diversity, wildlife habitat and soil ecosystem services within a single round. Though limitations exist there, cover crops, being young, temporary vegetative covers, can make excellent quality forage. In addition, the details, time and skill required to manage both crops and livestock are obvious adoption barriers to seeing cover crops as pasture. It might be the next step, however, in the re-discovery of the benefits lying within our undomesticated and native ecosystems. So, will we be equally willing to deal with the complexities of cycling at least a portion of it locally, including the ruminant animals within the system? The answer to this question will define the future of the spread of cover crops as forage across the world.

References

Aalhus, J. and Mandell, I. (1999). Nutritional effects of beef quality. pp. 48–59. *In*: Korver, D. and Morrison, J. (ed.). Proc. Western Nutrition Conf., 20th Calgary, AB, Canada, 16–17 September 1999, Alberta Agric., Food and Rural Dev., Edmonton, AB, Canada.

Addis, P.B. and Romans, J. (1989). The omega-3 fatty acid story. Proc Annu. Reciprocal Meat Conf., Am. Meat Sci. Assoc., 42: 41–47.

Agostini, M.A., Studdert, G.A., Martino, S.S., Costa, J.L., Balbuena, R.H., Ressia, J.M., Mendivil, G.O. and Lazaro, L. (2012). Crop residue grazing and tillage systems effects on soil physical properties and corn (*Zea mays* L.) performance. J. Soil Sci. Plant Nutr., 12: 271–282.

Allen, V.G. (2011). An international terminology for grazing lands and grazing animals. Grass and Forage Sci., 66: 2–28.

Allison, J.R. and Ott, S.L. (1987). Economics of using legumes as a nitrogen source in conservation tillage system. *In*: Power, J.F. (ed.). The Role of Legumes in Conservation Tillage Systems, Soil Cons. Soc. Am., Ankeny, Iowa.

Alves, A.L., Denardin, L.G.O., Martins, A.P., Anghinoni, I., Carvalho, P.C.F. and Tiecher, T. (2019). Soil acidification and P, K, Ca and Mg budget as affected by sheep grazing and crop rotation in a long-term integrated crop-livestock system in southern Brazil. Geoderma, 351: 197–208.

Army, T.J. and Hide, J.C. (1959). Effects of green manure corps on dryland wheat production in the Great Plains area of Montana. Agron. J., 51: 196–198.

Arnold, G.W. (1982). Grazing behavior. *In*: World Animal Science, BI: Grazing Animals, Elsevier, New York, N.Y.

Bade, D. and Dorsett, D.J. (2002). Forages for beef cattle, Texas A and M Agri-life Extension.

Barrow, N.J. (1987). Return of nutrients by animals. *In*: Ecosystems of the World, I78, Managed Grassland Analytical Studies, Elsevier, Amsterdam, The Netherlands.

Beck, B.F., Asmussen, L. and Leonard, R. (1985). Relationship of geology, physiography, agricultural land use and ground-water quality in southwest Georgia. Ground Water, 23: 627–634.

Beck, P.A., Stewart, J.B., Phillips, J.M., Watkins, K.B. and Gunter, S.A. (2007). Effect of species of cool-season annual grass inter seeded into Bermuda grass sod on the performance of growing calves. Journal of Animal Sci., 85: 536–44.

Bjorkman, T. and Chase, L. (2019). Cornell Co-operative Extension.

Blanco, H., Drewnoski, M.E., Burr, C., Lesoing, G., Williams, T., Redfearn, D.D. and Parsons, J. (2017). Does grazing or harvesting of cover crops affect soils and crop production? Assessment in Different Soil Types and Management Scenarios.

Blanco-Canqui, H., Holman, J.D., Schlegel, A.J., Tatarko, J. and Shaver, T. (2013). Replacing fallow with cover crops in a semiarid soil: Effects on soil properties. Soil Sci. Soc. Am. J., 77: 1026–1034.

Blanco-Canqui, H., Hergert, G.W. and Nielsen, R.A. (2015). Cattle manure application reduces soil's susceptibility to compaction and increases water retention after 71 years. Soil Sci. Soc. Am. J., 79: 212–223.

Boehncke, E. (1985). The role of animals in a biological farming system. *In*: Edens, T., Fridgen, C. and Battenfield, S. (eds.). Sustainable Agriculture and integrated Farming Systems, Mich. State Univ. Press, East Lansing.

Brandt, R.T. and Klopfenstein, T.J. (1986). Evaluation of alfalfa-corn cob association action. Comparative tests of alfalfa hay as a source of ruminal degradable protein. J. Anim. Sci., 63: 902–910.

Cameron, K.C. and Haynes, R.J. (1986). Retention and movement of nitrogen in soils. *In*: Mineral Nitrogen in the Plant-Soil System. Academic Press, Inc., Orlando, Fla.

Clark, J.T., Russell, J.R., Karlen, D.L., Singleton, P.L., Busby, W.D. and Peterson, B.C. (2004). Soil surface property and soybean yield response to corn stover grazing. Agron. J., 96: 1364–1371.

Commoner, B. (1970). Threats to the integrity of the nitrogen cycle: Nitrogen compounds in soil, water, atmosphere and precipitation. *In*: Singer, S.F. (ed.). Global Effects of Environmental Pollution, Springer Verlag, New York, N.Y.

Cox, G.W. and Atkins, M.D. (1979). Agricultural Ecology, W.H. Freeman and Co., San Francisco, CA.

Drewnoski, M., Parsons, J., Blanco, H., Redfearn, D., Hales, K. and MacDonald, J. (2018). Forages and pastures symposium: Cover crops in livestock production: Whole-system approach. Can cover crops pull double duty: Conservation and profitable forage production in the Midwestern United States? J. Ani. Sci., 28: 3503–3512.

Entz, M.H., Baron, V.S., Carr, P.M., Meyer, D.W., Jr. Smith, S.R. and McCaughey, W.P. (2002). Potential of forages to diversify cropping systems in the Northern Great Plains. Agron. J., 94.

Fae, G.S., Sulc, R.M., Barker, D.J., Dick, R.K. and Eastridge, M.L. (2009). Integrating winter annual forages into a no-till corn silage system. Agron. J., 101: 1286–1289.

Falkowski, P.G., Fenchel, T. and Delong, E. (2008). The microbial engines that drive Earth's biogeochemical cycles. Sci., 320: 1034–1039.

Farney, J.K., Sassenrath, G.F., Davis, C. and DeAnn, P. (2018). Growth, forage quality, and economics of cover crop mixes for grazing. Kansas Agricultural Experiment Station Research Reports, 4(3).

Forwood, J.R., Stypinski, P. and Paterson, J.A. (1989). Forage selection by cattle grazing orchard grass-legume pastures. APN J., 81: 409–414.

Franchin, M.F., Modolo, A.J., Adami, P.F. and Trogello, E. (2014). Effect of grazing intensities and seed furrow openers on corn development and yield in a crop-livestock system. Maydica, 59: 42–48.

Franzluebbers, A.J. and Stuedemann, J.A. (2008a). Soil physical responses to cattle grazing cover crops under conventional and no tillage in the Southern Piedmont USA. Soil Tillage Res., 100: 141–153.

Franzluebbers, A.J. and Stuedemann, J.A. (2008b). Early response of soil organic fractions to tillage and integrated crop-livestock production. Soil Sci. Soc. Am. J., 72: 613–625.

Franzluebbers, A.J. and Stuedemann, J.A. (2015). Does grazing of cover crops impact biologically active soil carbon and nitrogen fractions under inversion or not tillage management? J. Soil Water Conserv., 70: 365–373.

Grace, P.R., Oades, J.M., Keith, H. and Hancock, T.W. (1995). Trends in yields and soil organic carbon in the Permanent Rotation Trial at the Waite Agricultural Institute, South Australia. Aust. J. Exp. Agric., 35: 857–864.

Guldan, S.J., Gardner, J.C., Schatz, B.G., Goldstein, W., Klein, R.N., Havlin, J.L. and Schlegel, A. (1990). Substituting legumes for fallow in the Central and Northern Great Plains, Abs., Am. Soc. Agron., Madison, WI.

Hancock, D.W., Curt Lacy, R. and Lawton Stewart Jr., R. (2014). Forage Systems for Stocker Cattle, The University of Georgia Bulletin, 1392.

Hashemi, M. (1996). Forage Management: Perennial Forage Species for Pasture and Hay. https://ag.umass.edu/crops-dairy-livestock-equine/fact-sheets/perennial-legumes-grasses.

Heins, B. (2015). https://extension.umn.edu/soil-management-and-health/research-cover-crops-grazing-systems.

Hilder, E.J. (1969). The effect of the grazing animal on plant growth through the transfer of plant nutrients. *In*: James, B.J.F. (ed.). Intensive Utilisation of Pastures, Angus and Robertson, Sydney, Australia.

Hill, G.M., Gates, R.N. and Burton, G.W. (1993). Forage quality and grazing steer performance from Tifton 85 and Tifton 78 bermuda grass pastures. J. Anim. Sci., 71: 3219–3225.

Hoorman, J., Sundermeier, A.P., Islam, K.R. and Reeder, R.C. (2009). Using cover crops to convert to no-till. Crops & Soils, 42(6): Nov-Dec., pp. 9–13.

Hoveland, C.S., Hill, N.S., Lowrey Jr., R.S., Fales, S.L., McCormick, M.E. and Smith, Jr., A.E. (1988). Steer performance on bird foot trefoil and alfalfa pasture in central Georgia. J. Prod. Agr., 1: 343–346.

Jiang, J., Bjoerck, L., Fonden, R. and Emanuelson, M. (1996). Occurrence of conjugated cis-9, trans 11-octadecadonic acid in bovine milk: Effects of feed and dietary regimen. J. Dairy Sci., 79: 438–445.

Johnson, K.D., Lechtenberg, V.L., Vorst, J.J. and Hendrix, K.S. (1982). Indirect estimation of grass-legume ratios in mixed sward. Agron. J., 74: 1089–1091.

Kierkegaard, J.A., Lilley, J.M., Hunt, J.R., Sprague, S.J., Ytting, N.K., Rasmussen, I.S. and Graham, J.M. (2015). Effect of defoliation by grazing or shoot removal on the root growth of field-grown wheat (*Triticum aestivum* L.). Crop Pasture Sci., 66: 249–259.

Koepf, H.H. (1981). The principles and practice of biodynamic agriculture. *In*: Stonehouse, B. (ed.). Biological Husbandry, Butterworths, London, Eng.

Liebig, M.A., Tanaka, D.L., Kronberg, S.L., Scholljegerdes, E.J. and Karn, J.F. (2011). Soil hydrological attributes of an integrated crop-livestock agro-ecosystem: Increased adaptation through resistance to soil change. Appl. Environ. Soil Sci.

Mandell, I.B., Buchanan-Smith, J.G. and Campbell, C.P. (1998). Effects of forage vs. grain feeding on carcass characteristics, fatty acid composition and beef quality in Limousin-cross steers when time on feed is controlled. J. Anim. Sci., 76: 2619–2630.

Manousidis, T., Parissi, Z.M., Kyriazopoulosc, A.P., Malesiosa, C., Koutroubasd, S.D. and Abasa, Z. (2018). Relationships among nutritive value of selected forages, diet composition and milk quality in goats grazing in a Mediterranean woody rangeland. Livestock Science, 218: 8–9.

Mapfumo, E., Naeth, M.A., Baron, V.S., Dick, A.C. and Chanasyk, D.S. (2002). Grazing impacts on litter and roots: perennial versus annual grasses. J. Range Manag., 55: 16–22.

Mathison, G.W. (1993). The beef industry. *In*: J. Animal Production in Canada, 34–75.

McCaughey, W.P. and Cliplef, R.L. (1995). Carcass and organoleptic characteristics of meat from steers grazed alfalfa/grass pastures and finished on grain. Can. J. Anim. Sci., 76: 149–152.

McMechan, J.R. Wright and Ohnesorg, W. (2017). Suspected wheat stem maggot damage in corn following cover crops, Copwatch, University of Nebraska-Lincoln.

McVay, K.A., Radcliffe, D.E. and Hargrove, W.L. (1989). Winter legume effects on soil properties and nitrogen fertiliser requirements. Soil Sci. Soc. Am. J., 53-1856-1862.

Myers, L.F. (1969). The effect of the grazing animal on plant growth through grazing. *In*: James, B.J.F. (ed.). Intensive Utilization of Pastures, Angus and Robertson, Sydney, Australia.

NRC (National Research Council). (1989). Alternative Agriculture, Nat. Acad. Press, Washington, D.C.

Nuru, S. and Ryan, J.G. (1985). Proceedings of the Nigeria-Australia Seminar on Collaborative Agricultural Research, Shika, Nigeria, 1983, ACIAR Proceedings Series, 4–145.

Owens, LB., Edwards, W.M. and VanKeuren, R.W. (1984). Nitrogen values in surface runoff from fertilized pastures. J. Environ. Qual., 13: 310–312.

Peterson, R.G., Lucas, H.L. and Woodhouse, Jr., W.W. (1956). The distribution of excreta by freely grazing cattle and its effect on pasture fertility: I. Excretal distribution. Agron. J., 48: 440–444.

Peterson, R.G., Woodhouse, Jr., W.W. and Lucas, H.L. (1956). The distribution of excreta by freely grazing cattle and its effect on pasture fertility: II. Effect of returned excreta on the residual concentration of some fertiliser elements. Agron. J., 48: 444–449.

Pionke, H.B. and Urban, J.B. (1985). Effect of agricultural land use on groundwater quality in a small Pennsylvania watershed. Groundwater, 23: 68–80.

Puckridge, D.W. and French, R.J. (1983). The annual legume pasture in cereal-ley farming system of Southern Australia: A review. Agr. Ecosys. Environ., 9: 229–267.

Radford, B.J., Yule, D.F., Braunack, M. and Playford, C. (2008). Effects of grazing sorghum stubble on soil physical properties and subsequent crop performance. Am. J. Agric. Biol. Sci., 3: 734–742.

Rouquette, F.M., Jr., Kerby, J.L., Nimr, G.H. and Ellis, W.C. (2003). Tifton 85, Coastal bermudagrass, and supplement for backgrounding fall born calves during the summer. Beef Cattle Research in TX, pp. 62–66.

Russell, E.W. (1913). Soil Conditions and Plant Growth, Wiley Publ., London, Eng.

Savory, A. (1988). Holistic Resource Management. Island Press, Covelo, California.

Sims, J.R., Koala, S., Ditterline, R.L. and Weisner, E.L. (1985). Registration of 'George' black medic. Crop Sci., 25: 709–710.

Sims, J.R. (1989). CREST farming: A strategy for dryland farming in the Northern Great Plains-Intermountain region. Am. J. Alternative Agr., 485–90.

Sims, J.R. and Slinkard, A.E. (1991). Development and evaluation of germplasm and cultivars of cover crops. pp. 121–129. *In*: Hargrove, W.L. (ed.). Cover Crops for Clean Water, Soil and Water Conserv. Soc., Ankeny, IA.

Smith. R., Donna A.P. and Jeff, L. (2017). Using Cover Crops for Grazing Cattle, Univ. of Kentucky.

Soussana, J.F., Loiseau, P., Vuichard, N., Ceschia, E., Balesdent, J., Chevallier, T. and Arrouays, D. (2004). Carbon cycling and sequestration opportunities in temperate grasslands. Soil Use Manage., 20: 219–230.

Thiessen-Martens, J. and Entz, M.H. (2001). Availability of late-season heat and water resources for relay and double cropping with winter wheat in Prairie Canada. Can. J. Plant Sci., 81: 273–276.

Tothill, J.C. (1978). Comparative aspects of the ecology of pastures. *In*: Wilson, J.R. (ed.). Plant Relations in Pastures, Commonwealth Sci. Indus. Res. Org., Melbourne. Australia.

Twerdoff, D.A., Chanasyk, D.S., Mapfumo, E., Naeth, M.A. and Baron, V.S. (1999). Impacts of forage grazing and cultivation on near-surface relative compaction. Can. J. Soil Sci., 79: 465–471.

USDA-NRCS Bulletin.(2013). Farming in the 21st Century: A Practical Approach to Improve Soil Health.

Utley, P.R. Hollis, Monson, W.G., Marchant, W.H. and McCormick, W.C. (1974). Coastcross-1 bermuda grass, coastal bermuda grass and Pensacola bahia grass as summer pasture for steers. J. Anim. Sci., 38: 490–495.

Utley, P.R., Monson, W.G., Burton, G.W. and McCormick, W.C. (1981). Evaluation of Tifton 44, coastal and Callie Bermuda grasses as pasture for growing beef steers 1,2. J. Anim. Sci., 52: 725–728.

Wall, D.A. and Stobbe, E.H. (1984). The effect of tillage on soil temperature and corn (*Ziamays* L.) growth in Manitoba. Can. J. Plant Sci., 6459–67.

Wallace, H. (1892). Clover Culture, Homestead Co., Des Moines, Iowa.

Warren, S.D., Nevill, M.B., Blackburn, W.H. and Garza, N.E. (986). Soil response to trampling under intensive rotation grazing. Soil Sci. Soc. Am. J., 50: 1336–1341.

Watkins, B.R. and Clements, R.J. (1978). The effects of grazing animals on pastures. *In*: Wilson, J.R. (ed.). Plant Relations in Pastures, Commonwealth Sci. Indus. Res. Org., Melbourne, Australia.

Webber, G.D. (1990). The extension of the ley farming system in South Australia: A case study. *In*: Osman, A.E., Ibrahim, M.H. and Jones, M.A. (eds.). The Role of Legumes in the Farming Systems of the Mediterranean Area, Kluwer Academic Publ., DeVenter, The Netherlands.

Williams, A., Filbert, M., Solberg, K., Huff, P., Spratt, E. and Vergin, K. (2018). Grazing Cover Crops: A How-to-Guide. Natural Resources Conservation Service, USDA.

Zhu, C.J., Gantzer, C.J., Anderson, S.H., Albens, E.E. and Beuselinck, P.R. (1989). Runoff, soil, and dissolved nutrient losses from no-till soybeans with winter cover crops. Soil Sci. Soc. Am. J., 53: 1210–1214.

9

Cover Crops' Effect on Soil Quality and Soil Health

MA Rahman

1. Introduction

Soil quality is one of the three primary components of environmental quality, along with water and air quality (Islam and Weil, 2000; Andrews et al., 2002). The term 'soil quality' is often confused with 'soil health'. Scientifically, soil quality refers to the integration of soil's quantifiable biological, chemical and physical properties to perform functions. In contrast, soil health is a popular term portraying soil as a complex living system that acts holistically to perform functions, rather than as a mixture of sand, silt and clay. While soil quality is defined as the soil's capacity to function within an ecosystem and land-use boundaries to sustain biological productivity, maintain environmental quality and to promote plant and animal health (Doran and Parkin, 1994), soil health describes the integrated condition of the soil system in a holistic manner.

Soil health functions include support for economic crop production, recycling of applied nutrients, improving air and water quality and sustaining animal nutrition and health, as well as food quality and public health. Soil health (or quality) has been defined in many ways and usually includes an integration of soil biological, chemical and physical properties (Fig. 1).

Soil is a very complex multifunctional polydisperse system, of which its health or quality depends on variations in natural and anthropogenic factors, such as parent material, climate and topography and management practices. These factors determine and regulate the biological, physical and chemical properties of the soil. While soil quality cannot be measured directly, its assessment

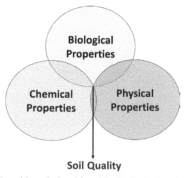

Fig. 1 Soil health (quality) and its relationship with physical, chemical, and biological properties.

The Ohio State University, Columbus, Ohio, USA.

relies on selected soil indicators or properties to quantify the management-induced changes in soils (Karlen et al., 1997; Islam and Weil, 2000; Weil et al., 2003).

A range of soil-quality indicators has been identified to estimate soil quality; some of the most important are those which are responsive to changes in soil management and these include soil structure, porosity, infiltration, soil rooting characteristics, plant available water, soil cover, soil acidity, electrical conductivity, plant nutrients, soil organic matter (SOM), microbial biomass and microbial diversity (Allen et al., 2011). Various studies have been conducted to evaluate soil-quality indicators under different land use types (Abbasi et al., 2007; Ishaq et al., 2014; Kalu et al., 2015). However, the most popular chemical indicators used to assess soil quality are soil organic carbon (OC), total nitrogen (TN) and soil acidity (pH). Increasing the amount of OC improves soil quality, as it contributes to many beneficial physical, chemical and biological processes in the soil ecosystem (Carson, 2013). Therefore, soil quality related to soil health is crucial because only healthy soil can ensure high crop yields. To determine the effects of management practices on soil quality, it is essential to identify easily measurable, sensitive and early and key indicators of soil's functionality in response to management practices (Table 1).

Cover cropping is one of the important components of sustainable agricultural practices that has received attention as a means of improving soil quality and is directly related to crop productivity and enhanced ecosystem services (Fig. 2). Cover crops are defined as the crops used to cover the ground as a living mulch or surface mulch to protect the soil and enhance soil productivity (Kaye

Table 1 Key indicators of soil functionality across agricultural management practices (*Source*: Islam and Well, 2000).

Ephemeral Changes within Days/weeks **(Very dynamic)**	Intermediate Subject to management Over several years **(Dynamic)**	Permanent Inherent to profile or site **(Inherent)**
Moisture content	Microbes/biodiversity	Depth/slope
Field respiration	Enzymes/basal respiration	Texture
pH and salts	Earthworm/nematode	Climate
Available nutrients	**Organic carbon (C)**	Stoniness
Bulk density	Total nitrogen	Fragipan
Cone index	Aggregate stability	Mineralogy
	Available water capacity	
‒ ‒ ‒ ‒ ‒ ‒ ‒ ‒ ➔	**Increasing Permanence**	‒ ‒ ‒ ‒ ‒ ‒ ‒ ‒ ➔

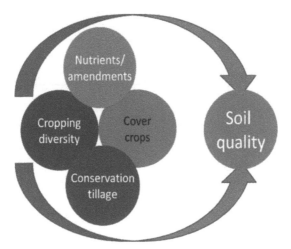

Fig. 2 Sustainable management practices and soil health quality.

and Quemada, 2017). Cover crops are either annual or perennial in nature and are planted between main crops (fallow) to improve soil quality and agricultural productivity by supporting biodiversity and efficiency, chemical buffering by accumulating SOM, regulating nutrient cycling, conserving soil moisture, reducing compaction with an associated increase in water filtration and adding soil physical stability (Hobbs et al., 2008).

Cover crops are grown to maintain soil quality by providing protection against soil erosion and nutrient leaching or runoff (Reeves, 1994), suppressing weeds, carbon sequestration, decreasing soil-borne pests and diseases, conserving soil moisture, reducing non-point source pollution, increasing water quality, etc. (Lal, 2015; Dabney et al., 2001).

Cover crops improve soil physical, chemical and biological properties by providing organic carbon content, cation exchange capacity, aggregate stability and water infiltration (Dabney et al., 2001). The schematic diagram for stepwise improvement of soil quality use of crops is presented in Fig. 3. Cover crops have been suggested to potentially improve soil nutrient nitrogen (N). Cover crops conserve nitrogen (N) by converting mobile nitrate into immobile plant protein (Weerasekara et al., 2017). Legume cover crops fix atmospheric N_2 and build up soil N, which benefits productivity and yield of subsequent cash crops while reducing N fertiliser requirements (Debney et al., 2010). Legume cover crops (e.g., alfalfa, vetches, clover) can fix nitrogen (N) biologically and increase SOM content (Lüscher et al., 2014). Non-legume cover crops (spinach, canola, flax) can absorb excess nitrate from the soil, increase crop biomass and improve soil quality (Finney et al., 2016; White et al., 2016). The SOM is a very important soil-quality indicator as cover crops have the potential to maintain higher levels of SOM, which might positively influence cation exchange capacity, water holding capacity, soil structure and microbial activity as well as might stabilise the soil (pH, which helps in plant nutrient uptake) (Sharma et al., 2018). In addition, SOM reduces compaction and crusting and binds soil particles together to reduce soil erosion.

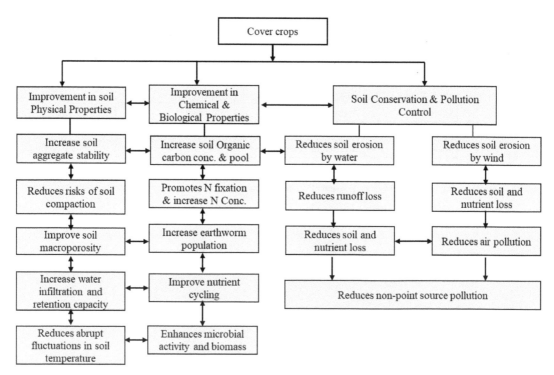

Fig. 3 Stepwise advantages of using cover crops and their effect on soil quality parameters (*Source*: Blanco-Canqui et al., 2015).

2. Cover Crops and Soil Biological Quality

There have been many types of cover crops commonly used all over the world. Legumes (soybean, peas, beans, crimson clover, hairy vetch, field peas, big flower vetch, red clover) and non-legumes (rye, wheat, barley, oats) are the most extensively used.

2.1 Soil Microbial Activity, Biodiversity and Efficiency

Use of cover crops is a widely growing strategy used to enhance soil microbial growth in agricultural systems. Reducing tillage and growing cover crops, both widely recommended practices for boosting soil health, are known to impact soil microbial communities. It is well established that soil microorganisms play a pivotal role in the function of the ecosystem through their contribution to SOM decomposition and biochemical cycling of nutrients (Bardgett and van der Putten, 2014). Soil microbes are the precursors and builders of SOM (Miltner et al., 2012).

Plant communities are a primary source of C that supports microbial growth and activity; therefore, crop management, such as cover crops, can be used to promote an increase in the size of the soil microbial community in an agricultural system. Microbial community can provide a broad range of functions to physical and chemical disturbance (Lehman et al., 2015). For example, arbuscular mycorrhizal fungi (AM fungi) enhance crop production by protecting host plants from pathogens, improving nutrient uptake and increasing host plant tolerance to environmental stresses, such as drought (Douds et al., 2005).

Cover crops, such as oat, cereal rye and winter wheat can increase AM fungi in agricultural soil (Lehman et al., 2012). Finney et al. (2017) reported that individual cover crop species favoured microbial functional groups. AM fungi were more abundant beneath oat and cereal rye cover crops. Non-AM fungi were positively associated with hairy vetch.

Generally, phospholipid fatty acid analysis (PLFA) is used to determine the size and structure of the soil microbial community. The presence of cover crops increased microbial biomass (indicated by total PLFA concentration) relative to the cover crops and control. Finney et al. (2017) studied the PLFA concentration on cover crops and control and reported that cover crops led to an average increase of 10.22 nmol g^{-1} relative to the control. The effects of cover crops on PLFA concentration is presented in Table 2. In the fall, abundances of gram-positive bacteria and actinomycetes were higher in Year 2 as compared to Year 1.

The establishment of cover crops provides a rhizosphere effect, whereby plant roots modify the soil habitats as they grow (Ramos et al., 2000), improving aeration and serving as a source of nutrients to microorganisms, leading to enhanced microbial growth and activity (Doran and Zeiss, 2000).

2.2 Soil Enzyme Activities and Decontamination

All soils contain a group of intracellular and extracellular enzymes with different origins that may be synthesised by plants, animals and microorganisms (Verdoucq et al., 2003). Enzymes play a vital role in agriculture and in nutrient cycling because they are constantly being synthesised, accumulated, inactivated and decomposed in the soil (Balota and Chaves, 2010). The choice to use enzymes to assess soil quality is based on their sensitivity to soil management, organic matter decomposition and relative ease of analysis (Balota and Chaves, 2010). The determination of soil fertility and plant yield using a single enzyme activity has been proven to be inappropriate (Nannipieri et al., 2012). This is because soil enzyme activities catalyse a particular reaction and therefore cannot be linked to the general soil-microbial activity, which comprises a wide range of different enzymatic reactions (Nannipieri et al., 2012). Enzymes respond to soil-management changes before other soil-quality indicator changes are detectable.

Soil enzymes play an important role in organic matter decomposition and nutrient cycling. Table 3 shows some common soil enzymes that can be used as biological soil quality indicators.

Table 2. Effects of cover crops on Phospholipid Fatty Acid (PLFA) concentration (nmol g^{-1}) in bulk soil in spring, approximately nine months following cover crop planting in 2011–2012 in Central Pennsylvania (*Source*: Finney et al., 2017).

Year/Crop	Total PLFA	Gram + Bacteria	Gram – Bacteria	Actinomycetes	Non-AM Fungi	AM Fungi	Protozoa
Year 1	100.1	25.14	31.01	13.32	2.43	4.19	1.05
Year 2	73.5	18.54	23.38	10.92	1.21	2.72	0.47
No cover crop	77.3	20.31	23.48	11.67	1.42	2.96	0.58
Sunn hemp (SH)	85.9	22.77	26.11	12.47	1.81	3.40	0.54
Soybean (SB)	82.8	21.64	25.24	12.06	1.70	3.26	0.68
Red Clover (RC)	86.2	21.24	27.47	11.78	1.90	3.40	0.67
Hairy Vetch (HV)	87.3	20.98	27.73	11.54	2.60	3.25	0.86
Forage radish (FR)	85.0	22.17	26.21	12.04	1.69	3.22	0.80
Oat (OA)	88.1	22.84	27.37	12.60	1.28	3.66	0.77
Canola (CA)	89.7	22.96	28.23	12.65	1.78	3.38	0.79
Cereal rye (CR)	88.0	21.73	27.27	12.16	1.79	3.64	0.82
FR + OA + CA + CR	85.0	21.38	26.66	12.11	1.40	3.48	0.75
SH + SB + FR + OA	89.2	22.65	27.83	12.57	1.91	3.48	0.78
RC + HV + CA + CR	88.0	20.66	28.38	11.55	2.41	3.54	0.77
SH + SB + CA + CR	86.26	21.54	27.19	11.98	1.52	3.56	0.74
RC + HV + FR + OA	89.26	21.76	28.56	11.99	1.95	3.68	0.90
8 species mix	94.47	22.95	30.20	12.67	2.11	3.92	0.94

Table 3 Soil enzymes as indicators of soil quality.

Soil Enzyme	Enzyme Reaction	Reaction Catalysed	Biological Indicators
Dehydrogenase	Electron transport system	$XH_2 + A \rightarrow X + AH_2$	C-cycling
β-glucosidase	Cellobiose hydrolysis	Glucoside + $H_2O \rightarrow$ ROH +glucose	C-cycling
Cellulase	Cellobiose hydrolysis	Hydrolysis of β-1, 4 – glucanbonds	C-cycling
Phenol oxidase	Cellulose hydrolysis	$A + H_2O_2 \rightarrow$ oxidised A + H_2O	C-cycling
Urease	Lignin hydrolysis	Urea $\rightarrow 2NH_3 + CO_2$	N-cycling
Amidase	Urea hydrolysis	Carboxylic acid amide + H_2O $\rightarrow$ Carboxylic acid + NH_3	N-cycling
Protease	N-mineralization	Proteins $\rightarrow$ Peptides/amino acids	N-cycling
Phosphatase	Release of PO_4^{3-}	Phosphate ester + $H_2O \rightarrow$ ROH phosphate	P-cycling
Arylsulfatase	Release of SO_4^{2-}	$ROSO_3^- + H_2O \rightarrow$ ROH SO_4^{-2}	S-cycling
Other soil enzymes	Hydrolysis	Hydrolysis	General organic Matter degradative enzyme activities

Source: Das and Varma (2010)

β-glucosidase is the most common, important and widely used soil quality indicator because it produces glucose, the C as energy source for the growth and activity of soil microbes as a final product (Merino et al., 2016). Cover crops such as vetch, canola and red clover led to an increase in the activity of β-glucosidase compared to oats (Mukumbareza et al., 2015). β-glucosidase activity increases because the soils amended with lower C:N crop residue favour its function, resulting in quick organic matter decomposition and nutrient release. Plants and microorganisms are the main sources of phosphatase enzymes in the soil. Plants use only inorganic P and a substantial amount of soil P is organically bound. When P is lacking in the soil, plant roots and microorganisms increase secretion of phosphatase to intensify the solubilisation and remobilisation of phosphate, therefore influencing the ability of the plant to cope with P-stressed conditions (Kai et al., 2002). Phosphatases are a good soil-quality indicator as their activity reflects the existing soil condition.

Legumes, such as chickpea, cowpea, Cyclopia and Aspalathus release more phosphatase enzymes than non-legumes (Maseko and Dakora, 2013). This is because legumes require more phosphorus in the symbiotic nitrogen fixation process than do cereals (Makoi and Ndakidemi, 2008). The increase in phosphatase activity in the legume roots and soils leads to a significant increase in plant available P (Makoi et al., 2010). Thus, phosphorus supply and assimilation can be estimated by acid and alkaline phosphatase activity in the low-P soils of legume crops (Maseko and Dakora, 2013). Mukumbareza et al. (2015) reported that rotation of maize with vetch and fertilised oat cover crops increased microbial biomass and the activities of phosphatase in soil.

The urease enzyme acts by aiding the hydrolysis of urea into CO_2 and NH_3, which leads to a rise in soil pH and nitrogen loss to the atmosphere through NH_3 volatilisation (Das and Varma, 2010). The presence of urea or an alternative N source activates urease production (Mobley et al., 1995). Studies of soil urease activity have been of great interest over the years and have been used as a good index of soil quality, because of the role of urease in the regulation of N supply to plants after urea fertilisation (Piotrowska-Dlugosz and Charzynski, 2014). Urease activity increased in a soil management system where maize was rotated with vetch and fertilised oat cover crops, which led to an increase in maize yield (Mukumbareza et al., 2015). The high urease activity recorded in soils treated with vetch and a combination of vetch and oats indicate a greater potential for N cycling through a lower C:N ratio.

Nevins et al. (2020) studied the synchrony of cover crop (cereal rye, hairy vetch and mixture) decomposition, enzyme activity and nitrogen availability in a corn agro-ecosystem in the midwest United States. They reported that corn planting occurred 12 days after cover crop termination in 2016, which was prior to soil β-glucoside activity significantly increasing at 39 days after termination for all cover crop treatments. For all treatments (2016) except hairy vetch, there was significantly greater β-glucosidase activity at the 53-day corn-growth stage. At 53-days corn-growth stage, among all treatments (cereal rye, hairy vetch, mixture), cereal rye and mixture treatments had significantly higher β-glucosidase activity compared to hairy vetch treatment (3.69 µmols para-nitrophenol release g dry soil^{-1} hr^{-1}) and the control (4.18 µmols para-nitrophenol release g dry soil^{-1} hr^{-1}) as shown in Fig. 4. In 2017, though not as pronounced as 2016, potential soil β-glucosidase activity increased in all cover crop treatments (Fig. 4). In 2016, potential soil urease activity increased with soil temperature in all cover crop treatments when averaged across tillage treatments until the final sampling date, when the corn was at physiological maturity, 162 days after cover crop termination (Fig. 5). The dynamics of potential soil urease activity in 2017 distinctly contrasted with the activity of 2016, as activity for all treatments decreased linearly with time. Specifically, in 2017, potential soil urease activity was not significantly different near termination as compared to the sampling date when the corn was at maturity as also cereal rye, hairy vetch and control treatments when averaged across the main effect of tillage.

A range of soil quality indicators are positively correlated with soil enzyme activities and labile SOM fractions (Veum et al., 2014). Thus, soil enzymes and their activities can be used as indicators for identifying cover crops' effect on soil quality. In the soil, enzymes play critical roles in microbially-mediated transformations of SOM and nutrient cycling. Cover crops can increase soil-microbial biomass and enzymatic activity (Hoorman, 2009; Nair and Ngouajio, 2012).

Cover crops can increase soil β-D-glucosaminidase and β-glucosidase activity, but the effects are not consistent, spatially or temporally (Fernandez et al., 2016). Bandick et al. (1999) conducted research using cover crops and reported that α-glucosidase, β-glucosidase, α-galactosidase, β-galactosidase, amidase, arylsulfatase, deaminase, fluorescein diacetate (FDA) hydrolase, invertase, cellulose and urease activities were greater in vegetable crop rotation systems with cover

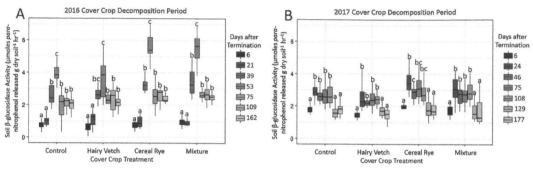

Fig. 4 Average measured level of potential soil β-glucosidase activity in each cover crop treatment during (A) 2016 and (B) 2017 corn-growing seasons (*Source*: Nevins et al., 2020).

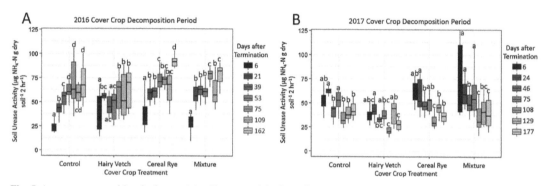

Fig. 5 Average measured level of potential soil urase activity in each cover crop treatment during (A) 2016 and (B) 2017 corn-growing seasons (*Source*: Nevins et al., 2020).

crops compared to fields without organic amendments. Ve Verka et al. (2019) conducted a study and reported that β-glucosidase levels were in the range of 35–105 μg pnp g⁻¹ dry soil h⁻¹ under cover crop treatment for two years.

Rankoth et al. (2019) carried out research with fall/winter cover crops like winter barley, winter cereal rye, winter triticale and winter oat in a corn-soybean rotation and found that β-glucosidase levels increase in 35–145 μg pnp g⁻¹ dry soil h⁻¹, which was six years after the initial cover crop establishment. Similarly, dehydrogenase levels changed from 8.5 μg TPF g⁻¹ dry soil h⁻¹ to 10–40 μg TPF g⁻¹ dry soil h⁻¹ within five years. In addition, β-glucosaminidase activity was significantly greater at 0–10 cm depth after three years of cover crop treatments. Overall, cover crops have a mixed effect on soil enzyme activities. β-glucoside is a commonly found, immobilised enzyme in soils (Mbuthia et al., 2015) that releases low molecular weight sugars (glucose), serving as an important energy source for soil microorganisms. Increasing β-glucosidase activity in soil accelerates the breakdown of plant materials in soil, thereby allowing greater availability of nutrients for the following cash crop (Mbuthia et al., 2015).

2.3 Cover Crops and Soil Chemical Quality

Nitrogen Dynamics

Legume cover crops (such as soybeans, clovers, sun hemp, cowpeas, Austrian winter pea, and field peas) can provide N to the following crop. Legume cover crops fix N from the air, adding up to 100–150 lb/acre of this essential N. Non-legume cover crops recycle leftover N from the soil, storing it in roots and aboveground plant material, where a portion will be available to the following crop. Every pound of N stored is a pound of N prevented from leaching out of the topsoil into streams. After treatment of cereal rye, there may not be enough N available early for the next crop; after a legume, the N will likely not be available until later in the growing season, depending upon when the crop decomposes. It all depends on the carbon-to-nitrogen (C:N). A C:N less than 20 allows the organic materials to decompose quickly while a C:N greater than 30 requires additional nitrogen and slows down decomposition (Table 4).

Cover crops are typically planted outside of the cash crop growing season to scavenge N from soil and prevent erosion and leaching (Schipanski et al., 2014). Then, when cover crops are killed, the N in their tissues can microbially mineralised to supply inorganic N to the subsequent cash crops (White et al., 2016). Cover crops that are good at scavenging N from soil (e.g., grasses) often have high C:N ratios when they are killed. During microbial decomposition of such cover crop residues, N is immobilised (White et al., 2014), reducing availability to cash crops to an extent that can limit yields in some cases (Finney et al., 2016). Conversely, legume cover crop tissues have low C:N ratios and thus, microbes decomposing their tissues mineralise N and increase N availability to cash crops. However, these legume cover crops can be poor scavengers for soil N and N leaching can be high under them (Thapa et al., 2018).

Table 4 The Carbon to Nitrogen (C:N) in Fall and Spring in all above-ground cover crop tissues.

Treatment	Cover Crop Biomass C:N			
	Between Wheat and Maize		Between Maize and Soybeans	
	Fall	*Spring*	*Fall*	*Spring*
Pea	11	8	11	10
Clover	11	10	11	11
Oat	33	NA	14	NA
Radish	18	NA	9	NA
Canola	21	19	10	12
Rye	19	33	13	35
3SppN	13	23	13	32
3SppW	30	29	13	35
4Spp	15	22	11	28
6Spp	21	24	12	31
Error	1	2	NA	NA

NA: Winter killed oat and radish. No spring data for these spp. (*Source*: Kaye et al., 2019).

Kaye et al. (2019) studied nitrogen levels in cover crops between wheat and maize. To study this effect, several cover crops were used between wheat and maize, and maize and soybean for monocultures and mixtures (3SppN: 3 species nitrogen, 3SppW: s species weed, 4Spp: 4 species and 6Spp: 6 species cover crops). Kaye et al. (2019) reported that between wheat and maize, fall oat stands had the highest C:N (33) and legume monocultures had the lowest (< 11), as presented in Table 4. The C:N of rye stands increased between fall sampling (19) and spring sampling (33), while the 3SppW mix had high C:N in both seasons because it was dominated by oat in the fall and rye in the spring. In contrast, canola stands had C:N of ~ 20 in both fall and spring. Legume stands had C:N of < 10 in spring, while all mixtures that contained legumes had C:N of ~ 22–24 in spring. Figure 6 shows the N content in cover crops between wheat and maize and indicates the N-scavenging capacity of the cover crops in different seasons.

2.4 Soil Organic Matter (Carbon) Dynamics

Soil organic carbon is a compositive indicator of soil quality and plays a critical role in food production and ecosystem services. It is a cornerstone of agro-ecosystem sustainability, as a driver

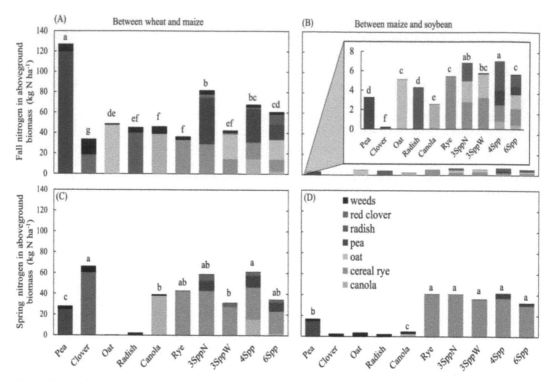

Fig. 6 Nitrogen in aboveground cover crop and weed tissues (Values are mean of 12 replicates). A and B: Fall biomass; C and D: Spring biomass (*Source*: Kaye et al., 2019).

of soil structure, nutrient cycling, water dynamics, microbial activity and biodiversity (Tautges et al., 2019). There is substantial evidence that cover crops increase C sequestration. Poeplau and Don (2015) conducted a meta-analysis and reported that cover crop treatments had a significantly higher C stock than the reference croplands.

A global meta-analysis of 30 studies found that cover crops increase C stock by 0.32 MG C ha⁻¹ year⁻¹ but was limited to the top 30 cm (Poeplau and Don, 2015). Soil texture with greater clay content, or those with low initial C concentration, may increase in OC more readily than sandy soils or those with high initial C concentration (Blanco-Canqui et al., 2015). Tautges et al. (2019) reported that in maize-tomato rotations, OC increased by 12.6 per cent with winter cover crops. A cover crop system improves the net ecosystem C balance of a cropland by replacing the bare fallow period (C source) with an additional period of C assimilation (Lal, 2001). Most importantly, C input comes from the root, which was found to contribute more effectively to the stable C pool than aboveground C input (Kätterer et al., 2011).

Cover crops provide additional biomass carbon input and soil cover, resulting in retainment or enhancement of OC. The OC retention or increment depends on management duration, soil texture, tillage, climate, cropping system and cover crop species. The importance of soil organic carbon is presented in Fig. 7. Reviews on OC and cover crops across different soil types, tillage systems and climates have reported that the cover crop can significantly increase OC from 0 to 3.50 Mg ha⁻¹ yr⁻¹ (Blanco-Canqui et al., 2015; Poeplau and Don, 2015).

The increment of SOC in soil by the cover crop is due to the addition of biomass C, improving soil aggregation to protect SOC (McVay et al., 1989; Villamil et al., 2006; Blanco-Canqui et al., 2015) and decreasing water and wind erosion potential (De Baets et al., 2011). Poeplau and Don (2015) reported that the longer a field is under a cover crop, the greater the SOC gain. Ruis et al. (2017) reported that there was a significant variability in SOC response to cover crop use and the mean annual SOC concentration gain was 0.49 ± 0.35 g kg⁻¹ for no-till, 0.11 ± 0.09 g kg⁻¹

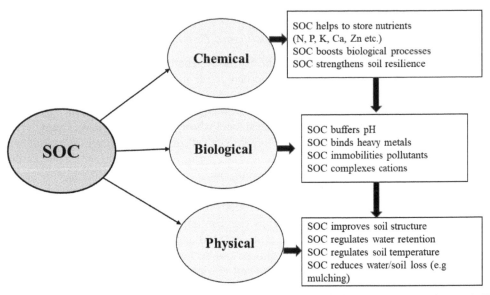

Fig. 7 Soil organic carbon (SOC) contributes to soil chemical, biological and physical properties (*Source*: Pham et al., 2018).

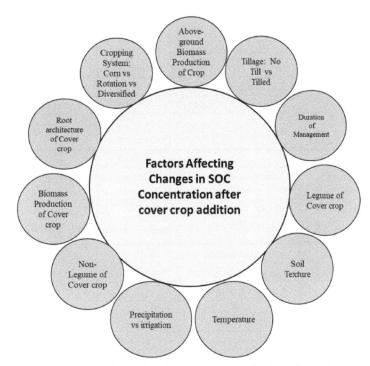

Fig. 8 Factors affecting soil organic carbon concentration after the addition of cover crops.

for conventional till and 0.47 ± 0.52 g kg^{-1} for other tillage practices. In addition, tillage does not affect SOC gain under a cover crop. The average annual SOC stock gain was 0.54 ± 0.17 for no-till, 0.29 ± 0.05 Mg ha^{-1}. More specifically, the average SOC concentration gains were 0.81 ± 0.75 g kg^{-1} yr^{-1} for *Brassicas*, 0.50 ± 0.32 g kg^{-1} yr^{-1} for legumes, and 0.61 ± 0.2 g kg^{-1} yr^{-1} for a mix. Therefore, cover crops have potential effects on SOC, but largely depend on cover crop species, tillage, temperature soil properties, etc. (Fig. 8).

3. Crop Crops and Soil Physical Quality

3.1 Soil Compaction (Bulk Density), Porosity and Erosion

Soil erosion is a major environmental problem. It causes negative impacts including the removal of nutrient-rich topsoil. Soil texture is an important factor in soil erosion. By providing a soil-stabilising root system and large quantities of crop residues on the soil surface, cover crops are considered a tool for improving soil quality and reducing soil erosion (Hargrove, 1991). Cover crops, like winter cereal grains (winter rye, triticale and wheat), annual grasses and summer cover crops produce high amounts of biomass, resulting in reduced soil erosion. Cover crop plant species cover the uniform and dense soil surfaces at times (i.e., spring) when soils are most susceptible to erosion and absorb kinetic energy of rain and wind, thus resulting in reduced soil erosion. Plant species of cover crops with abundant root biomass can anchor the soil, promote soil aggregation and increase water infiltration and SOM concentration, which reduce soil erosion. In addition, an increase in SOC concentration with cover crops can also improve soil aggregate stability, which directly reduces soil erosion (Blanco-Canqui et al., 2015). Blanco-Canqui et al. (2015) reported that runoff can decrease up to 80 per cent, and sediment loss between 40–96 per cent with cover crops. Therefore, there is a direct relationship between soil erosion with soil aggregation. The increased soil aggregation improves soil hydraulic properties, thus reducing soil erosion (Table 5).

Table 5 Effect of cover crops on soil physical quality – erosion.

Cover Crops	Species	Aliases	pH Preferred	Erosion Fighter	Weed Fighter
Non-Legumes	Annual Ryegrass	Italian ryegrass	6.0–7.0	Very good	Very good
	Barley		6.0–8.5	Excellent	Very good
	Oats	Spring oats	4.5–7.5	Very good	Excellent
	Rye	Winter, cereal, or grain rye	5.0–7.0	Excellent	Excellent
	Wheat		6.0–7.5	Very good	Very good
	Buckwheat		5.0–7.0	Fair	Excellent
	Sorghum-sudan	Sudax	6.0–7.0	Excellent	Very good
Brassicas	Mustards	Brown, oriental white, yellow	5.5–7.5	Very good	Very good
	Radish	Oilseed, Daikon, forage, radish	6.0–7.5	Very good	Excellent
	Rapeseed	Rape, canola	5.5–8	Very good	Very good
Legumes	Berseem Clover	BIGBEE, multi-cut	6.2–7.0	Very good	Excellent
	Cowpeas	Crowder peas, southern peas	5.5–6.5	Excellent	Excellent
	Crimson clover		5.5–7.0	Very good	Very good
	Field peas	Winter peas, Hairy vetch	6.0–7.0	Very good	Good
	Hairy vetch	Winter vetch	5.5–7.5	Good	Good
	Medics		6.0–7.0	Good	Very good
	Red clover		6.2–7.0	Good	Very good
	Subterranean clovers	Subcover	5.5–7.0	Very good	Excellent
	Sweet clovers		6.5–7.5	Very good	Very good
	White clover	White Dutch ladino	6.0–7.0	Very good	Very good
	Woolly pod vetch	Lana	6.0–8.0	Good	Excellent

Source: Clark (2012).

The soil aggregate stability in soil with cover crops is responsible for the reduction of soil erosion. Table 5 shows the capability of cover crops to protect against soil erosion. Results indicated that cover crops have root density ranging from 1.02 for phacelia and 2.95 kg m^{-3} for rye grass. The benefits of cover crops for reducing water erosion are widely recognised. The magnitude by which cover crops reduce water erosion is a function of biomass production and the cover crop species. Kaspar et al. (2001) reported that rye reduced runoff by 10 per cent in one year of the three-year study; however, oat did not reduce runoff. Across the three years, rye and oat reduced rill erosion by 54 per cent and 89 per cent, respectively. Crop species with a fibrous root system (e.g., rye grass, rye, oats) show high potential to control soil erosion, while cover crops with thick roots (e.g., white mustard, fodder radish) are less effective in preventing soil erosion (De Baets et al., 2011).

3.2 Soil Aggregate Stability and Water Infiltration

Soil aggregates are granules or clumps of soil made up of sand, silt and clay glued together by organic matter (Fig. 9). Soil structure refers to the size and shape of soil aggregates and the pore space between them. Stable aggregates protect SOM and improve soil structure, water-holding capacity and drought resistance. While unstable aggregates lead to reduced infiltration and surface sealing, the stable aggregates improves water infiltration (Fig. 9).

Cover crops are significantly impacted by the indexes of wet soil aggregate stability including the percentages of water stable soil aggregates (Table 6). Blanco-Canqui et al. (2019) studied the long-term effects of cover crops on soil aggregate properties and reported that grass cover crops improved soil aggregation and organic matter concentration. About 31–45 per cent of large water-stable aggregates (2–8.0 mm) increased in the 0–15 cm depth with treatment of grass cover crops. However, legume cover crops did not change soil physical properties, such as soil aggregation in the silty clay loam, after 12 years.

From the results (Table 7), it was found that cover crops increase the percentage of soil aggregates (< 0.25 mm diameter) at 0–7.5 cm depth compared to the control, while soil aggregate diameter (1.0–2.0 mm) increased at both the 0–7.5 and 7.5–15.0 cm depths. Grass cover crops increased the proportion of large aggregates (2–4.75 mm) by 31 per cent and reduced the proportion of microaggregates (< 0.25 mm) by 20 per cent as compared to the control. Similarly, grass cover crop increased mean weight diameter of wet aggregates by 34 per cent relative to both legume cover crops and control treatments (Fig. 10A). On the other hand, soil organic matter significantly increased under grass cover crop treatments compared to with no-cover crop (Fig. 10B). However, legume cover crop had no effect. Grass cover crop increased organic carbon by 11 per cent.

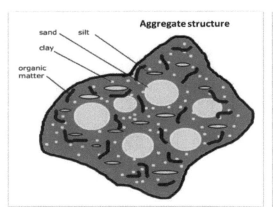
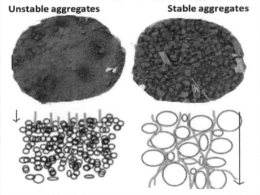

Fig. 9 Aggregate structure and components; unstable aggregate and stable aggregate (*Source*: Moebius-Clune et al., 2016).

Table 6 Cover crops' effect on soil bulk density and wet aggregate stability.

Soil Texture	Tillage	Crop	Time (Year)	Depth of Soil (cm)	Cover Crop	Bulk Density (Mg m⁻³)	Water Stable Aggregates (%)	Reference
Clay loam	No till	Corn	3	0–2.5	No CC		56.3	McVay et al. (1989)
					Crimson clover		55.0	
					Hairy vetch		58.2	
					Wheat		65.1	
Sandy clay loam	Conventional tillage	Grain sorghum	3	0–2.5	No CC		28.9	McVay et al. (1989)
					Crimson clover		37.9	
					Hairy vetch		36.7	
					Wheat		32.6	
Loamy sand	No-till	Sweet corn	3	0–7.6	No CC	1.73		Hubbard et al. (2013)
					Sunn hemp	1.71		
Find sandy loam	No-till	Corn	3	2.5–10	No CC	1.44		Wagger and Denton (1989)
					Wheat	1.52		
					Hairy vetch	1.47		
Silt loam	No-till	Corn-soybean	5	0–5 (bulk density) 0–15 (aggregate Stability)	No CC	1.32	38	Villamil et al. (2006)
					Rye	1.24	41	
					Hairy vetch	1.23	43	
					Rye+hairy vetch	1.23	44	
Silty clay loam	Conventional tillage	Soybean	1	0–50	No CC		82.6	Acuña and Villamil (2014)
					Cover crops		84.2	
Silt loam	No-till	Corn-soybean	5	0-5(bulk density) 0–15 (aggregate Stability)	No CC	1.32		Villamil et al. (2006)
					Rye	1.24		
					Hairy vetch	1.23.2020		
					Rye+hairy vetch	1.23		

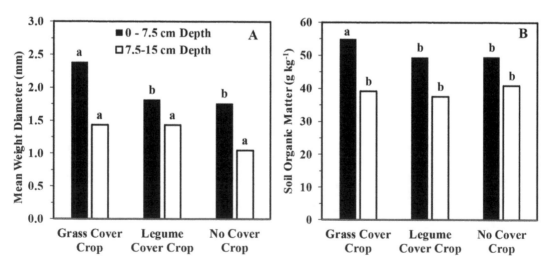

Fig. 10 Change in (a) wet soil aggregate stability expressed as mean weight diameter of water stable aggregates and (b) soil organic matter concentration with three cover crops treatment (winter wheat-corn-soybean) rotation on a silty loam in eastern Nebraska after 12 years of cover crop management practice (*Source*: Blanco-Canqui et al., 2019).

Table 7 Water stable soil aggregates (%) under cover crops and crop rotation phases for different aggregate sizes (mm) for two soil depths under winter wheat-corn-soybean rotation on a silty loam in eastern Nebraska after 12 years of cover crop management (*Source*: Blanco-Canqui et al., 2019).

	Water-Stable Aggregates (Mean %)					
	< 0.25 mm	0.25–0.5 mm	0.5–1 mm	1–2 mm	2–4.75 mm	4.75–8 mm
Effect of cover crop						
			0 to 7.5 cm depth			
Control	23	14	19	16	17	12
Legume	25	13	18	16	14	14
Grass	20	10	15	16	17	22
			7.5 to 15 cm depth			
Control	30	20	22	14	9	4
Legume	24	18	22	16	12	8
Grass	24	17	22	16	13	8
Effect of rotation						
			0 to 7.5 cm depth			
Soybean	34	15	16	13	13	8
Corn	15	10	18	19	21	18
Winter wheat	19	12	18	15	14	22
			7.5 to 15 cm depth			
Soybean	26	18	21	14	11	10
Corn	26	18	24	17	12	3
Winter wheat	26	19	22	15	12	6

Water infiltration is a key component of the water cycle, influencing how much precipitation becomes available to plants as opposed to what is lost through other pathways, such as runoff and evaporation. Alternative practices include the presence of livestock, crop residue, continuous plant roots and crop diversity (Fig. 11). These alternatives can change the infiltration rates through a range of physical, chemical and biological processes. Basche and DeLonge (2019) compared

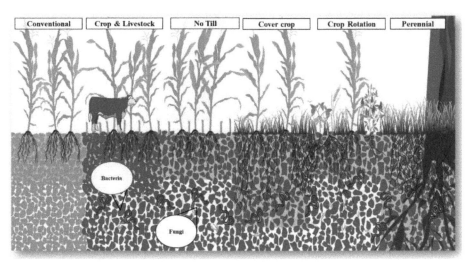

Fig. 11 Different agricultural practices and their impact on infiltration rates. Soil physical changes are represented by differences in porosity, compaction, and aggregation as represented in the size and distribution of soil aggregates (*Source*: Basche and DeLonge, 2019).

the infiltration rates in soil between conventional and alternative farming (no-till, cover crop, crop rotation, perennial, etc.) methods through meta-analysis. They found that perennial or cover crops led to the largest increase in infiltration rates (mean responses of 59.2 per cent and 34.8 per cent, respectively).

4. Cover Crops and Soil Quality Index

Over the years, several approaches were developed to quantify soil quality in response to management practices (Huddleston, 1984; Wymore, 1993; Islam, 1996; Andrews et al., 2004; Aziz et al., 2013). Islam (1996) used inductive additive approach 'considering higher values of soil properties are better indicator of soil quality'. To assess the soil quality, he emphasised on 'key indicator properties' of soil or crops that are sensitive and early indicators of soil's functionality, easily identifiable and precisely measurable, consistent changes in response to management practices and complementary to soil and crop properties associated with crop production and animal nutrition. In contrast, Andrews et al. (2004) developed the soil management assessment framework (SMAF) approach to calculate soil quality indices (SQI) in response to macrosystem management practices. This framework outline comprised of three basic parts: (1) indicator selection; (2) indicator interpretation and index integration; and (3) scoring functions to get overall soil quality index (Andrews et al., 2002). Example:

a) *Cornell Soil Health Test (CSHT)*: By adopting the SMAF, Cornell University soil testing laboratory developed CSHT, which is an integrative soil quality assessment tool using physical, chemical and biological properties of the soil. There are 42 potential indicators and given weightage are required for this CSHT scoring (Moebius-Clune et al., 2016).

b) *Alabama Soil Quality Index*: This is similar as CSHT; however, a slight modification was done that is site-specific. A difference was that a weight was assigned to each factor based on the judgement of the scientists' panel instead of the unweighed average in CSHT (Bosarge, 2015).

c) *Haney's Soil Health Test*: It is quite distinct compared to other soil-quality assessment methods. This test uses a unique set of parameters that are related to soil microbial activity and functions. This test uses water-extractable organic C and water-extractable organic N contents ratio, which was a sensitive indicator of soil microbial activity (Haney et al., 2012).

d) *OSU Soil Quality Test*: The Ohio State University soil quality test is a highly simplified method in which a dilute buffered reagent is used to react with active fractions of SOM, changing the deep purple colour of the solution to a light pink colour or colourless (Fig. 12).

Ohio State University soil quality test results and recommendations

Soil quality	Poor Soil	Fair Soil	Good Soil	Excellent Soil
Soil organic matter [%]	> 0 - 1	> 1 - 2.5	> 2.5 - 4.5	> 4.5
Active organic matter [kg/ha]	> 0 - 400	> 400 - 800	> 800 - 1600	> 1600
Available nitrogen [kg/ha]	> 0 - 12	> 12 - 26	> 26 - 40	> 40
Microbial biomass [kg/ha]	> 0 - 300	> 300 - 630	> 630 - 1280	> 1280
Aggregate stability [%]	> 0 - 25	> 25 - 40	> 40 - 70	> 70

Fig. 12 The Ohio State University soil quality test (Islam and Sundermeier, 2008).

The lighter the colour of the suspension after reacting with soil, the greater the amount of active organic matter content and the better the quality of the soil (Islam and Sundermeier, 2008). This field test instantly estimated total SOM, active SOM, available N contents, microbial biomass and soil aggregate stability.

A comparative evaluation on soil health tests (Fig. 13) related to their cost, depth and simplicity of analysis, turnaround, cost and other characteristics were evaluated (Spiegel, 2017).

Several research studies have evaluated and/or measured the impacts of cover crops with other factors on soil quality, in terms of calculated soil-quality indices or ratings (Islam and Weil, 2000). Islam (2010) has calculated soil quality index (≥ 0 to ≤ 1) based on additive integration of several indicator properties (inductive approach) in response to tillage x crop rotation with and without cover crops over time (2004–2009). The soil-quality properties were: (1) biological properties – total and active microbial biomass, basal and specific maintenance respiration rates, potentially mineralisable carbon and nitrogen, urease and dehydrogenase enzyme activities, and biodiversity; (2) chemical properties – total organic C and N, active C and N, particulate organic matter, particulate organic carbon and nitrogen, C and N lability; and (3) physical properties – bulk density and porosity, penetration resistance (compaction), aggregate stability and aggregation indices and plant available water capacity.

Results showed that crop rotation with conventional tillage, no-till had a significant impact on soil physical, chemical and biological quality at different soil depths (Fig. 14). Among crop rotations, no-till corn-soybean-wheat (NT-CSW) performed the best in improving soil-quality properties and soil quality over time (2004–2009). The calculation of overall soil quality showed that no-till with cover crop treatments increased soil quality from 60 per cent to 80 per cent at 0–7.5 cm depth.

Further studies have shown that the impact of cover crops significantly influenced soil quality under tillage x cropping diversity systems (Fig. 15). From the results, it was found that soil quality significantly increased with cover crop treatments over time (2004–2014), which directly impacted crop yield (%). As a result, cover crop treatment in a no-till cropping system will undoubtedly increase the soil quality (%). The results imply that multiple cropping systems, along with cover crops, could be more effective in enhancing soil quality than cropping systems alone.

Demir et al. (2019) studied the effects of different cover crop treatments on soil quality parameters in an apricot orchard. The treatment was carried out with hairy vetch, Hungarian vetch,

SOIL HEALTH TESTS
CAN YOU MEASURE WHAT CANNOT BE SEEN?

Soil Health Test	Price	Turnaround	Features
Soil Health Assessment, Cornell University	$95.00	8 weeks	10-page report includes a wealth of information including pH, aggregate stability, organic matter, active carbon, soil respiration, and nutrient availablity. An overall quality score is offered, plus recommendations for improving the soil health.
Phospholipid Fatty Acid (PLFA) Test, Ward Laboratories	$59.50	2 to 3 weeks	2-page report shows the presence of phospholipid fatty acids, a unique way to measure soil microbial activity. Indicators include total bacteria, total fungi, and protozoa.
Haney Soil Health Analysis, Ward Laboratories	$49.50	2 to 3 weeks	2-page report includes soil pH, organic matter, Solvita CO_2 Burst test, and several organic mineral results. The Soil Health Calculation includes cover crop recommendations and potential N savings.
Regular Soil Health Test, Woods End Laboratories	$55.00	2 to 3 weeks	1-page report featuring Solvita CO_2 Burst and SLAN tests, plus aggregate stability and organic matter. Includes fertility and soil health scores, with recommendations.
Soil Quality Field Test Kit, Ohio State University	$30.00	Instant	Easy-to-use kit includes chemistry, vials, directions, and color-coded chart indicating active organic matter and available nitrogen.

Successful Farming at Agriculture.com | **January 2017**

Fig. 13 Soil health tests comparison (*Source*: Spiegel, 2017).

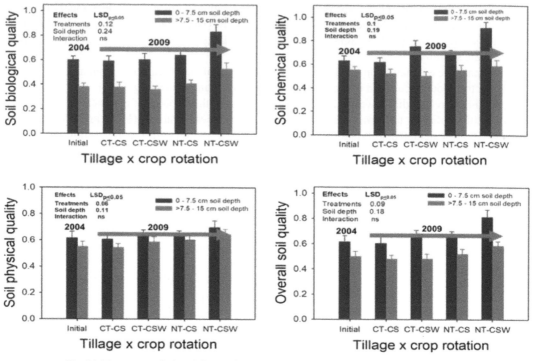

Fig. 14 Management-induced changes in temporal soil quality changes (*Source*: Islam, 2010).

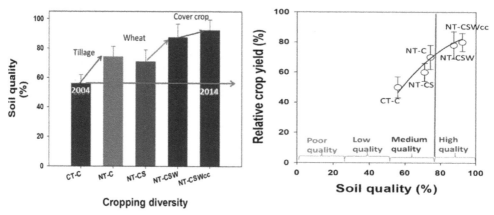

Fig. 15 Changes of soil quality (%) and crop yield (%) with cropping diversity.

a mixture of Hungarian vetch (70 per cent) and triticale (30 per cent) and lacy phacelia winter cover crops. This study reported that cover crops enhanced soil quality parameters like SOM, TN, electrical conductivity, soil basal respiration, structural stability index, aggregate stability, saturated hydraulic conductivity, bulk density, permanent wilting point, available water capacity and field moisture capacity. It was found that hairy vetch increased the SOM by 63.5 per cent, hydraulic conductivity by 248.7 per cent, available water capacity by 19.4 per cent, and structural stability index by 9.4 per cent in the 0–20 cm soil depth. The SOM contents increased followed the sequence as control plot < mechanically cultivated plot < lacy phacelia < Triticale < Hungarian vetch < hairy vetch.

Mbuthia et al. (2015) studied the effect of long-term tillage, cover crops and fertilisation on microbial community structure and activities that were linked to improvements in soil quality. This study characterised the impact of long-term (31 years) tillage (till and no-till), cover crops (Hairy vetch – *Vicia villosa* and winter wheat – *Triticum aestivum*) and a no-cover control on soil microbial community structure, activity and resultant soil quality calculated using the SMAF scoring index under continuous cotton (*Gossypium hirsutum*) production on a Lexington silt loam in western Tennessee. Soil quality indices were calculated based on the SMAF (Andrews et al., 2004). Seven of the 13 indices with scoring algorithms that are currently available under the SMAF quality scoring algorithms were used for this study. The selection of the SMAF indices is based on their role in certain soil functions that can be used as measurements for attaining specific management goals. MBC, pH, P, K and β-glucosidase activity are indices selected for their role in nutrient cycling. The SOC and bulk density were selected for their role in soil-water relations, aggregate stability, as well as filtering and buffering. All the selected indices are measures used for the assessment of crop productivity and ecosystem functioning (Andrews et al., 2004). The scores obtained from each indicator are then integrated into a soil quality index by dividing their sum by the total number of indicators used, then multiplying that number by 100. In Table 8, the extractable nutrients P, K, pH, microbial biomass and bulk density resulted in the highest soil quality scores from 0.85 to 1.00; meanwhile, SOC, calculated from soil C and β-glucosidase, had scores below 0.50 resulting in an overall soil quality index (SQI) ranging between 61–71 per cent. The soil quality score of SOC was highest for vetch cover crop treatment as compared to no-cover crop treatments. Cover crop also had a significant effect on P and β-glucosidase scores, which were significantly greater in the vetch cover crop as compared to no cover crop and wheat. The hairy vetch cover crop resulted in a significantly greater SQI as compared to no-cover crop and wheat.

Jokela et al. (2009) studied the effect of cover crops and liquid manures on soil quality indicators in a corn silage system. In this study, corn was grown for four years on a Bertrand silt loam in rotation with a number of crops and SQI was determined. Table 9 shows the change of soil quality indicators. They reported that cover crop treatment increased the active or labile carbon significantly and showed a good relationship with aggregate stability and microbial biomass. Overall, the use of cover crops with companion crops appears beneficial for corn silage systems. The SQI, a composite of scores, based on five soil parameters, ranged from 80–87 in the 0–5 cm depth and 73–83 in the 5–15 cm depth. The SQI differences were mainly due to differences in scores for total organic carbon, water stable aggregates and bulk density, even though differences in treatments were not significant for some of the individual sores (Table 9).

Table 8 Tillage and cover crop effect on soil quality indicators (average) based on the scores as determined by soil-management assessment framework (SMAF) (*Source*: Mbuthia et al., 2015).

Treatment	TOC	P	K	pH	BD	BG	MBC	SQI (%)
C Crop								
Vetch	0.44	0.98	0.99	0.95	0.97	0.24	0.88	70.6
Wheat	0.38	0.95	1.01	0.97	0.94	0.19	0.83	68.2
No cover	0.35	0.95	1.01	0.96	0.95	0.24	0.78	67.5
Tillage								
No-till	0.44	0.93	1.00	0.96	0.93	0.27	0.79	68.5
Till	0.34	0.99	1.01	0.97	0.97	0.17	0.86	69.1
N-rate								
0	0.31	0.91	1.03	0.99	0.93	0.21	0.93	66.2
34	0.39	0.97	1.02	0.98	0.95	0.27	0.84	69.7
67	0.36	1.00	0.99	0.96	097	0.21	0.90	69
101	0.49	1.00	1.00	0.92	0.97	0.19	0.96	70.2

TOC: total organic carbon; P-phosphorous; K-potassium; BD-soil bulk density; BG: β-glucosidase; MBC: microbial biomass C; Nitrogen fertilisation rate (N-rate) – 0, 34, 67 and 101 N kg/ha; SQI: soil quality index (an integration of all the individual quality scores).

Table 9 Soil quality index (soil management-assessment framework) as affected by cover crops and liquid manure application (*Source*: Jokela et al., 2009).

Treatment	TOC	AGG	pH	STP	BD	SQI
			Scores			
			0–5 cm depth			
KC-F	0.83	0.91	0.99	1.00	0.60	86.7
KC-C	0.84	0.94	0.95	1.00	0.61	86.9
RC-F	0.75	0.89	0.99	1.00	0.50	82.3
RC-C	0.83	0.91	0.95	1.00	0.60	85.7
IR-C	0.82	0.89	0.97	1.00	0.68	87.2
WR-C	0.72	0.95	0.92	1.00	0.65	84.6
NC-Man	0.66	0.88	0.98	1.00	0.48	80.0
NC-FN	0.79	0.96	0.94	1.00	0.58	85.3
			5–15 cm depth			
KC-F	0.60	0.84	0.96	1.00	0.50	78.0
KC-C	0.62	0.95	0.99	1.00	0.57	82.6
RC-F	0.59	0.79	0.94	1.00	0.46	75.5
RC-C	0.61	0.92	0.98	1.00	0.55	81.1
IR-C	0.60	0.90	0.91	1.00	0.53	78.7
WR-C	0.55	0.84	0.97	0.99	0.51	77.0
NC-Man	0.48	0.81	0.93	1.00	0.44	73.2
NC-FN	0.62	0.91	0.91	1.00	0.62	82.5

KC-C, Kura clover-corn; KC-F, Kura clover forage; RC-C, red clover-corn; RC-F, red clover forage; IR-C, Italian ryegrass-corn; WR-C, winter rye-corn; NC-FN, no-cover, N fertilised corn; NC-Man, no-cover, manured corn. TOC, total organic carbon; AGG, aggregate stability (from Macro-All); pH, soil pH; STP, soil test P (Bray-1); BD, bulk density; SQI, soil quality index (Soil Management-assessment Framework).

5. Conclusion

Improved soil quality or health is very important to support global food security. Cover crops improve soil quality and soil health by increasing biodiversity and efficiency, reducing soil erosion, improving soil aggregate stability and hydraulic properties, decreasing soil compaction, and increasing SOM, microbial activity and nutrient recycling. Soil-quality improvement is related to the characteristics of cover crops and their varieties, such as legume, non-legume, grasses, *Brassicas*, etc. These blends or individual characteristics of the cover crops could influence the effectiveness of the soil functions associated with soil quality and soil health. However, cover crops are highly site-specific and dependent on climate, soil texture, management practices and cover crop species and their integration in cropping systems. In addition, cover crop species that have a leafy biomass and fibrous root system are found to be more effective in protecting soil erosion when compared to a tap root system. The findings from this chapter emphasise the importance of long-term farmer field studies with standardised management protocols to confirm cover crops' effects on soil microbial enzyme activities and other quality parameters.

References

Abbasi, M.K., Zafar, M. and Khan, S.R. (2007). Influence of different land-cover types on the changes of selected soil properties in the mountain region of Rawalakot in Azad Jammu and Kashmir. Nutr. Cycl. Agroecosyst., 78: 97–110. Springer. doi:10.1007/s10705-006-9077-z.

Acuña, J.C.M. and Villamil, M.B. (2014). Short-term effects of cover crops and compaction on soil properties and soybean production in Illinois. Agron. J. 106: 860–870. DOI:10.2134/agronj13.0370.

Allen, D.E., Singh, B.P. and Dalal, R.C. (2011). Soil health indicators under climate change: a review of current knowledge. pp. 25–45. *In*: Singh, B.P., Cowie, A.L. and Chan, K.Y. (eds.). Soil Health and Climate Change. Springer-Verlag, Berlin, Heidelberg.

Andrews, S.S., Karlen, D.L. and Mitchell, J.P. (2002). A comparison of soil quality indexing methods for vegetable production systems in Northern California. Agric. Ecosyst. Environ., 90: 25–45, Elsevier, DOI:10.1016/S0167-8809(01)00174-8.

Andrews, S.S., Karlen, D.L. and Cambardella, C.A. (2004). The soil management assessment framework. Soil Sci., Soc. Am. J., 68: 1945–1962. doi:10.2136/sssaj2004.1945.

Aziz, I., Mahmood, T. and Islam, K.R. (2013). Effect of long-term no-till and conventional tillage practices on soil quality. Soil Tillage Res., 131: 28–35.

Balota, E.L. and Chaves, J.C.D. (2010). Enzymatic activity and mineralisation of carbon and nitrogen in soil cultivated with coffee and green manures. Rev. Bras. Cienc. do Solo, 34: 1573–1583. doi:10.1590/s0100-06832010000500010.

Bandick, A.K. and Dick, R.P. (1999). Field management effects on soil enzyme activities. Soil Biol. Biochem., 31: 1471–1479. DOI:10.1016/S0038-0717(99)00051-6.

Bardgett, R.D. and Van Der Putten, W.H. (2014). Belowground biodiversity and ecosystem functioning. Nature, 515: 505–511. doi:10.1038/nature13855.

Basche, A.D. and DeLonge, M.S. (2019). Comparing infiltration rates in soils managed with conventional and alternative farming methods: A meta-analysis. PLOS One, 14. DOI:10.1371/journal.pone.0215702.

Blanco-Canqui, H., Shaver, T.M., Lindquist, J.L., Shapiro, C.A., Elmore, R.W., Francis, C.A. and Hergert, G.W. (2015). Cover crops and ecosystem services: Insights from studies in temperate soils. Agron. J., 107: 2449–2474. DOI:10.2134/agronj15.0086.

Blanco-Canqui, H. and Jasa, P.J. (2019). Do grass and legume cover crops improve soil properties in the long term? Soil Sci. Soc. Am. J., 83: 1181–1187. DOI:10.2136/sssaj2019.02.0055.

Bosarge, T. (2015). A Soil Quality Index for Alabama. DOI:10.1017/CBO9781107415324.004.

Carson, Jennifer. (2013). How much Carbon can Soil Store? New South Wales Department of Primary Industries, New South Wales. (Accessed on 19 August 2020). http://soilquality.org.au/factsheets/how-much-carbon-can-soil-store-nsw.

Clark, A. (2012). Managing cover crops profitably (3rd ed.), Handbook Series Book 9. Sustainable Agriculture Research & Education. https://www.sare.org/wp-content/uploads/Managing-Cover-Crops-Profitably.pdf.

Dabney, S.M., Delgado, J.A. and Reeves, D.W. (2001). Using winter cover crops to improve soil and water quality. Commun. Soil Sci. Plant Anal., 32: 1221–1250. DOI:10.1081/CSS-100104110.

Dabney, S.M., Delgado, J.A., Meisinger, J.J., Schomberg, H.H., Liebig, M.A., Kaspar, T., Mitchell, J. and Reeves, W. (2010). Using cover crops and cropping systems for nitrogen management. pp. 231–82. *In*: Delgado, J.A. and Follett, R.F. (eds.). Advances in Nitrogen Management for Water Quality. Ankeny, IA: Soil & Water Conserv. Soc.

Das, S.K. and Varma, A. (2010). Role of Enzymes in Maintaining Soil Health, pp. 25–42, Springer, Berlin, Heidelberg. DOI:10.1007/978-3-642-14225-3_2.

De Baets, S., Poesen, J., Meersmans, J. and Serlet, L. (2011). Cover crops and their erosion-reducing effects during concentrated flow erosion. Catena, 85: 237–244. DOI:10.1016/j.catena.2011.01.009.

Demir, Z., Tursun, N. and Işık, D. (2019). Effects of different cover crops on soil quality parameters and yield in an apricot orchard. Intl. J. Agric. Biol., 21: 399–408. DOI:10.17957/IJAB/15.0000.

Doran, J.W. and Parkin, T.B. (1994). Defining and assessing soil quality. pp 3–21. *In*: Defining Soil Quality for a Sustainable Environment, Proc. Symposium, Minneapolis, MN, 1992, SSSA/ASA; special publication, 35. DOI:10.2136/sssaspecpub35.c1.

Doran, J.W. and Zeiss, M.R. (2000). Soil health and sustainability: Managing the biotic component of soil quality. Appl. Soil Ecol., 15: 3–11. DOI:10.1016/S0929-1393-067-6.

Douds, D.D., Nagahashi, G., Pfeffer, P.E., Kayser, W.M. and Reider, C. (2005). On-farm production and utilisation of arbuscular mycorrhizal fungus inoculum. Can. J. Plant Sci., 85: 15–21. DOI:10.4141/p03-168.

Fernandez, A.L., Sheaffer, C.C., Wyse, D.L., Staley, C., Gould, T.J. and Sadowsky, M.J. (2016). Associations between soil bacterial community structure and nutrient cycling functions in long-term organic farm soils following cover crop and organic fertilizer amendment. Sci. Total Environ., 566-567: 949–959. DOI:10.1016/j.scitotenv.2016.05.073.

Finney, D.M., White, C.M. and Kaye, J.P. (2016). Biomass production and carbon/nitrogen ratio influence ecosystem services from cover crop mixtures. Agron. J., 108: 39–52. DOI:10.2134/agronj15.0182.

Finney, D.M., Buyer, J.S. and Kaye, J.P. (2017). Living cover crops have immediate impacts on soil microbial community structure and function. J. Soil Water Conserv., 72: 361–373. DOI:10.2489/jswc.72.4.361.

Haney, R.L., Franzluebbers, A.J., Jin, V.L., Johnson, M.-V., Haney, E.B., White, M.J. and Harmel, R.D. (2012). Soil organic C:N vs. water-extractable organic C:N. Open J. Soil Sci., 02: 269–274. DOI:10.4236/ojss.2012.23032.

Hargrove, W.L. (1991). Cover Crops for Clean Water, The proceedings of an international conference West Tennessee Experiment Station, April 9–11, 1991, Jackson, Tennessee. DOI:10.2307/3451509.

Hobbs, P.R., Sayre, K. and Gupta, R. (2008). The role of conservation agriculture in sustainable agriculture. Philos. Trans. R. Soc. B Biol. Sci., 363: 543–555. DOI:10.1098/rstb.2007.2169.

Hoorman, J.J. (2009). Cover Crops to Improve Soil and Water Quality. [Online] Available: http://www.mccc.msu.edu/states/Ohio/OH_CoverCrops_to_Improve_Soi&Water_Quality.pdf [2020 Aug. 16].

Hubbard, R.K., Strickland, T.C. and Phatak, S. (2013). Effects of cover crop systems on soil physical properties and carbon/nitrogen relationships in the coastal plain of southeastern USA. Soil Tillage Res., 126: 276–283. DOI:10.1016/j.still.2012.07.009.

Huddleston, J.H. (1984). Development and use of soil productivity ratings in the United States. Geoderma, 32: 297–317.

Islam, K.R. (1996). Active carbon as a measure of soil quality, Ph.D. dissertation, Dept. of Natural Resources and Landscape Architecture, University of Maryland, College Park, MD, USA.

Islam, K.R. and Weil, R.R. (2000). Land use effects on soil quality in a tropical forest ecosystem of Bangladesh. Agric. Ecosyst. Environ., 79: 9–16, Elsevier. DOI:10.1016/S0167-8809(99)00145-0.

Islam, K.R. and Sundermeier, A.P. (2008). Soil Quality Test Kit—A simple test for active organic matter as a measure of soil quality. CFAES Agriculture and Natural Resources Factsheet SAG-4. https://ohioline.osu.edu/factsheet/SAG-4.

Islam, K.R. (2010). Cover Crops Impact on Soil Quality and Crop Yield, ASA-CSA-SSA International Meetings, October 31–November 4, Long Beach, Ca.

Ishaq, S., Begum, F., Ali, K., Ahmed, S., Ali, S., Karim, R., Ali, H. and Durrani, S.A. (2014). Assessment of soil quality under different land use practices in Altit Valley, Hunza Nagar, Gilgit-Baltistan. Proc. Int. Conf. For. Soil Rural Livelihoods: A Change. in Clim., Kathmandu, Nepal, 27–30 Sept. 2014: 43–54, Kathmandu University.

Jokela, W.E., Grabber, J.H., Karlen, D.L., Balser, T.C. and Palmquist, D.E. (2009). Cover crop and liquid manure effects on soil quality indicators in a corn silage system. Agron. J., 101: 727–737. DOI:10.2134/agronj2008.0191.

Kai, M., Takazumi, K., Adachi, H., Wasaki, J., Shinano, T. and Osaki, M. (2002). Cloning and characterisation of four phosphate transporter cDNAs in tobacco. Plant Sci., 163: 837–846. DOI:10.1016/S0168-9452(02)00233-9.

Kalu, S., Koirala, M., Raj Khadka, U. and C, A.K. (2015). Soil quality assessment for different land use in the Panchase area of Western Nepal. Int. J. Environ., 5: 38–43. doi:10.5963/IJEP0501006.

Karlen, D.L., Mausbach, M.J., Doran, J.W., Cline, R.G., Harris, R.F. and Schuman, G.E. (1997). Soil quality: a concept, definition and framework for evaluation (A Guest Editorial). Soil Sci. Soc. Am. J., 61: 4–10, Wiley. DOI:10.2136/sssaj1997.03615995006100010001x.

Kaspar, T.C., Radke, J.K. and Laflen, J.M. (2001). Small grain cover crops and wheel traffic effects on infiltration, runoff and erosion. J. Soil Water Conserv., 56.

Kätterer, T., Bolinder, M.A., Andrén, O., Kirchmann, H. and Menichetti, L. (2011). Roots contribute more to refractory soil organic matter than above-ground crop residues, as revealed by a long-term field experiment. Agric. Ecosyst. Environ., 141: 184–192. DOI:10.1016/j.agee.2011.02.029.

Kaye, J.P. and Quemada, M. (2017). Using cover crops to mitigate and adapt to climate change. A review 2. Agron. Sustain. Dev., 37: 4. DOI:10.1007/s13593-016-0410-x.

Kaye, J., Finney, D., White, C., Bradley, B., Schipanski, M., Alonso-Ayuso, M., Hunter, M., Burgess, M. and Mejia, C. (2019). Managing nitrogen through cover crop species selection in the U.S. Mid-Atlantic. PLOS One, 14: e0215448. DOI:10.1371/journal.pone.0215448.

Lal, R. (2001). World cropland soils as a source or sink for atmospheric carbon. Adv. Agron., 71: 145–191. DOI:10.1016/s0065-2113(01)71014-0.

Lal, R. (2015). Soil carbon sequestration and aggregation by cover cropping. J. Soil & Water Conserve., 70: 329–39.

Lehman, R.M., Taheri, W.I., Osborne, S.L., Buyer, J.S. and Douds, D.D. (2012). Fall cover cropping can increase arbuscular mycorrhizae in soils supporting intensive agricultural production. Appl. Soil Ecol., 61: 300–304. DOI:10.1016/j.apsoil.2011.11.008.

Lüscher, A., Mueller-Harvey, I., Soussana, J.F., Rees, R.M. and Peyraud, J.L. (2014). Potential of legume-based grassland-livestock systems in Europe: A review. Grass Forage Science, 69(2): 206–228. https://doi.org/10.1111/gfs.12124.

Makoi, J.H.J.R. and Ndakidemi, P.A. (2008). Selected soil enzymes: Examples of their potential roles in the ecosystem. African J. Biotechnol., 7: 181–191. doi:10.5897/AJB07.590.

Makoi, J.H.J.R., Chimphango, S.B.M. and Dakora, F.D. (2010). Elevated levels of acid and alkaline phosphatase activity in roots and rhizosphere of cowpea (*Vigna unguiculata* L. Walp.) genotypes grown in mixed culture and at different densities with sorghum (*Sorghum bicolor* L.). Crop Pasture Sci., 61: 279. doi:10.1071/CP09212.

Maseko, S.T. and Dakora, F.D. (2013). Rhizosphere acid and alkaline phosphatase activity as a marker of P nutrition in nodulated *Cyclopia* and *Aspalathus* species in the Cape fynbos of South Africa. South African J. Bot., 89: 289–295. doi:10.1016/j.sajb.2013.06.023.

Mbuthia, L.W., Acosta-Martínez, V., DeBryun, J., Schaeffer, S., Tyler, D., Odoi, E., Mpheshea, M., Walker, F. and Eash, N. 2015. Long term tillage, cover crop and fertilisation effects on microbial community structure, activity: Implications for soil quality. Soil Biol. Biochem., 89: 24–34. DOI:10.1016/j.soilbio.2015.06.016.

McVay, K.A., Radcliffe, D.E. and Hargrove, W.L. (1989). Winter legume effects on soil properties and nitrogen fertiliser requirements. Soil Sci. Soc. Am. J., 53: 1856–1862. DOI:10.2136/sssaj1989.03615995005300060040x.

Merino, C., Godoy, R. and Matus, F. (2016b). Soil enzymes and biological activity at different levels of organic matter stability. J. Soil Sci. Plant Nutr., 16: 14–30. DOI:10.4067/S0718-95162016005000002.

Miltner, A., Bombach, P., Schmidt-Brücken, B. and Kästner, M. (2012). SOM genesis: Microbial biomass as a significant source. Biogeochemistry, 111: 41–55. doi:10.1007/s10533-011-9658-z.

Mobley, H.L., Island, M.D. and Hausinger, R.P. (1995). Molecular biology of microbial ureases. Microbiol. Mol. Biol. Rev., 59.

Moebius-Clune, B., Moebius-Clune, D.J., Gugino, B., Idowu, O., Schindelbeck, R., Ristow, A., van Es, H., Thies, J., Shayler, H.A., McBride, M.B., Wolfe, D. and Abawi, G. (2016). Improving Aggregate Stability, Agronomy Factsheet 95, Cornell University Cooperative Exten. http://nmsp.cals.cornell.edu/publications/factsheets/factsheet95.pdf[2020 Aug. 16].

Moebius-Clune, B., Moebius-Clune, D.J., Gugino, B., Idowu, O., Schindelbeck, R., Ristow, A., van Es, H., Thies, J., Shayler, H.A., McBride, M.B., Wolfe, D. and Abawi, G. (2016). Comprehensive Assessment of Soil Health. http://www.nysaes.cornell.edu [2020 Aug. 16].

Mukumbareza, C., Muchaonyerwa, P. and Chiduza, C. (2015). Effects of oats and grazing vetch cover crops and fertilization on microbial biomass and activity after five years of rotation with maize. South Afr. J. Plant Soil, 32: 189–197. DOI:10.1080/02571862.2015.1025446.

Nair, A. and Ngouajio, M. (2012). Soil microbial biomass, functional microbial diversity and nematode community structure as affected by cover crops and compost in an organic vegetable production system. Appl. Soil Ecol., 58: 45–55. DOI:10.1016/j.apsoil.2012.03.008.

Nannipieri, P., Giagnoni, L., Renella, G., Puglisi, E., Ceccanti, B., Masciandaro, G., Fornasier, F., Moscatelli, M.C. and Marinari, S. (2012). Soil enzymology: Classical and molecular approaches. Biol. Fertil. Soils, 48: 743–762. DOI:10.1007/s00374-012-0723-0.

Nevins, C.J., Lacey, C. and Armstrong, S. (2020). The synchrony of cover crop decomposition, enzyme activity and nitrogen availability in a corn agro-ecosystem in the Midwest United States. Soil Tillage Res., 197. DOI:10.1016/j.still.2019.104518.

Piotrowska-Długosz, A. and Charzyński, P. (2014). The impact of the soil-sealing degree on microbial biomass, enzymatic activity and physicochemical properties in the Ekranic Technosols of Toruń (Poland). J. Soils Sed., 15: 47–59. DOI:10.1007/s11368-014-0963-8.

Pham, T.G., Nguyen, H.T. and Kappas, M. (2018). Assessment of soil quality indicators under different agricultural land uses and topographic aspects in Central Vietnam. Int. Soil Water Conserv. Res., 6: 280–288. DOI:10.1016/j.iswcr.2018.08.001.

Poeplau, C. and Don, A. 2015. Carbon sequestration in agricultural soils via cultivation of cover crops—A meta-analysis. Agric. Ecosys. Environ., 200: 33–41. DOI:10.1016/j.agee.2014.10.024.

Ramos, C., Mølbak, L. and Molin, S. (2000). Bacterial activity in the rhizosphere analysed at the single-cell level by monitoring ribosome contents and synthesis rates. Appl. Environ. Microbiol., 66: 801–809, Amer. Soc. Microb. DOI:10.1128/AEM.66.2.801-809.2000.

Rankoth, L.M., Udawatta, R.P., Veum, K.S., Jose, S. and Alagele, S. (2019). Cover crop influence on soil enzymes and selected chemical parameters for a claypan corn-soybean rotation. Agric., 9: 125. DOI:10.3390/agriculture9060125.

Reeves, D.W. (1994). Cover crops and rotations. pp. 125–172. *In*: Hatfield, J.L. and Stewart, B.A. (eds.). Crop Residue Management. Boca Raton, FL: Lewis Publishers, Soil Sci. Soc. Amer.

Ruis, S.J. and Blanco-Canqui, H. (2017). Cover crops could offset crop residue removal effects on soil carbon and other properties: A review. Agron. J., 109: 1785–1805. DOI:10.2134/agronj2016.12.0735.

Schipanski, M.E., Barbercheck, M., Douglas, M.R., Finney, D.M., Haider, K., Kaye, J.P., Kemanian, A.R., Mortensen, D.A., Ryan, M.R., Tooker, J. and White, C. (2014). A framework for evaluating ecosystem services provided by cover crops in agro-ecosystems. Agric. Syst., 125: 12–22. DOI:10.1016/j.agsy.2013.11.004.

Sharma, V., Irmak, S. and Padhi, J. (2018). Effects of cover crops on soil quality: Part I. Soil chemical properties—organic carbon, total nitrogen, pH, electrical conductivity, organic matter content, nitrate-nitrogen and phosphorus. J. Soil Water Conserv., 73: 637–651. DOI:10.2489/jswc.73.6.637.

Spiegel, B. (2017). Soil health tests—Can you measure what cannot be seen? Buyers' Guide, Successful Farming at Agriculture.com, January 2017, pp. 52–54.

Tautges, N.E., Chiartas, J.L., Gaudin, A.C.M., O'Geen, A.T., Herrera, I. and Scow, K.M. (2019). Deep soil inventories reveal that impacts of cover crops and compost on soil carbon sequestration differ in surface and subsurface soils. Glob. Chang. Biol., 25: 3753–3766. DOI:10.1111/gcb.14762.

Thapa, R., Mirsky, S.B. and Tully, K.L. (2018). Cover crops reduce nitrate leaching in agro-ecosystems: a global meta-analysis. J. Environ. Qual., 47: 1400–1411. DOI:10.2134/jeq2018.03.0107.

Verdoucq, L., Czjzek, M., Moriniere, J., Bevan, D.R. and Esen, A. (2003). Mutational and structural analysis of aglycone specificity in maize and sorghum β-glucosidases. J. Biol. Chem., 278: 25055–25062. DOI:10.1074/jbc.M301978200.

Veum, K.S., Goyne, K.W., Kremer, R.J., Miles, R.J. and Sudduth, K.A. (2014). Biological indicators of soil quality and soil organic matter characteristics in an agricultural management continuum. Biogeochemistry, 117: 81–99. DOI:10.1007/s10533-013-9868-7.

Ve Verka, J.S., Udawatta, R.P. and Kremer, R.J. (2019). Soil health indicator responses on Missouri claypan soils affected by landscape position, depth and management practices. J. Soil Water Conserv., 74: 126–137. DOI:10.2489/jswc.74.2.126.

Villamil, M.B., Bollero, G.A., Darmody, R.G., Simmons, F.W. and Bullock, D.G. (2006). No-till corn/soybean systems including winter cover crops. Soil Sci. Soc. Am. J., 70: 1936–1944. DOI:10.2136/sssaj2005.0350.

Wagger, M.G. and Denton, H.P. (1989). Influence of cover crop and wheel traffic on soil physical properties in continuous no-till corn. Soil Sci. Soc. Am. J., 53: 1206–1210. DOI:10.2136/sssaj1989.03615995005300040036x.

Weerasekara, C.S., Udawatta, R.P., Gantzer, C.J., Kremer, R.J., Jose, S. and Veum, K.S. (2017). Effects of cover crops on soil quality: selected chemical and biological parameters. Commun. Soil Sci. Plant Anal., 48: 2074–2082. DOI:10.1080/00103624.2017.1406103.

Weil, R.R., Islam, K.R., Stine, M.A., Gruver, J.B. and Samson-Liebig, S.E. (2003). Estimating active carbon for soil quality assessment: A simplified method for laboratory and field use. Am. J. Alter. Agri., 18(1): 3–17. DOI:10.1079/AJAA2003003.

White, C.M., Kemanian, A.R. and Kaye, J.P. (2014). Implications of carbon saturation model structures for simulated nitrogen mineralisation dynamics. Biogeosciences, 11: 6725–6738. DOI:10.5194/bg-11-6725-2014.

White, C.M., Finney, D.M., Kemanian, A.R. and Kaye, J.P. (2016). A model-data fusion approach for predicting cover crop nitrogen supply to corn. Agron. J., 108: 2527–2540. DOI:10.2134/agronj2016.05.0288.

Wymore, A.W. (1993). Model-based Systems Engineering, first ed., CRC Press, Boca Raton.

10

Cover Crops for Orchard Soil Management

Biswajit Das,* BK Kandpal and *H Lembisana Devi*

1. Introduction

An orchard is a demarcated area where fruit trees are planted in a systematic pattern and managed to maximise productivity in a sustainable manner. Orchard management practices (OMPs) begin with the transformation of natural vegetation into an orchard and continues throughout the fruit-bearing phases. An ideal integrated orchard management practice (IOMP) comprises of pit management at the time of planting, annual schedule of integrated nutrient management (INM), management of tree basins and tree canopies, orchard floor management (OFM) and residue recycling. The INM is generally formulated with organic manures, composts, chemical fertilisers, biofertilisers, lime and gypsum amendments, green manuring or cover crops and growth-stimulating nutrient compositions. The IOMPs vary with the crop type, agro-climates, growing season, soil properties, topography, fruit tree age, etc. Based on agro-climatic zones, fruit trees are classified as either temperate or tropical.

Temperate fruit trees include pome fruits—apple (*Malus* × *domestica*) and pear (Japanese: *Pyrus pyrifolia* and European: *Pyrus communis*); stone fruits—peach (*Prunus persica*), nectarine (*Prunus persica* var. *nucifera*), apricot (*Prunus armeniaca*), plum (Japanese: *Prunus salicina* and European: *Prunus domestica*), cherry (sweet cherry: *Prunus avium* and sour cherry: *Prunus cerasus*); nut fruits—walnut (*Juglans regia*), pecannut (*Carya illinoinensis*), almond (*Prunus amygdalus*), hazelnuts (*Corylus avellana*), pistachio nut (*Pistachia vera*), chestnut (*Castanea sativa*), etc.; berries and currents—blueberries (*Vaccinium* spp.), blackberry (*Rubus fruticosis*), raspberry (European: *Rubus idaeus* and American: *Rubus strigosus*), strawberry (*Fragaria* × *ananassa*), black currant (*Ribes nigrum*), red currant (*Ribes rubrum*), white currant (*Ribes sativum*), cape gooseberry (*Physalis peruviana*); and other vine crops—grapes (*Vitis vinifera*) and kiwi (*Actinidia chinensis*).

Tropical fruit trees include: subtropical fruits—mango (*Mangifera indica*), litchi (*Litchi chinensis*), banana (*Musa* spp.), citrus fruits (*Citrus* spp.), pomegranate (*Punica granatum*), guava (*Psidium guajava*), persimmon (*Diospyros kaki*), jackfruit (*Artocarpus heterophyllus*), pineapple (*Ananas comosus*), starfruit (*Averrhoa carambola*), papaya (*Carica papaya*), custard apple (*Annona squamosa*), sapodilla (*Manilkara zapota*), mangosteen (*Garcinia mangostana*), rambutan (*Nephelium lappaceum*), durian (*Durio zibethinus*), loquat (*Eriobotrya japonica*), date palm (*Phoenix dactylifera*), coconut (*Cocos neucifera*), longan (*Dimocarpus longan*) and many other minor fruit crops.

World fruit production has increased many folds over the past few decades and there has been a substantial increase in the area of fruit-tree plantations. Total estimated global fruit production at

ICAR-Research Complex for North-Eastern Hill Region, Tripura Center, Lembucherra-799210, Tripura, India.
* Corresponding author: biswajitsom_dr@yahoo.co.in

present is 865.6 million tons over a coverage area of 65.2 million ha, with a mean productivity of 13.3 ton ha^{-1} (FAO, 2017). This is around 200 million tons more than the 27.2 million ha area with a mean productivity of 7.3 ton ha^{-1} reported in 1961. An increasing trend in fruit production has been followed in all countries globally (Table 1). This was possible only because of the development of improved varieties and advanced production technologies, which constitute the IOMPs that are more soil fertility-enriching and productive than conventional orchard systems.

Cover crops represent an important component of OFM and are widely adopted in modern fruit production systems. This has helped to increase the productivity and quality of fruits, though area expansion was not significant for some fruit crops in certain countries. In orchards, cover crops overcome the disadvantages of clean cultivation, as under clean cultivation the orchard floor (i.e., interrow spaces) is kept barren. Orchard soils under clean cultivation become prone to soil-quality degradation due to accelerated surface runoff and windstorms and are exposed to direct solar radiations, resulting in higher soil temperature, loss of soil moisture, and surface drying, cracking and compaction. Loss of soil organic matter (SOM) and essential nutrients, as well as reduced soil microbial populations and lack of biodiversity, are major problems for clean cultivation.

Cover crops are those that are grown to cover the ground surface in the field of a regular or main crop without any interference to the growth and development of the regular crop. The cultivation time of these crops may be decided as per the agro-climatic zone, land conditions, nature of the regular crop and its growing season and the benefits of such crops. Considering the role of cover crops in a sustainable farming system—compared to mono-cropping, clean cultivation, deep tillage, and off-season fallow land—diversified cover crop cultivation has been considered an integral component of ecofriendly and conservation agricultural practices (CAP) to support ecosystem services (Sharma et al., 2018). Thus, the use of cover crops is a suitable component for ecological

Table 1 Increase in fruit production and productivity in some major fruit-producing countries (FAO, 2017).

Country	Crop	Area (000 ha)		Production (x 000 ton)		Productivity (ton ha^{-1})	
		1961	2017	1961	2017	1961	2017
USA	Apple	184.8	130.7	2,584.0	5,173.7	14.0	39.6
	Grapes	207.0	405.0	2,952.8	6,679.2	14.3	16.5
	Oranges	225.7	214.4	4,583.5	4,615.8	20.3	21.5
	Pears	35.8	18.8	598.0	678.0	16.7	36.1
	Peaches/ Nectarines	142.0	45.3	1,759.0	775.2	12.4	17.1
	Almonds	36.1	404.7	60.2	1,029.7	1.7	2.5
India	Apple	44.5	305.0	185.0	2,265.0	4.2	7.4
	Mango	850.0	2,212.0	6,988.0	19,506.0	8.2	8.8
	Peaches/ Nectarines	10.0	40.6	43.0	290.0	4.3	7.1
	Banana	165.0	860.0	2,257.0	30,477.0	13.7	35.4
	Grapes	4.4	137.0	70.0	2,922.0	16.0	21.3
China	Apple	90.0	2,220.4	167.0	41,391.5	2.0	18.6
	Pears	120.1	957.2	482.0	16,528.0	4.0	17.3
	Grapes	10.2	778.6	71.8	13,160.8	7.0	17.0
	Banana	13.2	381.3	177.7	11,423.0	13.5	30.0
	Oranges	16.4	511.7	43.6	8,685.8	2.7	17.0
	Peaches/ Nectarines	60.3	781.9	431.4	1,429.5	7.2	18.3

intensification approaches where anthropogenic inputs are replaced by ecofriendly management approaches without causing any damage to the existing ecosystem and productivity is enhanced by interventions based on CAP, which serve the ecological services (Wittwer et al., 2017). Planting of cover crops in orchards has been practiced since early times and objectives and cultivation practices of cover crops are different from traditional agricultural field operations (Ellenwood and Gorley, 1937; Lipecki and Berbec, 1997; Ingels et al., 2003; Jannoyer et al., 2011).

2. Cover Crops in Orchard

Cover crops in any orchard are cultivated or maintained to protect the surface soil, increase soil organic matter (SOM) content and create a congenial microenvironment for soil microbes, thereby, improving soil quality. These types of crops comprised a broader group of plants with special characteristics that make them suitable for growing in orchards. Cover crops may be cultivated in orchards as a seasonal crop or as permanent sod culture. Choice of crops depends on the agro-climatic zone, type and age of orchard, season, soil properties (especially pH) and purpose.

Type of crops may include vegetable/cereals/pulses/oil seed crops, green manuring crops, fodder crops, or permanent grass sods. The main differences in cover crop systems in an orchard versus agricultural land is: (1) with orchards, cover crops are grown in the inter-spaces (drive rows/ alleyways) of fruit trees as seasonal or permanent culture; and (2) in agricultural fields, cover crops are grown in the off-season as companion crops, relay crops, or to fill the time gap between main cropping seasons. These short duration crops' role in an agricultural field is to keep the soil active in terms of soil biodiversity, fertility, physical properties and to harvest some financial benefit instead of leaving the soil barren for a considerable period. However, in orchards, cover crop cultivation is a year-round continuous process throughout the orchard's life span.

3. Advantages of Cover Crops in Orchards

Cover crops may be annual or perennial in nature, comprising of grasses, legumes, vegetables, or agricultural crops that provide multifaceted benefits in the orchard ecosystem (Evans et al., 1988; Ingels et al., 1998; Sullivan, 2003; Sanchez et al., 2007; Kaspar and Singer, 2011; Martin, 2012; Steyn et al., 2014). These advantages include:

1. Protect soil erosion by stabilising the soil surface and building soil resistance to excessive runoff and windstorms.
2. Cover crops are selective crops that suppress the growth of noxious weeds in the orchard floor.
3. Prevent the soil from drying and cracking from excessive evaporation under strong solar radiation, or in response to prolonged drought.
4. Minimise nutrient (especially nitrate) leaching by enhancing recycling and utilisation by non-leguminous crops, and symbiotic N_2 fixation by leguminous cover crops.
5. Improve soil quality by improving biodiversity and efficiency, accumulating SOM and soil balancing and through soil physical stability.
6. Long-term cover crop cultivation makes the soil more friable and improves soil aggregate stability with a reduction in compaction (bulk density).
7. Improves hydraulic conductivity and water-holding capacity for greater soil moisture storage and availability to succeeding crops.
8. Act as mulches to maintain favorable soil temperature and aeration.
9. Apart from soil microbes, earthworm populations and other beneficial soil organisms also increase.
10. Eliminates the harmful effects of the herbicide application and minimizes weeds and barren soil management cost.

11. Compliance with the Environmental Quality Incentives Programme (NRCS, 2019).

12. Use of cover crops on the orchard floor is the best conservation horticultural practice with no-till or minimum tillage.

13. Flowering cover crops act as insectary crops by attracting pollinators, such as honeybees, bumble bees, syrphid flies, wasps, and as a host for hibernating many other predatory insects.

14. Cover crops' ultimate benefit is better fruit plant growth and higher production.

15. Serve as human food such as cereals, oil and pulses, as well as fodder for animals.

16. Act as a catch crop or filler crop and provide extra income apart from imparting soil health benefits during the juvenile period of the fruit trees, as well as during the fruit harvesting off-season of the year.

17. Help in establishing a rich and sustainable orchard ecosystem based on a conservation concept comprised of diversified flora and fauna with main fruit trees as the leader.

18. Lower C:N in leguminous crops facilitate faster decomposition and better recycling of nutrients for growing plants (SARE, 2012).

19. Legumes are compatible and synergistic with fruit crops and companion crops, and can be mixed or intercropped with other cover crops.

20. Eliminate loads of insects and pathogens in the orchards.

21. Legume crops emit less greenhouse gases compared to other filed crops (Jeuffroy et al., 2013), and together with fruit trees can sequester a significant amount of carbon in the plant biomass and soil.

22. Cultivation of these crops always enriches the aesthetic value of any orchard, and creates a better place for ecosystem services.

23. These crops, along with fruits trees, together act as a major player in carbon sequestration.

4. Suitable Cover Crops for Orchards

Cover crops comprise of a vast, but selective, group depending on their purpose for growing (Evans et al., 1988; Ingels et al., 1998; Jannoyer et al., 2011; Barney, 2012).

Grasses and Cereals: Several non-legume grasses and cereals, individually or in mixtures or blends, are grown as cover crops in orchard alleyways (Ingels et al., 1998; Tworkoski and Glenn, 2012). The erect and stout-growing grasses and cereals have clumps of tillers clasped in basal leaf sheath, whereas many of the creeping grasses grow horizontally through stems with numerous aboveground stolons and underground rhizomes covering the soil surface. The main feature of these grasses and cereals is their fibrous and shallow root system that binds and holds the soil firmly and acts as a soil conditioner through the incorporation of root excretion, decomposed roots and leaf biomass into the soil. Permanent grass sods are grown in the interrow spaces and maintained similarly to any lawn; however, proper distance is kept from trees. Tree rows and basins are kept bare or mulched to facilitate scheduled application of manures, fertilisers and irrigation, and other intercultural operations. Cultivation of forage grasses in orchards becomes necessary to meet the fodder demand of domestic animals. Under temperate conditions (Ahmad et al., 2017), tall fescue + clover and orchard grass + clover combinations in apple orchards were found to be better in terms of green fodder (26.2 and 23.7 ton ha^{-1}, respectively) and dry fodder production (12.1 and 10.7 ton ha^{-1}). Several of the grasses and cereals that are suitable as cover crops in orchards are described below (Evans et al., 1988; Ingels et al., 1998; Barney, 2012):

1. Ryegrass (annual ryegrass (*Lolium multiflorum*), perennial ryegrass (*Lolium perenne*), darnel (*Latium temulentum*) and winter rye (*Secale cereal*)): These are cool season grasses. Among them, annual ryegrass is fast-growing, non-spreading, bunchy, tillering type and may attain up

to 130 cm height with an extensive fibrous root system. Annual ryegrass may act as a nurse crop to cover the ground for a short duration in fall and spring seasons, followed by warm season cover crops regenerating from the dormancy stage in the summer. This is an excellent orchard cover crop as a weed suppressor because of its allelopathic nature, it functions as an erosion protector and soil builder with a higher rate of OM incorporation and is highly drought tolerant. It is a scavenger for nutrients and its higher N-use efficiency tends to draw excess leftover N from the soil, thereby reducing leaching (Clark, 2007). In cooler regions, it shows biennial nature and overwinters. It is also an excellent fodder crop. Perennial type is densely tufted, profusely tillering and attains a height up to 60 cm. Its adventitious shallow root system acts as a good soil stabiliser. Favourable growing period is fall and spring, whereas growth is suppressed and remains dormant during the hot summer months (Hannaway et al., 1999). Cutting is preferred when seeds are soft for fodder purpose. Winter ryegrass is a fast-growing annual, winter hardy and also called cereal or grain rye. It is tall (90–190 cm) with a fibrous shallow root system. Grains contain carbohydrate, gluten and are commonly used for bread and whiskey. Under unfavourable winter climate, it is the best cover crop sown in the fall and grows in spring to early summer. It is also a very good soil nitrogen and exchangeable potassium capture crop (Clark, 2007). However, the problem of ergotism may arise if infection of ergot fungus becomes severe.

2. Bermuda grass (*Cynodon dectylon*): This is a stoloniferous and rhizomatous creeping perennial grass. It is very hardy and suitable for hot, humid, sunny tropical and sub-tropical conditions. This type of sod is established by seeding, dribbling of stolons/rhizomes and turfing along the interrow spaces. Regular mowing is required to maintain the shape and height of the sod and Bermuda grass is very drought tolerant.

3. Orchard grass (*Dactylis glomerata*): This is a temperate perennial, bunch type, tall-growing grass (50–120 cm). Its non-rhizomatous roots system may go to deep soil. Seeds or seedlings are used for planting. Regular cutting of mature grasses is essential to restrict overgrowth, which may otherwise act as a hibernating place for pests and insects and affects the scheduled orchard cultural operations. It is drought tolerant and used as fodder.

4. Bent grass (*Agrostis* spp.): There are many species of bent grasses and some of those suitable for lawns and orchards are creeping bent grass (*Agrostis palustris*), colonia bent grass (*A. capillaries*), velvet bent grass (*A. canina*), brown bent grass (*A. vinealis*), and red top bent grass (*A. gigantea*). It is a perennial creeping, cool season grass, slender stem with stolons/rhizomes. It spreads very quickly by stolons or rhizomes and forms a dense mat. Bent grass roots are generally fibrous and shallow.

5. Fescus (*Festuca* spp.): Many types of fescue grasses are suitable for orchards with slender or strong creeping and tall-growing habits. They are cool season grasses. Under the fine fescue group (creeping and non-creeping types), there are creeping red fescue/chewing fescue (*Festuca rubra*), Idaho fescue (*Festuca idahoensis*), foxtail fescue (*Vulpia myuros* var. *hirsute*), and sheep fescue/hard fescue (*Festuca ovina*), whereas tall fescue (*Festuca arundinacea*) is a bunch type. Tall fescue grasses are coarse, bunch-type tillering clumps with erect growth, have leaves that are stout and flat, a dense root system, are shade loving and drought tolerant. Both tall and dwarf varieties of tall fescue are available. Fine fescues are low-growing and form short-statured clumps. Creeping fescue is a perennial grass that spreads by rhizomes and stolons, and its leaves are fine.

6. Timothy (*Phleum pretense*): Timothy is a perennial, cool season, bunch-type grass that may grow to around 100 cm in height. Its produces erect tillers in clumps with fibrous roots that are shallow and compact.

7. Oats (*Avena sativa*): Cool season, annual cereal crop with erect and stout clumps and fibrous shallow roots. It is a very efficient catch crop to uptake excess or leaching prone nutrients, such as nitrate and acts as a smother crop to restrict growth of weeds during the growing season. Its

allelopathic effect prevents the germination of weed seeds. Oats also act as a nurse crop for fall-sown legumes to help the legume crop survive the winter season (SARE, 2012).

8. Barley (*Hordeum vulgare*): Barley is an annual, cool season cereal crop that produces erect profuse tillers in clumps with fibrous shallow roots. With its fast-growing nature, it suppresses weeds and obnoxious plants. Its grains are used for commercial purpose and plant biomass is a good source of fodder.

9. Wheat (*Triticum aestivum*) and *wheat x rye* (*Triticale x Triticosecale*): These two crops are used for cereal grain and fodder. While mainly cultivated as field crops, they are also suitable as cover crops in orchards. They are very popular fodder as fresh and concentrate. These are cool season crops, have a bunchy erect growth habit and produce clumps of tillers with fibrous, shallow root systems.

Other grasses are California brome (*Bromus carinatus*), blue wild rye (*Elymus glaucus*), meadow barley (*Hordeum brachyantherum* ssp. *brachyantherum*), California barley (Hordeum brachyantherum ssp. *californicum*), California melic (*Melica californica*), nodding needlegrass (*Nassella cernua*), purple needlegrass (*Nassella pulchra*), pine bluegrass (*Poa secunda* ssp. *secunda*), creeping bluegrass (*Poa reptans*), Supina bluegrass (*Poa supine*), Kentucky bluegrass (*Poa pratensis*), buckwheat (*Fagopyrum esculentum*), Japan grass (*Zoysia japonica*), Korean grass/velvet grass/carpet grass (*Z. tenuifolia*), Manila grass (*Z. matella*), crabgrass (*Digitaria sanguinalis*), quackgrass (*Elytrigia repens*), buffalo grass/St. Augustine grass (*Stenotaphrum secundatum*), chain grass (*Sporobolus tremulus*), kikuyu grass (*Pennisetum clandestinum*), goosegrass (*Eleusine indica*), seashore grass (*Paspalum vaginatum*), sudangrass (*Sorghum sudanense*), sorghum-sudangrass (*Sorghum vulgare* × *S. sudanense*), soft chess (*Bromus hordeaceus* ssp. *Molliformis*), nut grass (*Cyperus rotundus*), love grass (*Chrysopogon aciculatus*), vetiver (*C. zizanioides*), chickweed (*Stellaria media*), shepherd's purse (*Capsella bursa-pastaris*), and scarlet pimpernel (*Anagallus arvensis*).

Some of the annual grasses are annual bluegrass, pearl millet, barley, corn and sorghum/sudan grass. Under the perennial group, orchard grasses are tall fescue, perennial ryegrass, Kentucky bluegrass, smooth brome grass, meadow foxtail, timothy, colonial bent grass, Bermuda grass, reed canary grass, wheat grass, big bluestem, switch grass and Indian grass. Sod-forming grasses have an extended network of stolons (overground stems) or/and rhizomes (underground stems) arising from adventitious buds at the crown tissues. This nature of extravaginally horizontal growth covers the soil surface and adventitious roots arise at the stolon nodes, which bind soils by clasping with dense and fibrous roots. Kentucky bluegrass, creeping foxtail, colonial bent grass, Bermuda grass, quackgrass, rough bluegrass, annual bluegrass and redtop are some examples of sod grasses. Most of the bunchgrasses do not form stolons or rhizomes, but produce profuse tillers. Several new plants arise from the adventitious buds at the basal crown tissue formed at the basal leaf sheath of the mother stem. Clumps of tillers, arising from the basal sheath, give erect extravaginally tufted growth. Perennial ryegrass, annual ryegrass, timothy, orchard grass, meadow foxtail, tall fescue, etc. are some bunch and erect growth habit grasses.

Legumes: This is a broad-spectrum group comprised of around 800 genera and 20,000 species under the family Fabaceae (Leguminosae) (Lewis et al., 2005). Most of them are annuals; however, perennial legumes such as clovers (yellow clover, Kura clover, etc.), sainfoin, milkvetch and alfalfa, as well as biennials, such as sweet clover, are widely grown as cover crops in orchards. They have podded fruits and a root nodule in association with *Rhizobium* bacteria (Graham and Vance, 2003). Growth habit may be dwarf, bushy, climbing vines, or tall-like pigeon peas. Some are cultivated for pulses/vegetables and many other species are used only as green fodder or manure purposes. Pulses contribute around 33 per cent of the dietary proteins and this is the second most important family after Poaceae (grass-cereal family) (Smykal et al., 2015). These crops are the most preferred cover crops in orchards for their special characteristics, such as wide adaptively, ability to fix atmospheric nitrogen, weed suppression, ability to produce a sufficient volume of biomass and are very good as an intercrop, relay intercrop, rotation and mixed crop. These crops are considered very good green manuring crops when incorporated into the soil by ploughing before flowering. Perennial clovers form very good permanent sods in the orchards without interfering with fruit tree growth and development, and orchard intercultural operations. Clovers are semi-erect dwarf and their dense, broad-obovate trifoliate leaves cover the soil surface with branching stems. Solitary flowers vary in colours depending on the species. Various legume cover crops include the following:

1. Tribe *Fabeae*: Grass pea/Indian pea (*Lathyrus sativus*), caley pea (*L. hirsutus*), sweet pea (*L. odoratus*), everlasting pea (*L. latifolius*), garden pea (*Pisum sativum*), Austrian winter pea (*P. sativum* subsp. *avense*), lentil (*Lens culinaris*), faba bean (*Vicia. faba*), purple vetch (*V. benghalensis*), hairy vetch (*V. villosa*), American vetch (*V. americana*), woolly pod vetch (*V. villosa* ssp. *dasycarpa*), tufted vetch (*V. cracca*) and Cahaba White vetch (*V. sativa* × *V. cordata*).

2. Tribe *Trifolieae*: Alfalfa *(Medicago sativa)*, bur clover (*M. polymorpha*), barrel medic (*M. truncatula*), white clover (*Trifolium repens*), red clover (*T. pratense*), alsike or Swedish clover (*T. hybridum*), meadow or zigzag clover (*T. medium*), hare's-foot trefoil (*T. arvense),* berseem clover (*T. alexandrinum),* rose clover (*T. hirtum*), subterranean clover (*T. subterraneum*), strawberry clover (*T. fragiferum*), crimson clover (*T. incarnatum*), arrow leaf (*T. vesiculosum*), ball clover (*T. nigresns*), Persian clover (*T. resupinatum*), Indian sweet clover (*Melilotus indicus*), white sweet clover (*M. alba*), yellow sweet clover (*M. officinalis*), Mediterranean sweet clover (*M. sulcatus*), volga sweet clover/Russian sweet clover (*M. wolgicus*) and meadow barley (*Hordeum brachyantherum* ssp. *brachyantherum*).

3. Tribe *Cicereae*: Cicer milkvetch (*Astragalus cicer*), liquorice milkvetch (*Astragalus glycyphyllos*), Chinese milkvetch (*Astragalus sinicus*) and chickpea (*Cicer arietinum*); some other species are *C. spongiosum, C. microphyllum, C. chorassanicum,* etc.

4. Tribe *Phaseoleae*: Common bean (*Phaseolus vulgaris*), tepary bean (*P. acutifolius*), runner bean (*P. coccineus*), year bean (*P. dumosus*), lima bean (*P. lunatus*), cowpea (*Vigna unguiculata*), yard long bean (*Vigna unguiculate* spp. *unguiculata*), green gram (*V. radiata*), adzuki bean (*V. angularis*), black gram (*V. mungo*), moth bean (*Vigna aconitifolia*), African gram/jungle mat bean (*V. trilobata*), bambara bean (*V. subterranean*) and soybean (*Glycine max*).

5. Tribe *Aeschynomeneae*: Groundnut (*Arachis hypogaea*) and forage peanut (*A. pintoi*).

6. Tribe *Genisteae*: Russel lupins (*Lupinus polyphyllus*), rainbow lupins (*L. regalis*), narrow leafed lupin (*L. angustifolius*), white lupin (*L. albus*) and yellow lupin (*L. luteus*).

Other legume crops are birds foot trefoil (*Lotus corniculatus*), velvet bean (*Mucuna* spp.), serradella (*Ornithopus sativus*), calopo/wild ground nut (*Calopogonium mucunoide*), sun hemp (*Crotalaria juncea*), showy rattle pod (*Crotalaria spectabilis*), jack bean (*Canavalia ensiformes*), pigeon pea (*Cajanus cajan*), black velvet bean (*Mucuna aterrina*), hyacinth bean (*Dolichos lablab*), sainfoin (*Onobrychis viciifolia*) and grey mucuna (*Mucuna conchinchinensis*).

***Brassicas and Forbs*:** This group includes broad-leaved, herbaceous, annual, cool season, non-leguminous plants. Mustard and rapeseeds under the genius *Brassica* comprise the group most used as cover crops in orchards (NRCS, 2015). The stem is erect and branches have a bushy growth habit (150–190 cm height) comprising of two types of leaves (i.e., pinnately cleft and lobed leaves at basal party and toothed or undivided on the slender stem). Flowers are borne on the spike in clusters—showy, yellow, or whitish (Ingels et al., 1998). Fruiting in capsules or siliques, profusely-seeded and seed colour varies in the range of brown, yellow, or black. In the temperate region, seeds are sown in spring with shallow tillage and harvested in fall, whereas in tropical-subtropical regions, seeds are sown in fall or early winter and harvested in February–March. The annual type *Brassica rapa* and biennial type *B. napus* can be grown in both winter and summer in Europe and North America (Gupta, 2016). Leaves are edible as a green vegetable and often used as fodder (Ayres and Clements, 2002).

Brassicas are the second largest oil seed crop. Several species are brown/Asian mustard (*Brassica juncea*); black mustard (*B. nigra*); toria/sarson/rapeseed/summer turnip, rape/Polish rape/ canola (*B. rapa* syn. *B. compestris*), Argentine rape/Swede rape/colza (*B. napus*); white mustard (*B. alba*) and Ethiopian or Abyssinian mustard (*B. carinata*). There are many other species under the family Brassicaceae used as vegetables, such as radish (*B. sativus*), turnip (*B. rapa* var. *rapa*) and all the cole crops, such as cauliflowers (*B. oleracea* var. *botrytis*), cabbage (*B. oleracea* var. *capitate*), broccoli (*B. oleracea* var. *italica*), knol-khol (*B. oleraea* var. *gongylodes*), brussels sprouts (*B. oleracea* var. *gemmifera*), kale (*B. oleracea* var. *acephala*), etc.

The main advantage, apart from soil benefits, of mustard and rapeseeds as cover crops is a solid patch of profuse flowering that attracts honeybees, as the flowers are bright and attractive in colour with sufficient pollen and nectar. Moreover, the higher plant density of these two crops makes the field a solid block. Presence of honeybees at the time of the blooming period of fruit crops enhances fruit set and yield. However, cole crops cultivated in orchards for vegetable purpose require proper cultural management. Green biomass of such crops may be recycled in orchards after proper composting. Hence, INM schedule should be adopted for cultivation of all these vegetable cole crops as intercrops in orchards, so that inter-competition between fruit trees and cover crops is eliminated.

5. Cover Crops and Orchard Floor Management

Orchard floor management (OFM) is very important for establishing a healthy orchard in terms of better soil, better biological, physical and chemical properties, proper fruit tree growth and higher yields. The ideal OFM practices include the adoption of the principles of conservation agriculture, so that input use efficiently is improved with the target of output maximisation in a sustainable manner. The main purpose of OFM is weed management, moisture conservation, soil temperature control, improved aeration and physical properties, enhanced activities of beneficial soil microbes and other organisms, optimisation of nutrient recycling, decrease of insect-pathogen load, better root development and to increase and facilitate nutrient uptake.

Cultivation of cover crops in the interrow spaces of orchards, along with mulching on the fruit tree rows, precision fertigation and addition of sufficient organic manures on the tree basins, and balanced tree canopy make an efficient integrated orchard-management module. An orchardist needs to select cover crops suitable for the orchard after considering factors (Ellenwood and Gorley, 1937; Nagy et al., 2010) such as, agro-climatic zone; orchard age; soil type; fruit crop type; purpose of growing (sod, forage, grain, vegetable, green manures, live mulch, etc.); perennial, biennial, or annual; growth habit and vigour (creeping, dwarf or semi erect, solitary stem, bushy, or bunch growth); root system (adventitious, tap root); regrowth; regeneration response to mowing; shearing; dormancy; grazing; herbicide application; sufficient biomass production; lower C:N and microbial decomposability and suitable season to grow. These crops should not become an obnoxious or invasive weed, non-competitive and non-allopathic to fruit trees.

Orchard ventilation and aeration should be properly managed. Access in the orchard should not be hindered and should not be a hibernation place or host for insect and pathogens. Some cover crops may be used as a trap crop for many insects and these trapped insects are killed by mechanical, chemical, or integrated methods. Clean cultivation by eliminating all types of plants from the orchard floor is often practiced in many orchards to minimise the cost of extra cultural management. This is done either by deep plowing, burning of weeds, or by a no-till approach, using herbicides. Other innovative methods are flame weeding and plastering, or using vinegar (acetic acid), especially in organic systems.

Under the flame weeding method, propane-fuelled flaming equipment is used to emit heat as high as 190°C to kill the plants by rapturing the plant cells and even by burning (Cisneros and Zandstra, 2008). In small, old mango orchards in West Bengal (India), where tree spacing is wider (10–12 × 10–12 m), orchardists keep the land clean by levelling and plastering with a mixture of fresh cowdung and clay (locally referred to as the *nikano* method) with the purpose of utilising the vacant space for grain threshing, hay storage and domestic cattle shelter (Mazumdar, 2004). However, clean cultivation and exposing the orchard floor surface for a long period has many disadvantages (Ames et al., 2004) and does not comply with the conservation agriculture concept. In deep-plowing, disturbed soil surface is prone to erosion, loss of moisture, depletion of organic matter, increased compaction, reduced infiltration and the extra cost of herbicide use with residual harmful effects. Orchard floor cultivation or permanent culturing of cover crops is the best approach for OFM (Ames et al., 2004; Mazumdar, 2004). There are various methods of cover crop management described in the following paragraphs.

A permanent strip of grass can be cultured along the orchard alleyways, as well as interspaces within rows, leaving the tree basin barren and the orchard floor covered under creeping grasses. However, in many orchards, tree rows are maintained as weed-free strips and sod is established along the interrow spaces with suitable grasses or legumes. Sods may be maintained as natural vegetation without any disturbances, pastures with grazing and orchard-floor mulching with mowed/chopped grasses. Sod may be maintained continuously throughout the orchard life span, or for two to three years followed by tilling and re-seeding. Most suitable crops for sod culture are creeping-type grasses, such as orchard grasses, fescues, timothy, Bermuda grass, *mutha*, Kentucky bluegrass, Korean grass and legumes, such as clovers and alfalfa. All these crops should be drought-tolerant, should not harbour nematodes and management cost should be at a minimum. Grass roots and mowed down aboveground plants parts are major contributors for SOM. Established sods should be easy and cost-effective in maintenance, able to survive under severe climate, tolerant to heavy machinery and farm vehicle movement and provide a lush, green sod mat for a better aesthetic look.

Grass alleys, along the inter-row with vegetation-free strips of 1.5–2 m within the tree rows, aid fruit trees by allowing proper root growth without any competition from the grasses (Roper, 2004). This type of sod culture is more efficient than solid grass covering the entire orchard floor. Organic mulch with straw, wood chips, black polyethylene sheet, or even mowed grasses on clean tree rows (including the tree basins of apple, pear, apricot, etc.) have been found to be very beneficial (Nagy et al., 2010). Generally, residue decomposition is slow when mulched with wood chips, which has high C:N, but lasts a long time. Other organic mulches add to SOM upon microbial decomposition and thereby improve soil aggregation, water infiltration and water-holding capacity and cation exchange capacity with optimum release of nutrients, better microbial activities and overall tree growth (Skroch and Sribbs, 1986; Nagy et al., 2010). Foxtail fescue sod in apricot orchards under Japanese conditions has been reported to suppress soil-borne pathogens by the release of volatile compounds from the roots and leaves, and increase the colonisation of bacteria, such as *Rosellinia necatrix, Fusarium oxysporum, Rhizoctonia solani* and *Pythium ultimum* that act as biocontrol agents. Moreover, these types of sods in an orchard create a favourable micro-environment for higher root colonisation and more hyphal growth of arbuscular mycorrhizal fungi (AMF) and P-solubilisation is greater due to root exodus by sod grass (Cruz and Ishii, 2012). The interaction between AMF and beneficial bacteria has been found to be beneficial for fruit tree growth (Linderman, 1992). Several studies (Merwin and Stiles, 1994; Atucha et al., 2011) reported that when different sod mixtures along with herbicide and management treatments were compared, the sod strips had shown lower soil compaction (1.2 to 1.3 g cm^{-3} bulk density), improved soil water availability and sorptivity (7.4 to 8.1 cm s^{-1}), higher cumulative infiltration (265 to 575 cm h^{-1}) and increased SOM content (5.2 to 5.6 g kg^{-1}) in apple orchards of the state of New York. The well-developed tap root systems of apple and peach trees under sod culture helped them to survive better and longer. However, under organic mulching, soil fertility as N, P, K, Ca, Mg and Mn contents was higher; however, grass sod soil also had better fertility compared to clean strips (Atucha et al., 2011).

Orchard vineyard nursery mix (OVN) of sod grasses (40 per cent perennial ryegrass, 30 per cent creeping red fescue, 30 epr cent chewing fescue) is the most preferred in modern orchards, such as high-density apple with tall spindle systems (Sazo et al., 2014). A modified low-growing mixture of hard fescues (50–55 per cent or more) with 20 per cent annual ryegrass was also found to be suitable for apple. Cover crop grasses, such as hairy vetch and a mixture of Hungarian vetch (70 per cent) + *Triticale* (30 per cent) and some quality of lacy phacelia grown in a Turkish apricot orchard in winter, followed by buckwheat in summer, increased soil quality parameters, such as SOM (2.3–2.6 per cent), total N (0.15–0.18 per cent), electrical conductivity (0.9–1.1 dSm^{-1}), soil respiration (36.0–41.5 mg CO_2 100 g^{-1}), structural stability index (54.0–55.4 per cent), aggregate stability (56.3–59.8 per cent), saturated hydraulic conductivity (0.9–1.3 cm h^{-1}), bulk density (1.0–1.1 g cm^{-3}), permanent wilting point (22.9–23.1 per cent), available water capacity (20.5–20.6 per cent), and field capacity (43.2–43.6 per cent) with increased fruit weight over the herbicide or bare control plots (Demir et al., 2019).

6. Methods of Sod Culture in Orchards

A good quality sod in the orchard can be established as per the procedures followed in the case of lawns. It is very essential to establish the sod methodologically for longevity, to create esthetic value and for ease of maintenance of the sod.

Soil Preparation: Orchard floor should be shallow-tilled, maintaining the present orchard floor gradient or levelled after removing all the stones and unwanted obnoxious grass roots. A 10–15 cm thick layer of fine weed seed/grass roots and a disinfectant mixture of garden soil, sand and farmyard manure (2:1:1) is spread uniformly over the orchard floor. The soil layer may be moistened by light irrigation with sprinklers to allow the germination of any weed-seeds, which may be immediately killed by any suitable herbicide. Spreading of soil mixture is done on the alleyways between the tree rows, leaving the cleaned/mulched tree rows, or may be applied within the tree row spaces.

Sod Establishment: There are a few specialised methods for planting sod grasses.

1. *Seeding*: Suitable sod grasses produce seed that is available from seed suppliers. Seed requirement is 25–30 kg ha^{-1} that is broadcasted after uniformly mixed with 200–250 kg of sand and saw dust/rice husk.

2. *Dribbling*: Small pieces of rooted stolons/rhizomes/suckers are planted at either a slanting or laying position at 8–10 cm spacing. Light sprinkler irrigation is given at intervals to keep the soil moistened. Planted grass propagules initiate above and below ground growth within 10–15 days, and by five to six months, the surface is covered with a radially expanded compact grass sod.

3. *Grass stem broadcasting*: Grown up rooted grasses are uprooted or procured from nursery suppliers and chopped into small pieces. The small rooted pieces are uniformly broadcasted/ sprinkled over the soil mixture during rainy days. Light rolling by use of a lawn roller is good to press the grass pieces gently on the surface to maintain better contact between grass roots and soil.

4. *Grass clump patches with soil*: Patches of grass clumps with roots intact with undisturbed soils are removed from the natural grasslands or pastures manually by spades or a sod cutter. Size of these patches may vary in dimension, such as 20–40 cm^2 in square/rectangle or 30–40 cm diameter in a circular shape. These small sod patches are carefully removed and transported to the orchards so that soil and roots are not disturbed. However, soil may fall from the fibrous root clutches, which does not affect the survival rate much. During the rainy season in tropical-sub-tropical areas, this method is very successful.

5. *Turfing*: Turfs of selected grasses are specially raised in lawn-grass nurseries. Turfs are compact sods of grass having rectangular dimensions of 75 cm^{-1} m width × 3–10 m long, or as per the requirement. These grass mats are raised on a soil or soilless medium and plastic or jute nets are placed at the base of this medium for ease of removing from the nursery, rolling, folding and transportation. These rolled turfs are placed as per the dimension of the alleyways in the orchard and unfolding is done carefully to spread the grass carpet properly over the orchard floor. Better

contact between the grass roots and orchard surface is ensured by gently pressing with a light roller. This is a very fast method of establishing grass sod in orchards.

6. *Turf plastering*: Chopped pieces (5–7 cm) of rooted stems of any hardy grasses such as Bermuda grass (*doob* grass) are mixed with garden soil, fresh cowdung and wood ash in a ratio of 2:1:1:$^1/_3$. This slurry is broadcasted over the orchard floor during the rainy season. These are rolled the next day for uniform spreading and contact with the orchard soil. After 10–15 days, new sprouts start and gradually over the next few months, radial growth of these grasses covers the orchard floor.

After Care: Maintenance of established sod is very critical to keep the growth of grass restricted. Otherwise, overgrown unmanaged grasses will invade the tree basin and make orchard intercultural operations difficult. Under such conditions, many insects and pathogens may harbour in the grasses. A uniform height of the grasses at 8–15 cm may be maintained with regular mowing. Rolling is done to keep stolon/rhizome growth horizontal and to maintain contact with the soil. Watering is done with sprinklers placed at specific distances to cover the entire area. Sometimes insecticide spray is required to kill the insects in order to avoid infestation on fruit trees. Fertiliser application is required for lush, green and healthy growth of these sod grasses and doses may vary depending on the agro-climatic zone and type of sod grasses. In general, urea or ammonium sulphate @ 1 kg 50 m^{-2} area during February–March, June–July and October–November is quite beneficial (TNAU, 2016) along with FYM @ 10 kg 10 m^{-2} area.

Green Manuring Crops: These crops are grown in the interrow spaces with the purpose to plough back whole plant biomass into the soil just before flower initiation or just after blooming, but before fruit set. The main objective of green manuring is to serve as a cover crop on the orchard floor during the growing period, so that it enriches the soil with organic matter and N, when ploughed back. Mostly annual or biennial leguminous crops with better N-fixing capacity (Table 2), or even some annual non-legume crops that can store sufficient N in biomass by capturing excess soil N, are grown as green manuring crops. This may be followed by the cultivation of vegetable or fodder crops the next season. Green leaves and even tender shoots of leguminous trees, may be spread over the orchard floor as green mulch during summer and mixed by shallow tilling for faster decomposing, and another crop may be grown the next season (Table 3) (Chandra, 2005; TNAU, 2016).

Table 2 *In situ* biological N fixation ability of leguminous green manuring crops.

Crop	Seed Rate (kg ha^{-1})	Green Manuring Duration (days)	N- fixation Potential (kg ha^{-1})	Nutrient (%) Dry Weight Basis			Reference
				N	P	K	
Sesbania spp. (*S. rostrata, S. aculeata, S. sesban*)	40–50	45–60	75–130	3.0	0.6	1.8	Islam et al. (2013); TNAU (2016)
Sunn hemp (*Crotalaria juncea*)	25–35	60–100	84–100	2.6	0.6	2.0	Islam et al. (2013); TNAU (2016)
Wild indigo (*Tephrosia purpurea*)	15–20	55–60	75–115	2.4	0.3	0.8	TNAU (2016)
Pillipesara (*Phaseolus trilobus*)	10–15	55–60	85–120	2.1	0.5	–	TNAU (2016)
Cowpea (*Vigna unguiculata*)	35–40	40–45	140–150	0.7	0.2	0.6	Chandra (2005)
Alfalfa (*Medicago* spp.)	13–20	70–80	200–250	3.0	0.2	2.6	Zhu et al. (1996); Koenig et al. (2009)

Table 3 Tree leaves nitrogen, phosphorus and potassium contents (Chandra, 2005; TNAU, 2016).

Tree/Crop	Nutrient Content (%) Dry Weight Basis		
	Nitrogen	Phosphorus	Potassium
Pongam oil tree/Indian beech (*Pongamia glabra* & *P. pinnata*)	3.2	0.3	2.4
Gliricidia (*Gliricidia sepium*)	2.8	0.3	4.6
Flame tree/gulmohur (*Delonix regia*)	2.8	0.5	0.5
Neem (*Azadirachta indica*)	2.8	0.3	0.4
Crown flower (*Calatropis gigantecum*)	2.1	0.7	3.6
Yellow Flame (*Peltophorum ferrugenum*)	2.6	0.4	0.5
Sesbania (*Sesbania speciose*)	2.7	0.5	2.2

Apart from enriching the soil with N and SOM, leaves of trees such as neem, when incorporated in the soil properly, can reduce the risk of soil-borne diseases like nematode and bacterial wilt problems from the field (Agyarko and Asante, 2005; Pontes et al., 2011). Moreover, many weeds growing on the orchard floor can be plowed back to add fresh biomass into the soil as green manure before they initiate seed set, such as parthenium (*Parthenium hysterophorus*), Trianthema (*Trianthema portulacastrum*), and Cassia (*Cassia fistula*) (Chandra, 2005; TNAU, 2016).

In a detailed study case by Litterick (2019) of apple plantations at Loddington Farms Ltd., UK, it was reported that multi-mixture of green manure grasses (pollinator mix) comprising five grass species, such as native grasses, Phacelia (quick growing hardy annual green manure forb with ability to enrich SOM) and eight legume species improved extractable P, K and Mg, pH, SOM, respiration and texture. Another soil-improving mixture comprised of five grass species and four legumes also performed better. Soils under green manure of Pinto peanut (*Arachis pintoi*) as mulch in a nectarine orchard had higher organic C (21.1 g kg^{-1}) in comparison to 14.8 g kg^{-1} in plots without any conservation measures (Wang et al., 2015). The beneficial effect of green manure crops in a Kent variety mango orchard was well reported under Brazilian conditions; a cultivation of mixture of 75 per cent leguminous crops (calopo, sunn hemp, showy rattle pod, jack bean, pigeon pea, hyacinth bean, black velvet bean, gray mucuna) in combination with 25 per cent non-legumes (sesame, sunflower, castor bean, pearl millet and sorghum) resulted in better soil fertility and mango tree growth (Mouco et al., 2015). Better mango tree growth was due to better soil fertility facilitating accumulation of fresh matter and nitrogen, phosphorus, potassium, calcium, magnesium, etc. in the leaves and branches. Moreover, such type of green manure cropping has the capacity to produce higher biomass (8.2 ton ha^{-1}) with higher content of organic C (SOC) in soil (3761.2 ton ha^{-1}), N (157 ton ha^{-1}), and K (219.6 ton ha^{-1}), as well as a higher rate of decomposition into the soil and better-released N, P, and K (Freitas et al., 2019). Alfalfa, as an alley crop in higher altitudes of temperate Indian conditions (Sofi et al., 2012), enriched the orchard soil with higher levels of particulate organic carbon (1.2 g kg^{-1}), labile SOC (4.9 g kg^{-1}), SOC (1.4 per cent), microbial biomass carbon (1.3 g kg^{-1}) and total SOC stocks (46.7 ton ha^{-1}).

The special characteristics and benefits of green manuring crops include the following (Rayns and Rosenfeld, 2010; TNAU, 2016):

i) Annual, herbaceous and fast growing.

ii) Tolerant to shade, drought, many other biotic and abiotic stresses and photo periodically insensitive.

iii) Ease of cultivation and management with minimum cost involvement.

iv) Wide agro-climatic adoptability and orchard compatibility.

v) Better water and fertiliser-use efficiency with minimum irrigation and manure-fertiliser application.

vi) Capable of better growth response, even under rain-fed conditions.

vii) Suitable and adaptive for a wide range of soils.

viii) Non-allelopathic and should not compete with fruit trees for input resources for growth and development.

ix) Leguminous crops should contain higher root nodule density with higher efficiency of *Rhizobium* synergism to provide N fertility. *In situ* biological N fixation right from the early growth stage by the leguminous crops, though may vary depending on the crop types (Table 2).

x) Non-legume green manure crops act as a capture crop by preventing N-leaching by capturing and utilising it in metabolism.

xi) Moreover, incorporation of green biomass adds up to a considerable amount SOM, N, P, and K, apart from improvement in soil biological and physical properties.

xii) Green manures are expected to decompose within a short period (15–30 days), so that soil is properly conditioned for the succeeding crop.

xiii) Double cropping in the orchard can be done with summer season green manure crops, followed by winter season green manure crops or vegetable/forage crops and vice-versa.

xiv) Mixed cropping with two or more green manure crops, with or without crops as companion crop in separate plots and relay cropping with other crops, can be practiced, increasing cropping intensity and productivity per unit of land.

xv) Green manure crops are the best option to optimise soil temperature, improve water retention, increase water infiltration and enrich orchard biodiversity.

Intercropping: Growing more than one crop together in the same field without any allelopathic effect or competition has been found to be sustainable to minimise crop failure risk, increase cropping intensity and increase productivity per unit area. However, in the orchard, intercropping of annual vegetable/spice crops in the interrow spaces is done beginning from the first year of orchard establishment until the attainment of fruit bearing stage, not only for soil fertility improvement, but also to generate income. Fruit trees are perennial in nature and have a long juvenile period of three years (citrus [*Citrus* spp.], stone fruits [*Prunus* spp.], guava [*Psidium guajava*], pomegranate [*Punica granatum*], etc.) to 10 years (mango [*Mangifera indica*], litchi [*Litchi chinensis*], walnut [*Juglans regia*] and pecan nut [*Carya illinoinensis*]) before they start bearing fruit. Moreover, income from the fruit comes only after harvesting, which may last for one month in a year. Hence, there is a long lean period annually in which the cost (in terms of income generation from fruit) of orchard inter-cultural operations and inputs is high.

Intercropping is the best-fit option for orchards, where sufficient interspaces are available due to wider tree spacing. Selection of a suitable intercrop may be based on only a single crop, a combination of several crops in separate rows, or plots or even a mixture of two or more crop seeds (Table 4). Tall-growing crops are not suitable as intercrops in a young orchard due to shade effect, incidence of insect-pathogen and competition with young fruit trees for nutrition and space. However, cereal grain crops such as millets, sorghum, wheat, maize, etc. may be grown in an orchard comprised of tall fruit-bearing trees such as mango, litchi, walnut, and even in apple trees where spacing is wider (7–12 m).

In cashew orchards, under tropical Indian conditions, annual crops such as tapioca (*Manihot esculenta*), pulses, turmeric (*Curcuma longa*), ginger (*Zingiber officinale*), yam (*Dioscorea* spp.), maize (*Zea mays*), groundnuts, horse gram, cowpea, beans and even pineapples (*Ananas comosus*) are very much suitable for inter cropping (Visalakshi et al., 2015). Intercropping in the orchard not only generates income, but the role of cover crops is very much evident in soil erosion protection, weed suppression, improvement in soil physico-chemical properties, overall soil fertility and fruit production (Linares et al., 2008; Gill et al., 2018). In a kinnow mandarin orchard, under Indian sup-tropical conditions, inter-cropping of cluster bean (summer) followed by wheat (winter), green

Table 4 Suitable intercrops in orchards.

Orchard	Intercrop Combination	Agro-climatic Zone	Reference
Apple, pear	Pea, red clover, French bean, cabbage, and strawberry	Temperate climate, India	Bhat et al. (2018)
Almond	1. Vegetable and oil seeds: Pea, saffron, pulses, onion, garlic, cole crops and mustard 2. Seed production: Turnip, knoll khol, carrot and mustard 3. Medicinal and aromatic plants: lavender, pepper mint, geranium and aloe	Temperate climate, India	Ahmed and Verma (2009)
Cherry	Pea, red clover, French, bean, strawberry, cabbage, oats and maize	Temperate climate, India	Bhat et al. (2015)
Mango	Cowpea, French bean, brinjal, tomato, onion, ginger, turmeric, coriander, elephant foot yam, *Colocasia*, radish, carrot, mustard, cole crops, chili, aloe, pigeon pea, periwinkle and pineapple	Tropical-sub-tropical climate, India	Ravitchandirane, and Haripriya. (2011); Singh et al. (2012); Tiwari and Bhadel (2014)
Kinnow	Soybean, pea, cowpea, gram, cluster bean, French bean, fenugreek, green gram and okra	Sub-tropical climate, India	Singh et al. (2012); Gill et al. (2018)
Ber	1. Spices: Black caraway (*Nigella sativa*), aniseed (*Pimpinella anisum*), carom/ajwain (*Trachyspermum ammi*), fenugreek (*Trigonella foenum-graecum*) and coriander (*Coriandrum sativum*) 2. Vegetables/legumes: Cowpea, cluster bean, black gram, okra and green gram	Semi-Arid-subtropical climate, India	Meena et al. (2017)
Cashew nut	1. Legumes: Groundnuts, horse gram, cowpea and beans. 2. Mixed crops: Tapioca, pulses, turmeric, ginger, yam, maize and pineapple	Tropical climate, India	Visalakshi et al. (2005)
Litchi	1. Potato (Nov–Feb)-amaranth (March-May)-cowpea (June–Aug.)-radish (Sept.–Oct.) 2. Cauliflower (Aug.–Nov.)-potato (Dec.–March)- hyacinth (April–July) 3. Cabbage (Sept.–Dec.)-brinjal (Jan.–May)-cucumber (June–Aug.) 4. Palak (Nov.–Jan.)-cluster bean (Feb.–May)-brinjal (June–Oct.) 5. Cow pea-potato/green gram-niger/black gram-Mustard/ green gram-gram/black gram-gram/cowpea/mustard/cow pea-gram	Sub-tropical climate, India	Singh et al. (2012)
Guava	Pea, cowpea, gram, beans, turmeric, *Colocasia* and elephant foot yam	Sub-tropical conditions, India	Singh et al. 2012); Singh et al. (2016)
Aonla	Elephant Foot Yam, aroids, turmeric and ginger, fennel, cluster bean, coriander, carom/ajwain, brinjal, moth bean and mustard		Hare Krishan et al. (2013); Arya et al. (2010); Singh et al. (2016)
Mango	Sequence: Maize-tepary bean/French bean/pigeon pea/	Tropical climate, Brazil	Agreda et al. (2006)
Mandarin orange	Brachiaria, Pearl millet and Jack bean	Northern Bahia, Brazil	Oliveira et al. (2016)
Mandarin Orange	Hyacinth bean, mucuna and cowpea	Coastal region, Kenya	Mulinge et al. (2018)

Table 4 Contd. ...

...Table 4 Contd.

Orchard	Intercrop Combination	Agro-climatic Zone	Reference
Nut orchards	1. Grasses and green manure crops: Crown vetch, crimson clover, bluegrass, hairy vetch, red clover and Sweet clover, alfalfa, annual ryegrass, phacelia and Sunn hemp. 2. Crops: Oats, soybean, winter wheat, buckwheat, canola and pea. 3. Pollination facilitator crops: Witchhazel (Jan.–Mar.), redbud (Mar.–Apr.), plums (Mar.–May), service berry (Mar.–May), pecan (Apr.–May), black raspberries (Apr.–Jun.), peaches, cherries, pawpaw (Apr.–May), sassafras, elderberry (Jun.–Jul.) and willow (Apr.–May)	American conditions	Sambeek (2017)
Apple	1. Grasses: Red fescue, hard fescue, perennial ryegrass and Kentucky grass. 2. Legumes: Dutch clover, red clover, white clover, subterranean clover, vetches and alfalfa.	American conditions	Pavek and Granatstein (2014)

gram (summer) + fenugreek (winter), and cluster bean (summer) + fenugreek (winter) were suitable in terms of kinnow tree growth, fruit quality, and higher additional income from the intercrops (Gill et al., 2018). In a mango orchard under Mexican conditions (Agreda et al., 2006), maize crop followed by French bean/Tepary bean/pigeon pea improved soil fertility, suppressed weeds and increased mango fruit yield (8–10 ton ha^{-1}). Under such systems, SOC (2.6–3.4 per cent), N (0.23–0.26 per cent), C:N (10–13) and other soil chemical parameters were also improved (Agreda, 2008). However, in high-to-medium density apple orchards, inter-cropping may negatively affect tree growth and fruit yield due to competition for space, over- and aboveground plant growth, light, moisture, nutrients, etc. (Gao et al., 2013). To eliminate such competition, appropriate distance should be maintained between fruit trees and intercrop strips, and additional management and manure-fertiliser schedule for intercrops may be adopted. Moreover, it is advisable to avoid heavy feeder intercrops during the fruit development period, which may otherwise negatively affect the fruit physico-chemical properties.

Intercropping may act as a pest reduction approach, such as in apple San Jose scale control by intercrop with *Phacelia* spp. and *Eryngium* spp., grape leaf hopper control by intercrop with wild blackberry. Grape Pacific mite control by intercrop with Johnson grass, grape Willamette mite control by intercrop with sudangrass and Johnson grass, grape aphid control by intercrop with kale, peach oriental fruit moth control by intercrops with strawberry, and walnut aphid control by intercrop with grassy ground cover (SARE, 2014) due to either harboring predators or chemical effect. Many other flowers and aromatic plants, such as marigold (*Tagetes* spp.), basil (*Ocimum basilicum*), lavender (*Lavandula angustifolia*), rosemary (*Rosmarinus officinalis*), mint (*Mentha* spp.), chrysanthemum (*Chrysanthemum indicum*), onion and garlic (*Allium* spp.), thyme (*Thymus vulgaris*), sage (*Salvia officinalis*), catnip (*Nepeta cataria*), fennel (*Foeniculum vulgare*), ageratum (*Ageratum houstonianum*), summer savory (*Satureja hortensis*), and lemon grasses (*Cymbopogon citratus*) may act as a repellent for insects (Catherine, 1997; Song et al., 2010; Song et al., 2019). This was also demonstrated in managing Tortricid spp. in apple, due to the chemical toxicity or repellency by these aromatic plants on insect feeding, oviposition, and absence of a hibernating place (Lu et al., 2008; Song et al., 2019).

Intercrop combination of mango + brinjal or bottle gourd under Indian sub-tropical conditions (Singh et al., 2015) was profitable with 19.4 ton ha^{-1} eggplant (brinjal) and 13.5 ton ha^{-1} bottle gourd (squash) yields and an increase in mango yield by 8.6 per cent. Improvement in soil quality was recorded in the intercropping system apple + red clover/French bean/peas under Indian temperate climate conditions with soil bulk density of 1.3 g cm^{-3}, total porosity 0.5 m^{3}m^{-3}, SOC content 14.7 g kg^{-1}, and total N 1.6 g kg^{-1}, which were significantly better than the apple monoculture or apple with paddy culture.

Under integrated farming systems (IFS), intercropping and mixed farming are integral parts of an efficient resource management system through the integration of interrelated, inter-dependent and compatible components comprising of fruit trees, fodder trees, vegetables, cereals and selective livestock, such as domestic animals, birds (poultry/duck) and fishery. Therefore, to develop different copping systems (such as agriculture + horticulture agriculture + horticulture + fishery + poultry, agriculture + horticulture + piggery + fishery, agriculture + horticulture + cattle + piggery + fishery, and agriculture + horticulture + silviculture), proper integration of all the compatible components is critical for the sustainability and profitability of such systems (Kumar et al., 2018; Lal et al., 2018; Panwar et al., 2019). Incorporation of organic residues and manures and recycling of inputs make the farm soil under IFS very fertile by increasing SOM content, physico-chemical properties and biodiversity, resulting in higher water infiltration (Walia and Kaur, 2013). Under a multiple-cropping approach, the cropping diversity has been successfully adopted in many agro-climatic zones comprised of horticultural and other compatible crops in the same orchard to get the benefit of maximum resource utilisation and minimising the mono-crop failure risk (Ghosh and Bandopadhyay, 2011). In this system, varying heights of perennial, semi-perennial, biennial and annual crops are geometrically planted, utilising the surface and aerial space. Salient features of a multi-storied system include:

i) tall crops with wider spacing – mango (spacing 10–12 × 10–12 m, tree height 6.0–15 m), cashew nut (spacing 10–12 × 10–12 m, tree height 5–10 m), coffee (*Coffee arabica*) (spacing 2.5 × 2.5 m, tree height 3–5 m), coconut (*Cocos nucifera*) (spacing 7.5 × 7.5 m, tree height 15–20 m), areca nut (*Areca catechu*) (spacing 2.7 × 2.7 m, tree height 15–20 m) and oil-palm (*Elaeis guineensis*) (spacing 9 × 9 m, tree height 10–15 m);

ii) climbing vines such as black pepper (*Piper nigrum*), Chinese potato (*Coleus rotundifolius*) and yams (*Dioscorea* spp.) are trailed on the trees, semi-erect low height crops, such as turmeric (*Curcuma longa*), ginger (*Zingiber officinale*), tapioca (*Manihot esculenta*), elephant foot yam (*Amorphophallus paeoniifolius*), American taro (*Xanthosoma* spp.), small cardamom (*Elettaria cardamomum*), large cradamom (*Amomum subulatum*), cluster bean, hyacinth bean, etc. and ground cover crops, such as cowpea, French bean, grams, pea, sweet potato, colocasia, pineapple and marigold, etc. are shade loving and non-allelopathic to other companion crops in the system;

iii) short height fruit crops, such as papaya (*Carica papaya*), banana (*Musa* spp.), guava, drumstick (*Moringa oleifera*), and koronda (*Carissa karonda*);

iv) weed suppressive and less susceptible to insect and disease infestation;

v) other tree spices, such as cinnamon (*Cinnamomum verum*), nutmeg (*Myristica fragrans*), allspice (*Pimenta dioica*), clove (*Syzygium aromaticum*) and some selective medicinal and aromatic plants, such as patchouli (*Pogostemon cablin*), Indian Snake root (*Rauvolfia serpentine*), and long pepper (*Piper longum*) are suitable (Nath et al., 2015; Nimbolkar et al., 2016).

However, integrated management practices estimating the fertiliser requirement of all the companion crops is essential to eliminate completion for nutrition. Fruit tree-based intercropping or multi-storied cropping improves the soil organic carbon pool and the carbon sequestration rate is high compared to monocropping fields (Sau et al., 2017). Intercropping in apple orchards with tall fescue, orchard grass and red and white clover under Indian conditions increased SOC to 0.9 per cent from 0.6 per cent under clean cultivation, along with an increase in available N (393.7–505.5 kg ha^{-1}), P (13.2–18 kg ha^{-1}), and K (412.8–454.1 kg ha^{-1}) (Ahmad et al., 2018). Cover crops in olive orchards under semi-arid conditions in Spain (Marquez-Garcia et al., 2013) reportedly reduce soil erosion by 80.5 per cent, SOC loss by 67.7 per cent and SOC sink was increased by 12.3 ton CO_2 ha^{-1} year^{-1} compared to plowing.

7. Land Use Change and Role of Cover Crops in Orchards

Land-use conversion from natural vegetation or grassland into orchards has been a common practice worldwide. Moreover, precision fruit plantation technology requires uniformly flat land, which is generally performed by bulldozer machines. Such land-levelling operations displace fertile topsoil, expose infertile subsoils, result in loss of SOC, destroy beneficial biological population and lead to the depletion of C-stock (Chen et al., 2011; Qin et al., 2014; Seyum et al., 2019). The top 15 cm depth of soil is the prime location for microbial activities, and where sufficient labile C-substrate is available from organic residues, cover crop roots, and litters (Hoorman and Islam, 2010).

The C and other nutrients are depleted, even if only 30–40 cm of the topsoil is removed or affected. All the soil physical properties are disturbed by increasing compaction, and reducing critical soil properties, such as aggregate stability, saturated hydraulic conductivity, permeability and water-holding capacity. Land levelling under natural vegetation may reduce SOC to 2.6 from 19.5 g kg^{-1} (Woodward, 1996; Yang et al., 2018). The SOC density of topsoil changed with the land conversion, such as in natural forest or woodlands, abundant croplands, orchards, natural grasslands, and agricultural cultivated lands with SOC density of 25.5, 20.7, 17, 19.4 and 19.1 ton ha^{-1} (Qin et al., 2014).

Under such degraded land conditions, a sustainable intensification approach is needed by producing more from the optimisation of resources in the agroecosystem (Lal, 2015). Therefore, scheduled IOMPs are adopted to increase SOM content and regain soil productivity, so that an equilibrium among soil fertility, input utilisation, tree growth and fruit productivity may be achieved. Fruit tree pit filling with sufficient garden soil and organic manure mixture, followed by INM comprised of organic manures, inorganic fertilisers, and bio-fertilisers in subsequent years constitute an ideal strategy for fruit tree nutrition (Das et al., 2016; Oliveira et al., 2016). However, in those depleted lands, orchard floor management is important for overall soil fertility restoration to establish a sustainable, productive orchard ecosystem.

Grass sods cover the barren surface very quickly, or even cultivation of other cover crops with minimum tillage and biomass recycling is advisable under such situations to restore soil fertility. Root residues of these crops gradually add SOC into the soil by microbial decomposition, even if aboveground biomass of these cover crops is incorporated into the soil as green manure, or taken away as fodder and grain/vegetable/seed produce, or left on the ground after chopping and mowing (Austin et al., 2017). Aboveground biomass, after recycling into the soil, contributes SOM and the release of nutrients, and such litter quality is dependent on C:N of the cover crops for optimum microbial decomposition. Low C:N crops with labile C-chemistry such as pea, lentil, cowpea, soybean, Sunn hemp, clovers, or even turnip, radish, canola, rape, and mustard are decomposed faster in comparison to high C:N crops with more recalcitrant C-chemistry crops such as corn, sorghum, sunflower, millet and wheat or other mulched straw residues (Cotrufo et al., 2013; NRCS, 2011).

Cropping sequence may be adjusted with fast decomposition green manure crops followed by slower decomposing crops, which may even be decomposed the next season. In general, C:N of legumes may range between 14–20 in roots and 18–24 in shoots, and usually contribute a total dry matter of 340.5 g kg^{-1} from shoot and 394.3 g kg^{-1} from roots (Talgre et al., 2017). Thus, such types of crops can contribute total carbon of around 457.8 mg g^{-1} by shoot and 425.5 mg g^{-1} by roots, and total nitrogen around 21.4 mg g^{-1} by shoot and 27.7 mg g^{-1} by roots. In citrus orchards under Brazilian conditions (Oliveira et al., 2016), cultivation of brachiaria (*Brachiaria decumbens*), pearl millet (*Pennisetum glaucum*), jack bean (*Canavalia ensiformis*), or spontaneous vegetation cover improved SOC status (6.7–13.3 g kg^{-1}). The INM on mango under sub-tropical Indian conditions comprised of 10 kg Vermicompost + 100 g N + 50 g P + 100 g K tree^{-1} year of age^{-1} + *Azotobacter* + PSM + *Trichoderma harzianum* + organic mulching increased SOC to 0.45 per cent from 0.3 per cent under an organic manure + inorganic fertiliser regime (Adak et al., 2014). Long-term soil restorative cover crop cultivation and IOMPs in the orchard can make land-use changes more productive.

References

Adak, T., Singha, A., Kumar, K., Shukla, S.K., Singh, A. and Singh, V.K. (2014). Soil organic carbon, dehydrogenase activity, nutrient availability and leaf nutrient content as affected by organic and inorganic source of nutrient in mango orchard soil. J. Soil Sci. Plant Nutri., 2: 394–406.

Agreda, F.J.M., Pohlan, J. and Janssens, M.J.J. (2006). Effects of Legumes Intercropped in Mango Orchards in the Soconusco, Chiapas, Mexico, Tropentag 2006, University of Bonn, October 11–13, 2006, Conference on International Agricultural Research for Development. http://www.tropentag.de/2006/abstracts/full/386.pdf.

Agreda, F.J.M. (2008). Sustainable Management of Fruit Orchards in the Soconusco, Chiapas, Mexico – Intercropping Cash and Trap Crops, Publisher: Shaker Verlag GmbH, Germany, pp. 122. http://hss.ulb.uni-bonn.de/2008/1424/1424.pdf.

Agyarko, K. and Asante, J.A. (2005). Nematode dynamics in a soil amended with neem leaves and poultry manure. Asian J. Plant Sci., 4: 426–428.

Ahmad, S., Khan, P.A., Verma, D.K., Mir, N.H., Sharma, A. and Wani, S.A. (2017). Forage production and orchard floor management through grass/legume intercropping in apple-based agroforestry system. Inter. J. of Chem. Stud., 6: 953–959.

Ahmad, S., Khan, P.A., Verma, D.K., Mir, N.H., Sharma, A. and Wani, S.A. (2018). Forage production and orchard floor management through grass/legume intercropping in apple-based agroforestry systems. Inter. J. of Chem. Stud., 6: 953–958.

Ahmed, N. and Verma, M.K. (2009). Scientific Almond Cultivation for Higher Returns, Central Institute of Temperate Horticulture, Indian Council of Agricultural Research, Srinagar, J&K, India, pp. 14.

Ames, G.A., Kuepper, G. and Guerena, M. (2004). Tree fruits: Organic production overview. Appropriate Technology Transfer for Rural Areas (ATTRA), Fundamentals of Sustainable Agriculture, National Sustainable Agriculture Information Service (NSAIS). National Centre for Appropriate Technology (NCAT), Fayetteville, Arkansas (P.O. Box 3657, Fayetteville, AR 72702), Butte, Montana and Davis, California. http://drcsc.org/resources/fruitover.pdf.

Arya, R., Awasthi, O.P., Singh J. and Arya C.K. (2010). Comparison of fruit-based multi-species cropping system under arid region of Rajasthan. Indian J. Agri. Sci., 80: 423–426.

Atucha, A., Merwin, I.A. and Michael, G.B. (2011). Long-term effects of four groundcover management systems in an apple orchard. Hort. Science, 46: 1176–1183.

Austin, E.E., Kyle, W., Mcdaniel, M.D., Robertson, G.P. and Grandy, A.S. (2017). Cover crop root contributions to soil carbon in a no-till corn bioenergy cropping system. GCB Bioenergy, 9: 1252–1263.

Ayres, L. and Clements, O.B. (2002). Forage Brassicas – quality crops for livestock production, Agfact p. 2.1.13, first edition, The State of New South Wales NSW Agriculture. https://www.dpi.nsw.gov.au/__data/assets/pdf_file/0003/146730/forage-brassicas-quality-crops-for-livestock-production.pdf.

Barney, D.L. (2012). Storey's Guide to Growing Organic Orchard Fruits, Storey Publishing, 210 MASS MoCA Way, North Adams, MA 01247, USA, pp. 286–324.

Bhat, R., Wani, W.M., Sharma, M.K. and Hussain, S. (2015). Influence of intercrops on cropping, quality, and relative economic yield of sweet cherry cv. Bigarreau Noir Grossa (Misri). Inter. J. Res. Eng. App. Sci., 5: 264–272.

Catherine, R.R. (1997). The potential of botanical essential oils for insect pest control. Integ. Pest Mgt. Rev., 2: 25–34.

Chandra, K. (2005). Organic Manures, Regional Centre of Organic Farming, No. 34, 5th Main Road, Hebbal, Banglaore-24, India, pp. 46. https:// ncof.dacnet.nic.in/Training manuals/Training manuals in English/Organic manures. pdf.

Chen, L., Qi, X., Zhang, X., Li, Q. and Zhang, Y. (2011). Effect of agricultural land use changes on soil nutrient use efficiency in an agricultural area, Beijing, China. Chin. Geogra. Sci., 21: 392–402.

Cisneros, J.J. and Zandstra, B.H. (2008). Flame weeding effects on several weed species. Weed Tech., 22: 290–295.

Clark, A. (2007). Managing Cover Crops Profitably, third ed., Sustainable Agriculture Network, Beltsville, MD. http://mccc.msu.edu/wp-content/uploads/2016/09/Managing Cover Crops Profitably_ Annual Ryegrass.pdf.

Cotrufo M.F., Wallenstein, M.D., Boot, C.M., Denef, K. and Paul, E. (2013). The Microbial Efficiency-Matrix Stabilization (MEMS) framework integrates plant litter decomposition with soil organic matter stabilisation: Do labile plant inputs form stable soil organic matter? Glob. Chang Biol., 19: 988–995.

Cruz, A.F and Ishii, T. (2012). Sod Culture management contributes to sustainable fruit growing through propagation of arbuscular mycorrhizal fungi and their hyper-microorganisms and decreasing of agricultural inputs. pp. 477–484. *In*: Mourao, I. and Aksoy, U. (eds.). Proc. XXVIIIth IHC – IS on Organic Horticulture: Productivity and Sustainability, Acta Hort., 933, ISHS.

Das, B., Harekrishna, Ranjan, J.K., Pragya, N. Ahmad and Attri, B.L. (2016). Integrated nutrient management and mulching for higher productivity of spur type apple (*Malus domestica*) cultivars. Ind. J. Agric. Sci., 86: 1016–1023.

Demir, Z., Tursun, N. and Işik, D. (2019). Effects of different cover crops on soil quality parameters and yield in an apricot orchard. Inter. J. Agri. Bio., 21: 399–408.

Ellenwood, C.W. and Gourley, J.H. (1937). Cultural Systems for the Apple in Ohio, Bulletin 580, Ohio Agricultural Experiment station, Wooster, Ohio, USA.

Evans, D.O., Joy, J.R. and Chia, C.L. (1988). Cover Crops for Orchards in Hawaii, Research Extension Series 094. 630 US ISSN 0271-9916, College of Tropical Agriculture and Human Resources, University of Hawaii. https://www.nrcs.usda.gov/Internet/FSE_ PLANTMATERIALS/publications/hipmctn806.pdf.

[FAO] Food and Agriculture Organisation of the United Nations. (2017). FAOSTAT. http://www.fao.org/faostat/en/#data/QC.

Freitas, M.S.C., Souto, J.S., Gonçalves, M., Almeida, L.E.S., Salviano, A.M. and Giongo, V. (2019). Decomposition and nutrient release of cover crops in mango cultivation in Brazilian semi-arid region. Rev Bras Cienc Solo., 43: e0170402.

Gao, L., Xu, H., Bi, H., Xi, W., Bao, B., Wang, X., Bi, C. and Chang, Y. (2013). Intercropping competition between apple trees and crops in agroforestry systems on the Loess Plateau of China. PLoS One, 8(7): e70739. DOI: 10.1371/journal.pone.0070739.

Ghosh, D.K. and Bandopadhyay, A. (2011). Productivity and profitability of coconut-based cropping systems with fruits and black pepper in West Bengal. J. Crop Weed, 7: 134–137.

Gill, H.K. and McSorley, R. (2011). Cover Crops for Managing Root-Knot Nematodes, ENY063, U.S. Department of Agriculture, UF/IFAS Extension Service, University of Florida, IFAS, Florida A & M University Cooperative Extension Program and Boards of County Commissioners Cooperating. Nick T. Place, Dean for UF/IFAS Extension. https://edis.ifas.ufl.edu/pdffiles/IN/IN89200.pdf.

Gill, M.S., Khehra, S. and Gupta, N. (2018). Impact of intercropping on yield, fruit quality and economics of young kinnow mandarin plants. J. Appl. Nat. Sci., 10: 954–957.

Graham, P.H. and Vance, C.P. (2003). Legumes: Importance and constraints to greater use. Plant Physiol., 131: 872–877.

Gupta, S.K. (2016). Brassicas. pp. 33–53. *In*: Gupta, S.K. (ed.). Breeding Oilseed Crops for Sustainable Production: Opportunities and Constraints, Academic Press.

Hannaway, D., Fransen, S., Cropper, J., Teel, M., Chaney, M., Griggs, T., Halse, R., Hart, J., Cheeke, P., Klinger, R. and Lane, W. (1999). Perennial Ryegrass (*Lolium perenne* L.), PNW 503 • April 1999, Oregon State University. file:///C:/Users/USER/Downloads/pnw503.pdf.

Hare Krishan, Singh, I.S., Bhargava, R. and Sharma, S.K. (2013). Fruit-based cropping systems for sustainable production. ICAR News, 9: 9.

Hoorman, J.J., Islam, R. and Sundermeier, A. (2009). Sustainable Crop Rotations with Cover Crops, Fact Sheet Agriculture and Natural Resources, SAG-9-09, The Ohio State University Extension. file:///C:/Users/USER/Downloads/Sustainable_Crop Rotations.

Ingels, C.A. and Klonsky, K.M. (1998). Historical and current uses. pp. 3–7. *In*: Ingels, C.A., Bugg, R.L., McCourty, G.T. and Christensen, L.P. (eds.). Cover Crops in Vineyards: A Grower's Handbook, University of California, Division of Agriculture and Natural Resources, Publication No. 3338.

Ingels, C.A., Bugg, R.L. and Thomas, R.L. (1998). Cover crop species and descriptions. pp. 8–26. *In*: Ingels C.A., Bugg, R.L., McCourty, G.T. and Christensen, L.P. (eds.). Cover Crops in Vineyards: A Grower's Handbook, University of California, Division of Agriculture and Natural Resources, Publication No. 3338.

Ingels, C.A., Prichard, T., Berry, A., Scow, K. and Whisson, D. (2003). Selection and effects of cover cropping in vineyards and orchards. Proc. California Weed Sci. Soc., 55: 811–84.

Jannoyer, M.L., Le Bellecb, F., Lavignea, C., Achardc, R. and Malezieuxd, E. (2011). Choosing cover crops to enhance ecological services in orchards: A multiple criteria and systemic approach applied to tropical areas. Procedia Env. Sci., 9: 104–112.

Jeuffroy, M.H., Baranger, E., Carrouee, B., Chezelles, E.D., Gosme, M. and Henault, C. (2013). Nitrous oxide emissions from crop rotations including wheat, oilseed rape and dry peas. Biogeosciences, 10: 1787–97.

Kaspar, T.C. and Singe, J.W. (2011). The Use of Cover Crops to Manage Soil, Publications from USDAARS/UNL Faculty. 1382. https://digitalcommons.unl.edu/usdaarsfacpub/1382.

Koenig, R.T., Horneck, D., Platt, T., Petersen, P., Stevens, R., Fransen, S. and Brown, B. (2009). Nutrient Management Guide for Dryland and Irrigated Alfalfa in the Inland Northwest, A Pacific Northwest Extension Publication, PNW0611, Washington State University, Oregon State University and University of Idaho, USA. http://agresearch.montana.edu/wtarc/producer info/agronomy-nutrient-management/Alfalfa/NutrientManagementGuide.pdf.

Kumar, S., Bhatt, B., Dey, A., Shivani, Kumar, U., Idris, M. and Mishra, J.S. (2018). Integrated farming system in India: Status, scope, and prospects in changing agricultural scenario. Indian J. Agril. Sci., 88: 1661–1675.

Lal, M., Patidar, J., Kumar, S. and Patidar, A. (2018). Different integrated farming system model for irrigated condition of India on basis of economic assessment: A case study: A review. Inter. J. Chem. Stud., 6: 166–175.

Lal, R. (2015). Restoring soil quality to mitigate soil degradation. Sustainability, 7: 5875–5895.

Lewis, G., Schrire, B., Mackinder, B. and Lock, M. (2005). Legumes of the World, Royal Botanic Gardens, Kew, UK.

Linares, J., Scholberg, J., Boote, K., Chase, C.A., Ferguson, J.J. and McSorley, R. (2008). Use of the cover crop weed index to evaluate weed suppression by cover crops in organic citrus orchards. Hort. Sci., 43: 27–34.

Linderman, R.G. (1992). VA mycorrhizae and microbial interactions. pp. 45–70. *In*: Bethlenfalvay, G.J. and Linderman, R.G. (eds.). VA Mycorrhizae in Sustainable Agriculture, ASA Special Publ. No. 54, Madison, WI.

Lipecki, L. and Berbec, S. (1997). Soil management in perennial crops: Orchards and hop gardens. Soil Tillage Res., 43: 169–184.

Litterick, A. (2019). Green manures improve soil health in apple orchards, Fact sheet, AHDB Horticulture, Stoneleigh Park, Warwickshire, UK. https://projectblue.blob.core.windows.net/media/Default/Imported%20 Publication%20Docs/AHDB%20Horticulture%20/GreenManuresInAppleOrchardCrops2407_WEB.pdf.

Lu, Y.H., Zhang, Y.J. and Wu, K.M. (2008). Host-plant selection mechanisms and behavioural manipulation strategies of phytophagous insects. Acta Ecol. Sinica., 28: 5113–5122.

Marquez-Garcia, F., Gonzalez-Sanchez, E.J., Castro-Garcia, S. and Ordonez-Fernande, R. (2013). Improvement of soil carbon sink by cover crops in olive orchards under semiarid conditions. Influence of the type of soil and weed. Spanish J. Agril Res., 11: 335–346.

Martin, O. (2012). Choosing and using cover crops in the home garden and orchard, News and Notes of the UCSC Farm & Garden, Issue 135, Fall 2012, Centre for Agroecology and Sustainable Food Systems. https://casfs.ucsc. edu/documents/for-the-gardener/choosing-cover-crops.pdf.

Mazumdar, B.C. (2004). Orchard Irrigation and Soil Management Practices, Daya Publishing House, New Delhi, India, pp. 86–89.

Meena, S.S., Lal, G., Mehta, R.S., Meena, R.D., Kumar, N. and Tripathi, G.K. (2017). Comparative study for yield and economics of seed spices based cropping system with fruit and vegetable crops. Int. J. Seed Spices, 7(1): 35–39.

Merwin, I.A. and Stiles, W.C. (1994). Orchard ground cover management impacts on apple tree growth and yield, and nutrient availability and uptake. J. Amer. Soc. Hort. Sci., 119: 209–215.

Mouco, M.A.C., Silva, D.J., Giongo, V. and Mendes, A.M.S. (2015). Green manures in 'Kent' mango orchard. pp. 179–184. *In*: Espinal, J.J. et al. (eds.). Proc. Xth Intl. Mango Symposium, Acta Hort., 1075, ISHS 2015.

Mulinge, J.M., Saha, H.M., Mounde, L.G. and Wasilwa, L.A. (2018). Effects of legume cover crops on orange (*Citrus sinensis*) fruit weight and brix. Inter. J. Plant Soil Sci., 21: 1–9.

Nagy, P.T., Kincses, I., Lang, T., Szoke, S.L., Nyeki, J. and Szabo, Z. (2010). Importance of orchard floor management in organic fruit growing (nutritional aspects). Inter. J. Horti. Sci., 16: 61–67.

Nath, J.C., Deka, K.K., Saud, B. and Maheswarappa, H.P. (2015). Intercropping of medicinal and aromatic crops in adult coconut garden under Brahmaputra valley region of Assam. J. Plant Crops, 43: 17–22.

Nimbolkar, P.K., Awachare, C., Chander, S. and Husain, F. (2016). Multi-storied cropping system in horticulture—A sustainable land use approach. Inter. J. Agri. Sci., 8: 3016–3019.

[NRCS] National Resources Conservation Service. (2011). Carbon to Nitrogen Ratios in Cropping Systems, USDA. East National Technology Support Centre, Greensboro. NC. https://www.nrcs.usda.gov/Internet/FSE_ DOCUMENTS/nrcseprd331820.pdf.

[NRCS] Natural Resource Conservation Service. (2015). Cool Season Cover Crop Species and Planting Dates and Techniques, Technical Note No: TX-PM-15-03, United States Department of Agriculture, USA. https://www. nrcs.usda.gov/Internet/FSE_PLANTMATERIALS/publications/etpmctn12683.pdf5.

[NRCS] National Resources Conservation Service. (2019). The Environmental Quality Incentives Programme, USDA. https://www.nrcs. usda.gov/wps/portal/nrcs/main/national/programs/financial/ eqip/.

Oliveira, F.E.R., Oliveira, J.M. and Xavier, F.A.S. (2016). Changes in soil organic carbon fractions in response to cover crops in an orange orchard. Revista Brasileria de Ciência do Solo, 40: e0150105. DOI: 10.1590/18069657rbcs20150105.

Panwar, A.S., Poonam, K., Natesan, R., Ashisa, P. and Mohammad, S. (2019). Horticulture-based integrated farming systems: a viable option for doubling farmers income. pp. 495–509. *In*: Chadha, K.L., Singh, S.K., Prakash, J. and Patel, B.P. (eds.). Shaping the Future of Horticulture. Kruger Brentt Publisher.

Pavek, P.L.S. and Granatstein, D.M. (2014). The potential for legume cover crops in Washington apple orchards: A discussion and literature review. Plant Materials Technical Note No. 22, USDA Natural Resources Conservation Service, Spokane, Washington.

Pontes, N.D.C., Kronka, A.Z., Moraes, M.F.H., Nascimento, A.S. and Fujinawa, M.F. (2011). Incorporation of neem leaves into soil to control bacterial wilt of tomato. J. Plant Path., 93: 741–744.

Qin, Y., Xin, Z., Yu, X. and Xiao, Y. (2014). Influence of vegetation restoration on topsoil organic carbon in a small catchment of the loess hilly region, China. PLoS One, 9: e94489. DOI: 10.1371/journal.pone.0094489.

Ravitchandirane, V. and Haripriya, K. (2011). Intercropping with medicinal plants in mango cv. Alphonso. Plant Archives, 11: 413–416.

Rayns, F. and Rosenfeld, A. 2010. Green manures—effects on soil nutrient management and soil physical and biological properties. Fact sheets 24/10. Soil grown crops. Projects FV 299 and 299a. The Horticulture Development Company. Warwickshire, UK. www.organicresearchcentre.com 09.09.2015.

Roper, T.R. (2004). Orchard Floor Management for Fruit Trees, A 3562, RP-08-2004(SR-07/95), University of Wisconsin-Extension, Co-operative Extension Publication, USA. http://www.uvm.edu/~fruit/treefruit/tf_horticulture/AppleHortBasics/Readings/WI_groundcover.pdf. (Accessed on 11/03/2019).

Sambeek, J.V. (2017). Cover crops to improve soil health and pollinator habitat in nut orchards. Missouri Nut Growers Association (MGNA) Newsletter, 17: 6–12.

Sanchez, E.E., Giayetto, A., Cichon, L., Fernandez, D., Aruani, M.C. and Curetti, M. (2007). Cover crops influence soil properties and tree performance in an organic apple (*Malus domestica* Borkh) orchard in northern Patagonia. Plant Soil, 292: 193–203. https://doi.org/10.1007/s11104-007-9215-7.

[SARE] Sustainable Agriculture Research and Education. (2012). Managing Cover Crops Profitably, Handbook Series Book 9, third ed., 1122 Patapsco Building University of Maryland College Park, MD 20742-6715. file:///C:/Users/USER/Downloads/Managing_Cover_Crops_Profitably.pdf.

[SARE] Sustainable Agriculture Research and Education. (2014). Recent Advances in Ecological Pest Management, USDA, 1122 Patapsco Building | University of Maryland | College Park, MD 20742. https://www.sare.org/Learning-Center/Books/Manage-Insects-on-Your-Farm/Text-Version/Recent-Advances-in-Ecological-Pest-Management.

Sau, S., Sarkar, S., Das, A., Saha, S. and Datta, P. (2017). Space and time utilisation in horticulture-based cropping system: An income doubling approach from same piece of land. J. Pharm. and Phytochem., 6: 619–624.

Sazo, M.M., Breth, D. and Tee, E. (2014). Establishing row-middle ground over options for high density apple orchards in western NY. New York Fruit Quart., 22: 15–20.

Seyum, S., Taddese, G. and Mebrate, T. (2019). Land use land cover changes on soil carbon stock in the Weshem Watershed, Ethiopia. Forestry Res. Eng.: Inter. J., 3: 24–30.

Sharma, P., Singh, A., Kahlon, C.S., Brar, A.S., Grover, K.K., Dia, M. and Steiner, R.L. (2018). The role of cover crops towards sustainable soil health and agriculture—A review paper. Amer. J. Plant Sci., 9: 1935–1951.

Singh, G., Nath, V., Pandey, S.D., Ray, P.K. and Singh, H.P. (2012). The Litchi, Food and Agriculture Organisation of the United Nations, New Delhi, India, pp. 126–135. https://midh.gov.in/tmnehs/writereaddata/A-Contents.pdf.

Singh, J., Arya, C.K., Bhatnagar, P., Jain, S. and Pandey, S.B.S. (2012). Intercropping in orchards is better option. Indian Hort., Jan.–Feb., pp. 5–7.

Singh, S.K., Sharma, M. and Singh, P.K. (2016). Combined approach of intercropping and INM to improve availability of soil and leaf nutrients in fruit trees. J. Chem. Pharm. Sci., 9: 823–829.

Singh, S.K., Sharma, M. and Singh, P.K. (2016). Intercropping—An approach to reduce fruit drop and improve fruit quality in guava. J. Chem. Pharm. Sci., 9: 3182–3187.

Singh, V.K., Singh, A., Soni, M.K., Singh, K. and Singh, A. (2015). Increasing profitability of mango (*Mangifera indica*) orchard through intercropping. Acta Hortic., 1066: 151–157.

Skroch, W.A. and Shribbs, J.M. (1986). Orchard floor management: An overview. Hort. Sci., 21: 390–393.

Smykal, P., Coyne, C.J., Ambrose, M.J., Maxted, N., Schaefer, H., Blair, M.W.B., Berger, J., Greene, S.L., Nelson, M.N., Besharat, N., Vymyslick, T.Y., Toker, C., Saxena, R.K., Roorkiwal, M., Pandey, M.K., Hu, J., Li, Y.H., Wang, L.X., Guo, Y., Qiu, L.J., Redden, R.J. and Varshney, R.K. (2015). Legume crops phylogeny and genetic diversity for science and breeding. Critical Rev. in Plant Sci., 34: 43–10.

Sofi, J.A., Rattan, R.K. and Datta, S.P. (2012). Soil organic carbon pools in the apple orchards of Shopian District of Jammu and Kashmir. J. Indian Soc. Soil Sci., 60: 187–197.

Song, B., Jiao, H., Tang, G. and Yao, Y. (2019). Combining repellent and attractive aromatic plants to enhance biological control of three Tortricid Species (Lepidoptera: Tortricidae) in an apple orchard. Florida Entomologist, 97: 1679–1689.

Song, B.Z., Wu, H.Y., Kong, Y., Zhang, J., Du, Y.L., Hu, J.H. and Yao, Y.C. (2010). Effects of intercropping with aromatic plants on diversity and structure of an arthropod community in a pear orchard. BioControl, 55: 741–751.

Steyn, J.N., Crafford, J.E., Louw, S. and Gliessman, S.R. (2014). The potential use of cover crops for building soil quality and as trap crops for stinkbugs in sub-tropical fruit orchards: knowledge gaps and research needs. African J. Agril. Res., 9: 1522–1529.

Sullivan, P. (2003). Overview of cover crops and green manures. Appropriate Technology Transfer for Rural Areas (ATTRA), Fundamentals of Sustainable Agriculture, National Sustainable Agriculture Information Service (NSAIS), National Centre for Appropriate Technology (NCAT), U.S. Department of Agriculture. Fayetteville, Arkansas (P.O. Box 3657, Fayetteville, AR 72702), Butte, Montana, and Davis, California. https://cpb-us-e1.wpmucdn.com/blogs.cornell.edu/dist/e/4211/files/2014/04/Overview-of-Cover-Crops-and-Green-Manures-19wvmad.pdf.

Talgre, L., Roostalu, H., Mäeorg, E. and Lauringson, E. (2017). Nitrogen and carbon release during decomposition of roots and shoots of leguminous green manure crops. Agron. Res., 15: 594–601.

Tiwari, T. and Baghel, B.S. (2014). Effect of intercropping on plant and soil of Dashehari mango orchard under low productive environments. The Asian J. Hort., 9: 439–442.

TNAU. (2016). Green manuring crop. TNAU agritech portal. Agriculture. https://agritech.tnau.ac.in/agriculture/agri_greenmanuring_agronomygreenmanures.html.

Tworkoski, T. and Glenn, D.M. (2012). Weed suppression by grasses for orchard floor management. Weed Tech., 26: 559–565. 10.2307/23264371.

Visalakshi, M., Jawaharlal, M. and Ganga, M. (2015). Intercropping in cashew orchards. Acta Hortic., 1080: 295–298.

Walia, S.S. and Kaur, N. (2013). Integrated farming system –an eco-friendly approach for sustainable agricultural environment—A review. Greener J. Agron. Forestry Hort., 1: 1–11.

Wang, K.H., Sipes, B.S. and Schmitt, D.P. (2002). Crotalaria as a cover crop for nematode management: A review. Nematropica, 32: 35–57.

Wang, Y.X., Weng, B.Q., Ye, J., Zhong, Z.M. and Huang, Y.B. (2015). Carbon sequestration in a nectarine orchard as affected by green manure in China. Eur. J. Hortic. Sci., 80: 208–215.

Wittwer, R.A., Dorn, B., Jossi, W. and van der Heijden, M.G.A. (2017). Cover crops support ecological intensification of arable cropping systems. Scientific Reports, 7: 41911.

Woodward, C.L. (1996). Soil compaction and topsoil removal effects on soil properties and seedling growth in Amazonian Ecuador. Forest Eco. Manag., 82: 197–209.

Yang, Y., Chen, Y., Li, Z. and Chen, Y. (2018). Land-use/cover conversion affects soil organic carbon stocks: A case study along the main channel of the Tarim River, China. PLoS One, 13(11): e0206903. https://doi.org/10.1371/journal.pone.0206903.

Zhu, Y.P., Sheaffer, C.C., Russelle, M.P. and Vance, C.P. (1996). Dinitrogen Fixation of Annual Medicago Species, North American Alfalfa Improvement Conference, 35, Oklahoma City, Oklahoma City, p.75.

11

Cover Crop Mixes for Diversity, Carbon and Conservation Agriculture

Reicosky, DC,[1,]* *Ademir Calegari,*[2] *Danilo Rheinheimer dos Santos*[3] and
Tales Tiecher[4]

1. Introduction

As world population increases and food demands rise, keeping our soil healthy and productive becomes exceedingly important in agriculture. The expanding global population, expected to reach nearly 10 billion by 2050 (United Nations, 2014; Sanderman et al., 2017), is exerting mounting pressure on the finite land area and resources for growing food. Traditional conventional agriculture, with an emphasis on intensive tillage and monoculture practices, has resulted in slow environmental degradation that may ultimately jeopardise our food security (Hatfield et al., 2017). Meeting population needs while minimising impacts on the environment (Foley et al., 2011) will require sustainable agriculture (Tilman et al., 2011; Garnett et al., 2013). Since over one third of arable land is in agriculture globally (World Bank, 2015), finding ways to increase soil carbon (C) in agricultural systems will be a major component of using soils as a sink (Lal, 2005). A number of agricultural management strategies sequester soil C by increasing C inputs into the soil and enhancing various soil processes that protect C from microbial turnover.

Farmers around the world have relied on intensive tillage for the last 10,000 years (Lal et al., 2007; Reusser et al., 2015). Although this approach has had many benefits, it also has resulted in some serious problems, notably the resulting susceptibility of soil to erosion. According to Pimentel et al. (1995), about 430 million ha—almost one-third of the global arable land area—has been lost to soil erosion. Efforts to control human-induced land degradation and soil erosion have been building on the ruins of the past tillage and monoculture concepts (Lal et al., 2007; Montgomery, 2007a,b). In many places, tillage for planting and cultivation loosens the soil, leaving it vulnerable to transport by wind or water. In sloping lands, the tillage process itself can transport soil downslope (Lindstrom et al., 1990). Consequently, erosion rates from conventionally plowed agricultural fields can be 1–2 orders of magnitude higher than rates of soil production (Montgomery, 2007b). We are losing soil and C faster than nature can make it because we are working against nature rather than working with nature.

[1] Soil Scientist Emeritus, USDA-ARS, Morris, MN, USA.
[2] Soil Scientist Senior, Agricultural Research Institute (IAPAR), Londrina, Parana State, Brazil.
[3] Professor, University Federal of Santa Maria, Rio Grande do Sul State. Researcher CNPq.
[4] Professor, Soil Department of Federal University of Rio Grande do Sul (UFRGS), Porto Alegre, Rio Grande do Sul, Brazil.
  Emails: ademircalegari@bol.com.br; danilonesaf@gmail.com; tales.tiecher@ufrgs.br
* Corresponding author: don.reicosky@gmail.com

Climate change poses a major threat to food security through its direct impact on long-term agriculture (Basche et al., 2016a). Climate extremes negatively affect crop, livestock, and fishery production through yield reductions, biological migration, and loss of ecosystem services, which ultimately lead to a reduction in agricultural incomes and an increase in food prices. Lobell et al. (2013) reported on the role of extreme heat, associated with increased vapor pressure deficit, impacting maize production in the United States; this contributes to water stress in two ways: (1) by increasing demand for soil water to sustain a given rate of C assimilation, and (2) by reducing the future supply of soil water by raising transpiration rates. Soil organic carbon (SOC) sequestration can partially support the mitigation of these issues while offering part of the solution to a warming climate (FAO, 2017).

Food security rests on our living soils. Meeting the twin challenges of growing more food and reducing environmental damage will require careful attention to the world's soils (Lal, 2014; Admunson et al., 2015; Hatfield et al., 2017). Soils are a critical component of terrestrial ecosystems, a good place to store atmospheric carbon dioxide (CO_2), and a fundamental constituent for sustaining all life on earth (Lal, 2004; Lal et al., 2013; Lal, 2014). We need climate-smart agriculture: a transition to farming practices that are better-suited to cope with impacts of climate extremes that jeopardise global food security and natural resources (Hatfield et al., 2011; Branca et al., 2011; Gattinger et al., 2011; Lal, 2004, 2014). The recent increase in the frequency and intensity of extreme climate events has encouraged the science community to view soil as a C storage location for greenhouse gases, primarily CO_2. The consequences of human degradation of soil resources are far ranging and related to loss of soil organic matter (SOM) (Lal, 2004; FAO and ITPS, 2015; Montanarella et al., 2016; Keesstra et al., 2016). Montanarella et al. (2016) and Hatfield et al. (2017) reported soil erosion was identified as a grave threat, leading to deteriorating water quality. At the same level of concern, the major soil processes threatening ecosystem services (ES) include soil organic C loss, soil contamination, soil acidification, soil salinisation, soil biodiversity loss, soil surface effects, soil nutrient status, soil compaction, and soil moisture conditions (FAO and ITPS, 2015). Estimating the SOC loss due to land use change and intensive agriculture has been difficult but is a critical step in understanding whether SOC sequestration can be an effective climate mitigation strategy (Lal, 2015c; Sanderman et al., 2017). There is optimism the continued loss of SOM in intensive agricultural production systems can be reversed with conservation agriculture (CA) systems utilizing diverse rotations and cover crop mixes with minimum soil disturbance and minimal environmental impact.

Carbon is a major player in the greenhouse effect and climate mitigation, in soil health and ecosystem services, and in our food security. The multiple synergistic benefits of cover crop mixes and C management are required for sustainable production. Our objectives in this chapter are to analyze cover crop management to optimise C input for all CA system ecosystem services, soil health, sustainable production, climate mitigation and our food security for future generations. Emphasis will be on plant biodiversity for optimum C management, nutrient cycling and protection, and utilisation and minimum soil disturbance that represent holistic and systems regeneration and thinking with a focus on climate mitigation and global food security.

2. Cover Crop Terminology Dilemma

Agriculture production systems require clear and explicit global communication. The term 'cover crops' was initially defined to include those crops that provide soil cover to prevent erosion. Cover crop is a term applied to a number of plant species that farmers, ranchers and landowners may plant to help manage soil erosion and fertility, fix nitrogen, preserve moisture content, and control weeds and diseases. Hartwig and Ammon (2002) provide a comprehensive overview of cover crops (and living mulches) planted into or after a cash crop, and then commonly destroyed before the next crop is planted. As production systems have evolved, other "service" crop names are being used for many multifaceted functions within ecosystems, and as a result, we come up with a different name depending on the expected impact of, and the reason for, using that crop in a production system. Names like green manure crops (Dutra, 1919; Cherr et al., 2006), carbon crops, catch crops, companion crops, nurse crops, rotational crops and forage crops, perennial crops, summer annual crops, winter annual crops, pulse crops, cereal crops, and inter-crops (Calegari et al., 1993; Miyazawa et al., 2014) are used. Intercropping increases SOM decomposition, presumably through reduced SOM recalcitrance resulting from lower C:N, higher litter input, and better N retention (Cong et al., 2015a,b). Green manure crops are planted between harvested crops as cover crops (Thorup-Kristensen et al., 2003). The green manure is young and succulent plant material that is often turned into the soil to improve the organic matter and nutrient content. Unfortunately, the benefits are short-lived because maximizing the soil residue contact results in rapid decomposition of the plant biomass with associated nutrient release. The interaction of tillage with the incorporation of the crop biomass is not nearly as good as managing the biomass maintained on the surface for protection and minimising soil residue contact.

Another term that could be used is 'soil health crops', for the way they contribute to improve nutrient cycling and biological activity, or 'soil energy crops' or 'soil function crops' for the way both the aboveground and belowground components contribute as an energy source for the soil biology. Additionally, they could also be called 'crops' because many of the benefits generated in the soil are directly related to the physical, chemical and biological benefits of C in the soil using cover crop cocktails, cover crop mixes, and cover crop blends. Another name is 'synergy crops', due to the synergistic benefits from minimum soil disturbance and diverse cover crop mixes (Calegari, 2016). The name or type of crop is not that important; however, it becomes important in communicating the intended role or function. The important point is growing a crop as long as biologically possible, capturing as much CO_2 from the atmosphere, and converting it into biochemical energy for our food production and maintenance of the soil biology functions. It is the continued C cycling of any plant that is important, not the name of the plant or a specific function (Janzen, 2015).

3. Soil Carbon Benefits

The soil is a fundamental foundation of our life, economy and environmental quality. Soil is alive and as vital to human survival as air, water and the sun; its protection and enrichment with energy and organic C are needed for the future sustainability of our planet. Soil C stands at the forefront of CA and soil health due to its critical importance in regulating physical, chemical and biological processes and properties of soils. Cover crops play a vital role in soil C accumulation (Lal, 2015a). As a living entity, soil is a diverse ecosystem with an estimated 10,000–50,000 different taxa in a teaspoon of soil, making it one of the most complex on the planet and essential for human life. The soil functions include food production, water purification, greenhouse gas reduction, and pollution cleanup, to name a few. Even though microbes in soil are essential for life on earth, scientists readily admit they still know relatively little about them. A gram of soil—about a quarter of a teaspoon—can easily contain a billion bacterial cells and several miles of fungal filaments. These numbers are difficult to estimate and keep getting adjusted as we learn more about the soil microbiome through research and sequencing efforts.

Carbon is the key component of SOM (~ 58 per cent C) that improves soil physical, chemical, and biological properties and processes (Corsi et al., 2012; Lal, 2015; Kane, 2015; FAO, 2017; Sanderman et al., 2017). Throughout this discussion, the terms C, SOC, SOM and organic residues may be used interchangeably with the understanding C is the key element with all aspects of structure and energy in the soil-plant-atmosphere system.

Soil organic matter is critical for the stabilization of soil structure, retention and release of plant nutrients, and maintenance of water-holding capacity, thus making it a key indicator for agricultural productivity and environmental resilience. The microbial decomposition of SOM further releases mineral

> Carbon is the framework and the fuel of every living thing!
> Bryan Jorgensen, no-till farmer, Ideal, SD

nutrients, thereby making them available for plant growth (van der Wal and de Boer, 2017), while better plant growth and higher productivity contribute to ensuring food security.

Plant C is transient with continuous movement through the soil food web, meaning that plant C is constantly changing as it is transformed into new organisms or converted into different compounds (Janzen, 2015; Kane, 2015). Soil scientists classify C into general or pools based on how long the C remains in the soil, a figure often referred to as 'mean residence time'. The most commonly used model of these pools includes three different groupings: the fast (or labile) pool, the slow pool, and the stable pool, based on the function of its physical and chemical stability (Jenkinson and Rayners, 1977; FAO and ITPS, 2015; O'Rourke et al., 2015). The fast pool (labile or active pool) of fresh organic C decomposition, resulting in a large proportion of the initial biomass being lost in one to two years. The intermediate pool is comprised of microbially processed organic C that is partially stabilised on mineral surfaces and/or protected within aggregates, with turnover times in the range 10–100 years. A slow pool (refractory or stable pool) with highly stabilised SOC enters a period of very slow turnover of 100 to > 1,000 years and ultimately into humus.

Soil organic carbon is dynamic, however, and anthropogenic impacts on soil can turn it into either a net sink or a net source of greenhouse gases (GHGs). Soil has become a focus for looking at ways to mitigate climate change (Hatfield et al., 2011; Lal, 2015a; Sanderman et al., 2017; Hatfield et al., 2017). Healthy soil is more resilient in a changing environment. It also nourishes plants and allows the plants to remove CO_2 from the atmosphere. Soil is also a massive repository for C; it houses the C compounds of decaying plants and animals as well as everything that lives within the soil, from microbes to worms to plant roots exudates essential for soil microbes to be able to flourish and continue C and nutrient cycling. The composition and breakdown rate of SOM affects the diversity and biological activity of soil organisms, plant nutrient availability, soil structure and porosity, water infiltration rate, and water-holding capacity. A list of soil C benefits are shown in Table 1. While each of the individual benefits may not seem critical, the combined benefits with the associated synergies becomes very demanding with respect to providing ecosystem services and sustainable food production.

Soil productivity is closely linked to SOM and its primary component, SOC. Sequestration of C in SOM is also an important approach for reducing the concentration of CO_2 in the atmosphere (Lal, 1999; Hatfield et al., 2011; Lal, 2015a; Sanderman et al., 2017). Soil organic matter, which includes soil humus and all the plant, animal, and microbial residues in the soil, is generally assumed to be 50–58 per cent C by mass (Nelson and Sommers, 1996). In general, SOC increases when inputs of plant residue C to the soil are greater than C losses through decomposition, erosion, and leaching (Paustian et al., 2016).

Plant biomass is a treasure trove of many forms of C that serves as the foundation for many other 'organic compounds' in agricultural production systems. Soil C storage and other conservation benefits require a detailed knowledge of plant biomass C production and management. Johnson et al. (2006) reviewed grain and biomass yield, harvest index data and root C/shoot C ratios for a few major crops, estimating total plant C input to the soil. Vegetative aboveground biomass is relatively easy to measure; however, the root system represents more of a challenge with the available organic source C inputs into the soil. Understanding the role of C translocated belowground is critical to understanding the soil C cycle. It is the reduced C (photosynthate) translocated below ground that supports root growth and maintenance; root biomass plus root rhizodeposition provides the energy inputs into the soil food web. Balesdent and Balabane (1992), Balesdent and Balabane (1996), Allmaras et al. (2004), Wilts et al. (2004) and Kätterer et al. (2011) noted the relative importance of root biomass C plus rhizodeposition (total root C) compared to shoot C. An accurate accounting of total root C sources is critical for assessing the overall plant-derived C inputs into the soil.

Redin et al. (2014a) studied root mineralisation of 20 different crops and four botanical families, and showed that the mineralisation of root C varied greatly in terms of kinetics and in the total amount of C mineralized (36–59 per cent of added C), and that mineralisation constant was negatively correlated with hemicelluloses and positively with N content. Moreover, Poaceae roots

Table 1 Benefits of soil organic carbon (SOC) in agricultural ecosystems (*Footnote*: Numerous Sources, notably Lal, 2015a).

Soil C Increases	Soil C Decreases
Crop drought resistance	Risks of drought crop yield losses
Water infiltration	Soil runoff
Transpiration	Erosion
Soil tilth	Evaporation
Soil structure	Sediments
Available water holding capacity	Temperature
Water use efficiency	Crusting
Water storage	Pollution
Root depth	Soil compaction
Soil biological activity and bio-pores	Desertification
Water quality	Fertiliser inputs
Root and worm bio-pores	Air pollution
Aeration	Water pollution
Nutrient: cycling, adsorption and protection	Chemical fertiliser inputs
Drainage	Energy costs
Root depth	Time and labour
Cation exchange capacity	Equipment wear
Soil buffer capacity	Air pollution
Biological activity (micro, meso, macro fauna and flora); natural enemies to control pests and soil root diseases	Desertification
Nutrient storage	Agriculture C footprint
Biodiversity of microflora and fauna	Soil degradation
Adsorption of pesticides	Salinisation
Soil aesthetic appeal	Nutrient losses
Capacity for manure and other wastes	Pests and diseases, nematodes population
Wildlife and pollinators	Soil and water toxicity
Species diversity	Water losses
Resilience	Soil imbalance

that combined high hemicelluloses, low cellulose and low total N showed low degradation rate and cumulative C mineralisation. This study demonstrated that the chemical composition of roots, as for the aboveground parts of plants, can correctly predict their rates of decomposition in soils.

The C content of vegetation is surprisingly constant across a wide variety of tissue types and species. Schlesinger (1991) noted that C content of biomass is almost always found to be between 45–50 per cent (by oven-dry mass) for woody biomass. Calegari et al. (1993) showed that most typical cover crop plants contain between 38–45 per cent C, with outliers of white lupin at 48 per cent and ryegrass at 59 per cent C. Redin et al. (2014b) also found C content ranging from 39–49 per cent in 20 different crops and cover crops, but with N content ranging from 0.3–5.2 per cent.

4. Benefits of Cover Crops

a. Energy Capture Efficiency

Life is energy-driven. Sunlight is the only renewable energy for life on earth. Agriculture is responsible for the capture and transfer of solar energy, plus some nutrients from the soil, into biochemical energy and food security. Cover crops can play a critical role in providing continuous living cover for as long as biologically possible. The living cover captures C and ultimately provides soil protection through living plants, dormant plants and dead crop residue. Carbon captured by plants is the energy that enables nutrient uptake and fuels our biological system and associated nutrient cycling. Through photosynthesis, the combination of CO_2 and H_2O, under the influence of sunlight, creates carbohydrates that serve as the foundation of plant structure and C management that enable crop growth and grain yield as our primary food source (Janzen, 2006, 2015; Reicosky and Janzen, 2018). Carbon management, either C storage or C cycling, starts with photosynthesis, where plants convert it into some form of soil C, where it may be stored, or it may be transformed/recycled through utilisation by biological activity and released as CO_2 emissions. This process has been going on for many billions of years, and unless the soil is disturbed by intensive tillage releasing CO_2 emissions back to the atmosphere, it reaches a natural equilibrium. Kuzyakov and Cheng (2004), using C^{13} and C^{14} tracers, determined that the process of photosynthesis is the limiting step in getting C into and through the root system. The tight coupling of the negative rhizosphere effect on SOM decomposition with photosynthesis suggests managers work to maximize photosynthesis in our production system. In other words, continuous living cover on the landscape, as long as biologically possible, should be our goal. Our C management must lead to production of roots and exudates (Bais et al., 2006; Haichar et al., 2008), grain, and stover for combining agronomic and cover crops for maximum C input in agricultural and ecosystem services.

With conventional intensive agriculture, virtually all fields have something green growing during the six-month summer growing season, then terminated with harvest and fall tillage. This leaves bare soil exposed to the elements without any plant material to capture additional C and supply nutrition and energy to the soil biology during the other six months. We must make sure we are using our resources (sun, soil, water, air) efficiently all the time.

We want living plants capturing C within the biological limits because that growth helps keep the soil microbes active. Plant-captured C is the energy that fuels this biological system, and the associated nutrient cycling providing our food security. We need to develop methods to extend the "C capture season" before and after the main crop for efficient utilisation of solar energy. Conservation agriculture and diverse cover crop mixes can lengthen the C capture season. Utilising cover crop spp. before and after the normal growing season extends the period of C capture for the ecosystem, while maintaining soil protection during these normally fallow periods. Our goal is to grow something as long as it is biologically possible within the temperature and water limits that nature provides. Even in northern climates with extremely low temperatures, the presence of dormant or dead cover crops is a readily available source of energy when the climate conditions are right, and also provide protection from wind and water erosion during this critical period. While there are many challenges doing this economically, with a little more research and some genetic manipulation, we may be able to get a combination of plants that can accumulate C and protect the soil 365 days per year.

Cover crops are multifaceted with respect to the soil functions they address, resulting in the numerous benefits listed in Fig. 1. Cover crops can improve soil tilth, control erosion and weeds (Teasdale et al., 2007) and maintain SOM content. Soil compaction and water infiltration, which may leach soil nutrients (especially N), can be reduced (Kaspar and Singer, 2011). Thus, cover crops can be used to retain and recycle plant nutrients, especially N, P, K and S, between cropping cycles. Cover crops also provide a habitat for pollinators and beneficial insects, as well as provide rotations to break plant disease cycles.

Cover crops provide agricultural production systems with many benefits, including soil and nutrient retention, resources and habitat for beneficial organisms and weed suppression (Fig. 1). All of these 'oddly' synergistic factors working together improve food security, environmental quality, our quality-of-life, and profitability. With a short growing season, cool temperatures can hinder the establishment of productive cover crops between cash crops; living mulch systems may provide growers with opportunities to establish cover crops earlier in the growing season, thereby increasing the duration of soil cover, whether it is dead, dormant, or alive. Cover crops can restore

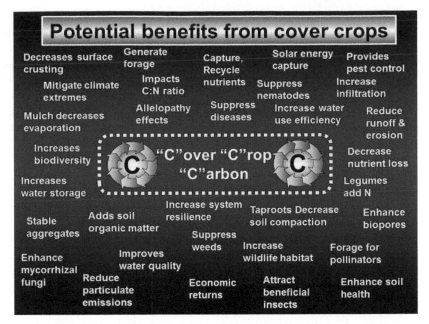

Fig. 1 Multiple benefits of cover crops.

soil quality and maintain soil health through improved C management (Reicosky and Forcella, 1998; Franzluebbers, 2005; Calegari et al., 2008; Florentin et al., 2010; Blanco-Canqui et al., 2015; Lal, 2015b; Raphael et al., 2016).

Mulching is applying plant residue, or other suitable organic material produced offsite, to the land surface (USDA NRCS, 2014a). The mulch mat was found to conserve soil moisture, reduce energy use with irrigation, control erosion, facilitate the establishment of plant cover, and improve soil health. Mulching generally has a positive impact on physical soil properties, e.g., bulk density (Kahlon et al., 2013) and porosity (Glab and Kulig, 2008). Kahlon et al. (2013) found that saturated hydraulic conductivity and mean aggregate size in the surface layer increased under ridge till, no-till, and plow-till systems. Kaye and Quemada (2017) reviewed cover crops to mitigate climate change by tallying the positive and negative impacts on the net global warming potential of agricultural fields. They found cover crop effects on GHGs fluxes, surface albedo changes due to cover cropping, and enabling climate change adaptation through reduced vulnerability to erosion from extreme rain and drought events, with increased soil water management options during droughts or periods of soil saturation, and retention of nitrogen mineralized due to warming (Kaye and Quemada, 2017). Many ecosystem services are traditionally expected from cover cropping but can now be promoted synergistically with services related to climate extremes. Farmers, consumers, and policymakers can now collaborate to expect cover cropping benefits, such as improved soil quality, water quality, and climate-change adaptation and mitigation, which are good for all of society.

b. Manage Biological Diversity

We are slowly understanding that soil is a natural living biological system where soil organisms are part of the soil (Garbeva et al., 2004; Wall and Nielsen, 2012; Bardgett and van der Putten, 2014; Van der Wal and de Boer, 2017), influencing soil physical, chemical, and biological properties such as hydrology, aeration and gaseous composition, all of which are essential for the primary production and the decomposition of organic residues and waste materials and nutrient protection and release. Ecosystem services depend on biodiversity in three areas: ecosystem, species, and genetic diversity. The combined diversity in the soil microbial species, along with the combined diversity in many different plant species, provides a powerful mix for strengthening ecosystem services. A diverse population of species provides 'functional redundancy' that may be needed during extreme climate events or to counter a disease/nematode or insect infestation. Wider range of species will likely have a few species that will cover for the impaired species within the soil system. Diversity in biology must be understood to manage soil for optimum ecosystem services and agricultural production. Brussaard et al. (1997) reviewed the current knowledge on biodiversity in soils, its role in ecosystem processes, its importance for human purposes, and its resilience against stress and disturbance to identify areas of future research. The diversity in soil biology must be supported by the diversity of C energy input, supplied by both agronomic and cover crop mixes, for the system to efficiently function by cycling nutrients into the soil plant system for ecosystem services as illustrated in Fig. 2. Note that C cycling is the energy supply located in the center of all the soil physical, chemical and biological properties and processes.

Cover crops add to the biodiversity in agricultural production systems (Anderson, 2008). All species of plants have their own unique characteristics, including how they interact with other plants (such as providing shade, competition, or fixing nitrogen) and organisms (such as attracting beneficial insects or repelling insects that could damage neighboring specimens). The cover crops can also attract wildlife to agricultural fields by providing habitat, feeding opportunities (on insects attracted by the plants, for instance), and protection from the elements and predators. There are six broad types of cover crops based on temperature, grass versus broadleaf, warm and cool season grasses, warm and cool season broad leaves and legumes and *Brassicas* that contribute to biodiversity. Farmers do not always need to plant expensive cover crop mixes when cheap, single species covers often work. However, by adding biodiversity, moisture management, exploring soil profile depth layers, less N loss and better SOM, and other multiple synergistic economic and environmental benefits are reasons to adopt cover mixes.

With more diverse crop rotation, farmers can diversify crop production to include pasture, small grains such as wheat and oats, and cover crops such as clover. Rotating many different crops builds the soil and naturally disrupts the pests and diseases that strike when a farmer grows only one or two crops. Moreover, cultivating a winter cover crop is a beneficial practice for enhancing soil microbial quality, soil organic C stock, and promoting the use of the microbial biomass as a substantial reserve of nutrients (Balota et al., 2014).

The power of diversity is strong in natural systems. "Agricultural biodiversity is the first link in the food chain, developed and safeguarded by indigenous people throughout the world, and it makes an essential contribution to feeding the world" stated Nakhauka (2009). The living soil system

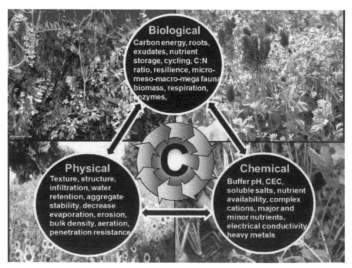

Fig. 2 Cover crop mixes provide enhanced biodiversity in biological, chemical and physical processes and properties with various forms of C and nutrients for soil organisms and subsequent crops.

includes a jungle of creatures, seen and unseen, working underfoot by performing countless critical functions (Wall and Nielsen, 2012; Bardgett and van der Putten, 2014; van der Wal and De Boer, 2017). Biodiversity is an element of sustainability and necessary for harmony and stability in nature (Nielsen et al., 2011; Orgiazzi et al., 2016); understanding and nurturing this biodiversity is needed to ensure that management practices sustain ecosystem services (Pereira et al., 2012; Costello et al., 2013; Pascual et al., 2015). Biodiversity plays a critical role connecting physical and chemical factors and processes in the soil, as illustrated in Fig. 2.

Soil biodiversity reflects the mix of living organisms in the soil. These organisms interact with one another, as well as with plants and small animals, forming a web of biological activity (Orgiazzi et al., 2016). Mycorrhizal fungi have a fundamental role in the crop root system, while in symbiosis, it will promote more available water, phosphorus, and other nutrients—and beyond that, promote higher fungi population in the rhizosphere that may control populations of nematodes and other pathogens (Balota et al., 2014). On the one hand, soil biodiversity contributes greatly to the formation of SOM from organic litter, thereby contributing to the enhancement of SOC content. On the other hand, the amount and quality of SOM (and consequently SOC) determines the number and activity of soil biota that interact with plant roots. Therefore, the soil microbial community structure is influenced largely by the quality and quantity of SOC, and to a lesser extent by plant diversity (Thiele-Brunh et al., 2012).

What makes a diverse system work is the ability of all of the parts to work together naturally. Soil biodiversity loss is a decline in the diversity of (micro- and macro-) organisms. In turn, this enhances the ability of soil to provide critical ecosystem services (Schnitzer et al., 2010; McDaniel et al., 2014; Tilman et al., 2014). Diversity and minimum disturbance are critically important in natural systems, however in recent agricultural ecosystems, "monoculture attitudes" have prevailed with limited plant species choices and mixtures (Cardinale et al., 2011; Thiele-Brunh et al., 2012). Diversity can reduce the risks incorporated in rotations with a minimum of three crops and at least one legume. Enhancing biodiversity enables a more holistic agriculture in which annual monocultures are replaced by mixtures of complementary crops that have the innate resilience and high biological productivity of natural ecosystems.

One way of enhancing the diversity of organisms in the soil is to grow a greater diversity of crops and by extending crop rotations (Poeplau and Don, 2015; Chatterjee et al., 2016). Introducing a variety of cover crops including cool season and warm season species, legumes and grasses, and shallow rooted and deep-rooted crops (Williams and Weil, 2004; Anderson, 2008; Burney et al., 2010; Kemper et al., 2011; Kätterer et al., 2011; Paustian et al., 2016) can also promote diversity. Crop rotation influences soil C dynamics, whereas rotating a larger number of crop species may influence soil C and N pools by affecting the proportions of the decomposable organic compounds returned to the soil, leading to potential gains in soil C (Hutchinson et al., 2007; Ogle et al., 2005; Schomberg et al., 2006). Increasing inter-specific diversity via CA also reduces vulnerability to damage from pests and extreme weather compared to monoculture systems. This resilience becomes important in the future, with greater potential for climate extremes and the incursion of new pests. Using multiple species also has synergistic benefits for nutrient uptake, nutrient cycling, wildlife habitat and pollinators. The robust and resilient nature of CA systems may also help to protect stored C from climatic disturbances, increasing the permanence and benefits of C.

> **The power of diversity is strong in natural systems. Diversity creates a dynamic mix of species and activity that provides multiple ecosystem services and enriches productivity.**

Schipanski et al. (2014) developed a framework to quantify ecosystem services provided by cover crops and to facilitate more multifunctional agricultural systems. They estimated that cover crops could increase eight out of 11 ecosystem services without negatively influencing crop yields. Thus, provisioning multiple ecosystem services from agricultural soil is largely accomplished by cover crops, fertiliser management, and no-till cropping systems (Syswerda and Robertson, 2014). Vukicevich et al. (2016) have demonstrated that (1) increasing plant diversity increases soil microbial diversity, minimizing the proliferation of soil-borne pathogens; (2) populations of beneficial microbes can be increased by increasing plant functional group richness (e.g., legumes, C_4 grasses, C_3 grasses, and non-leguminous forbs); (3) *Brassicas* suppress fungal pathogens and promote disease-suppressive bacteria; (4) native plants may further promote beneficial soil microbiota; and (5) frequent tillage, herbicide use, and copper fungicides can harm populations of beneficial microbes and, in some cases, contribute to greater crop decline.

c. Carbon Content and Cycling

Soil organic matter contains a large portion of the world's C and plays an important role in maintaining productive soils and water quality, however, a consensus on the nature of SOM is lacking. Lehmann and Kleber (2015) argue that SOM should no longer be seen as large, persistent, and chemically unique substances, but as a continuum of progressively decomposing organic compounds. An emerging alternate view suggests "SOM is a continuum of progressively decomposing organic compounds" and that recalcitrant 'humic substances' do not exist in soils (Lehmann and Kleber, 2015). The complexity and interactions of these segments of the C cycle contribute to the contentious nature of SOM (Lehmann and Kleber, 2015).

Plants capture CO_2, and then release it by respiration as part of the C cycle (Wagger et al., 1998; Nielsen et al., 2011; Frasier et al., 2016). Changes to either of these processes in response to climate change have profound implications on how much ecosystems soak up CO_2 emissions from burning fossil fuels. Carbon cycling is fundamental to all life on Earth, with storage of C in the soil being important for regulating climate, enhancing soil fertility to sustain crops, and capturing and retaining water (Kay, 1998; Lal, 2015b). But scientific capability to predict soil C stocks in terrestrial ecosystems is limited, partly due to the dynamic nature of the yet-unknown impact of diverse species (Nielsen et al., 2011). However, anthropogenic impacts on soil C can turn soil into either a net sink or source of GHGs.

Plant C capture in photosynthesis is the most efficient way of taking CO_2 out of the atmosphere and has been working for centuries. The simple C cycle showing the flow of C from photosynthesis to the roots, into SOM, through the soil, and back to the atmosphere demonstrates complexities of understanding rapid C cycling within an agricultural system. The C found in various pools (stable SOM, decomposing tissue, soil organisms, free organic C, exudates, etc.) all contribute to SOC. Photosynthate provides the energy for plant (above and below ground) growth and maintenance, but a significant fraction, as much as 40 per cent (Johnen and Sauerbeck, 1977; Bais et al., 2006), provides food for the belowground soil biology. A unit of C captured, reduced, and moved through plant and soil organisms may traverse several trophic levels before returning to the atmosphere as CO_2, and only a small fraction becomes stabilised SOM. Stabilised SOM is not immune to eventual decomposition, but decomposes at a slower rate.

The conversion of natural ecosystems to agricultural ecosystems with tillage disturbs the soil ecological balance, soil processes, organic C, and biotic C pools. Extractive farming practices, low external inputs, and soil-degrading land use all deplete terrestrial C pools (Sá et al., 2001). Manna et al. (2016) reviewed C sequestration as an effective strategy to mitigate climate change. Globally, there is a C crisis in soil, especially in tropical and subtropical ecosystems, because of increased CO_2 emissions from soil. There is a challenge to maintain the soil C status. Cover crops, less intensive tillage, and nutrient management in CA are viable options for soil C sequestration and the mitigation of C emissions.

All the transformations discussed in this chapter—decomposition, mineralisation, immobilization, denitrification, and nutrient cycling—are at least, in part, dependent on soil organisms. The same factors that regulate the SOM pools, especially precipitation and temperature, also control the activity and community composition of soil organisms. In fact, SOM and soil organisms are so interdependent that it is difficult to discuss one without the other.

In many no-tillage (NT) systems, crop residue input may be low and fallow periods excessive, as a result, soil degradation occurs. Cover crops, properly managed, can increase the building of organic matter even faster than no-till alone (Nielsen et al., 2011). Frasier et al. (2016) evaluated sorghum (*Sorghum bicolour* Moench.) monoculture, and with rye and vetch monoculture, as well as a mix of both species as cover crops. Cover crops increased plant litter, root biomass, total C, microbial biomass C, and microbial biomass N. Litter cover improved soil moisture to 45–50 per cent water-filled pore space and soil temperatures not exceeding 25°C during the warmest month. Their findings support the view that cover crops, specifically legumes in NT systems, can increase soil ecosystem services related to water and C storage, habitat for biodiversity, and nutrient availability (Schipanski et al., 2014).

Cover crop cultivars can be bred and adapted to the unique conditions inherent in CA to realize their full potential as a high-yielding alternative to conventional tillage agriculture. Stavi et al. (2012) studied the impact of winter cover crops in monoculture and a mixed culture on soil properties. They evaluated Austrian winter peas (*Pisum sativum* L.), radishes (*Raphanus sativus* L.), and a mix of both cultivars during the spring season to determine their impact on the soil quality at 0–5 and 5–10 cm depths. The concentrations of soil organic C and total N increased with radishes < Austrian winter peas < the mix of the two, respectively. The Austrian winter peas, the mix of the two, and radishes resulted in the highest, intermediate, and lowest water stable aggregates, mean weight diameter of aggregates, and saturated hydraulic conductivity, respectively, in terms of suitability of the cover crops for improving soil quality.

Two global meta-analyses document significant increases in C when a cover crop is a component of a crop rotation. McDaniel et al. (2014) found that including cover crops in crop rotations led to an average 8.5 per cent increase in total C concentration, and Poeplau and Don (2015) calculated an average increase of 0.32 Mg C ha^{-1} yr^{-1}.

The amount of SOC stored in a given soil is dependent on the equilibrium between the amount of C entering the soil and the amount of C leaving the soil as C-based respiration gases resulting from microbial mineralisation and, to a lesser extent, leaching from the soil as dissolved organic C. Locally, C can also be lost or gained through soil erosion or deposition, leading to the redistribution of soil C at local, landscape, and regional scales. Levels of SOC storage are mainly controlled by managing the amount and type of organic residues that enter the soil (i.e., the input of organic C to the soil system) and minimising the soil C losses (Kane, 2015; FAO and ITPS, 2015).

Increasing organic matter content of the soil has many physical, chemical, and biological benefits. Organic matter will not only increase the cation exchange capacity (CEC) of the soil, but the decomposition of added and residual organic compounds will release available N for plant uptake. Management practices to increase SOM and total N in soils include cover crops (Kaspar and Singer, 2011), green manure, crop rotations, crop residue management (Turmel et al., 2015; Murphy et al., 2016), addition of animal manure and compost (D'Hose et al., 2016), biochar addition (Lehmann, 2007), adopting no-till or minimum tillage systems (Mazzoncini et al., 2016; Sapkota et al., 2017), or combinations of these practices (Bulluck et al., 2002; Raphael et al., 2016; Wei et al., 2016; Ghafoor et al., 2017; Mulvaney et al., 2017).

Only about one third of crop residue C remains in the soil after one year (Angers and Chenu, 1977; Jenkinson and Rayner, 1977; Voroney et al., 1989; Powlson et al., 2011) and only 10–20 per cent remain after two years (Buyanovsky and Wagner, 1977; Broder and Wagner, 1988). Just a small

> "Organic matter functions mainly as it is decayed and destroyed. Its value lies in its dynamic nature."
> W. Albrecht, 1938

fraction of the C added to soil enters non-ephemeral pools (Paul et al., 1997; Lal et al., 2003; Franzluebbers et al., 1998; Robert, 2001), and even that 'stabilised' C is not locked away permanently, but remains vulnerable to gradual release over periods of decades to centuries (Janzen, 2015). As a result, SOM cannot be a 'storehouse' of C permanently locked away; instead, it may be better visualised as a 'stream' of C flowing in and out of the soil at various rates, and back to atmospheric CO$_2$. The latter perspective, based on the endless cycling of C through soil, reflects the transient, thermodynamically unstable nature of soil C, continually decayed by relentless microbial activity in soil (Kleber and Johnson, 2010; Janzen, 2015; Lehmann and Kleber, 2015).

Soil management can have a significant effect on organic matter levels in the soil. The composition and breakdown rate SOM affect the diversity and biological activity of soil organisms, plant nutrient availability, soil structure and porosity, water infiltration rate, and water-holding capacity. Intensive tillage is a practice that has negative impact on all soil properties important to ecosystem services. Franzluebbers and Arshad (1996) measured the impact of canola residue incorporation, maximising residue soil contact, which increased the decomposition rate when compared to surface applied crop residue. Franzluebbers (2004) found the portion of total N remaining as lignin-bound N increased in both treatments, with no difference between the incorporated residue or surface applied residue, suggesting the need for more research in this area. Franzluebbers et al. (1999) showed that the depth distribution of soil C significantly decreased in the surface layer after four years of conventional disc harrow tillage (15 cm-depth) compared to no-tillage. Other studies reported serious impacts of intensive tillage on soil degradation (Doran, 1987; Reicosky and Lindstrom, 1993, 1995; Ellert and Janzen, 1999; Mathew et al., 2012; Poffenbarger et al., 2015; Zuber and Villami, 2016). In fact, Franzluebbers (2005) found no-tillage plus cover crops provided two times more C storage than no-tillage alone, reflecting the combination of minimum soil disturbance and cover crops increasing soil C, and developing the foundation for the principles of conservation agriculture and soil health. Calegari et al. (2013a), in a Rhodic Hapludox in southern Brazil after 19 years of studies, found that the soil disturbance by plowing enhanced the macroporosity, decreased the microporosity, and promoted the formation of smaller aggregate sizes compared to no-till. Apart from the soil management, all winter species (black oat, blue lupin, radish, hairy vetch, and wheat) increased aggregate size classes, mean weight diameter, geometric mean diameter, and stability index compared to the fallow treatments. In the no-till treatments, most of sequestered C was stored in the soil aggregates.

Recent research on soil C dynamics and its influence on the global C cycle has been driven in part by increasing awareness of: (1) the importance of SOC for microbial C turnover that extends beyond a depth of 20 cm (Schimel and Schaeffer, 2012; Vogel et al., 2014); (2) the link between microbial communities and the dynamic soil properties in relation to the C cycle and its interaction with other biogeochemical cycles; and (3) the influence of plant diversity in increasing soil microbial activity and soil C storage (Nielsen et al., 2011; Lange et al., 2015).

The composition of the microbial community (e.g., bacteria: fungi) may be affected by tillage, and also has preferential decomposition of certain compounds (Strickland and Rousk, 2010; Heggason et al., 2007, 2010; Mathew et al., 2012; Mbuthia et al., 2015). Fungi have fine delicate filaments, or hyphae, that extend into places where plant roots cannot access nutrients and water in the soil. As a result, fungi are considered to be favored in no-tillage systems (Frey et al., 1999; Young and Ritz, 2000; Andrade et al., 2003). A low bacteria-to-fungi ratio also contributes to more C and N storage in no-tillage systems (Bailey et al., 2002; Wilson et al., 2009; Six et al., 2006; Jiang et al., 2011) and in forested systems (Clemmensen et al., 2013). A comparison of an intensive wheat-cropping system with a bacteria-dominated food web to a managed grassland with a fungi-dominated food web demonstrated that tillage lowered the resistance and resilience of the soil food web to drought (Vries et al., 2012). A study on the impact of red clover, chicory, and ryegrass grown before cereals showed an increase in the number of soil fungi spp. with the more biological diverse rotation; interestingly, the same study found that a tillage treatment increased pathogenic fungi vs. using NT (Detheridge et al., 2016).

Knowledge of changes during decomposition of differentially placed crop residue is important in understanding the potential effects of tillage systems on soil quality properties, including C and N conservation and soil tilth. A major factor contributing to the dynamics of soil C cycling is seen in the role of tillage incorporating the surface crop residue (Wilson and Hargrove, 1986; Ghidey and Alberts, 1993; Franzluebbers et al., 1996). Franzluebbers et al. (1996) found surface-placed canola residue had 57 to 30 per cent more mass remaining than buried residue. A fair amount of C in the crop can be accumulated during the growing season, and with intensive tillage, more than 70 per cent of the C will be returned to the atmosphere through microbial decomposition. However, the remaining plant residue on the soil surface, coupled with minimum soil contact, results in the prolonged protection of the surface soil and slow release of nutrients and C to the biological community.

Plant root exudates and plant-microbe interactions can also influence certain species or classes of microorganisms in the soil, with subsequent effects on other plants. For example, the glucosinolates and isothiocyanates released by crops and weeds in the crucifer family (such as *Brassica* crops, wild mustards, and yellow rocket) can inhibit soil fungi, including some pathogens (Haramoto and Gallandt, 2004). Crucifers and other nonmycorrhizal host plants, while not directly toxic to mycorrhizae, do not support the high populations of active mycorrhizal fungi often found in the soil after strong-host species like most legumes.

Soil C dynamics is a key component of crop production systems and provides various ecosystem services (Schipanski et al., 2014). Ecosystem C dynamics is the soil component that involves SOC storage through the C sequestration process and potential release of C as CO_2 into the atmosphere (Heimann and Reichstein, 2008; Schipanski et al., 2014) related to gain and loss of C by the soil. SOC can be recycled through the aboveground assimilatory processes (plant photosynthesis) and belowground heterotrophic respiratory processes through decomposition by soil microbial activity and respiration by animals and other organisms in the soil.

In eastern Australia, Liu et al. (2016) showed that SOC changes under crop-pasture rotation varies from northern to southern sites. They reported that temperature and rainfall play an important role and noted that this drove the dynamics of SOC and its interactions among farming management practices. A study conducted in Illinois, United States, on a silty clay loam soil by Zuber et al. (2015) showed that after 15 years, the corn–soybean–wheat (3-year) rotation increased SOC compared to that under continuous corn (2-year) or continuous soybean (2-year). However, in some short-term studies, the benefits of crop rotation in improving SOC have also been observed. Zhu et al. (2014) found that under a rice–wheat rotation system in China, where residues of rice and wheat were added into the soil, improved SOM. Data from this study showed that rice–wheat crop rotation built up SOM because of the leftover residue. A 49-year study in Ohio by Kumar et al. (2012) reported that SOC stocks under continuous corn and corn–soybean rotation was almost the same.

Another study in China, under winter wheat–summer corn crop rotation, showed an increase in SOC by 12 per cent in 0–5 cm, 17 per cent in 5–10 cm and 6 per cent in 10–20 cm, as well as a decrease in SOC by 7 per cent in 20–30 cm soil depth affected by crop rotations and tillage systems (Zhao et al., 2015). Shrestha et al. (2013), in semi-arid southwestern Saskatchewan, Canada, compared SOC stocks and rate of SOC change under one continuous crop and four three-year fallow-containing crop rotations managed with a NT system, to two fallow-containing crop rotations under minimum-tillage (MT). After 11 years, they observed that the SOC (0-to-15-cm depth) was 0.2 Mg C ha^{-1} greater under continuous crop compared to the fallow-containing systems. They concluded there were no significant differences in SOC and rate of SOC change among fallow-containing rotations, or between MT and NT. Aziz et al. (2011) evaluated the impact of crop rotations on soil quality and showed that corn–soybean–wheat (2002–07) had a significant impact on the SOC and TN contents of these crops. Various studies conducted across the world involving diverse crop rotations and their impacts on SOC, N, and crop production show that diverse crop rotations have significant impacts on improving SOC (Derpsch et al., 1986; Calegari and Alexander, 1998; Andrade et al., 2003; Calegari et al., 2008; McDaniel et al., 2014; Poeplau and Don, 2015; Chatterjee et al., 2016).

In subtropical and tropical conditions, many Brazilian studies show great potential of CA in C and N storage. Calegari et al. (2008), studying a long-term experiment in Pato Branco, southwestern Paraná in a Rhodic Hapludox in southern Brazil, found after 19 years of different cropping sequences and tillage management, NT management sequestered 6.84 Mg ha^{-1} more organic C compared to CT (64.6 per cent) at the 0-to-10 cm soil depth, 29.4 per cent more at the 0-to-20 cm soil depth, but equivalent amounts as CT at the 20-to-40 cm soil depth (Table 2). Greater amounts of SOC were found within the 0-to-20 cm depth (i.e., the moldboard plow layer). The NT system sequestered organic C at a rate of 1.24 Mg ha^{-1} yr^{-1}, while the CT system sequestered organic C at a rate of 0.96 Mg ha^{-1} yr^{-1}. Independent of soil management, winter fallow stored the lowest organic

Table 2 Total soil organic carbon affected by soil management and cropping systems after 19 years of cultivation.

Soil depth	Soil man	Winter Treatments						Aver.	CV	LSD (P < 0.05)		Forest
		Fallow	Vetch	Wheat	Radish	Oat	Lupine			Crop	man	
cm		Mg ha^{-1}							%	Mg ha^{-1}		
0–5	NT	19.20	24.37	22.73	25.27	26.46	24.43	23.74				
	CT	15.12	17.16	17.66	18.21	18.26	19.93	17.72				
	Aver.	17.16	20.77	20.20	21.74	22.36	22.18		9.02	2.24	1.29	80.64
5–10	NT	15.14	18.98	16.16	18.64	16.21	16.95	17.01				
	CT	14.61	16.42	15.18	17.59	17.39	17.17	16.39				
	Aver.	14.88	17.70	15.67	18.12	16.80	17.06		9.34	1.87	ns	41.57
10–20	NT	26.29	28.04	28.23	29.33	27.45	29.06	28.07				
	CT	28.10	30.06	30.55	33.00	32.15	32.69	31.09	8.62	3.05	1.76	31.76
	Aver.	27.20	29.05	29.39	31.17	29.80	30.88					
20–30	NT	21.29	22.64	20.83	22.12	22.71	26.48	22.68				
	CT	21.58	24.60	24.10	23.91	26.64	22.62	23.91				
	Aver.	21.44	23.62	22.47	23.02	24.68	24.55		8.25	2.30	1.33	26.27
30–40	NT	17.73	18.71	17.74	18.60	18.54	20.85	18.70				
	CT	18.45	17.25	17.73	17.96	20.22	16.93	18.09				
	Aver.	18.10	17.98	17.74	18.28	19.38	18.89		11.0	ns	ns	22.16
40–60	NT	29.88	29.15	30.62	30.49	30.94	36.04	31.19				
	CT	28.86	29.53	29.25	28.77	34.07	29.80	30.05				
	Aver.	29.37	29.34	29.94	29.63	17.01	32.92		9.66	ns	ns	19.19

C at all soil depths (0–10, 10–20, 20–40, 0–20, 0–40 cm). When winter cover crops such as blue lupin, hairy vetch, black oat (*Avena strigosa* Schreb.), oilseed radish (*Raphanus sativus* L.), and winter wheat (*Triticum aestivum* L.) were used with NT, in general, greater amounts of organic C were sequestered. The one exception was in the 10–20 cm depth increment for the moldboard plow treatment at a slightly higher C content, apparently related to inversion tillage incorporating surface crop residue (Table 2). Continuous NT management combined with winter cover crops resulted in the greatest amount of SOM in the surface soil and was the only cropped treatment that approached the undisturbed forest condition. Thus, the NT system serves as a management model for sustaining the productivity of Oxisols in tropical and sub-tropical regions of the world, one to be emulated by Brazilian farmers and others who are managing similar soil types. Considering the surface 40 cm of soil, the NT system sequestered only 6.7 Mg C ha^{-1} higher compared to CT. The rates of C sequestration were 1.24 and 0.96 Mg ha^{-1} yr^{-1} to NT and CT systems, respectively.

In contrast, results by Roscoe and Buurman (2003), after 30 years of cultivation, found no differences in SOC levels between NT and CT in a Dark Red Latosol (Typic Haplustox) in the cerrado area of Minas Gerais State, Brazil. The authors attributed this lack of difference to the high clay contents and Fe + Al oxi-hydroxides concentrations and physicochemical protection of organic C. Working in the tropical central savanna region of Brazil, Centurion et al. (1985) and Corazza et al. (1999) found higher soil C stocks under NT than CT in the surface 0-to-20 and 0-to-30 cm soil layers, but when the evaluation was extended to 100 cm soil depth, these differences disappeared due to lower C content in the 30–100 cm layer under NT. Also, Freitas et al. (2000) did not observe changes in C stocks (0–40 cm) in a clayey Dark Red Latosol (Oxisol) after 25 years of conventional cultivation (vegetables, rice, corn, and beans), when compared to natural cerrado vegetation. Corazza et al. (1999), in an Oxisol under tropical native savanna vegetation of central Brazil, also observed a relative increase of 26.2 Mg ha^{-1} SOC in soil under crop rotation and NT for 12 years, compared to conventional tillage, representing an annual accumulation rate of 2.18 Mg ha^{-1} yr^{-1}. These results are showing the high diversity of soil conditions (mineral, organic, texture, edaphoclimatic aspects) that contribute to achieve different results.

Because cover crops are normally grown during fallow periods, the addition of cover crops to a cropping system can increase total residue C inputs to soil, and has the potential to increase SOC (Karlen and Cambardella, 1996; Lal, 1999; Jarecki and Lal, 2003; Mitchell et al., 2013). Similarly, the rate at which cover crop residues decompose also affects the balance between losses and inputs of C into soil. Cover crop residue decomposition depends primarily on temperature, water content, biochemical constituents, residue quantity, C to N ratio, and soil contact. Kuo et al. (1997) reported that SOC half-lives for rye, hairy vetch, and annual ryegrass were similar and averaged 31 days and 57 days in two years when residues were buried 15 cm below the soil surface in mesh bags before planting corn. They attributed the slower decomposition in one of the two years to wetter soils and lower temperature. Although hairy vetch had lower shoot C to N ratios than ryegrass or rye, this did not affect the observed decay rate, which probably indicates that N was not limiting and that other factors, such as lignin concentration of residues or environmental conditions, limited the decomposition rate. In contrast, Ruffo and Bollero (2003) reported hairy vetch decomposed more rapidly than rye and that the decomposition rate of both responded to water content and temperature.

Black oat and oilseed radish grown in southern Brazil and Paraguay and could be good cover crops in the Southeastern United States. Black oat dry matter production is much like cereal rye; and like rye, the residues have a large carbon to nitrogen ratio (C:N) that can influence soil N mineralization–immobilization (Derpsch, 1990; Bolliger et al., 2006). Bauer and Reeves (1999) found that cotton (*Gossypium hirsutum* L.) yields were 120 kg ha^{-1} greater following black oat than following rye on a coastal plain soil in South Carolina, United States. Greater cotton yield following black oat may have been due to greater N availability because C:N of black oat is lower than that of rye (Ceretta et al., 2002). Oilseed radish has been used in southern Brazil as a fodder crop and cover crop. It grows rapidly in the fall and spring, and can scavenge significant quantities of N; however, the residues decompose very rapidly due to their low C:N. In addition to its influence on N cycling, oilseed radish contains glucosinolates (thioglucoside-N-hydroxysulfates), the precursors of isothiocyanates, chemicals known for fungicidal, bactericidal, nematocidal, and allelopathic properties and are the focus of medical research because of their potential cancer chemoprotective attributes (Fahey et al., 2001).

Schomberg et al. (2006) evaluated four cover crops: crimson clover, oilseed radish, black oat, and rye on a Cecil soil (sandy clay loam, fine, kaolinitic, thermic Typic Kanhapludult, with 2–3 per cent slope) typical of the southern U.S. Piedmont landscape. The experimental area was in Bermuda grass (*Cynodon dactylon* L. Pers.) hay production the previous four years. Soybean (*Glycine max* L. Merr.) was grown in 1999, but due to excessive deer (*Odocoileus virginanus* Zimm.) damage, cotton was grown in 2000, 2001 and 2002. Schomberg et al. (2006) showed that black oat and oilseed radish cover biomass production was less than cereal rye, but like crimson clover, and the amount of N contained in the residues was greater than in cereal rye with the effects of black oat and oilseed radish residues on N mineralization more similar to crimson clover than to cereal rye. Because N mineralisation measurements associated with these two cover crops did not indicate net N immobilisation during the summer season, there should be no reason to alter N fertilisation recommendations. The effects of cover crop biomass produced by these species were sufficient to help control soil erosion in conservation tillage systems; cold-hardiness may need to be addressed through breeding or selection to increase the geographic range for black oat. The large biomass produced by radish in the fall and early spring could be useful in rotations where earlier planting dates are desired and for preventing leaching of residual N. Producers in the Southeastern United States should consider black oat and oilseed radish as alternative cover crops to gain additional benefits associated with cover crop rotation, like disease and pest reduction, while maintaining soil N availability to summer cash crops (Schomberg et al., 2006).

Cover crops have been used successfully to increase soil C, especially in locations with mild winters that allow substantial cover crop growth (Beale et al., 1955; Patrick et al., 1957; Utomo et al., 1990; Kuo et al., 1997; Nyakatawa et al., 2001; Sainju et al., 2002). In some of these studies, cover crop residues were incorporated with tillage (Beale et al., 1955; Patrick et al., 1957; Kuo et al., 1997; Sainju et al., 2002). As mentioned earlier, more intensive tillage can increase the relative rate of cover crop decomposition and reduce the retention of C. Beale et al. (1955) reported that SOM was 28 per cent higher after 10 yr of a vetch and rye cover crop with mulch tillage compared to cover crops and moldboard plow tillage. Despite this evidence for increases in soil C and residue C inputs with cover crops, it is difficult to measure a change in SOC in cropping systems to which cover crops have been added. This is partly because it is difficult to measure small changes in SOC in field soils with relatively high background SOC levels and large variations in SOC with depth and terrain (Kaspar et al., 2008).

Additionally, cover crops may not produce large amounts of biomass in some locations or climates and may be a relatively small percentage of the total biomass produced in cropping systems like continuous corn. For example, Eckert (1991) in Ohio was not able to detect an increase in soil C using rye as a cover crop in no-till continuous corn or corn–soybean rotations. Duiker and Hartwig (2004) reported similar SOC levels in a crown vetch (*Coronilla varia* L.) living mulch treatment and the control after 13 years and concluded that severe suppression of the crown vetch living mulch to reduce competition with the corn crop had also reduced soil C inputs and benefits. Similarly, Utomo et al. (1990) observed no change in soil C using rye in either no-till or conventional tillage, but measured an increase with a hairy vetch cover crop in no-till, which produced more biomass than rye in a zero N treatment. Mendes et al. (1999) found that red clover or triticale (×Triticosecale Witt mack) winter cover crops did not increase soil C in a tilled vegetable production system, suggesting a negative effect of tillage.

The contribution of NT management to mitigate climate change by C sequestration is perceived to be low because: (1) the capacity for soil C sink is finite (Sommer and Bossio, 2014; Adenle et al., 2015; Corbeels et al., 2016; Powlson et al., 2016;); (2) diverse crop sequences or combinations with worldwide adoption of NT promote variable effects of NT on crop yields at global scale (Pittelkow et al., 2014); (3) difficulty of obtaining credible estimates of SOC on landscape scale and requiring a complex framework encompassing a wide range of climate, soils (texture, mineralogy), crops and cropping systems that exacerbate uncertainties in assessing C sequestration (Sá et al., 2013; Sommer and Bossio, 2014; Adenle et al., 2015; Lam et al., 2013); (4) high risks of re-emission of SOC sequestered because even a single tillage event in a long-term NT soil may negate previous gains in SOC stock (Sá et al., 2014; Stockfisch et al., 1999); (5) a high variation and uncertainties of the C sequestration rates in fields under NT involving three CA principles (FAO, 2014; Kassam et al., 2015) already practiced on < 15 per cent of the global cropland; and (6) low input of biomass-C return because of extreme weather events (e.g., long dry period or excessive rainfall). While there are many concerns about the use of CA principles to mitigate climate change, numerous research needs are easily identified and need further support.

Kuzyakov et al. (2000) and Kuzyakov (2010) discussed priming effects of SOM decomposition and C cycling. Priming occurs when fresh substrates accelerate the rate of SOM decomposition (Jenkinson, 1971) and when green manure crops are incorporated with tillage. An actively growing cover crop, with plentiful root exudates readily available to stimulate microbial activity, is an example of priming the soil biology. When the readily available supply runs out, the microbes must then turn to the more recalcitrant forms of SOM as an energy source (Kuzyakov and Cheng, 2004). The net effect of this priming is an overall decrease in SOM unless careful management is taken to keep the SOM inputs and outputs in balance. These dynamics in the induced priming effect indicates that the microbial response to exudates, which is released diurnally, is fast. This short activation of microorganisms by exudates leads to the change of SOM decomposition (Kuzyakov and Cheng, 2004). Any factor affecting photosynthesis, or substrate supply to roots and rhizosphere microorganisms, is an important determinant of root-derived CO_2 efflux, and thereby, total CO_2 efflux from soils.

d. Nutrient Content, Capture, and Cycling

Key cover crop challenges for N include: (1) identifying management practices that increase SOM content; (2) accurately estimating the contribution of organic N to the nutrition of plants in a cropping season; and (3) reduce the many possible loss reactions in the N cycle (Doran and Smith, 1991). Adoption of practices by farmers to increase SOM can be slow due to added expense and lack of knowledge.

Recognising the importance of biodiversity in natural systems, utilising cover crop mixes provides variable nutrient concentrations that also enable management options for the farmers to correct specific nutrient deficiencies on problem soils. Calegari et al. (1993) measured the nutrient concentration for a select group of 17 cover crops and found the C content of most cover crops in their study ranged between 37 and 45 per cent C, with two outliers being white lupin at 47.5 per cent and ryegrass at 52.9 per cent C. As a result, efforts to maximise C input o maximizing plant biomass and selecting the cover crop best adapted to the soil and environmental conditions. The N content ranged from a minimum of 0.77 per cent N for wheat to a maximum of 3.82 per cent N for hairy vetch, reflecting a fivefold variation in N. The combined diversity with a wide range of C and N contents provides ample opportunity to manage the C:N for optimum soil cover and N fixation. The P content ranged from a minimum of 0.05 per cent for white oat to a maximum of 0.30 per cent P for hairy vetch, reflecting a six fold variation in P content within the group of plants studied (Calegari et al., 1993). Similarly, the K content ranged from a minimum of 1.15 per cent in wheat and a maximum of 3.55 per cent *Ornithopus sativus*, reflecting a threefold variation in K content. Similar variations in the nutrient concentrations for the other nutrients were noted. The wide variation in the nutrient concentration within this group of cover crops suggest a benefit of biodiversity where one species with a high concentration of a specific nutrient can offset the low concentration of a companion crop. This wide variation in nutrient concentration partially explains the better performance of multi-species cover crop mixes relative to mono cover crop species (Hargrove and Frye, 1987; Havlin et al., 1990).

Cover crops grown during fallow periods can increase total N in the soil by changing the annual patterns of N uptake and mineralization, reducing NO_3 leaching, retrieving NO_3 from deep soil layers, and fixing atmospheric N_2. If the cover crops are legumes, they increase SOM and provide a living mulch and forage for livestock (Kaspar and Singer, 2011; Chatterjee and Clay, 2016). Forage production is highly dependent on climate, planting date, and termination dates with greater amounts of N_2 fixation occurring under warm conditions and earlier termination dates (Kaspar and Singer, 2011). Replacing fallow in a no-till winter wheat-fallow rotation with winter and spring cover crops for five years in the semiarid central Great Plains in the United States reduced runoff loss of sediment, total P and NO_3 (Blanco-Canqui et al., 2013). Appropriate plant species is an important consideration to obtain desired benefits as nitrate-leaching reductions with cover crops vary greatly.

Diversity in cover crop mixes can be useful to modify or adapt to soil properties in no-tillage systems. Tiecher et al. (2018) evaluated soil tillage systems and winter cover crops' impact on soil acidity, nutrient availability, and P and K budget on grain production in a highly weathered subtropical Oxisol from Southern Brazil. The native soil, a Rhodic Hapkudox, was characterized by low soil pH, high Al saturation, 73 per cent clay (mainly kaolinite), and high iron oxides. Additional work by Rheinheimer et al. (2018a,b) showed that in the NT system, soil acidity was very low in surface layers and it increased with soil depth due to the superficial applications of fertilisers and limestone, together with the deposition of crop residues on the surface without their incorporation due to low soil disturbance. The previous lime and fertilisers incorporated in the conventional system maintained low Al saturation and adequate levels of nutrients in the subsurface layer. The adoption of cover crops and no-tillage system improved and stabilized the grain production yields with 'normal' P, K and N fertiliser rates. This combined with high plant residues on the soil surface increased slowly over the 23 years of the study, enabling the accumulation of organic C and nutrients (Fig. 3).

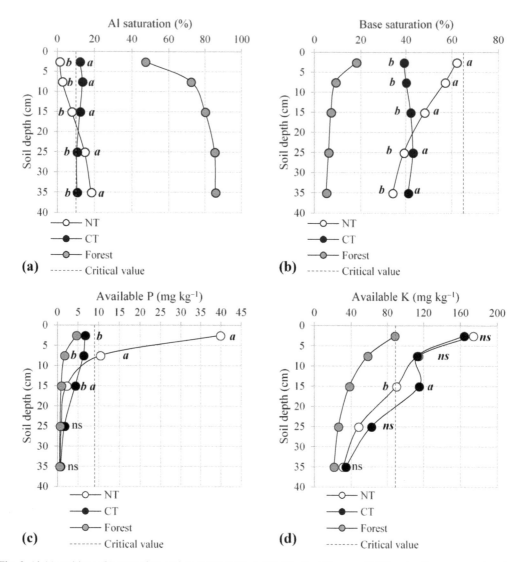

Fig. 3 Al (a) and base (b) saturation, and available P (c) and K (d) after 23 years of different soil management systems. Letters in bold and italic indicate difference between the forest soil and the cultivated soil under NT or CT by Mann-Whitney U test at p < 0.05 (adapted from Tiecher et al., 2018).

Results show that NT system builds up a nutrient availability gradient, with a higher concentration of nutrients on the soil surface layers, abruptly decreasing with soil depth (Tiecher et al., 2018).

Growing cover crops in the winter is effective for increasing P and K availability through plant cycling (Tiecher et al., 2018). Nutrient cycling by winter cover crops reduces P and K losses, especially when the soil is not plowed. Among the cover crops tested, black oat stood out due to its greater production of biomass, resulting in higher P and K availability in the soil surface. Lupine resulted in a greater cycling of P, possibly due to its ability to absorb P from less labile forms in the soil. Fallow in the winter decreases the P and K use efficiency. Growing cover crops is also an alternative to redistribute nutrients in the soil profile and avoid nutrient accumulation in the soil surface, which can be easily lost by runoff and soil erosion in conventional tillage systems. Franchini et al. (2004) demonstrated that *Vicia sativa* was the most efficient cover crop species as a P carrier into the roots from the surface layer to lower layers.

The negative budget of P during the 23 years of the study (Tiecher et al., 2018) was estimated because a large soil sink of P (Fe and Al oxi-hydroxides and kaolinite clay) exists, which is the main P adsorbent in highly weathered subtropical soils (Bortoluzzi et al., 2015). However, the less negative P budget in NT (-260 ± 21 kg ha^{-1}) compared to CT (-305 ± 21 kg ha^{-1}) shows that the efficiency of P use increases in NT systems (Table 3), related to minimum soil disturbance. Plowing in the CT system expose the P-added fertilisers to the functional groups adsorbing P, while in NT there is saturation of the most active sites by P, and subsequent accumulation of P in more available forms (Rheinheimer and Anghinoni, 2001). Minimum soil disturbance in NT systems also allows C accumulation in the surface layer, contributing to enhanced soil structure. The adsorption of functional groups onto Fe-oxides can increase P availability by altering surface charges, boosting competition for adsorption sites and replacing adsorbed anions (Fink et al., 2016). Growing a winter cover crop also reduced soil P and K losses, demonstrating the importance of cover plants in the cycling of these elements. Fallow in NT alone results in K losses, while with winter cover crops, the K-budget varies from null (for wheat) to $+223$ kg ha^{-1} with black oat, demonstrating the nutrient-cycling potential of black oat and its high biomass production.

The use of cover crops with high soil–root contact is important for P and K uptake because they have low mobility in the soil. In the long-term, the lack of any recycling of crop residues in mono-cropping systems with low residue production might cause severe K depletion from the soil K reserve (non-exchangeable K) (Srinivasarao et al., 2014), or K losses by leaching (Alfaro et al., 2006). Different crop rotation systems can alter soil P availability due to access of soil organic P by increasing exudation of acid enzyme phosphatase (Kunze et al., 2011; Cui et al., 2015; Chavarría et al., 2016), or by the exudation of organic acids with low molecular weight that promotes P mobilization by ligand exchange or by occupying P adsorption sites (Neumann and Römheld, 1999; Bayon et al., 2006). Moreover, the mycorrhizal or non-mycorrhizal character of cover crops have a regulatory effect on enzyme activity linked to organic P mineralization in the soil (Kunze et al., 2011). The use of non-mycorrhizal plants as lupin and oilseed radish (*Raphanus sativus* L.) increase the phosphatases enzymes activity in the soil compared to vetch (*Vicia sativa* spp.) and black oat (*Avena strigosa* Schreb) (mycorrhizal plants) (Dalla Costa and Lovato, 2004). Nutrient losses can be reduced by using cover crops that promote P and K accumulation, maintaining the nutrient in the soil-plant system, and avoiding K losses by leaching or P losses by soil-fixation (Rosolem and Calonego, 2013).

The data in Table 4 show that crop residues and tillage regime caused significant alteration and redistribution of nutrients within the soil profile of highly weathered soils (Tiecher et al., 2018). There are also likely to be effects in nutrient cycling, and certainly soil physical and biological properties that were not evaluated in this study. Although no-till system caused a nutrient concentration on the surface layer, this was important and not a disadvantage for corn development. Thus, the no-tillage system promoted better soil conditions for C and P, and consequently, P and N became more available for uptake in grain (corn, soybean, wheat, bean) production (Table 4).

Nature has been evolving for 3.8 billion years and has learned to recycle everything. One way of enhancing nutrient cycling in the soil is to grow a diverse set of cover crops, thereby extending crop rotations and minimising nutrient export (Poeplau and Don, 2015; Chatterjee et al., 2016). Nutrient cycling is driven by C cycling as the primary energy source for the soil biology depicted in Fig. 4. The efficiency of the nutrient cycling is a function of the diversity of the soil biology and plant biomass diversity as a source of nutrients. The wide variation in the diversity within the soil biology species, and a corresponding variation in the diversity of the nutrient concentration and cover crop biomass, provides ample potential for optimising the system for efficient uptake and crop production. When the SOC content falls below a critical level, the soil is endangered, as the soil aggregates become destabilised and soil nutrient cycling is compromised. In addition to nutrient cycling (Kell, 2011; Manna et al., 2016), increasing C in the soil should still be pursued for improving soil structure and decreasing atmospheric CO_2 (Lal, 2013, 2015a; Hatfield et al., 2011, 2017).

Tully and Rebecca (2017) reviewed metrics of farm management strategies that tighten nutrient cycles and maintain yields. Metrics for efficient nutrient cycling in agro-ecosystems included reduced runoff and erosion, reduced leaching, improved soil C storage, enhanced microbial

Table 3 Phosphorus and potassium budget after 23 years of different soil management systems and winter cover crops (Adapted from Tiecher et al., 2018).

Soil Management System	Winter Cover Crops	Yield (Mg ha⁻¹)			Removed by Harvest (kg ha⁻¹)		Available in the 0–40 cm Soil Layer (kg ha⁻¹)				Added by Fertilisation (kg ha⁻¹)		Net Nutrient Input (kg ha⁻¹)	
							Initial		23-yrs After					
		Corn	Soybean	Wheat	P	K	P	K	P	K	P	K	P	K
NT	Fallow	52.4	28.3	-	356	731	8.3	413.7	23.8	254.5	659	669	-287	-97
	Vetch	58.4	28.0	-	375	755	8.3	413.7	26.1	338.1	659	669	-266	11
	Wheat	52.3	27.8	12.8	408	786	8.3	413.7	24.8	298.6	659	669	-234	1
	Radish	53.7	28.9	-	364	747	8.3	413.7	26.7	401.2	659	669	-277	65
	Black oat	54.7	29.2	-	370	757	8.3	413.7	43.1	548.5	659	669	-255	223
	Blue lupin	59.8	29.1	-	387	780	8.3	413.7	41.1	368.6	659	669	-240	66
CT	Fallow	49.6	25.0	-	326	662	8.3	413.7	16.0	394.8	659	669	-325	-26
	Vetch	50.8	25.6	-	334	678	8.3	413.7	14.9	473.2	659	669	-318	69
	Wheat	50.2	25.5	12.8	387	737	8.3	413.7	15.4	469.4	659	669	-265	123
	Radish	52.5	25.9	-	342	691	8.3	413.7	15.9	359.6	659	669	-310	-32
	Black oat	52.0	26.4	-	343	698	8.3	413.7	14.4	418.7	659	669	-310	34
	Blue lupin	53.5	26.4	-	348	704	8.3	413.7	16.3	346.6	659	669	-303	-32
Mean	NT fallow	52.4	28.3	0	356	731	8.3	413.7	23.8	254.5	659	669	-287	-97
	NT cover crops	55.7	28.6	13	381	765	8.3	413.7	32.3	391.0	659	669	-254	73
	CT fallow	49.6	25.0	0	326	662	8.3	413.7	16.0	394.8	659	669	-325	-26
	CT cover crops	51.8	26.0	13	351	701	8.3	413.7	15.4	413.5	659	669	-301	32

Table 4 Total aboveground dry biomass yield over 23 years in different winter treatments and soil management systems (NT = No-Tillage; CT = Conventional Tillage) (adapted from Tiecher et al., 2018).

Winter Treats	Above-ground Dry Biomass Yield (Mg ha⁻¹)							
	Winter Cover Crops		Summer Crops Residues		Total		Corn Production 2009	
	NT	CT	NT	CT	NT	CT	NT	CT
Fallow	42.5	31.0	93.1	91.7	135.5	122.7	8.3	8.0
Wheat	75.0	68.6	89.9	86.0	164.9	154.6	8.0	8.1
Radish	84.9	69.5	99.0	97.1	183.9	166.5	7.4	7.2
Vetch	85.6	73.3	97.7	91.0	183.3	164.3	9.5	8.1
Lupin	87.5	76.0	97.2	91.1	184.7	167.1	9.1	7.5
Oat	99.9	87.5	95.4	93.6	195.3	181.1	8.4	7.0

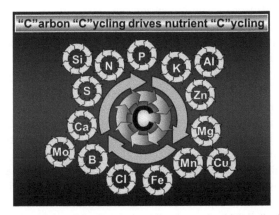

Fig. 4 Schematic representation of the C cycle providing the energy driving all soil nutrient cycling.

biomass, low greenhouse gas emissions, improved water-holding capacity, and high yields. Cropping systems' multiple services and synergistic benefits gives us insight into how to match practices to environmental goals, and where the uncertainties and opportunities exist for resilient agro-ecosystem management.

Tonitto et al. (2006) use a meta-analysis to evaluate crop yield, nitrate leaching, or soil nitrate between conventional (receiving inorganic fertiliser with a winter bare fallow) and diversified systems managed using either a non-legume over-wintering cover crop (amended with inorganic fertiliser) or a legume over-wintering cover crop (no additional N fertiliser). Yields under non-legume cover crop management were not significantly different from those in the conventional, bare fallow systems, while leaching was reduced by 70 per cent. On average, nitrate leaching was reduced by 40 per cent in legume-based systems relative to conventional fertiliser-based systems. Post-harvest soil nitrate status, a measure of potential N loss, was similar in conventional and green manure systems, suggesting that reductions in leaching losses were largely due to avoidance of bare fallow periods. Furthermore, as agriculture increasingly dominates the landscape, it is important for agricultural land to provide ecological needs currently met by unmanaged habitat. The meta-analysis of Tonitto et al. (2006) suggests diversified cropping systems have the potential to achieve these goals, reflecting the importance of a diverse landscape and elimination of bare fallow periods.

Residue carbon:nitrogen – Plant material has a variable C to N ratio (C:N ratio) depending on species, growth stage, and the environmental conditions during growth (Ågren and Weih, 2012). In comparison, the soil microbial biomass has a narrower C:N; soil microbial biomass has been found to have an average C:N:P of 60:7:1 globally (Hartman and Richardson, 2013). A range of plant, soil physical and chemical properties, and fixed environmental factors, together with residue management practices (e.g., incorporation method), interact to determine the decomposition of added organic materials (Kumar and Goh, 2000).

The C:N ratio of cover crop residue is a good indicator of whether immobilisation or mineralization will occur. Values exceeding C:N of ~ 30:1 are generally expected to immobilise N during the early stages of the decomposition process (Finney et al., 2016). Residue incorporation by tillage also increases N release. Biochemical composition and environmental conditions were the primary factors affecting short-term N mineralisation from warm season legume and grass cover crops in organic farming systems (O'Connell et al., 2015). Greater potential N mineralization was found in legume cover crops, but even cover crops with a C:N ratio of > 40:1 had net N mineralisation (O'Connell et al., 2015). Use of legumes as a summer fallow crop in small grain systems resulted in higher potentially mineralisable C and N, and microbial biomass C (O'Dea et al., 2015). Legumes may increase long-term no-till system resilience and sustainability in the northern Great Plains by increasing the available N supply by 26–50 per cent compared to wheat-only systems, thereby reducing the need for N fertiliser for subsequence crops. The half-life of potentially mineralisable C was shortest in intensified systems and was longest in the legume systems, potentially mitigating the negative effects of soil organic matter losses from summer fallow (O'Dea et al., 2015). Liebman (2012) showed that red clover had a N fertiliser replacement value for corn of 87–184 kg N ha^{-1} and alfalfa supplied corn with equivalent 70–121 kg N ha^{-1} in Iowa. Ultimately, N management using cover crops requires that N availability be synchronised so that inorganic N is readily available during periods of active uptake by cash crops and minimally available during periods when cash crops are not growing, to reduce losses of N to air and water (Kaspar and Singer, 2011). This may not be easily achievable and is an important area of research on which to focus, so as to utilise nutrients released from cover crops efficiently.

The C:N of small grain residues is mostly dependent on time of termination. Early termination of grass cover crops results in a narrower C:N ratio, typical of young plant tissue. If killed too early, this narrower C:N ratio results in rapid decomposition of a smaller amount of residue, reducing ground coverage. The N contribution from small grain cover crops depends on N availability during the cover crop growing period, the total amount of biomass produced, and the growth stage when the cover is terminated. Because of the need for residue in conservation agriculture systems (CAS), small grain cover crops are often allowed to grow as long as possible. Termination date depends on crop rotation and climate. When small grain cover crops are killed at flowering, the C:N ratio is usually greater than 30:1. Delaying the rye termination date from early to late boot stage increased average aboveground dry matter accumulation with no negative effect on corn yield (Duiker and Curran, 2005).

The C:N ratio of mature legume residues varies from 25:1 to 9:1 and is typically well below 20:1, the guideline threshold where rapid mineralization of the N in the residue occurs. Residues on the soil surface decompose more slowly than those incorporated in conventional tillage systems. Consequently, in no-tillage systems with crop residue on the soil surface, legume-residue N may not be readily available during the early part of the growing season (Franzluebbers and Arshad, 1996; Franzluebbers, 2004).

In North Carolina, delaying the kill date of crimson clover two weeks beyond 50 per cent bloom, and hairy vetch two weeks beyond 25 per cent bloom, increased the biomass of clover by 41 per cent and vetch by 61 per cent. Corresponding increases in N content were 23% for clover and 41 per cent for vetch (Wagger, 1989). In Maryland, hairy vetch fixed about 2.24kg Nha^{-1}day^{-1} from April 10 to May 5, resulting in an additional 67.3 kg Nha^{-1} in aboveground biomass (Clark et al., 1995; Clark et al., 1997a,b).

Cover Crop Characteristics – Perennial plant communities are critical regulators of ecosystem functions, such as water management and C and N cycling. Glover et al. (2007) reviewed the traits of perennials with their roots commonly exceeding 2 m. Deep roots, however, mean resilience (Kell, 2011). Perennial crops would transform the process of farming and its environmental effects by using resources more effectively, thereby being less dependent on human inputs and more productive for a longer time. Perennials also anchor and support the ecosystem that nourishes them, whereas short-lived and short-rooted annuals allow water, soil, and nutrients to be lost (Kaspar et al., 2008). The more resilient perennials are also expected to fare better than annuals in a warming climate. Greenhouse gases released into the atmosphere by conventional crop production inputs and tillage, minus C sequestered in soil, is negative for perennial crops (Glover et al., 2007). Establishing a perennial cover crop root system will give farmers more choices in what they can grow, while sustainably producing food.

Plant diversity loss in conventional agriculture impairs ecosystem functioning, including important effects on soil. Gould et al. (2016) showed that high plant diversity in grassland systems increases soil aggregate stability and that plant root traits play a major role in determining diversity impacts and benefit essential soil physical properties. Roots lose metabolites to the soil at rates significant to soil organisms, and we need to know if the mechanisms of passive diffusion identified in hydroponics apply in soil, and whether other, active mechanisms complement them (Farrar et al., 2003).

Root type and root distribution pattern play an important role in the understanding and estimating of soil C allocation and the effect of crop roots C input on soil C balance in agro-ecosystems (Kell, 2011; Chen and Weil, 2011; Fan et al., 2016). A database of 96 profiles was compiled and a root distribution pattern was fitted to a modified logistic dose response curve for 11 temperate crops (Fan et al., 2016). Roots contribute more to refractory SOM than aboveground crop residues, as revealed by a long-term field experiment. A large part of the C input from cover crop is added as roots, which was found to contribute more effectively to the relatively stable C pool than aboveground C input (Balesdent and Balabane, 1992, 1996; Allmaras et al., 2004; Wilts et al., 2004; Kätterer et al., 2011).

Oilseed radish contributes biodiversity and has about the same absorption capacity for N as wheat, and it develops a very deep root mass (Chen and Weil, 2011). It is an excellent nutrient scavenger. This combination enables the cover crop to capture maximum N from deep in the soil profile to feed the following corn crop. Deep-rooted cover crops like oilseed radish can help reverse the traditional theory of N stratification. Nitrogen allowed to concentrate deep in the soil requires a deep-rooted crop to capture that N (Kell, 2011).

Cover crops with shallow fibrous root systems build soil aggregation and alleviate compaction in the surface layer. Cover crops with deep tap roots can help breakup compacted layers, bring up nutrients from the subsoil to make them available for the following crop, and provide access to the subsoil for the following crop via root channels left behind (Kell, 2012). Cover crops can, thus, recycle nutrients that would otherwise be lost through leaching during off-season periods. Leguminous cover crops can also fix atmospheric nitrogen that then becomes available to the following crop. Benefits from cover crops include protection of the soil from water and wind erosion, improved soil aggregation and water storage, suppression of soil-borne pathogens, support beneficial microbial activity, increased active and SOM, and C sequestration.

The cover crop root and shoot is another critical parameter to be considered (Kell, 2011). Johnson et al. (2006) used previously determined root and shoot to estimate the total source C from crop residues, roots and rhizo deposits by the growing crop. Balesdent and Balabane (1996) found that the contribution of root-derived C to SOM was 1.5 times that of stalks + leaves, whereas the corresponding ratio of biomasses was less than 0.5. They attributed this to a high belowground production, and a relatively slow biodegradation of root-derived material.

5. Conservation Agriculture Systems (CAS) for Healthy Soils

Conservation agriculture systems and associated C management has the potential for meeting climate mitigation and food demands, while minimising environmental damage, and to examine how improvements in C cycling and energy flow might advance the effectiveness of CA systems in meeting these challenges (Janzen, 2006, 2015; Reicosky and Janzen, 2018). Conservation agriculture emphasizing soil health principles has been highlighted as a key route toward this urgent goal (Calegari et al., 2005; Hobbs et al., 2008; Kassam et al., 2009; Kassam et al., 2015; Erenstein et al., 2012; Kassam and Friedrich, 2012; Corsi et al., 2012; Pretty and Bharucha, 2014; Lal, 2015b; Reicosky and Janzen, 2018).

True conservation is more about plant management than soil management because of the importance of C and C cycling and because plants are the C energy conduit from the atmosphere to the soil (Janzen, 2015). Crop rotations usually increase SOM content when compared with monocultures (Odell et al., 1984; Wagner, 1989; Dick et al., 1986a,b; Johnston, 1986; Hargrove and Frye, 1987; Havlin et al., 1990), suggesting numerous benefits. Crop rotation also influences soil C dynamics by affecting the diversity and proportions of the decomposable organic compounds returned to the soil, leading to potential gains in soil C (Hutchinson et al., 2007; Ogle et al., 2005). The guiding CA principles are that nature manages soil better with biodiversity and when it is left alone. The benefits of this system are that we can dramatically reduce synthetic inputs (fertilisers and pesticides) while increasing SOM and improving soil biology. This, in turn, reduces CO_2 emissions from the soil and from fossil fuels, enhances environmentally quality and is more sustainable.

Conventional agriculture with intensive tillage is one of the most destructive forces against biodiversity, whereas CA is a new, more natural system where humans can produce food and energy sustainably (Fig. 5). The CA system mimics principles of 'perennial-based agriculture' (Basche and Edelson, 2017) or 'continuous living cover' (Basche and De Longe, 2017) to maintain ecosystem services within biological limits. With CA, environmental benefits include less erosion, better water conservation, improvement in air quality due to lower GHGs emissions, and a chance for larger biodiversity in a given area (Erenstein et al., 2012; Kassam and Friedrich, 2012; Corsi et al., 2012; Pretty and Bharucha, 2014; Lal, 2015b; Reicosky and Janzen, 2018). Conservation agriculture is not either/or for any core principle, but the simultaneous interaction and integration of the three core principles synergistically producing more economic and environmental benefits than the sum of the parts (Reicosky and Janzen, 2018).

The principles of CA are universal, but the solutions are local and revolve around C cycling using a systems approach (Kassam and Friedrich, 2012; Lal, 2015b). Conservation agriculture is a broad term to describe: (1) continuous soil protection with crop residue cover; (2) continuous biodiversity with diverse agronomic and cover cropping systems; and (3) continuous minimum soil disturbance (no-till). While each of the principles may be considered a separate entity, the fluid integration of all three principles and their supportive soil health practices as a system are keys to effective CA as shown schematically in Fig. 5. While definitions of soil health are still evolving (Doran, 2002; Cornell, 2009; FAO, 2014; USDA NRCS, 2014b), they all have three principles in

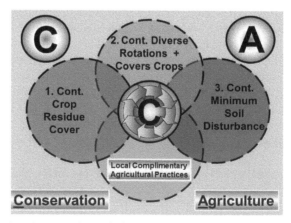

Fig. 5 Schematic representation of conservation agriculture systems with three principles and complementary agricultural practices encompassing C cycling (Reicosky and Janzen, 2018).

common like CA. The natural interactions and diversity within the cropping system contribute to numerous synergistic economic and environmental benefits (Lal, 2015b).

The first principle is to preserve either live cover or dormant crops, or dead crop residues accumulating as mulches on the soil surface (Fig. 5). Retaining mulch between crops provides better protection against erosion and can also maintain higher soil moisture in dry regions, enrich the soil with SOM, and, if the mulch is sufficiently dense, prevent the regrowth of weeds.

Basche et al. (2016b), using a meta-analysis, found soil water improvement with a long-term cover crop in the system. The cover crop increased the water retained in the soil at water potentials associated with field capacity by 10–11 per cent, as well as increasing plant-available water by 21–22 per cent. They concluded that the rye, if managed properly, could improve soil water dynamics without impacting cash crop growth and yield.

The second principle addresses biodiversity (Fig. 5), ideally consisting of at least three different plant species including one legume in the rotation (Anderson, 2011; Chatterjee et al., 2016). To incorporate more diversity, some farmers are using up to 14 species in cover crop mixes.

The third principle of CA systems is minimal soil disturbance (Fig. 5), typically achieved using no-tillage or direct seeding techniques. Planting perennial crops such as forages which, by definition, involve no cultivation for the duration of their growth. The additional component of complementary agricultural practices allows the farmer to adjust operation to economic and site-specific needs.

> **The synergistic simplicity of Conservation Agriculture with minimum soil disturbance (minimizes C and soil loss) and the use of diverse crop rotations and cover crop mixes (maximizes soil protection and C input) for soil biodiversity protection and soil health regeneration benefits for the environment, the farmer, and all of society through food security.**

The three CA principles should be integrated and applied continuously for improved C management and long-term sustainability. Conservation agriculture derives many of its multiple benefits from the synergistic simplicity of NT (minimises C and soil loss) and the use of diverse rotations and cover crop mixes (maximises soil coverage and C input) for soil diversity protection and regeneration. Today, however, CA practitioners aim for greater soil cover because of additional protective and nutritional benefits of crop residue. Cover crops are critical to capturing off season C and producing surface residue protection and minimum disturbance benefits.

Conservation agriculture has been adopted in more and more areas since the 1970's due to improvements in planting equipment, herbicides, and other technologies (Kassam et al., 2009; Kassam et al., 2015). Several long-term, incremental benefits of CA have emerged (Table 5), and more recently, the contributions of diverse cover crop mixes with an emphasis on maximizing C input. The most important benefits have been attributed to the SOM accumulation at the surface for erosion protection, enhanced water infiltration and storage, and efficient nutrient cycling.

The economic benefits provided by conversion from CT to CA can only be for provided qualitatively. Individual differences in farm operations, soil types, the rate of transition to CA, and a host of other factors contribute to this challenge. Some farmers expressed concern over the expense of new seeding equipment and the cost of cover crop mixes, however, the early adapters with innovative skills have made management decisions that demonstrate economic savings. Anecdotal data from a few early adapters of CA suggest input savings ranging from \$245–\$500 ha^{-1} yr^{-1} (Mitchell et al., 2012), depending on the farm and many personal assumptions involved.

Multiple economic benefits accrue mainly due to reductions in tillage, diesel fuel, size of equipment required, equipment maintenance, and labour listed in Table 5. The savings associated with reduced synthetic N fertiliser are substantial and contribute to decreasing the C footprint of CA. Incorporating the principles of biodiversity reduces pesticide and insecticide costs that may provide chemical management challenges in the transition from conventional to CA. The economic incentives associated with soil C storage with CA in the form of C credits, offsets, and/or taxes are still being evaluated. The economics of soil health is the subject of definition, measurement, research, and education programs to help minimise potential ecosystem disturbance while maximising nutrient cycling, which can lead to greater profitability for sustainable production.

Table 5 Summary of the benefits of conservation agriculture systems (CAS).

Benefits of Conservation Agriculture:

Anecdotal economic benefits decreased input costs.

1. Fuel ~ 50%
2. Labor ~ 50%
3. Equipment ~ 40-50%
4. Repair and maintenance ~ 40%
5. Nitrogen fertilizer > 50%
6. Pesticides > 50%
7. Water Management >30%

Ecosystem Benefits with CA systems

1. **Climate resiliency and minimum water, wind and tillage erosion** (keep the soil in place because erosion loses soil faster than nature can make it)
2. **Maintain "continuous living crop" or crop residue cover and carbon input** (manage crop residue for use protective soil blanket with carbon and nutrient cycling)
3. **Keep available water in the root zone** (decrease runoff and increase infiltration and increased carbon content and water holding capacity, decreased nutrient leaching loss)
4. **Enhance soil fauna habitat and activity** (increased earthworm population, deeper root penetration and bio-pores, better balance of bacteria and fungi)
5. **Decrease fossil fuel use and carbon footprint** (less diesel required for tillage, fewer passes over the field, and lower repair and the maintenance costs; legumes fixed nitrogen decreasing the need for synthetic fertilizer)
6. **Manage diverse crop rotations and cover crop mixes to control weeds and break up disease and pest cycles** (requires less fertilizer, insecticides and herbicides)
7. **Minimizes soil carbon loss** (low soil disturbance minimizes CO_2 loss, disturbance of microbial and fungal activity important in organic matter decomposition and nutrient cycling)
8. **Enhanced economic profitability and environmental quality** (reduced input costs with improved soil, water, and air quality with significant, but unknown economic value)
9. **Harvest the maximum amount of solar energy** (required for photosynthesis, carbon capture and food production, provides optimum energy utilization released in respiration as part of the carbon cycle)
10. **Enables a better balance of natural diversity** (provides aesthetically pleasing habitat for songbirds, pollinators and wildlife)

Cover crops provide major contributions to the environmental benefits of CA systems by decreasing erosion, increasing WUE, enhancing nutrient capture and cycling, in addition to enhancing C cycling, as listed in Table 5. Cover crop biomass is a source of SOM that stimulates soil biological activity. Soil organic matter and cover crop residues improve soil physical properties (Calegari et al., 2008), resulting in greater water infiltration due to direct effects of the residue coverage, or to changes in soil structure, greater soil aggregation, or tilth, as well as in better nutrient (Calegari et al., 2013a,b) and moisture management and less surface sealing. This is because residue intercepts rain drops, reducing the dispersal of clay particles during a rainfall or irrigation event and greater soil porosity due to macropores formed as roots grow, die, and decompose. Improvements in soil physical properties depend on soil type, crops grown, and residue management system, as well as temperature and rainfall. Grasses and *Brassicas* are better than legumes at reducing N leaching (Dabney et al., 2001; Kremen and Weil, 2006; Meisinger et al., 1991). Cereal rye is very effective at reducing N leaching because it is cold-tolerant, has rapid growth, and produces a large quantity of biomass (Delgado, 1998). Winter annual weeds do not effectively reduce N losses. Regardless of soil type, however, tillage will very quickly negate cover crop benefits associated with increased SOM. Simply put, any tillage breaks down SOM much faster than NT, hence the need for continuous minimum soil disturbance.

Diverse crop rotations provide numerous benefits to any cropping system. Crop rotations usually increase SOM content when compared with monocultures (Odell et al., 1984; Dick et al., 1986a,b; Johnston, 1986; Hargrove and Frye, 1987; Havlin et al., 1990). It is critical in reducing the incidence of diseases and pests and is also credited with improving nutrient use and reducing weeds. Cover crops increase the complexity and intensity of crop rotations, effectively increasing crop rotation benefits that may require higher level management to minimise the chance that cover crops can adversely affect other crops in the rotation.

Factors influencing farmers' conservation management choices in cropping and soil management include short or long-term economic, social norms, family and social relationships, convenience, management experience, perceptions, and beliefs. With the many positive attributes of CA systems, the relatively slow acceptance by farmers around the world is still puzzling. While CA is more knowledge and management intensive than conventional tillage agriculture, CA is a practical, agro-ecological approach to achieving sustainable agriculture intensification (Garnett et al., 2013; Kassam and Friedrich, 2012). Dumanski et al. (2014) stated the biggest challenge is managing the ecology of the systems to enhance environmental quality, while optimizing economic returns. Perceived risks of individual farm production practices, environmental constraints, and perception problems often limit the adoption of many sustainable practices. Forgetting traditional tillage agriculture and learning no-tillage techniques, weed control problems, and equipment expense have contributed to slow acceptance. At the same time, erosion seriously degraded soil systems, causing the farmers to be self-motivated and to communicate with researchers, and the importance of farmer-run associations were the major contributing factors for increasing acceptance of CA. Other studies have shown a variety of reasons for the slow acceptance of CA (Featherstone and Goodwin, 1993; Drost et al., 1996; Knowler and Bradshaw, 2007; Coughenour, 2003). After reviewing recommendations for increasing the adoption of CA practices, Carlisle (2016) found a complementary approach combining education, research, policy, measures to overcome equipment barriers, and efforts to address farm and food system context hold the most promise.

Carlisle (2016) presents another emergent theme of CA and soil health practices adoption literature, is the interaction among different practices. Multiple studies found that farmers who engaged in one conservation practice were more likely to engage in others (Bergtold et al., 2012; Lichtenberg, 2004; Ryan et al., 2003; Singer et al., 2007; Upadhyay et al., 2003; Wilson et al., 2014). Those who rotated crops, or had longer rotations, likely would use other conservation practice (Vitale et al., 2011; Wu and Babcock, 1998). Both agronomic and sociological explanations were offered: many rotation crops provide more time for a winter cover crop; management skills associated with one soil health practice can be transferrable to other practices; and use of soil health practices may indicate or even help cultivate a 'conservation mindset', which would increase the likelihood of a farmer using other soil health practices.

Ingram et al. (2014) found many European farmers are unconvinced of the economic benefits of practices for managing soil C, suggesting incentives are needed, either as subsidies or as evidence of the cost effectiveness of practices. They identified key barriers to the uptake of CA practices including perceived scientific uncertainty about the efficacy of practices, lack of real life 'best practice' examples to show farmers, difficulty in demonstrating the positive effects of soil C management practices and economic benefits over a long-time scale, and advisors being unable to provide suitable advice due to inadequate information or training. All new measures and advice should be integrated into existing programs to avoid a fragmented policy approach.

6. Summary and Conclusion

We, as a society, must curtail environmental destruction caused by conventional tillage agriculture and change our stewardship of the Earth and life, if human misery is to be avoided. We are on a collision course with the natural world, exceeding our natural resource capacity. Concerns about current, impending, or potential damage on planet Earth include freshwater availability, marine life depletion, ocean dead zones, forest loss, soil erosion, water and air pollution, biodiversity destruction, climate extremes, ozone depletion, and continued human population growth. Fundamental changes are urgently needed to avoid the consequences of our present conventional course. Conservation agriculture systems, with an emphasis on biodiversity and C management utilising cover crop mixes, shows a lot of promise for food security.

Cover crops build SOM, protect against soil erosion, cycle nutrients, reduce compaction, improve soil structure and all soil attributes (chemical, physical, and biological), capture C from the atmosphere, and build overall soil health and make soil more resilient to extreme weather. Cover crops and minimum soil disturbance, as well as crop rotation systems, allow stable soil N and C concentrations and microbial community composition and promote higher biodiversity of communities of fungi, bacteria, and actinomycetes, resulting in a balanced, healthy community and minimizing the feast or famine phenomenon experienced in monoculture systems. Healthy soil biology is intrinsic to the provision of life support services, environmental goods and services, and food security. If soil health, along with water quality and soil biodiversity, is allowed to deteriorate, this will adversely impact the quantity and quality of food produced from the soil, and ultimately, the quality of life on the planet. Inversion tillage (conventional agriculture) is a major biotic disturbance that seriously disrupts and damages soil biological constituents, while also contributing to climate change by rapidly oxidising soil C and releasing CO_2 into the atmosphere. Inversion tillage and high levels of synthetic inputs most noticeably impacts the larger soil biota, like earthworms and filamentous fungi, slicing and dicing them, and decreasing their effectiveness. Scientists and many experienced farmers all over the world have noted other important benefits of soil management, including increased yields from improved soil fertility and better water-holding capacity, which also support farmers in adapting to climate change.

The benefits provided by cover crop mixes and innovative cover crop management provide options for many soil types and geographic locations. The way we use cover crops to make agriculture sustainable and climate-resilient, as well as economically profitable and environmentally friendly, is left to our creative innovation. Large numbers of winter cover crop species provide flexibility and biodiversity, enabling a more natural production system providing C capture and soil protection during the 'offseason'. Benefits of soil C management for agricultural ecosystems start with C capture in photosynthesis, followed by C flow through the system and eventually back to the atmosphere, completing the carbon cycle. The importance of soil C management and the role of diverse cover crop mixtures for C balance cannot be overstated. Cover crops are multifaceted with respect to the soil functions they address, resulting in numerous ecosystem benefits. Cultivating cover crops, in winter and/or summer/autumn/spring, is beneficial for enhancing soil microbial quality, SOM stocks, and promoting microbial activity to recycle a substantial reserve of nutrients for subsequent agronomic crops. Ecosystem services traditionally expected from cover crops can now be promoted with other practices to enable synergistic benefits to cope with climate extremes. Farmers, consumers, and policymakers can now collaborate to expect cover cropping benefits for soil, water, air quality, and climate-change adaptation and mitigation for the good of society. Nutrient losses by leaching can be reduced by cover crops that also promote P and K accumulation, maintain nutrients in the soil-plant system, and avoid K losses by leaching or P losses by soil fixation.

Cover crops, as soil-building assets, can also help control weeds; however, this requires increased management. Cover crops affect weed management in CA by competing for light, water and nutrients, using cover crop residue to suppress seed germination, and releasing allelopathic compounds; cover crop residue with high C:N's also lasts longer. Root type and root distribution pattern plays an important role in the understanding and estimating of soil C allocation, and the effect of crop roots C input on soil C balance in agro-ecosystems. Growing cover crops is also an alternative way to redistribute nutrients in the soil profile and avoid nutrient accumulation in the soil surface, which can be easily lost by runoff and soil erosion in conventional tillage systems.

The process of growth and decay balances C in natural systems. The conversion of natural ecosystems to agricultural ecosystems with tillage disturbs the soil ecological balance, soil processes, organic C, and biotic C pools. Carbon capture has been going on for many billions of years, and unless the soil is disturbed by intensive tillage releasing CO_2 emissions back to the atmosphere, it reaches a natural equilibrium. Tillage results in the loss of SOM primarily through three mechanisms: (1) mineralisation of C due to the breakdown of soil aggregates and changes in temperature and moisture regimes; (2) leaching of organic C; and (3) accelerated rates of erosion. Soil erosion caused by excessive tillage is the most visual example of humankind's influence on soil function and degradation. Continuous NT management, combined with winter cover crops, results in the greatest amount of SOM in the surface soil and was the only cropped treatment that approached the adjacent undisturbed forest. Thus, the NT system serves as a management model for sustaining the productivity of Oxisols and other 'problem soils' in tropical and sub-tropical regions of the world. Many of the impacts of cover crops are related to C input and the way it affects the soil system, particularly soil aggregation, where all soil aggregation parameters were enhanced under the NT system. All these C forms can also have a direct or indirect effect on soil structure for efficient infiltration and water-use efficiency, and contribute to enhanced water-holding capacity. Cover crops can do a lot to increase WUE in our production systems, with plant C as our best water management tool. The adoption of cover crops and a no-tillage system improved and stabilised grain production yields with 'normal' P, K, and N fertiliser rates.

Conservation agriculture and soil health system concepts need to be implemented on all producing landscapes if we are going to have food security for future generations. Soil must be considered a 'living system', and as such, must be nurtured and protected. Understanding the soil is a habitat for a living biological community, soil health is an intuitive, appealing concept that leads to a clear understanding and faster action from a management perspective. Conservation agriculture and soil health system concepts introduced in the United States by the Natural Resources Conservation Service (USDA NRCS, 2014a,b) are closely related. Soil health (SH), sometimes known as soil quality, includes the first three components (expressed a little differently) of CA and adds a fourth component: Keep living roots in the soil for as long as biologically possible. While both CA and SH allow for the integration of animals into the ecosystem services and soil C, recent discussions place more emphasis on including the impact of SH on animal health in production systems. Conservation agriculture systems incorporate the combination of physical, chemical, and biological properties into the three primary CA principles, allowing soil to function as a living organism that supports plants, animals, and humans. While there are some concerns about the use of CA principles to mitigate climate change, numerous research needs are easily identified and need further support. There must be a strong partnership among all sectors to promote the adoption and success of these conservation approaches. Emphasis is on the scientific foundation of CA, and its value to our global society is paramount if we expect food security for future generations.

References

Adenle, A.A., Stevens, C. and Bridgewater, P. (2015). Global conservation and management of biodiversity in developing countries: An opportunity for a new approach. Environ. Sci. Policy, 45: 104–108.

Admunson, R.L., Berhe, A.A., Hopmans, J.W., Olson, C., Sztein, A.E. and Sparks, D.L. (2015). Soil and human security in the 21st century. Science, 348: 635–647.

Ågren, G.I. and Weih, M. (2012). Plant stoichiometry at different scales: Element concentration patterns reflect environment more than genotype. New Phytologist, 194: 944–952.

Alfaro, M.A., Jarvis, S.C. and Gregory, P.J. (2006). Factors affecting potassium leaching in different soils. Soil Use Management, 20: 182–189.

Allmaras, R.R., Linden, D.R. and Clapp, C.E. (2004). Corn-residue transformations into root and soil carbon as related to nitrogen, tillage and stover management. Soil Science Society of America Journal, 68: 1366–1375.

Anderson, R.L. (2008). Diversity and no-till: keys for pest management in the U.S. Great Plains. Weed Science Society of America, 56: 141–145.

Anderson, R.L. (2011). Synergism: A rotation effect of improved growth efficiency. Advances in Agronomy, 112: 205–223.

Andrade, D.S., Colozzi-Filho, A.K. and Giller, E. (2003). The soil microbial community and soil tillage. pp. 51–81. *In*: Titi, I.A.E. (ed.). Soil Tillage in Agro-ecosystems. CRC Press, Boca Raton, FL.

Angers, D.A. and Chenu, C. (1977). Dynamics of soil aggregation and C sequestration. pp.199–206. *In*: Lal, R. (ed.). Soil Processes and the Carbon Cycle. CRC Press, Boca Raton. FL.

Aziz, I., Asharf, M., Mahmood, T. and Islam, K.R. (2011). Crop rotation impact on soil quality. Pak. J. Bot., 43: 949–960.

Bailey, V.L., Smith, J.L. and Bolton Jr, H. (2002). Fungal-to-bacterial ratios in soils investigated for enhanced C sequestration. Soil Biology and Biochemistry, 34(7): 997–1007.

Bais, H.P., Weir, T.L., Perry, L.G., Gilroy, S. and Vivanco, J.M. (2006). The role of root exudates in rhizosphere interactions with plants and other organisms. Ann. Rev. Plant Biol., 57: 233–266.

Balesdent, J. and Balabane, M. (1992). Maize root-derived soil organic carbon estimated by natural 13^C abundance. Soil Biology and Biochemistry, 24: 97–101.

Balesdent, J. and Balabane, M. (1996). Major contribution of roots to soil carbon storage inferred from maize cultivated soils. Soil Biology and Biochemistry, 28: 1261–1263.

Balota, E.L., Calegari, A. and Nakatani, A.S. (2014). Benefits of winter cover crops and no-tillage for microbial parameters in a Brazilian Oxisol: A long-term study. Agric. Ecosys. Environ., 197: 31–40.

Bardgett, R.D. and Van der Putten, W.H. (2014). Belowground biodiversity and ecosystem functioning. Nature, 515: 505–511.

Basche, A.D., Archontoulis, S.A., Kaspar, T.K., Jaynes, D.B., Parkin, T.B. and Miguez, F.E. (2016a). Simulating long-term impacts of cover crops and climate change on crop production and environmental outcomes in the Midwestern United States. Agric. Ecosys. Environ., 218: 95–106.

Basche, A.D. and Edelson, O. (2017). Improving water resilience with more perennially-based agriculture. Agroecology and Sustainable Food Systems, 41: 799–824.

Basche, A.D. and DeLonge, M. (2017). The impact of continuous living cover on soil hydrologic properties: A meta-analysis. Soil Science Society America Journal, 81: 1179–1190.

Bauer, P.J. and Reeves, D.W. (1999). A comparison of winter cereal species and planting dates as residue cover for cotton grown with conservation tillage. Crop Science, 39: 1824–1830.

Bayon, R.C., Weisskopf, L., Martinoia, E., Jansa, J., Frossard, E., Keller, F., Föllmi, K.B. and Gobat, J.M. (2006). Soil phosphorus uptake by continuously cropped *Lupinus albus*: A new microcosm design. Plant Soil, 283: 309–321.

Beale, O.W., Nutt, G.B. and Peele, T.C. (1955). The effects of mulch tillage on runoff, erosion, soil properties, and crop yield. Soil Set Soc. Amer. Proc., 19: 244–247.

Bergtold, Jason, S., Patricia, A. Duffy, Diane, Hite and Randy L. Raper. (2012). Demographic and management factors affecting the adoption and perceived yield benefit of winter cover crops in the southeast. Journal of Agricultural and Applied Economics, 44(1): 99–116.

Blanco-Canqui, H., Shapiro, C.A., Wortmann, C.S., Drijber, R.A., Mamo, M., Shaver, T.M. and Ferguson, R.B. (2013). Soil organic carbon: The value to soil properties. Journal of Soil and Water Conservation, 68: 129–134.

Blanco-Canqui, H., Shaver, T.M., Lindquist, J.L., Shapiro, C.A., Elmore, R.W., Francis, C.A. and Hergert, G.W. (2015). Cover crops and ecosystem services: Insights from studies in temperate soils. Agronomy Journal, 107: 2449–74.

Bolliger, A., Magid, J., Amado, T.J.C., Skóra Neto, F., Santos, R., Bona, F.D., Bayer, C., Bergamaschi, H. and Dieckow, J. (2006). Carbono orgânico no solo em sistemas irrigados por aspersão sob plantio diretoe preparo convencional. Revista Brasileira de Ciência do Solo, 30: 911–919.

Bortoluzzi, E.C., Pérez, C.A.S., Ardisson, J.D., Tiecher, T. and Caner, L. (2019). Occurrence of iron and aluminum sesquioxides and their implications for the P sorption in subtropical soils. Appl. Clay Sci., 104: 196–204.

Branca, G., McCarthy, N., Lipper, L. and Jolejole, M.C. (2011). Climate-smart agriculture: A synthesis of empirical evidence of food security and mitigation benefits from improved cropland management. Mitigation of Climate Change in Agriculture, Series No. 3, Food and Agriculture Organisation of the United States.

Broder, M.W. and Wagner, G.H. (1988). Microbial colonisation and decomposition of corn, wheat and soybean residue. Soil Sci. Soc. Am. J., 52: 112–117.

Brussaard, L., Behan-Pelletier, V.M., Bignell, D.E., Brown, V.K., Didden, W., Folgarait, P., Fragoso, C., Freckman, D.W., Gupta, V.V.S.R., Hattori, T., Hawksworth, D.L., Klopatek, C., Lavelle, P., Malloch, D.W., Rusek, J., Söderström, B., Tiedje, J.M. and Virginia, R.A. (1997). Biodiversity and ecosystem functioning in soil. Ambio, 26: 563–570.

Bulluck, L.R. and Ristaino, J.B. (2002). Synthetic and organic amendments affect southern blight, soil microbial communities and yield of processing tomatoes. Phytopathology, 92: in press.

Burney, J.A., Davis, S.J. and Lobell, D.B. (2010). Greenhouse gas mitigation by agricultural intensification. Proc. National Acad. Sci. U.S. Amer., 107: 12052–12057.

Buyanovsky, G.A. and Wagner, G.H. (1997). Crop residue input to soil organic matter on Sanborn field. pp. 73–83. *In*: Paul, E.A. et al. (eds.). Soil Organic Matter in Temperate Agro-ecosystems: Long-term Experiments in North America. CRC Press, Boca Raton, FL.

Calegari, A., Mondardo, A., Bulisani, E.A., Wildner, L. do P., Costa, M.B.B., Alcântara, P.B., Miyasaka, S. and Amado, T.J.C. (1993). Adubação verde no sul do Brasil, second ed., AS-PTA, Rio de Janeiro, 346 pp.

Calegari, A., Ashburner, J. and Fowler, R. (2005). Conservation Agriculture in Africa, FAO, Regional Office for Africa, Accra, Ghana. ISBN: 9988-627-04. 91p.

Calegari, A., Hargrove, W.L., Rheinheimer, D.S. Ralisch, R., Tessier, D., Tourdonnet, S. and Guimarães, M.F. (2008). Impact of long-term no-tillage and cropping system management on soil organic carbon in an oxisol: a model for sustainability. Agronomy Journal, 100: 1013–1019.

Calegari, A., Rheinheimer, D.S., Tourdonnet, S., Tessier, D., Hargrove, W.L., Ralisch, R., Guimarães, M.F., Tavares and Filho. J. (2013a). Soil physical properties affected by soil management and crop rotation in a long-term experiment in Southern Brazil. Commun. Soil Sc. Plant Analysis, 104(13): 2019–2031.

Calegari, A., Tiecher, T., Hargrove, W.L., Ralisch, R., Tessier, D., Tourdonnet, S., Guimarães, M.F. and Rheinheimer, D. (2013b). Long-term effect of different soil management systems and winter crops on soil acidity and vertical distribution of nutrients in a Brazilian Oxisol. Soil and Tillage Research, 133: 32–39.

Calegari, A. (2016). Plantas de cobertura, Manual Técnico. Fev., Penergetic. Uberaba, MG, Brasil.

Cardinale, B.J., Matulich, K.L., Hooper, D.U., Byrnes, J.E., Duffy, E., Gamfeldt, L. and Gonzalez, A. (2011). The functional role of producer diversity in ecosystems. American Journal of Botany, 98(3): 572–592.

Carlisle, L. (2016). Factors influencing farmer adoption of soil health practices in the United States: A narrative review. Journal Agro-ecology and Sustainable Food Systems, 40(6): 583–613.

Centurion, J.F., Dematte, J.L.I. and Fernandes, F.M. (1985). Efeitos de sistemas de preparo nas propriedades químicas de um solo sob cerrado cultivado com soja. Revista Brasileira Ciência do Solo, 9: 267–270.

Ceretta, C.A., Basso, C.J., Herbes, M.G., Poletto, N. and Silveira, M.J. (2002). Produção e decomposição de fitomassa de plantas invernais de cobertura de solo e milho, sob diferentes manejos da adubação nitrogenada. Ciência Rural, 32: 49–54.

Chatterjee, Amitava, Cooper, K., Klaustermeier, A., Awale, R. and Cihacek, L.J. (2016). Does crop species diversity influence soil carbon and nitrogen pools? Agron. J., 108: 427–432.

Chavarría, D.N., Verdenelli, R.A., Serri, D.L., Restovich, S.B., Andriulo, A.E., Meriles, J.M. and Vargas-Gil, S. (2016). Effect of cover crops on microbial community structure and related enzyme activities and macronutrient availability. Eur. J. Soil Biol., 76: 74–82.

Chen, G. and Weil, R.R. (2011). Root growth and yield of maize as affected by soil compaction and cover crops. Soil Tillage Res., 117: 17–27.

Cherr, C.M., Scholberg, J.M.S. and McSorley, R.M. (2006). Green manure approaches to crop production: a synthesis. Agronomy Journal, 98: 302–319.

Clark, A.J., Decker, A.M., Meisinger, J.J., Mulford, F.R. and McIntosh, M.S. (1995). Hairy vetch kill date effects on soil water and corn production. Agronomy Journal, 87: 579–585.

Clark, A.J., Decker, A.M., Meisinger, J.J. and McIntosh, M.S. (1997a). Kill date of vetch, rye, and vetch-rye mixture: I. Cover crop and corn nitrogen. Agronomy Journal, 89: 427–434.

Clark, A.J., Meisinger, J.J., Decker, A.M. and Mulford, F.R. (2007b). Effects of a grass-selective herbicide in a vetch–rye cover crop system on corn grain yield and soil moisture. Agron. J., 99: 43–48.

Clemmensen, K.., Bahr, A., Ovaskainen, O., Dahlberg, A., Ekblad, A., Wallander, H., Stenlid, J., Finlay, R.D., Wardle, D.A. and Lindahl, B.D. (2013). Roots and associated fungi drive long-term carbon sequestration in boreal. Forestry Science, 339: 1615–1618.

Cong, W.F., Hoffland, E., Li, L., Janssen, B.H. and Werf, W. (2015). Intercropping affects the rate of decomposition of soil organic matter and root litter. Plant Soil, 391: 399–411.

Cong, W.F., Hoffland, E., Long, L., Six, J., Sun, J.H., Bao, X.G., Zhang, F.S. and Werf, W. (2015). Intercropping enhances soil carbon and nitrogen. Global Change Biology, 21: 1715–1716.

Corazza, E.J., Da Silva, J.E., Resck, D.V.S. and Gomes, A.C. (1999). Comportamento de diferentes sistemas de manejo como fonte ou depósito de carbono em relação à vegetação de Cerrado. Revista Brasileira de Ciência Solo, 23: 425–432.

Corbeels, M., Neto, M.S., Marchao, R.L. and Ferreira, E.G. (2016). Evidence of limited carbon sequestration in soils under no-tillage systems in the Cerrado of Brazil. Sci. Rep., 6: 21450.

Cornell. (2009). Cornell Soil Health Assessment Training Manual. http://soilhealth.cals.cornell.edu., Ithaca, NY: Cornell University.

Corsi, S., Friedrich, T., Kassam, A., Michele, M. and Sà, J.M. (2012). Soil Organic Carbon Accumulation and Greenhouse Gas Emission Reductions from Conservation Agriculture: A literature review, Integrated Crop

Management, vol. 16, 89 pp, Plant Production and Protection Division, Food and Agriculture Organisation of the United Nations, Rome, 2012.

Costello, M.J., May, R.M. and Stork, N.E. (2013). Can we name Earth's species before they go extinct? Science, 339: 413–416.

Coughenour, C.M. (2003). Innovating conservation agriculture: The case of no-till cropping. Rural Sociology, 68(2): 278–304.

Cui, H., Zhou, Y., Gu, Z., Zhu, H., Fu, S. and Yao, Q. (2015). The combined effects of cover crops and symbiotic microbes on phosphatase gene and organic phosphorus hydrolysis in subtropical orchard soils. Soil Biol. Biochem., Elsevier, 82: 119–126.

Dabney, S.M., Delgado, J.A. and Reeves, D.W. (2001). Using winter cover crops to improve soil and water quality. Communications in Soil Science and Plant Analysis, 32: 1221–1250.

Dalla Costa, M. and Lovato, P.E. (2004). Fosfatases na dinâmica do fósforo do solo sob culturas de cobertura com espécies micorrízicas e não micorrízicas. Pesquisa Agropecuária Brasileira, 39: 603–605.

D'Hose, T., Ruysschaert, G., Viaene, N., Debode, J., Vanden Nest, T., Van Vaerenbergh, J. et al. (2016). Farm compost amendment and non-inversion tillage improve soil quality without increasing the risk for N and P leaching. Agric. Ecosyst. Environ., 225: 126–139.

Delgado, J.A. (1998). Sequential NLEAP simulations to examine effect of early and late planted winter cover crops on nitrogen dynamics. J. Soil Water Cons., 53: 241–244.

Derpsch, R., Sidiras, N. and Roth, C.H. (1986). Results of studies made from 1977 to 1984 to control erosion by cover crops and zero tillage techniques in Paraná, Brazil. Soil Till. Res., 8: 253–263.

Derpsch, R. (1990). Do crop rotation and green manuring have a place in the wheat farming systems of the warmer area? pp. 284–299. *In*: Saunders, D.A. (ed.). Wheat for the Nontraditional Warm Areas, Proceedings of a Conference, CIMMYT, Mexico, D.F.

Detheridge, A.P., Brand, G., Beechen, R., Crotty, F.V., Sanderson, R., Griffith, G.W. and Marley, C.L. (2016). The legacy effect of cover crops on soil fungal populations in a cereal rotation. Agriculture, Ecosystems and Environment, 228: 49–61.

Dick, W.A., Van Doren, D.M., Jr., Triplett, G.B., Jr. and Henry, J.E. (1986a). Influence of long-term tillage and rotation combinations on crop yields and selected soil parameters: I. Mollic Ochraqualf Res. Bull., 1180, Ohio Agric. Res. Dev. Ctr., Ohio State Univ., Wooster, OH.

Dick, W.A., Van Doren, D.M., Jr., Triplett, G.B., Jr. and Henry, J.E. (1986b). Influence of long-term tillage and rotation combinations on crop yields and selected soil parameters: II. Typic Fragiudalf. Communications in Soil Science and Plant Analysis, 22: 19–20.

Doran, J.W. (1987). Microbial biomass and mineralisable nitrogen distributions in no-tillage and plowed soils. Biology and Fertility of Soils, 5: 68–75.

Doran, J.W. and Smith, M.S. (1991). Role of cover crops in nitrogen cycling. *In*: Hargrove, W.L. (ed.). Cover Crops for Clean Water. Ankeny: Soil and Water Conservation Society.

Doran, J.W. (2002). Soil health and global sustainability: Translating science into practice. Agriculture, Ecosystems and Environment, 88: 119–127.

Drost, D., Long, G., Wilson, D., Miller, B. and Campbell, W. (1996). Barriers to adopting sustainable agricultural practices. Journal of Extension, 34(6): 1–30.

Duiker, S.W. and Curran, W.S. (2005). Rye cover crop management for corn production in the northern. Mid-Atlantic Regional Agronomy Journal, 97: 1413–1418.

Duiker Sjoerd, W. and Nathan L. Hartwig. (2004). Living mulches of legumes in imidazolinone-resistant corn. Agron. J., 96(4): 1021–1028.

Dumanski, J., Right, D.C.R. and Peiretti, R.A. (2014). Pioneers in soil conservation and conservation agriculture. International Soil and Water Conservation Research, 2: 107.

Dutra, G.R.D. (1919). Adubos verdes: Sua produção e modo de emprego. Campinas: Instituto Agronômico, 76p.

Eckert, D.J. (1991). Chemical attributes of soils subjected to no-till cropping with rye cover crops. Soil Sci. Soc. Am. J., 55: 405–409.

Ellert, B.H. and Janzen, H.H. (1999). Short-term influence of tillage on CO_2 fluxes from a semi-arid soil on the Canadian Prairies. Soil and Tillage Research, 50: 21–32.

Erenstein, O., Sayre, K., Wall, P., Hellin, J. and Dixon, J. (2012). Conservation agriculture in maize- and wheat-based systems in the (sub)tropics: Lessons from adaptation initiatives in South Asia, Mexico and Southern Africa. J. Sust. Agric., 36: 180–206. 10.1080/10440046.2011.620230.

Fahey, J.W., Zalcmann, A.T. and Talalay, P. (2001). The chemical diversity and distribution of glucosinolates and isothiocyanates among plants. Phytochemistry, 56: 5–51.

Fan, J., McConkey, B.G., Wang, H. and Janzen, H.H. (2016). Root distribution by depth for temperate agricultural crops. Field Crops Research, 189: 68–74.

FAO. (2014). The State of Food Insecurity in the World 2014: Strengthening the Enabling Environment for Food Security and Nutrition, FAO. Rome.http://www.fao.org/soils-portal/soil-degradation-restoration/global-soil-health-indicators-and-assessment/global-soil-health/en/.

FAO. (2017). Soil Organic Carbon: The Hidden Potential, Food and Agriculture Organisation of the United Nations Rome, Italy, p. 90.

FAO and ITPS. (2015). Status of the World's Soil Resources (SWSR) – Main Report, Food and Agriculture Organisation of the United Nations and Intergovernmental Technical Panel on Soils, Rome, Italy.

Farrar, J., Hawes, M., Jones, D. and Lindow, S. (2003). How root control the flux of carbon to rhizosphere. Ecology, 84: 827–837.

Featherstone, A.M. and Goodwin, B.K. (1993). Factors influencing a farmer's decision to invest in long-term conservation improvements. Land Economics, 69(1): 67–81.

Fink, J.R., Inda, A.V., Tiecher, T. and Barrón, V. (2016). Iron oxides and organic matter on soil phosphorus availability. Ciência e Agrotecnologia, 40(4): 369–379.

Finney, D.M., Eckert, S.E. and Kaye, J.P. (2016). Drivers of nitrogen dynamics in ecologically based agriculture revealed by long-term, high frequency field measurements. Ecological Applications, 25: 2210–2227.

Florentin, M.A., Peñalva, M., Calegari, A. and Derpsch, R. (2010). Green manure/cover crops and crop rotation in conservation agriculture on small farmers. Integr. Crop Manag., 12: 11–109.

Foley, A.J., Ramankutty, N., Brauman, K.A., Cassidy, E.S., Gerber, J.S., Johnston, M., Mueller, N.D., O'Connell, C., Ray, D.K., Oeste, P.C., Balzer, C., Bennett, E.M., Carpenter, S.R., Hill, J., Monfreda, C., Polasky, S., Rockström, J., Sheehan, J., Siebert, S., Tilman, D. and Zaks. D.P.M. (2011). Solutions for a cultivated planet. Nature, 478: 337–342.

Franchini, J.C., Pavan, M.A. and Miyazawa, M. (2004). Redistribution of phosphorus in soil through cover crop roots. Brazilian Archives of Biology and Technology, 47: 381–386.

Franzluebbers, A.J. and Arshad, M.A. (1996). Soil organic pools during early adoption of conservation tillage in northwestern Canada. Soil Sci. Soc. Amer. J., 60: 1422–1427.

Franzluebbers, A.J., Hons, F.M. and Zuberer, D.A. (1998). *In situ* and potential CO_2 evolution from a fluventic Ustochrept in southcentral Texas as affected by tillage and crop management. Soil Till. Res., 47: 303–308.

Franzluebbers, A.J., Langdale, G.W. and Schomberg, H.H. (1999). Soil carbon, nitrogen, and aggregation in response to type and frequency of tillage. Soil Sci. Soc. Amer. J., 63: 349–355.

Franzluebbers, A.J. (2004). Tillage and residue management effect on soil organic matter, ch. 8. pp. 227–268. *In*: Magdoff and Weil (eds.). Soil Organic Matter in Sustainable Agriculture. CRC Press, Boca Raton, Fl.

Franzluebbers, A.J. (2005). Soil organic carbon sequestration and agricultural greenhouse gas emissions in the southeastern USA. Soil and Tillage Res., 83: 120–147.

Frasier, L., Quiroga, A. and Noellemeyer, E. (2016). Effect of different cover crops on C and N cycling in sorghum NT systems. Science of the Total Environment, 562: 628–639.

Frey, S.D., Elliott, E.T. and Paustian, K. (1999). Bacterial and fungal abundance and biomass in conventional and no tillage ecosystems along to climatic gradients. Soil Biol. Biochem., 31: 573–585.

Garbeva, P., Van Veen, J.A. and Van Elsas, J.D. (2004). Microbial diversity in soil: Selection of microbial populations by plant and soil type and implications for disease supressiveness. Annual Review of Phytopathology, 42: 243–270.

Garnett, T., Appleby, M.C., Balmford, A. et al. (2013). Sustainable intensification in agriculture: Premises and policies. Science, 341: 33–34.

Gattinger, A., Jawtusch, J., Muller, A. et al. (2011). No-till agriculture—A climate smart solution? Climate Change and Agriculture Report No. 2, Misereor e.V., Aachen, Germany.

Ghafoor, A., Poeplau, C. and Kätterer, T. (2017). Fate of straw- and root-derived carbon in a Swedish agricultural soil. Biol. Fertil. Soils, 53: 257–267.

Ghidey, F. and Alberts, E.E. (1993). Residue type and placement effect on decomposition: field study and model evaluation. Trans. ASAE, 36: 1611–1617.

Glover, J.D., Cox, C.M. and Reganold, J.P. (2007). Future farming: A return to roots? Scientific American, 297: 83–89.

Gould, L.J., John, N.Q., Weigelt, A., Deyn, G.B.D. and Bardgett, R.D. (2016). Plant diversity and root traits benefit physical properties key to soil function in grasslands. Ecology Letters, 19: 1140–1149.

Haichar, F.E.Z., Marol, C., Berge, O., Rangel-Castro, J.I., Prosser, J.I., Balesdent, J., Heulin, T. and Achouak, W. (2008). Plant host habitat and root exudates shape soil bacterial community structure. ISME Journal, 2: 1221–1230.

Haramoto, E.R. and Gallandt, E.R. (2004). *Brassica* cover cropping for weed management: A review. Renewable Agriculture and Food Systems, 19: 187–198.

Hargrove, W.L. and Frye, W.W. (1987). The need for legume cover crops in conservation tillage production. pp. 1–4. *In*: Power, D.F. (ed.). The Role of Legumes in Conservation Tillage System, Soil Conserv. Soc., AM, Ankeny, IA.

Hartman, W.H. and Richardson, C.J. (2013). Differential nutrient limitation of soil microbial biomass and metabolic quotients (qCO_2): Is there biological stoichiometry of soil microbes? PLoS One, 8: 1–14.

Hartwig, N. and Ammon, H.U. (2002). Cover crops and living mulches. Weed Sci., 50: 688–600.

Hatfield, J.L., Boote, K.J. and Kimball, B.A. (2011). Climate impacts on agriculture: Implications for crop production. Agronomy Journal, 103: 351–370.

Hatfield, J.L., Thomas, J.S. and Cruse, R.M. (2017). Soil: The forgotten piece of the water, food, energy nexus. Advances in Agronomy, 143: 1–46.

Havlin, J.L., Kissel, D.E., Maddux, L.E., Classen, M.M. and Long, J.H. (1990). Crop rotation and tillage effects on soil organic carbon and nitrogen. Soil Sci. Soc. Amer. J., 54: 448–452.

Heimann, M. and Reichstein, M. (2008). Terrestrial ecosystem carbon dynamics and climate feedbacks. Nature, 45: 289–292. doi:10.1038/nature06591.

Hobbs, P.R., Sayre, K. and Gupta, R. (2008). The role of conservation agriculture in sustainable agriculture. Philosophical Transactions of the Royal Society B, 363: 543–555.

Hutchinson, J.J., Campbell, C.A. and Desjardins, R.L. (2007). Some perspectives on carbon sequestration in agriculture. Agricultural and Forest Meteorology, 142(2): 288–302.

Janzen, H.H. (2006). The soil carbon dilemma: Shall we hoard it or use it? Soil Biology and Biochemistry, 38: 419–424.

Janzen, H.H. (2015). Beyond carbon sequestration: Soil as conduit of solar energy. European Journal of Soil Science, 66(1): 19–32.

Jarecki, M.K. and Lal, R. (2003). Crop management for soil carbon sequestration. Crit. Rev. Plant Sci., 22: 471–502. https://doi.org/10.1080/713608318.

Jenkinson, D.S. (1971). Studies on the decomposition of C14 labeled organic matter in soil. Soil Science, 111: 64–70.

Jenkinson, D.S. and Rayner, J.H. (1977). The turnover of soil organic matter in some of the Rothamsted classical experiments. Soil Science, 123: 298–305.

Jiang, X., Wright, A., Wang, X. and Liang, F. (2011). Tillage-induced changes in fungal and bacterial biomass associated with soil aggregates: A long-term field study in a subtropical rice soil in China. Applied Soil Ecology, 48(2): 168–173.

Johnen, G. and Sauerbeck, D. (1977). A tracer technique for measuring growth, mass and microbial breakdown of plant roots during vegetation. pp. 366–373. *In*: Lohm, V. and Persson, T. (eds.). Soil Organisms as Components of Ecosystems, Proc. Int. Soil Zoological Colloquium, sixth Ecol. Bull., Stockholm, Sweden.

Johnson, J.M.F., Allmaras, R.R. and Reicosky, D.C. (2006). Estimating source carbon from crop residues, roots and rhizodeposits using the national grain-yield database. Agron. J., 98: 622–636.

Johnston, A.E. (1986). Soil organic matter, effects on soil and crops. Soil Use Manag., 2: 97–105.

Kahlon, M., Singh, R. and Merrie, A. (2013). Twenty-two years of tillage and mulching impacts on soil physical characteristics and carbon sequestration in Central Ohio. Soil and Tillage Research, 126: 151–158.

Kane, D. (2015). Carbon Sequestration Potential on Agricultural Lands: A Review of Current Science and Available Practices. In association with National Sustainable Agriculture Coalition Breakthrough Strategies and Solutions. https://sustainableagriculture.net/wp-content FK.

Karlen, D.L. and Cambardella, C.A. (1996). Conservation strategies for improving soil quality and organic matter storage. pp. 395–420. *In*: Carter, M.R. and Stewart, B.A. (eds.). Structure and Organic Matter Storage in Soils. Lewis Publ., CRC Press, Boca Raton, FL.

Kaspar, T.C., Kladivko, E.J., Singer, J.W., Morse, S. and Mutch, D.R. (2008). Potential and Limitations of Cover Crops, Living Mulches and Perennials to Reduce Nutrient Losses to Water Sources from Agricultural Fields in the Upper Mississippi River Basin, Society of Agricultural and Biological Engineers, 127–148.

Kaspar, T.C. and Singer, J.W. (2011). The use of cover crops to manage soil. pp. 321–337. *In*: Hatfield, J.L. and Sauer, T.J. (eds.). Soil Management: Building a Stable Base for Agriculture, Madison, WI: American Society of Agronomy and Soil Science Society of America.

Kassam, A., Friedrich, T. and Shaxson, F. (2009). The spread of conservation agriculture: justification, sustainability and uptake. Intern. J. Agric. Sustainability, 7: 292–320.

Kassam, A. and Friedrich, T. (2012). An ecologically sustainable approach to agricultural production intensification: Global perspectives and developments, Field Actions Science Reports, Reconciling Poverty Eradication and Protection of the Environment, special issue 6, Institut Veolia Environnement, France. http://factsreports.revues.org/1382.

Kassam, A., Friedrich, T., Derpsch, R. and Kienzle, J. (2015). Overview of the Worldwide Spread of Conservation Agriculture, Field Actions Science Reports [online], vol. 8, online since 26 September 2015, connection on 01 December 2017.

Kätterer, T., Bolinder, M.A. and Andrén, O. (2011). Roots contribute more to refractory soil organic matter than above-ground crop residues as revealed by a long-term field experiment. Agriculture, Ecosystems and Environment, 141: 184–192.

Kay, B. (1998). Soil structure and organic carbon: A review. Soil Processes and the Carbon Cycle, CRC Press, Boca Raton, 169–197.

Kaye, P. and Quemada, M. (2017). Using cover crops to mitigate and adapt to climate change. Agronomy for Sustainable Development, 37: 41–47.

Keesstra, S.D., Bouma, J., Wallinga, J., Tittonell, P., Smith, P. and Fresco, L.O. (2016). The significance of soils and soil science towards realisation of the United Nations Sustainable Development Goals. Soil, 2(2): 111–128.

Kell, D.B. (2011). Breeding crop plants with deep roots: Their role in sustainable carbon, nutrient and water sequestration. Annals of Botany, 108(3): 407–418.

Kell, D.B. (2012). Large-scale sequestration of atmospheric carbon via plant roots in natural and agricultural ecosystems: Why and how. Philosophical Transactions of the Royal Society B: Biological Sciences, 367(1595): 1589–1597.

Kemper, W., Doral, N., Schneider, N. and Sinclair, T. (2011). No-till can increase earthworm populations and rooting depths. Journal of Soil and Water Conservation, 66(1): 13A–17A.

Kleber, M. and Johnson, M.G. (2010). Advances in understanding the molecular structure of soil organic matter: implications for interactions in the environment. Adv. Agron., 106: 78–142.

Knowler, D. and Bradshaw, B. (2007). Farmers' adoption of conservation agriculture: A review and synthesis of recent research. Food Policy, 32(1): 25–48.

Kremen, A. and Weil, R.R. (2006). Monitoring nitrogen uptake and mineralisation by *Brassica* cover crops in Maryland. 18th World Congress of Soil Science, 155–40.

Kumar, K. and Goh, K.M. (2000). Crop residues and management practices: Effects on soil quality, soil nitrogen dynamics, crop yield, and nitrogen recovery. Adv. Agron., 68: 197–319.

Kumar, S., Kadono, A., Lal, R. and Dick, W. (2012). Soil hydrological properties as influenced by 50 years of tillage and cropping systems of two contrasting soils in Ohio. Soil Science Society of America Journal, 76: 1798–1809.

Kunze, A., Costa, M.D., Epping, J., Loffaguen, J.C., Schuh, R. and Lovato, P.E. (2011). Phosphatase activity in sandy soil influenced by mycorrhizal and non-mycorrhizal cover crops. Revista Brasileira Ciência do Solo, 35: 705–711.

Kuo, M.H., Nadeau, E.T. and Grayhack, E.J. (1997). Multiple phosphorylated forms of the *Saccharomyces cerevisiae* Mcm1 protein include an isoform induced in response to high salt concentrations. Mol. Cell Biol., 17(2): 819–32.

Kuzyakov, Y. (2010). Priming effects: Interactions between living and dead organic matter. Soil Biology Biochemistry, 43: 1363–1371.

Kuzyakov, Y., Friedel, J.K. and Stahr, K. (2000). Review of mechanisms and quantification of priming effects. Soil Biology and Biochemistry, 32: 1485–1498.

Kuzyakov, Y. and Cheng, W. (2004). Photosynthesis controls of CO_2 efflux from maize rhizosphere. Plant and Soil, 263: 85–99.

Lal, R. (1999). Soil management and restoration for C sequestration to mitigate the accelerated greenhouse effect. Environmental Science and Technology, 1: 307–326.

Lal, R. (2004). Soil carbon sequestration impacts on global climate change and food security. Science, 304: 1623–1627.

Lal, R. (2005). Soil erosion and carbon dynamics. Soil and Tillage Research, 81(2): 137–142.

Lal, R. (2007). Carbon Sequestration. Philosophical Transactions of the Royal Soc., 363: 815–830.

Lal, R. (2013). Intensive agriculture and the soil carbon pool. pp. 59–72. *In*: Kang, M.S. and Banga, S.S. (eds.). Combating Climate Change: An Agricultural Perspective. Taylor and Francis, Boca Raton.

Lal, R. (2014). Climate strategic soil management. Challenges, 5(1): 43–74.

Lal, R. (2015a). A soil-carbon sequestration and aggregation by cover cropping. Journal of soil and Water Conservation, 70: 329–339.

Lal, R. (2015b). A system approach to conservation agriculture. J. Soil Water Conser., 70: 82A–88A.

Lal, R. (2015c). Cover cropping and the 4 per thousand proposals. J. Soil Water Conser., 70: 141A.

Lam, S.K., Chen, D., Mosier, A.R. and Roush, R. (2013). The potential for carbon sequestration in Australian agricultural soils is technically and economically limited. Scientific Reports, 3: 2179.

Lange, M., Eisenhauer, N., Sierra, C.A., Bessler, H., Engels, C., Griffiths, R.I., Mellado-Vázquez, P.G., Malik, A.A., Roy, J., Scheu, S., Steinbeiss, S., Thomson, B.C., Trumbore, S.E. and Gleixner, G. (2015). Plant diversity increases soil microbial activity and soil carbon storage. Nature Communications, 6: 6707.

Lehmann, J. (2007). Bioenergy in the black. Frontiers in Ecology and the Environment, 5: 381–387.

Lehmann, J. and Kleber, M. (2015). The contentious nature of soil organic matter. Nature, 528(7580): 60–68.

Liebman, M., Graef, R.L., Nettleton, D. and Cambardella, C.A. (2012). Use of legume green manures as nitrogen sources for corn production. Renewable Agric. Food Sys., 27: 180–191.

Lichtenberg, Erik. (2004). Cost-responsiveness of conservation practice adoption: A revealed preference approach. Journal of Agricultural and Resource Economics, 29(3): 420–435.

Lindstrom, M.J., Nelson, W.W. and Schumacher, T.E. (1990). Soil movement by tillage as affected by slope. Soil and Tillage Research, 17(3-4): 255–264.

Liu, L., O'Leary, J.G., Ma, Y., Cowie, A., Li, F.Y., McCaskill, M., Conyers, M., Dalal, R., Robertson, F. and Dougherty, W. (2016). Modelling soil organic carbon 2. Changes under a range of cropping and grazing farming systems in eastern Australia. Geoderma, 265: 164–175.

Lobell, D.B., Hammer, G.L., McLean, G., Messina, C., Roberts, M.J. and Schlenker, W. (2013). The critical role of extreme heat for maize production in the United States. Nature Climate Change, 3: 497–501.

Manna, M.C., Muneshwar, Singh, Wanjari, R.H., Mandal and Patra, A.K. (2016). Soil nutrient management for carbon. Encyc. Soil Science, DOI: 10.1081/E-ESS3-120052914.

Mazzoncini, M., Antichi, D., Benec, C., Risaliti, R., Petrid, M. and Bonarie, E. (2016). Soil carbon and nitrogen changes after 28 years of no-tillage management under Mediterranean conditions. European Journal of Agronomy, 77: 156–165.

Mathew, R.P., Feng, Y., Githinji, L., Ankumah, R. and Balkcom, K.S. (2012). Impact of no-tillage and conventional tillage systems on soil microbial communities. Appl. Environ. Soil Sci.

Mbuthia, L.W., Acosta-Martínez, V., DeBruyn, J., Schaeffer, S. and Tyler, D. (2015). Long-term tillage, cover crop and fertilisation effects on microbial community structure, activity: Implications for soil quality. Soil Biology. Biochemistry, 89: 24–34.

McDaniel, M., Tiemann, L. and Grandy, A.S. (2014). Does agricultural crop diversity enhance soil microbial biomass and organic matter dynamics? A meta-analysis. Ecol. Appl., 24: 560–570.

Meisinger, J.J. (1991). Effects of cover crops on groundwater quality. pp. 57–68. *In*: Hargrove (ed.). Cover Crops for Clean Water, Soil and Water Conservation Society.

Mitchell, D.C., Castellano, M.J., Sawyer, J.E. and Pantoja, J.L. (2013). Cover crop effects on nitrous oxide emissions from a maize-based cropping system: role of carbon inputs. Soil Science Society of America Journal, 77: 1765–1773.

Mitchell, J., Carter, L. and Munk, D. (2012). Conservation tillage systems for cotton advance in the San Joaquin Valley. California Agriculture, 66: 108–15.

Miyazawa, K., Takeda, M., Murakami, T. and Murayama, T. (2014). Dual and triple intercropping: potential benefits for annual green manure production. Plant Prod. Sci., 17: 194–201.

Montanarella, L., Pennock, D.J., McKenzie, N. and Badraoui, N. (2016). World's soils are under threat. Soil, 2: 79–82.

Montgomery, D.R. (2007a). Dirt: The Erosion of Civilisations, Berkeley, CA: University of California, 285.

Montgomery, D.R. (2007b). Soil erosion and agricultural sustainability. Proceedings of the National Academy of Sciences of the United States of America, 104: 13,268–13,272.

Mulvaney, M.J., Wood, C.W., Balkcom, K.S., Kemble, J. and Shannon, D.A. (2017). No-till with high biomass cover crops and invasive legume mulches increased total soil carbon after three years of collard production. Agroecology and Sustainable Food Systems, 41(1): 30–45.

Murphy, F., Ewins, C., Carbonnier, F. and Quinn, B. (2016). Wastewater treatment works (WwTW) as a source of microplastics in the aquatic environment. Environ. Sci. Technol., 50: 5800–5808.

Nakhauka, E.B. (2009). Agricultural biodiversity for food and nutrient security: The Kenyan perspective. International Journal of Biodiversity and Conservation, 1(7): 208–214.

Neumann, G. and Römheld, V. (1999). Root excretion of carboxylic acids and protons in phosphorus-deficient plants. Plant Soil, 211: 121–130.

Nelson, D.W. and Sommers, L.E. (1996). Methods of soil analysis, Part 3, chemical methods. Soil Science Society of America Book Series no. 5: 961–1010.

Nielsen, U.N., Ayres, E., Wall, D.H. and Bardgett, R.D. (2011). Soil biodiversity and carbon cycling: A review and synthesis of studies examining diversity-function relationships. Global Change Biol., 62: 105–116.

Nyakatawa, E.Z., Reddy, K.C. and Sistani, K.R. (2001b). Tillage, cover cropping and poultry litter effects on selected soil chemical properties. Soil Till. Res., 58: 69–79.

O'Dea, Justin K., Clain A. Jones, Catherine A. Zabinski, Perry R. Miller and Ilai N. Keren. (2015). Legume, cropping intensity and N-fertilisation effects on soil attributes and processes from an eight-year-old semiarid wheat system. Nutr. Cycl. Agroeco., 02: 179–194.

O'Connell, K., Jinks-Robertson, S. and Peter, T.D. (2015). Elevated genome-wide instability in yeast mutants lacking RNase H activity. Genetics, 201(3): 963–75.

O'Rourke, S.M., Angers, D.A., Holden, N.M. and Mcbratney, A.B. (2015). Soil organic carbon across scales. Global Change Biology, 21: 3561–3574.

Odell, R.T., Melsted, S.W. and Walker, W.M. (1984). Changes in organic carbon and nitrogen of Morrow plot's soil under different treatments. Soil Science, 137: 160–171.

Ogle, S.M., Breidt, F.J. and Paustian, K. (2005). Agricultural management impacts on soil organic carbon storage under moist and dry climatic conditions of temperate and tropical regions. Biogeochemistry, 72: 507–513.

Orgiazzi, A., Bardgett, R.D., Barrios, E., Behan-Pelletier, V., Briones, M.J.I., Chotte, J-L., De Deyn, G.B., Eggleton, P., Fierer, N., Fraser, T., Hedlund, K., Jeffery, S., Johnson, N.C., Jones, A., Kandeler, E., Kaneko, N., Lavelle, P., Lemanceau, P., Miko, L., Montanarella, L., Moreira, F.M.S., Ramirez, K.S., Scheu, S., Singh, B.K., Six, J., van der Putten, W.H. and Wall, D.H. (2016). Global Soil Biodiversity Atlas, Luxembourg: European Commission, Publications Office of the European Union.

Patrick, W., Haddon, C. and Hendrix, J. (1957). The effect of longtime use of winter cover crops on certain physical properties of commerce loam. Soil Sci. Soc. Am. J., 21: 366–368.

Paul, E.A., Follett, R.F., Leavitt, S.W., Halvorson, A.D., Peterson, G.A. and Lyon, D. (1997). Radiocarbon dating for determination of soil organic matter pool sizes and dynamics. Soil Science Society of America Journal, 61: 1058–1067.

Paustian, K., Lehmann, J., Ogle, S., Reay, D., Robertson, G.P. and Smith, P. (2016). 'Climate-smart' soils: A new management paradigm for global agriculture. Nature, 532(7597): 49–57.

Pereira, H.M., Navarro, L.M. and Martins, I.S. (2012). Global biodiversity change: The bad, the good, and the unknown. Annual Review of Environment and Resources, 37: 25–50.

Pimentel, D., Harvey, C., Resosudarmo, P. et al. (1995). Environmental and economic cost of soil erosion and conservation benefits. Science, New Series, 267: 1117–1123.

Poeplau, C. and Don, A. (2015). Carbon sequestration in agricultural soils via cultivation of cover crops—A meta-analysis. Agriculture, Ecosystems and Environment, 200: 33–41.

Poffenbarger, H., Mirsky, S.B., Weil, R.R., Kramer, M., Spargo, J.T. and Cavagelli, M.A. (2015). Legume proportion, poultry litter, and tillage effects on cover crop decomposition. Agronomy Journal, 107: 2083–2096.

Powlson, D.S., Whitmore, A.P. and Goulding, K.W.T. (2011). Soil carbon sequestration to mitigate climate change: A critical re-examination to identify the true and the false. European Journal of Soil Science, 62(1): 42–55.

Powlson, D.S., Stirling, C.M., Thierfelder, C., White, R.P. and Jat, M.L. (2016). Does conservation agriculture deliver climate change mitigation through soil carbon sequestration in tropical agro-ecosystems? Agric. Ecosyst. Environ., 220: 164–174.

Pretty, J. and Bharucha, Z.P. (2014). Sustainable intensification in agricultural systems. Annals of Botany, 114: 1571–1596.

Raphael, J.P.A., Calonego, J.C., Milori, D.M.B.P. and Rosolem, C.A. (2016). Soil organic matter in crop rotations under no-till. Soil and Tillage Research, 155: 45–53.

Redin, M., Guénon, R., Recous, S., Schmatz, R., Freitas, L.L., Aita, C. and Giacomini, S.J. (2014a). Carbon mineralisation in soil of roots from twenty crop species, as affected by their chemical composition and botanical family. Plant Soil, 378: 205–214.

Redin, M., Recous, S., Aita, C., Dietrich, G., Skolaude, A.C., Ludke, W.H., Schmatz, R. and Giacomini, S.J. (2014b). How the chemical composition and heterogeneity of crop residue mixtures decomposing at the soil surface affects C and N mineralisation. Soil Biol. Biochem., 78: 65–75.

Reicosky, D.C. and Lindstrom, M.J. (1993). Fall tillage method: Effect on short-term carbon dioxide flux from soil. Agron. J., 85: 1237–1243.

Reicosky, D.C. and Lindstrom, M.J. (1995). The impact of fall tillage on short-term carbon dioxide flux. pp. 177–187. *In*: Lal, R. et al. (eds.). Soils and Global Change. Lewis Publ., Chelsea, MI.

Reicosky, D.C. and Forcella, F. (1998). Cover crop and soil quality interactions in agro-ecosystems. J. Soil and Water Conservation, 53(3): 224–229.

Reicosky, D.C. and Janzen, H.H. (2018). Conservation agriculture: maintaining land productivity and health by managing carbon flows, chap. 4. *In*: Lal, R. and Stewart, B.A. (eds.). Soil and Climate. Advances in Soil Science, Taylor and Francis Group, LLC (in press).

Reusser, P., Bierman, P. and Rood, D. (2015). Quantifying human impacts on rates of erosion and sediment transport at a landscape scale. Geology, published online. DOI: 10.1130/G36272.1.

Rheinheimer, D.S. and Anghinoni, I. (2001). Distribuição do fósforo inorgânico em sistemas de manejo de solo. Pesq. Agropec. Bras., 36: 151–160.

Rheinheimer, D.S., Tiecher, T., Gonzatto, R., Santanna, M.A., Brunetto, G. and Silva, L.S. (2018a). Long-term effect of surface and incorporated liming in the conversion of natural grassland to no-till system for grain production in a highly acidic sandy-loam Ultisol from South Brazilian campos. Soil and Tillage Research, 180: 222–231.

Rheinheimer, D.S., Tiecher, T., Gonzatto, R., Zafar, M. and Brunetto, G. (2018b). Residual effect of surface-applied lime on soil acidity properties in a long-term experiment under no-till in a Southern Brazilian sandy Ultisol. Geoderma, 313: 7–16.

Robert, M. (2001). Soil-carbon sequestration for improved land management. World Soil Resources Reports. 96(ISSN 0532-0488), FAO, Rome, Italy.

Roscoe, R.A. and Buurman, P. (2003). Tillage effects on soil organic matter in density fractions of a Cerrado Oxisol. Soil Tillage Res., 70: 107–119.

Rosolem, C.A. and Calonego, J.C. (2013). Phosphorus and potassium budget in the soil – plant system in crop rotations under no-till. Soil Tillage Res., 126: 127–133.

Ruffo, M.L. and Bollero, G.A. (2003). Residue decomposition and C and N release rates prediction based on biochemical fractions using principal component regression. Agron. J., 95: 1034–1040.

Ryan, R.M. and Brown, K.W. (2003). The benefits of being present: mindfulness and its role in phychological well-being. Journal of Personality and Social Phychology, 84(4): 822–848.

Sainju, U., Whitehead, W. and Singh, B. (2002). Cover crops and nitrogen fertilisation effects on soil aggregation and carbon and nitrogen pools. Canadian J. Soil Science, 83: 155–165.

Sanderman, Jonathan, Tomislav Hengl and Gregory J. Fiske. (2017). Soil carbon debt of 12,000 years of human land use. PNAS, 114(36): 9575–9580.

Sapkota, T.B., Jat, R., Singh, R.G., Jat, M., Stirling, C., Jat, M. et al. (2017). Soil organic carbon changes after seven years of conservation agriculture in a rice-wheat system of the eastern Indo-Gangetic Plains. Soil Use and Management, 33: 81–89.

Schimel, J. and Schaeffer, S. (2012). Microbial control over carbon cycling in soil. Frontiers in Microbiology, 3(348): 155–165.

Schipanski, M., Barbercheck, M., Douglas, M.R., Finney, D.M., Haider, K., Kaye, J.P., Kemanian, A.R., Mortensen, D.A., Ryan, M.R., Tooker, J. and White, C.M. (2014). A conceptual framework for evaluating ecosystem services provided by cover crops in agro-ecosystems. Agric Syst., 125: 12–22.

Schlesinger, W.H. (1991). Biogeochemistry, An Analysis of Global Change. New York, USA, Academic Press.

Schnitzer, S.A., Klironomos, J.N., Hilleris Lambers, J., Kinkel, L.L., Reich, P.B., Xiao, K. and Scheffer, M. (2010). Soil microbes drive the classic plant diversity – productivity pattern. Ecology, 92(2): 296–303.

Schomberg, H.H., Endale, D.M., Calegari, A., Peixoto, R.T.G., Miyazawa, M. and Cabrera, M.L. (2006). Influence of cover crops on potential nitrogen availability to succeeding crops in a Southern Piedmont soil. Springer, Berlin/Heidelberg, Biol. Fert. Soils, 42: 299–307.

Shrestha, B.M., Singh, B.R., Sitaula, K., Lal, R. and Bajracharya, R.M. (2007). Soil aggregate- and particle-associated organic carbon under different land uses in Nepal. Soil Sci. Soc. Amer. J., 71: 1194–1203.

Singer, J.W., Nusser, S.M. and Alf, C.J. (2007). Are cover crops being used in the U.S. corn belt? Journal of Soil and Water Conservation, 62: 353–358.

Six, J., Frey, S.D., Thiet, R.K. and Batten, K.M. (2006). Bacterial and fungal contributions to carbon sequestration in agro-ecosystems. Soil Science Society of America Journal, 70: 555–569.

Sommer, R. and Bossio, D. (2014). Dynamics and climate change mitigation potential of soil organic carbon sequestration. J. Environ. Manage., 1, 144: 83–87.

Srinivasarao, C., Kundu, S., Ramachandrappa B.K., Reddy, S., Lal, R., Venkateswarlu, B., Sahrawat, K.L. and Naik, R.P. (2014). Potassium release characteristics, potassium balance, and finger millet (*Eleusine coracana* G.) yield sustainability in a 27-year long experiment on an Alfisol in the semi-arid tropical India. Plant Soil, 374: 315–330.

Stavi, I., Lal, R., Jones, S. and Reeder, R.C. (2012). Implications of cover crops for soil quality and geodiversity in a humid-temperate region in the Mid-western USA. Land Degradation Development, 23: 322–330.

Stockfisch, N., Forstreuter, T. and Ehlers, W. (1999). Ploughing effects on soil organic matter after twenty years of conservation tillage in Lower Saxony, Germany. Soil Till. Res., 52: 91–101.

Strickland, M.S. and Rousk, J. (2010). Considering fungal, bacterial dominance in soils—Methods, controls and ecosystem implications. Soil Biology and Biochemistry, 42: 1385–1395.

Syswerda, S.P. and Robertson, G.P. (2014). Ecosystem services along a management gradient in Michigan (USA) cropping systems. Agriculture Ecosystems and Environment, 189: 28–35.

Teasdale, J.R., Brandsaeter, L.O., Calegari, A. and Skora Neto, F. (2007). Cover crops and weed management. pp. 49–64. *In*: Mahesh K. Upadhyaya and Robert E. Blackshaw (eds.). Non-chemical Weed Management: Principles, Concepts and Technology. Reading, UK.

Thiele-Brunh, S., Bloem, J., de Vries, F.T., Kalbitz, K. and Wagg, C. (2012). Linking soil biodiversity and agricultural soil management. Environmental Sustainability, 4: 523–528.

Thorup-Kristensen, K., Magid, J. and Jensen, L.S. (2003). Catch crops and green manures as biological tools in nitrogen management in temperate zones. Adv. Agron., 79: 227–302.

Tiecher, T., Calegari, A., Caner, L. and Rheinheimer, D.D.S. (2018). Soil fertility and nutrient budget after 23 years of different soil tillage systems and winter cover crops in a subtropical Oxisol. Geoderma, 308: 78–85.

Tilman, D., Balzer, C., Hill, J. and Befort, B.L. (2011). Global food demand and the sustainable intensification of agriculture. Proceedings of the National Academy of Sciences of the United States of America, 108: 20260–20264.

Tilman, D., Isbell, F. and Cowles, J.M. (2014). Biodiversity and ecosystem functioning. Annual Review of Ecology, Evolution and Systematics, 45: 471–493.

Tonitto, C., David, M.B. and Drinkwater, L.E. (2006). Replacing bare fallows with cover crops in fertilizer-intensive cropping systems: A meta-analysis of crop yield and N dynamics. Agriculture, Ecosystems and Environment, 112: 58–72.

Turmel, M., Spearatti, A., Baudron, F. and Verhulst, N. (2015). Crop residue management and soil health: A systems analysis. Agricultural Systems, 134: 6–16.

Tully, K. and Rebecca, R. (2017). Nutrient cycling in agro-ecosystems: Balancing food and environmental objectives. Agroecology and Sustainable Food Systems, 41: 761–798.

Upadhyay, B.M., Young, D.L., Wang, H.H. and Wandschneider, P. (2003). How do farmers who adopt multiple conservation practices differ from their neighbours? Amer. J. Altern. Agric., 18: 27–36.

United Nations. (2014). Concise Report on the World Population Situation, 2014, United Nat ions Department of Economic and Social Affairs, Washington D.C.

USDA NRCS. (2014a). Natural Resources Conservation Service, National Handbook of Conservation Practices.

USDA NRCS (Natural Resources Conservation Service). (2014b). Soil Health Home. http://www.nrcs.usda.gov/wps/portal/nrcs/main/soils/health/.

Utomo, M., Frye, W.W. and Blevins, R.L. (1990). Sustaining soil nitrogen for corn using hairy vetch cover crop. Agron. J., 82: 979–983.

van der Wal, A. and De Boer, W. (2017). Dinner in the dark: Illuminating drivers of soil organic matter decomposition. Soil Biology and Biochemistry, 105: 45–48.

Vitale, J.D., Godsey, C., Edwards, J. and Taylor, R. (2011). The adoption of conservation tillage practices in Oklahoma: Findings from a producer survey. J. Soil Water Conser., 66: 250–264.

Vogel, C., Mueller, C.W., Höschen, C., Buegger, F., Heister, K., Schulz, S., Schloter, M. and Kögel-Knabner, I. (2014). Submicron structures provide preferential spots for carbon and nitrogen sequestration in soils. Nature Communications, 5: 2947.

Voroney, R.P., Paul, E.A. and Anderson, D.W. (1989). Decomposition of wheat straw and stabilization of microbial products. Can. J. Soil Sci., 69: 63–77.

Vukicevich, E., Lowery, T., Bowen, P. and Úrbez, T. (2016). Cover crops to increase soil microbial diversity and mitigate decline in perennial agriculture. Agron. Sust. Dev., 36.

Wagger, M.G. (1989). Cover crop management and N rate in relation to growth and yield of no-till corn. Agronomy Journal, 81: 533–538.

Wagger, M.G., Cabrera, M.L. and Ranells, N.N. (1998). Nitrogen and carbon cycling in relation to cover crop residue quality. Journal Soil Water Conservation, 53: 214–218.

Wall, D.H. and Nielsen, U.N. (2012). Biodiversity and ecosystem services: Is it the same below ground? Nature Education Knowledge.

Williams, S.M. and Weil, R.R. (2004). Crop cover root channels may alleviate soil compaction effects on soybean crop. Soil Science Society of America Journal, 68(4): 1403–1409.

Wilson, D.O. and Hargrove, W.L. (1986). Release of nitrogen from Crimson Clover residue under two tillage systems. Soil Science Society of America Journal, 50: 1251–1254.

Wilson, G.W.T., Rice, C.W., Rillig, M.C., Springer, A. and Hartnett, D.C. (2009). Soil aggregation and carbon sequestration are tightly correlated with the abundance of arbuscular mycorrhizal fungi. Ecology Letters, 452–461.

Wilts, A.R., Reicosky, D.C., Allmaras, R.R. and Clapp, C.E. (2004). Long-term corn residue effects: Harvest alternatives, soil carbon turnover, and root-derived carbon. Soil Science Society of America Journal, 68: 1342–1351.

World Bank. (2015). Agricultural Land (% of Land Area). Available at http://data.worldbank.org/indicator/AG.LND.AGRI.ZS/countries?display=graph.

Wu, J.J. and Babcock, B.A. (1998). The choice of tillage, rotation, and soil testing practices: economic and environmental implications. Amer. J. Agric. Econ., 80: 494–511.

Young, I.M. and Ritz, K. (2000). Tillage, habitat space and function of soil microbes. Soil Tillage Research, 53: 201–213.

Zhao, X., Xue, J.F., Zhang, X.Q., Kong, F.L., Lal, R. and Zhang, H.L. (2015). Stratification and storage of soil organic carbon and nitrogen as affected by tillage practices in the North China Plain. PLoS One. DOI:10.1371/journal.pone.0128873.

Zhu, L., Hu, N., Yang, M., Zhan, X. and Zhang, Z. (2014). Effects of different tillage and straw return on soil organic carbon in a rice-wheat rotation system. PLoS One. DOI.org/10.1371/journal.pone.0088900.

Zuber, Stacy M. and Villamil, M.B. (2015). Meta-analysis approach to assess effect of tillage on microbial biomass and enzyme activities. Soil Biology and Biochemistry, 97: 176–187.

Zuber, S. and Villami, M.B. (2016). Meta-analysis approach to assess effect of tillage on microbial biomass and enzyme activities. Soil Biology and Biochemistry, 97: 176–187.

12

Cover Crops and Soil Nitrogen Cycling

Nitu, TT, UM Milu and *MMR Jahangir**

1. Introduction

Current agricultural practices, such as mono-cropping, deep plowing and reactive agrochemicals usage are associated with increasing environmental and economic concerns, including air and water pollution, decrease in biodiversity, greenhouse gas emissions, soil degradation and food insecurity. Cover cropping is an emerging agricultural practice playing a major role in enhancing sustainable agriculture and supporting ecosystem services (Groff, 2015; Schipanski et al., 2014). Cover crops are planted between two main crops and their residue is incorporated into the soils before the main crop is planted. They have multiple uses in agricultural systems, such as mulch, green manure, catch crop, etc. When included in crop rotation, a cover crop improves soil health via the addition of SOM and essential nutrients, improves biodiversity and efficiency, affects stoichiometrically-linked N, P, K and S dynamics, reduces weed and soil-borne diseases, supports soil balancing and consequently improves the system productivity (Dabney et al., 2001; Brust et al., 2014; Cordeau et al., 2015). If they are precisely grown, cover crops protect the soil from erosion and prevent the loss of nutrients through leaching and surface runoff (Kuo and Sainju, 1998; Kaye and Quemada, 2017). However, cover crop impacts on crop performance, soil quality and nutrient dynamics depend on factors, such as type, use and management. Evidence of 65 years of research in the United States and Canada showed that winter grasses as cover crops neither increase nor decrease corn yields, although corn grown for grain yielded relatively higher than silage corn after grass cover crops. Legume cover crops resulted in subsequently higher (30–33 per cent) corn yields when N fertiliser rates were low, or the tillage system shifted from conventional tillage (CT) to no-tillage (NT). A mixture or blend of cover crops increased corn yields by 30 per cent when the cover crop was terminated later (Marcillo and Miguez, 2017).

Though N is the most important essential nutrient for plant growth in terrestrial ecosystems, the availability of N is one of the major limiting factors in primary productivity (Cole et al., 2008). There has been a dramatic increase in the use of chemical N fertilisers over the last 50 years to meet the agricultural needs of a global population (Conant et al., 2013). The common scenario is that increased application of N fertiliser gives higher crop yield, but crop plants can use only 30–50 per cent of applied N efficiently (Prasad and Datta, 1979). Between the years 2000 and 2050, global N withdrawal by crops is projected to increase by 60 per cent, total N inputs in arable land by 40 per cent and the surplus by 28 per cent. This indicates that the increase in nutrient recovery cannot balance the impressive increase in nutrient demand to achieve increased crop production (Bouxman et al., 2013).

Dept. of Soil Science, Bangladesh Agricultural University, Mymensingh, Bangladesh.
* Corresponding author: mmrjahangir@bau.edu.bd

Using cover crops has the positive impact of enhancing the availability and uptake of N by plants. In the case of nutrient cycling, cover crops enhance soil N dynamics and microbiological function of N mineralisation, nitrification and denitrification as measured by assays of potential microbial activity (Steenwerth and Belina, 2008a). Cover crops impact the denitrification process, especially greenhouse gas emissions (GHG) significantly, as the intermediate product of denitrification, N_2O, is a potent GHG gas. Further research is needed to understand the reliability of N release from cover crop sources compared to fertiliser N, as N release from cover crop residues is possibly controlled by parameters that vary regionally, seasonally and with management strategies (Kaye and Quemada, 2017). Therefore, understanding of N cycle processes under varying management systems with cover crops will help illustrate the impact of cover crops on soil N pools and dynamics, N budget, N-use efficiency and surplus.

2. Cover Crop Impacts on Soil Nitrogen Dynamics

One of the prime impacts of cover crops is to provide the SOM, which enhances soil biological activity. Long-term cover cropping can increase SOM that can stimulate soil microbial communities, nutrient dynamics and availability (Schmidt et al., 2018). Conversely, Van der Linden et al. (1987) observed a minor change in SOM content following 20 years of green manure application as a cover crop. Similarly, a 15-year NT and cover crop system increased the SOM content from 1.82–1.83 per cent at 0–30 cm depth, and the system with cover crops increased the biomass index by 15 per cent compared to the system without any cover crops (Sapkota et al., 2012). A study reported that cover cropping has positive effects on plant production, leading to an increase in crop yields up to 24 per cent in a reduced tillage organic system and only 2 per cent in a conventional system with tillage (Wiltwer et al., 2017). Cover crops can modify soil properties by increasing SOM content, with a legume-based mixture and thus increase the nutrient availability for succeeding crops (Hubbard et al., 2013). However, understanding how cover crop management impacts soil biochemical properties and how these interactions might affect soil nutrient cycling, especially N, is still limited (Romdhane et al., 2019).

Legume cover crops fix atmospheric N_2 whereas non-legume cover crops scavenge and recycle soil N for their growth. Non-legumes also improve N-use efficiency by reducing NO_3^- loss from the soil-plant ecosystem. Generally, cover-crop-residue decomposition depends on the C:N; for example, C:N is 68:1 for rye and 10:1 for hairy vetch. The detrimental impacts of NO_3^- loss from the soil have toxicological implications for animals and humans (Camarguo and Alonso, 2006). Recently, the use of cover crops with minimum tillage has become popular in conservation agricultural systems as an effective way of reducing N loss through N cycling (Jahangir et al., 2014; Premrov et al., 2014). Cover crop impacts on soil C and N dynamics are linked to multiple factors, including cover type and management. Ramdhane et al. (2019) reported higher total N and C content when cover crops were killed by frost, compared to rolling and glyphosate termination (Fig. 1), while cover crop biomass was positively correlated to soil carbon and C:N.

Several past researches reported that cover crops significantly affected nutrient uptake patterns. Fageria (2014) reported that macronutrient uptake by plants was in the order of nitrogen > calcium > potassium > magnesium, while phosphorus (P) and micronutrient uptake pattern was in the order of iron > manganese > zinc > copper. In the case of micronutrients (Zn, Cu, Fe and Mn), the nutrient-use efficiency of plants varies among cover crops, as well as nutrients. Studies conducted in tropical and subtropical systems showed significant effects of cover crops on soil and plant P contents (Samson et al., 1990).

Cover crops are also used to improve the P-use efficiency of added organic or mineral fertilisers by increasing soil biological activity or uptake and protection of soluble mineral P in strongly P-fixing soils (Kamh et al., 1998; Kuo et al., 2005). It enhances the soil microbial community by providing a legacy of increased mycorrhizal abundance, microbial biomass P and phosphatase activity (Hallama et al., 2018). But a long-term soybean-ruzigrass (*Brachiaria ruziziensis*) rotation

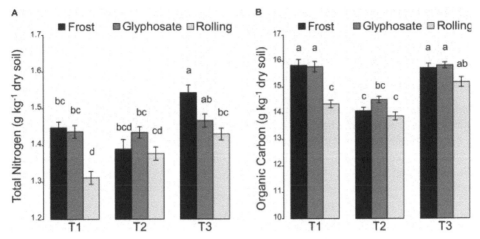

Fig. 1 Effects of the interaction between sampling time (T1, T2 and T3) and cover crop termination by frost, glyphosate, or rolling on soil (A) total nitrogen and (B) organic carbon. All values are means ± standard errors (Romdhane et al., 2019).

resulted in lower soil P availability due to decreased P mobility, regardless of P rates, by reducing P diffusion and resupply from the soil solid phase, which may have a significant impact on crop production (Almeida, 2019).

Nitrogen is an essential element for plant growth and an adequate amount of N supply improves crop yield. The surface depth of mineral soil naturally contains 0.02–0.5 per cent of nitrogen. The SOM is considered the largest N pool in the rhizosphere and most of the soil N is tied to SOM and some of N bound with silicate clays, which resist the rapid microbial breakdown. Mineralisation is the only process for most of the plants to use this N, except legumes. Point to be noted here is that atmospheric N_2 fixation by nodulation is a special characteristic of leguminous plants and use of this type of plant in crop rotation is one of the processes of N input in farming systems. Mineralisation is a slow process since the decomposition of organic compounds is complex and time consuming. Only 2–3 per cent of organic nitrogen is mineralised under natural conditions, otherwise the plant-available N immobilises. Plants may acquire N from the soil as NH_4^+ (Chaillou and Lamaze, 1997), NO_3^-, NO_2^- (Darwinkel, 1975), or in simple organic forms (Lipson and Monson, 1998). The quantity of plant-available forms of N, as nitrate and ammonium, is not more than 1–2 per cent of the total N in soil. There are many pathways N can follow in becoming lost from the system—it can be fixed by clay temporally, leached out by drainage, or lost by denitrification and volatilisation. These are relatively quicker processes than mineralisation of the plant-available form of N. The interaction of various forms of N in soil, plant, animal and the atmosphere constitute a total N dynamic. The N supplied to the soil faces many transformations before it is removed from the system. This cycle is affected by the plant selection, nature of applied organic sources, rate of microbial activity, physical and chemical properties of soil, favourable climatic conditions and the adapted agricultural system of any region. Changes in soil properties, due to cover-crop management rather than the composition of cover-crop mixtures, were related to changes in the abundance of ammonia oxidisers and denitrifiers, while there was no effect on the total bacterial abundance (Ramdhane et al., 2019).

Agricultural production is one of the primary human activities to alter N dynamics in terrestrial ecosystems, generally by increasing the N mobility remarkably. Considering uptake, ammonification, runoff, leaching and nitrification-denitrification, its surpluses the N pool by 3.08×10^4 tons (Zhao et al., 2010). Farmers are adapting various modern techniques to increase N inputs that can satisfy plant demand, especially the use of excess N fertilisers. By adapting alternative farming systems, the overall N budget of an agricultural field is already being researched for reducing chemical fertilisation. Unravelling the underlying processes by which cover-crop management shapes soil properties and biology is of utmost importance in helping to select optimised agricultural practices to sustain global food security (Ramdhane et al., 2019).

3. N Cycle Processes in Soils under Cover Crop

3.1 Biological N Fixation

Biological N_2 fixation is a biochemical process through which the elemental N_2 combines into organic form. This process is mediated by a number of microorganisms, including several species of bacteria, a few actinomycetes and blue-green algae. By far the largest amount of terrestrial N fixation is carried out biologically (globally 90–130 Tg N/yr) (Galloway, 1995). The reaction involved in the fixation is represented by:

$$N_2 + 6H + e^- \xrightarrow[Fe, Mo]{nitrogenous} 2\,NH_3$$

Ammonia combines with organic acid to form amino acid and then proteins.

$$NH_3 + \text{organic acid} \longrightarrow \text{amino acid} \longrightarrow \text{protein}$$

The reduction of N is carried out by nitrogenase enzyme.

The symbiosis association of *Rhizobium* and legume cover crop in one sense preserves atmospheric N_2 for later use by the next crop in the rotation. *Rhizobium* invades the root heir and cortical cell and forms a nodule. The host plant provides carbohydrates to the bacteria for energy and in return, the bacteria stimulate the plant to fix the N from the atmosphere. Throughout the entire growing period, the legumes conserve N and use it for cellular structure. Some N passes through the root and nodule into the soil and some of it mineralises into NH_4^+ and NO_3^- so that any plant can later use it. In an alternative way to increase the organic matter content and improve soil aggregate structure, the legume cover crops are incorporated with the soil. Their C:N is not so high and during the growing period, the main plant can utilise the preserved N through its quick decomposition and mineralisation.

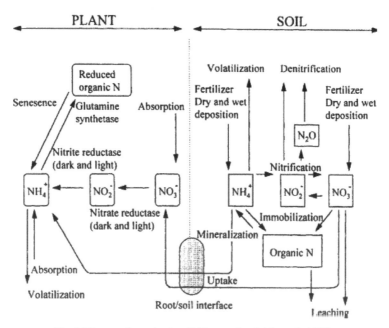

Fig. 2 Nitrogen dynamics in soil (*Source*: Stockdale et al., 1997).

Table 1 Typical levels of N fixation from legume cover crops (*Source*: Brady, 1996).

Legume Crops	Associated Organisms	Typical Levels of N Fixation (kg N/ha yr^{-1})
Alfalfa (*Medicago sativa*)	*Rhizobium*	150–200
Clover (*Trifolium pretense* L.)		100–150
Soybean (*Glycine max* L.)		50–150
Cowpea (*Vigna unquiculata*)		50–100
Lupin (*Lupinus*)		50–100
Vetch (*Vicia vilbosa*)		50–125
Bean (*Phaseolus vulgaris*)		30–50

3.2 Mineralisation

The conversion of an organic form of N into an inorganic form is called mineralisation. The whole process involves two reactions: (1) ammonification and (2) nitrification. The total reaction is as follows:

$$RNH_2 \longleftrightarrow ROH + NH_4^+ \longleftrightarrow NO_2^- + 4H+ \longleftrightarrow NO_3^-$$

3.2.1 Ammonification

Ammonification is the biological conversion of organic N to ammonia. The N in soils is present mostly within organic compounds that make it insoluble and unavailable for use by higher plants. Much of the N is present as amine groups ($R-NH_2$) in proteins, or as part of humic compounds. Soil microorganisms, particularly exoenzymes, deaminate these to simpler amino-acids (monomers), then to NH_4^+. Ramdhane et al. (2019) suggested that changes in soil C and N dynamics are a function of changes in the abundance of ammonia oxidisers and denitrifiers, while there was no effect on the total bacterial abundance. Ammonium concentration in soil increased from early winter to mid-spring and then decreased. Another study shows, in summer, increases in NH_4^+–N corresponded to decreases in microbial biomass N (MBN) and increases in dissolved organic N and C (Steenwerth and Belina, 2008b), suggesting that MBN turnover and cover crop-derived organic matter contributed to these C and N pools. The ammonification reaction is represented by:

$$R\text{-}NH_2 + H_2O + H^+ \rightarrow R\text{-}OH + NH_4^+$$
$$R\text{-}NH_2 + H_2 \rightarrow NH_4^+ + \text{Energy}$$

Nitrogen concentration and C:N increased over time through cover-crop cultivation. Most of the symbiotic N is used for legume growth (Peoples et al., 1995) and is therefore accumulated in organic matter. Some of this N can be used later as animal feed in the form of protein in herbage (Kramberger et al., 2007), while the rest of the accumulated N can be taken up by subsequent crops after ploughing-in and mineralisation of the organic matter (Voughan et al., 2000; Andraski and Bundy, 2005). After mineralisation of ploughed-in cover crops, the accumulated N is available for subsequent crops. When a legume cover crop is incorporated into the soil, a substantial amount of N is usually mineralised within a few weeks. Nitrogen continues to mineralise in the ensuing weeks as the organic matter decomposes (Snapp et al., 2005).

3.2.2 Nitrification

Nitrification is a chemolitho-autotrophic oxidation of ammonia (NH_3) to NO_2^- to NO_3^-. Nitrification can also produce N_2O. This process is carried out by a small group of microbes (e.g., *Nitrosomonas, Nitrospira, Nitrosococcus*) that convert NH_3 to NO_2^- and *Nitrobacter* that convert NO_2^- to NO_3^-. These bacterial genera use ammonia or NO_2^- as an energy source and molecular oxygen as an electron acceptor, while carbon dioxide is used as a carbon source. Potential N mineralisation is an indicator of potential soil N availability. The greenhouse gas N_2O is biologically produced in soil via nitrification and denitrification. This process occurs strictly under aerobic conditions (Lee et al., 2009), but soil N availability, temperature, SOM, temperature, pH, alkalinity, inorganic C source, moisture, microbial abundance, NH_4^+, and DO control the entire microbial process. Cover crop-cultivated soil supports greater potential nitrification, denitrification and mineralisation than cultivated soil. It indicates that microbial activity and enzymatic activity is higher in cover crop-cultivated soil.

The nitrification process is enhanced by temperature (Cookson et al., 2002) and is curtailed below pH 6 (Paul and Clark, 1989) and inhibited at greater than pH 8 (Whitehead, 1995). In general, it is assumed that, N mineralisation is low under low soil temperatures (i.e., if soil temperature is under 5°C) (Thorup-Kristensen et al., 2003). High soil temperatures and rainfall or irrigation in the spring provide favourable conditions for rapid decomposition of cover-crop residue, particularly legumes with narrow C:N's in their biomass. Both N mineralisation and nitrification have been found to occur rapidly following woolly pod vetch incorporation, as shown by a brief peak in soil ammonium levels within 10–15 days after incorporation, followed by a rapid increase in soil nitrate (Stivers and Shennan, 1991).

Cover crop can improve soil quality by increasing SOM levels through the input of crop biomass over time. The incorporation of cover-crop biomass into soil also provides additional OM inputs into the system, which can, in turn, lead to improved soil OM content, physical properties and water infiltration characteristics (Bolton et al., 1985; Macrae and Mehuys, 1985; Patrick et al., 1957; Smith et al., 1987; Tisdall and Oades, 1982; Williams, 1966).

In another study, the incorporated plant residues tended to reduce soil pH. Changes in soil pH due to plant residue incorporation depend on the quality of residues, rate of application of the residues and the initial soil pH (Wong et al., 1998; Paul et al., 2001; Xu and Coventry, 2003). When cover crop is incorporated with soil, the decomposition process is started and produces organic acid. In the nitrification process, for each NH_4^+ molecule, $2H^+$ are produced and thus the pH drops. Therefore, liming can be helpful for neutralising the acidity developed in the nitrification process (Ahn, 2006). The whole reaction can be shown as follows:

$$NH_4^+ + 1.5\ O_2 \rightarrow NO_2^- + H_2O + 2H^+$$
$$NO_2^- + 0.5\ O_2 \rightarrow NO_3^-$$
$$2NH_4^+ + 3O_2 \rightarrow 2NO_2^- + 2H_2O + 4H^+ + \text{Energy};$$
$$2NO_2^- + O_2 \rightarrow 2NO_3^- + \text{Energy}$$

3.3 Immobilisation

This process opposes mineralisation and describes the conversion of Nr (NH_4^+ and NO_3^- formed by mineralisation and nitrification) into organic forms and, ultimately, biomass. The extent of immobilisation is determined by the amount and quality of C and N in soils and sediments, but in groundwater, dissolved organic C and N is generally low. Mineralisation, nitrification and immobilisation occur simultaneously, resulting in the transformations of N from organic to inorganic forms and vice versa. Therefore, a net mineralisation term is generally used to determine the amount of NO_3^- and NH_4^+ readily available to be fed from the soil to groundwater and then groundwater to receptors.

$$NO_3^- + 2e^- \rightarrow NO_2^- + 6e^- \rightarrow NH_4^+$$
$$NH_4^+ + R\text{-}OH \rightarrow R\text{-}NH_2 + H_2O + H^+$$

In annual ecosystems, cover crops have been used to augment the SOM content, thereby offsetting tillage-induced reductions in SOM and increasing soil N retention between crop rotations (Jackson et al., 2004). In a conservational agricultural system of adapting grass-based cover crops, relatively little amounts of N release for the cash crop during subsequent growing seasons and N immobilisation in some cases (Snapp et al., 2005; Thorup-Kristensen et al., 2003).

Soil with greater SOM content demonstrates a higher potential to immobilise and retain N (Barrett and Burke, 2000), resulting in greater potential N availability. They can support higher microbial biomass, thus serving to reduce N loss through immobilisation (Jackson, 2000), but higher soil labile C has also been associated with an increased capacity for denitrification (Drury et al., 1991). The dynamics of net N mineralisation or N immobilisation depend on the C:N, cellulose and lignin concentrations and neutral detergent fibre concentration in organic matter (Jensen et al., 2005). The C:N of the cover crop residues determines the general timing of release of N from the residues. With high C:N ratios (greater than 25:1), the decomposing cover crop residues will first immobilise N from the soil or recent fertiliser N additions. Only after some period of time will the N start to be released or mineralised. If the cover crop gets into the reproductive phase before termination, there will be more N immobilisation after termination and a good application of starter N fertiliser should be used.

Cover crop management, likewise, influences the potential for N accumulation and release. Any management activity that influences cover-crop biomass accumulation and/or C:N ratio may affect subsequent N release. The application of fall fertiliser or organic amendments to a grass or broadleaf cover crop may increase N accumulation and potential release. Cover crop planting can affect N release because earlier planting in the fall generally maximises biomass and thus, nitrogen accumulation. Spring cover crop termination timing can, likewise, affect cover crop nitrogen release. Legumes that are terminated early have less time to fix N and so have less N to potentially provide to a following cash crop. Cover crop species with high C:N at maturity may be terminated early (i.e., at tillering, rather than anthesis) to decrease the risk of nitrogen immobilisation in the soil. Finally, climate influences N releases because of its effect, both on cover crop growth and N accumulation and because it affects the microbes that break down cover crop biomass to release nitrogen. Most microbes thrive in warm, damp, but not saturated soils. Under such conditions, N release will occur more quickly than under cold, dry conditions.

3.4 Denitrification

Denitrification is a microbially-mediated, multi-step process that converts NO_3^- to di-nitrogen (N_2) in low-oxygen environments. The intermediate products of denitrification are NO_2^-, nitric oxide (NO), and N_2O. Nitrous oxide is an obligate intermediate product of denitrification that is a potent GHG, but the stable end product of denitrification is environmentally benign. In denitrification, NO_3^- is used as a terminal electron acceptor to produce N_2 or N_2O (Starr and Gillham, 1993). This process is carried out by heterotrophs (*Pseudomonas, Micrococcus, Achromobactor* and *Bacillus*) and autotrophs. Heterotrophs are microbes that need organic substrates to obtain their C source for growth and evolution and obtain energy from organic matter. In contrast, autotrophs utilise inorganic substances as an energy source and CO_2 as a C source (Rijn et al., 2006). Denitrification occurs under anoxic conditions (because DO can suppress the enzyme systems required for this process) where electron donors (organic C) are available. Therefore, in soil microsites a rapid gradient of DO can be established, which often allow both nitrification and denitrification to occur in sequence (Lee et al., 2009).

Denitrification rate is influenced by NO_3 concentration, microbial flora, type and quality of organic C source, hydroperiods, plant residues, DO, redox potential, soil type and moisture, temperature, pH, presence of denitrifiers and the depth of overlying water (Golterman, 2004; Sirivedhin and Gray, 2006). Denitrification process is represented as below:

$$2NO_3^- \rightarrow 2NO_2^- \rightarrow 2NO \rightarrow N_2O \rightarrow N_2$$

Each of the steps is conducted with the help of a reductase enzyme and the reaction can stop in any of the stages. The gaseous product of that specific stage can release to the atmosphere. Biological denitrification can best be described by the following reaction:

$$4NO_3^- + 5CH_2O + 4H+ \rightarrow 2N_2 + 5CO_2 + 7H_2O$$

The NO_3^- reduction reaction can be written as a half-equation that illustrates the role of electron (e^-) transfer in the process and is non-specific to the electron donor (Tesoriero et al., 2000):

$$2NO_3^- + 12H^+ + 10e^- \rightarrow N_2 + 6H_2O$$
$$NO_3^- + 1.25\,(CH_2O) \rightarrow 0.5\,N_2 + 0.75\,H_2O + 1.25\,CO_2 + OH$$
$$NO_3^- + (CH_2O) \rightarrow 0.5\,N_2O + 0.5\,H_2O + CO_2 + OH$$

Adaptation of cover crops in a cropping pattern has the potential to enhance the denitrification rate. Firstly, winter cover crops conserve water by reducing the evaporation rate with the help of a canopy and this has the potential to recharge groundwater. Secondly, cover crops are one of the main sources of soil organic carbon, which increases the microbial activity of denitrifying bacteria. It has been found that certain types of cover crops affect the decomposition rate of organic matter (highest in saturated soil and depends on C:N of the crop residue) and mineralisation of N, hence the denitrification capacity of soil.

Ramdhane et al. (2019) showed significant variations in denitrifiers' functional genes under varying cover crop management systems (Fig. 3). Some studies show the denitrification rate is higher in the upper soil horizon (Clement et al., 2002; Cosandey et al., 2003; Kustermann et al., 2010) depending on the extent of soil moisture content (Khalil and Baggs, 2005). From

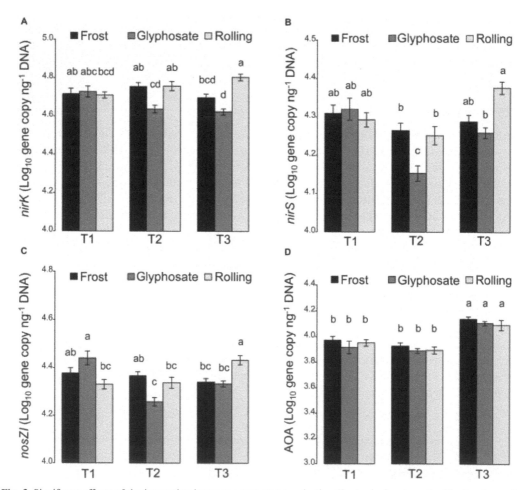

Fig. 3 Significant effects of the interaction between cover crop termination (frost, glyphosate and rolling) and sampling time (T1, T2 and T3) on the abundance of N-cycling communities. (A) *nirK* denitrifiers, (B) *nirS* denitrifiers, (C) *nosZI* denitrifiers and (D) ammonia oxidising *archaea* (AOA) (Log_{10} gene copy ng⁻¹ DNA). All values are means ± standard errors (*Source*: Romdhane et al., 2019).

the findings of Khalil and Richard (2010), it is clearly understood that subsoil horizon has the small denitrification capacity of a grazed pasture; but in the presence of cover crop, the potential of denitrification in subsoil significantly increases, especially for rye grass compared to clover grass. Under field conditions, nitrous oxide and elemental N_2 are lost in the highest quantity, but nitrous oxide emission is dominant. Non-legume cover crops, such as winter cereal, contribute to the reduction of N_2O emissions due to their deep roots, which allow them to extract soil nitrate more efficiently than legumes (Kallenbach et al., 2010). Legume systems are estimated to emit 0.4 Mt N_2O-N annually, around 10 per cent of total agricultural N_2O emissions (Stehfest and Bouwman, 2006). Some examples of total N_2O emissions from field-grown legumes, N fertilised grass pastures and crops, or un-fertilised soils in North and South America, Europe, South Asia, East Asia, Australia and New Zealand are presented in Table 2 (Jensen et al., 2012).

Winter season is particularly prone to N_2O emissions in temperate regions, though soil microbial activity slows during the non-growing season (Wagner-Riddle and Thurtell, 1998; Dörsch et al., 2004; Ellert and Janzen, 2008; Hao, 2015). Cover crops release labile C and N through root exudates and rhizo deposition during their growth phase and freeze-thaw cycles, which can stimulate microbial activity and increase N_2O emissions (Petersen et al., 2011; Gul and Whalen, 2013; Mitchell et al., 2013). In the non-growing season, thaw events are associated with the greatest N_2O fluxes and denitrification is considered the dominant N_2O production pathway (Wagner-Riddle et al., 2008; Risk et al., 2014). Denitrification may be enhanced in the non-growing season as freezing and thawing alter water availability, which directly controls O_2 diffusion and stimulates microbial activity by increasing substrate solubility (Skogland et al., 1988). For GHG emission, specifically N_2O emission, there are knowledge gaps about the effect of plant species selection and cover crop management (retention, incorporation, and removal) (Basche et al., 2014).

Table 2 Various cover crop management impacts on N_2O emissions (*Source*: Jensen et al., 2012).

Category and Species	Number of Site-years	Total N_2O Emission per Growing Season of Year (kg N_2O-N ha^{-1})	
Pure legume stands	14	Range	Mean
Alfalfa	3	0.67–4.57	1.99
White clover		0.50–0.90	0.79
Mixed pasture sward			
Grass-clover	8	0.10–1.30	0.54
Legume crops			
Faba bean	1		0.41
Lupin	1		0.05
Chickpea	5	0.03–0.16	0.06
Field pea	6	0.38–1.73	0.65
Soybean	33	0.29–7.09	1.58
Mean of all legumes	37		1.29
N-fertilised pasture			
Grass	19	0.03–18.16	4.49
N-fertilised crops			
Wheat	18	0.09–8.57	2.73
Maize	22	0.16–12.67	2.72
Canola	8	0.13–8.60	2.65
Mean of fertilised systems	67		3.22
Soil			
No N fertilizer of legume	33	0.03–4.80	1.20

Agricultural soils account for 60 per cent of N_2O emissions globally and 56 per cent in the United States (USEPA, 2018). The N_2O is a potent greenhouse gas with a long lifespan in the atmosphere; the 100-year warming potential of a gram of N_2O is 298 times greater than a gram of CO_2 when carbon-climate feedbacks are accounted for (CO_2e from N_2O emissions = 298 × N_2O emission rate; (IPCC, 2013). Agricultural fields are the main source for N_2O emissions. Basche et al. (2014) conducted a meta-analysis about cover crop effects on N_2O; the key finding was that over an entire growing season, there was no identical difference of N_2O fluxes between cover crop plots and no-cover crop controls. Field studies in a Spanish case illustrate that legume cover crops had higher N_2O emissions than non-legumes when the cover crops were growing. In contrast, after the cover crops were killed and maize had been planted, plots with a non-legume cover crop history tended to have higher N_2O emissions than plots with a legume cover crop history (Guardia et al., 2016). The N_2O mitigation potential for winter cover crops ranges from 0.2 to 1.1 kg N_2O ha^{-1} yr^{-1} (Ussiri and Lal, 2013). Since cover crops are the source of organic matter biomass, the decomposition of organic matter stimulates the microbial activity, which results in CO_2 emissions in the soil. Increased CO_2 emissions under cover crops may be particularly relevant during the cover crop growing period, and not after cover crop termination (Sanz-Cobena et al., 2014; Guardia et al., 2016). Cover crop effects on GHG fluxes typically mitigate warming by ~ 100 to 150 g CO_2e/m^2/yr (Kaye and Quemada, 2017). According to Guardia et al. (2016), there is no significant effect of cover crop systems on CH_4 emissions.

Shallow groundwater is a potent source of N_2O emissions. Denitrification occurs under a local anaerobic condition here within the microsites in particulate organic matter. According to an experiment by Jahangir et al. (2014), a lower concentration of NO_3^--N was detected in the groundwater below a spring barley-mustard cover crop than in a no-crop system, which tends the higher denitrification in presence of a mustard cover crop. Figure 4 shows the total denitrification in two different cropping systems (spring barley-mustard cover crop and no-cover crop system). Further chemical reduction of N_2O to N_2 under anaerobic conditions may be possible for upward movement through the soil profile with a presence of an adequate amount of labile organic C (Elmi et al., 2003; Castle et al., 1998). This process can protect the shallow groundwater from NO_3^- contamination.

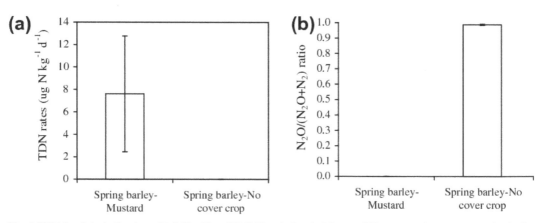

Fig. 4 TDN (total denitrification = N_2O-N + N_2 and N_2O-N mole fraction) in two different cropping systems: spring barley with a mustard cover crop and a no-cover crop system (*Source*: Jahangir et al., 2014).

3.5 Nitrogen Leaching

In agriculture, leaching or nutrient leaching is the downward movement of dissolved nutrients in the soil profile with percolating water due to rain and irrigation. Nutrient leaching refers to nitrate leaching. Water added in excess of the soil's water-holding capacity will carry nitrate and other salts downward. Nutrients that are leached below the rooting zone of the vegetation are at least temporarily lost from the system, although they may be recycled if roots grow deeper. Controlling nitrate leaching can be a challenge for farmers because it requires simultaneous management of two essentials of plant growth; N and soil moisture (such as rainfall, irrigation, evaporation and transpiration). In general, more water infiltration results in nitrate moving deeper into the profile.

Excess N after cash crop uptake is prone to leaching and can increase NO_3^- concentrations in ground- and surface-water bodies (Quemada et al., 2013; Zhao et al., 2016; Russo et al., 2017; Thapa and Chatterjee, 2017). Increased NO_3^- levels in aquatic ecosystems may result in eutrophication (i.e., algal blooms) which degrades aquatic habitats and harms aquatic species (Carpenter et al., 1998; McIsaac et al., 2001). In another study, NO_3^- leaching indirectly contributed to greenhouse gas emissions through nitrous oxide emissions resulting from incomplete denitrification in the subsurface (Jahangir et al., 2012).

Baker (2001) reported that fine-tuning farm management practices such as crop rotation, no-tillage and N management for greater N use efficiency (i.e., application of the '4Rs') can collectively reduce NO_3^- leaching by 25–30 per cent. However, adoption of these practices alone does not reduce NO_3^- leaching to acceptable levels because most NO_3^- leaching occurs during the fallow (late fall, winter and early spring) period, when there is no crop present to take up surplus N after cash crop harvest (Dinnes et al., 2002). Winter annual cover crops have been recognised as an effective means to capture surplus N and reduce NO_3^- leaching (Meisinger et al., 1990; Dinnes et al., 2002; Tonitto et al., 2006; Quemada et al., 2013; Tully and Ryals, 2017).

The use of cover crops for reducing NO_3^- leaching has been proven. Winter cover crops, especially non-legumes, such as grasses and broadleaf species, can reduce NO_3^- leaching by 3570 per cent depending on intrinsic (soil and climate) and extrinsic factors (management) (Tonitto et al., 2006; Quemada et al., 2013). Besides reduction of NO_3^- leaching, cover crops also influence soil N, soil water and weed dynamics, thereby affecting subsequent cash crop yields (Marcillo and Miguez, 2017). During the rainy season, cultivation of cover cropping reduces leaching losses, increases water infiltration rate and recycling of nutrients to the subsequent cash crops following residue incorporation.

Deep-rooted cover crops reduce the amount of nitrate N that can be lost to leaching over the fall, winter and spring. For retaining N over winter, seeding a grass or cereal, such as cereal rye, can be very effective. Plants with deep taproots, such as radish, can also capture N from depth though in some cases, the timing of nutrient release may occur early in the spring before uptake by a crop such as corn. The most effective cover crops for reducing nitrate loss are those that gain a reasonable amount of biomass in the fall and/or spring and have both dense and deep root systems.

Following a period of bare fallow with a cover crop also has the potential to reduce leaching losses of mobile nutrients, such as nitrate, and thus reduce groundwater contamination (Muller et al., 1987). This is important when there is sufficient rainfall to cause percolation of water down the soil profile and when the root zone either contains significant amounts of residual NO_3^- or higher temperatures to support continued N mineralisation, especially nitrification, during the fallow period. Recently, emphasis has been placed on the uptake of soil mineral N, which could otherwise be leached as NO_3^- into deeper soil layers and groundwater (Shipley et al., 1992; Logsdon et al., 2002), lost during nitrification as NO and N_2O_3, or denitrified as N_2 (Jenkinson, 2001).

So, cover crop cultivation or crop rotation with cover cropping reduces the risk of nitrate leaching to fix N from the air, to conserve soil N and to transfer N to the following crops. Non-legume crops are very effective in scavenging nutrients (Clark, 2012), especially grasses, which usually provide rapid ground cover when sown in the autumn (Stobart, 2015). In addition, grasses are considered to be good at scavenging excess nutrients, especially N, left in the soil after harvest (Clark, 2012). Grasses also tend to root at a shallower depth compared to other cover crop types such as *Brassicas*, but the shallow rooting system on heavy land is beneficial for opening up the soil surface (Stobart, 2015). Typical grasses include rye, oats and sorghum-sudangrass. White et al. (2016) reported that *Brassicas*, such as mustard and radishes, when integrated in crop rotation, improved soil structure and reduced NO_3 leaching.

3.6 *Ammonia Volatilisation*

Ammonia (NH_3) formation in the soil from manure and crop residue decomposition, as well as from hydrolysis of N fertilisers, can be volatilised in substantial quantities under certain environmental conditions. The NH_3 volatilisation from agricultural sources have contributed significantly to air pollution, soil acidification water eutrophication, biodiversity loss and declining public and animal health. After the incorporation of organic manure and fertilisers into the topsoil, 25–75 per cent of N was reduced by NH_3 volatilisation. Several factors, especially high soil pH (> 9) and high soil temperature, enhance the NH_3 volatilisation rate. The clay and labile SOM contents are vital in influencing the NH_3 emission rates because both have the exchangeable sites for NH_4^+ adsorption. Volatilisation of NH_3 is also highest in dry conditions.

During cover crop growth, low soil temperatures and N availability likely contributed to low gaseous loss for all treatments (Parsons et al., 1991). Thus, NO_3 leaching is the primary N loss pathway during this time. Karlen (1990) mentions that while fertiliser-use efficiency may be lower initially after the termination of cover crop due to increased immobilisation under reduced tillage, soil and fertiliser N will be conserved as SOM is built up and fertiliser requirements may decrease over time. Precipitation most likely induced fertiliser mineralisation, increasing soil NH_3 concentrations and, as soil moisture declined, volatilisation occurred (Rochette et al., 2009). In addition, N loss pathways are also strongly influenced by the type of fertiliser used. Fertilisers that provide a C source often stimulate heterotrophic microbial activity and release N for nitrification (Mitchell et al., 2013) and those that hydrolyse rapidly prior to assimilating into the soil profile are prone to volatilization (Rochette et al., 2009).

Quemada and Cabrera (1997) found a maximum NH_3 loss of about 6 per cent of residue N within three weeks. These data suggest that N losses as NH_3 escaping from surface-placed cover crops is minimal during dry periods, while during wet periods, NH_3 is leached into and held in the soils or subsoils. Therefore, only small losses of cover crop N from NH_3 volatilisation are likely in a no-till system, but volatilisation loss of N due to a cover crop in an arable system under various soil and climatic conditions are still not clear.

3.7 *Anammox*

The Anammox is a biochemical process in which NH_4^+ is oxidised to elemental N_2 via a chain of reactions under anaerobic conditions. It was a recent discovery and a very important step in the N cycle. The process is carried out by anammox bacteria, which use nitrite as the electron acceptor. Anammox process is critically important in the marine N cycle and the relative contribution of the anammox process to the total production of di-nitrogen gas (N_2) has been estimated to be 50 per cent in the ocean (Etten, 2009). Besides the marine and freshwater ecosystem, anammox bacteria have also been detected in terrestrial ecosystems (Humbert, 2010). The anammox process was reported to account for 4–37 per cent of the nitrogen loss in agricultural soils (Hu et al., 2011).

The ecological distribution of anammox bacteria and their contribution to N loss in natural ecosystems are influenced by some local environmental conditions: the organic content, NOx-concentration, environmental stability, salinity and temperature have been described as major influencing factors. Only the cultivation practices of a paddy field provide favourable environmental conditions for the anammox bacteria (Zhu et al., 2010), which includes waterlogged, oxygen-limiting conditions, high temperature and coexistence of ammonium and nitrate ion. The significant loss of N due to anammox occurring in paddy fields is similar to that for NH_3 volatilisation (up to 40 per cent), leaching (9–15 per cent), runoff (5–7 per cent) and denitrification (up to 40 per cent) (De Datta et al., 1991; Xing and Zhu, 2000; Zhao et al., 2009). This observation has changed the concept of N cycle in paddy fields.

From slurry manure plus ammonia fertiliser applied to the soil of a 25-year paddy-wheat rotation, about 23 per cent of applied fertiliser is lost via this pathway (Zhao et al., 2009). Though there is no direct correlation of cover crop residue and anammox reaction, it can be guessed that the incorporation of cover crops may increase the anammox side-by-side denitrification at the same time, if the anammox bacteria is present in this system. This is because crop residue increases the organic matter content of topsoil and NH_4^+ and NO_3^- concentration in soil through mineralisation.

Nitrogen as NO_3^- and NH_4^+ are among the most ubiquitous groundwater contaminants due to widespread use in agriculture as fertilisers, as unintentional discharge in septage and effluent (Harter et al., 2002). The high abundance of anammox and associated nitrifying and denitrifying bacteria may remediate excess NO_3^- and NH_4^+ in groundwater (or mineralised) to N_2 by the endogenous microbiota. Carbon source from decomposed crop residue enhances the activity of nitrifying and denitrifying coupled with anammox bacteria. Grassy cover crops, especially cereal rye or annual ryegrass, directly uptake NO_3^- from a greater soil depth. For the limited NO_3^- anammox, bacteria may not be able to oxidise the NH_4^+ to N_2 at shallow depth.

4. Conclusion

The problem of N balance in an agricultural field is twofold: firstly, the maintenance of adequate N supply in soil and secondly, regulation of the soluble form of N to ensure it is readily available to meet crop demand. Using cover crops in crop rotation is the exact solution, whereas it provides an organic source for N and has a great impact in overall N dynamics in an agricultural system. Cover crops enrich the microbial abundance and stimulate microbial activity by providing a source of energy for the microbes. Through enhancing mineralisation and atmospheric N_2-fixation, cover crops are able to provide an adequate amount of N for plant growth. By trapping NO_3^- from the soil, grass cover crops reduce the mobility of N toward groundwater. Denitrification is one of major pathways that contribute to N loss and the emission of N_2O into the atmosphere. In fallow periods, grass cover crops reduce the denitrification rate in the field and sometimes this denitrification can be a way to prevent NO_3^- contamination in surface water and groundwater. It reduces volatilisation of NH_3. Anammox is a process to reduces N_2O emissions in groundwater as well as after the incorporation of cover crop residue. Therefore, cover cropping can be effective in reducing N mobility, which contributes to the N balance for higher crop yields, and protection of the environment from contamination.

References

Ahn, Y.H. (2006). Sustainable nitrogen elimination biotechnologies: A review. Process Biochem., 41: 1709–1721.

Almeida, D.S., Menezes-Blackburnb, D., Zhangb, H., Haygarthb, P.M. and Rosolema, C.A. (2019). Phosphorus availability and dynamics in soil affected by long-term ruzigrass cover crop. Geoderma, 337: 434–443.

Andraski, T.W. and Bundy, L.G. (2005). Cover crop effects on corn yield response to nitrogen on an irrigated sandy soil. Agron. J., 97: 1239–1244.

Barrett, J.E. and Burke, I.C. (2000). Potential nitrogen immobilisation in grassland soils across a soil organic matter gradient. Soil Biol. Biochem., 32: 1707–1716.

Basche, A., Miguez, F.E., Kaspar, T. and Castellano, M.J. (2014). Do cover crops increase or decrease nitrous oxide emissions? A meta-analysis. J. Soil Water Conserv., 69: 471–482.

Baumhardt, R.L. and Lascano, R.J. (1996). Rain infiltration as affected by wheat residue amount and distribution in ridged tillage. Soil Sci. Soc. Amer. J., 60: 1908–1913.

Berhe, A.A., Harte, J., Harden, J.W. and Torn, M.S. (2007). The significance of erosion-induced terrestrial carbon sink. Bioscience, 57: 337–346.

Bolton, H., Elliott, L.F., Papendick, R.I. and Bezdicek, D.F. (1985). Soil microbial biomass and selected soil enzyme activities: Effect of fertilisation and cropping practices. Soil Biol. Biochem., 17: 297–302.

Bouwman, L., Goldewijk, K.K., Van Der Hoek, K.W., Beusen, A.H.W., Van Vuuren, D.P., Willems, J., Rufino, M.C. and Stehfest, E. (2013). Exploring global changes in nitrogen and phosphorus cycles in agriculture induced by livestock production over the 1900–2050 period. Proc. NCfat. Acad. Sci., 110: 21195.

Brady, N.C. (1996). The Nature and Properties of Soil, tenth ed., Prentice Hall International Inc. Englewood Cliffs.

Bruce, R.R., Langdale, G.W., West, L.T. and Miller, W.P. (1992). Soil surface modification by biomass inputs affecting rainfall infiltration. Soil Sci. Soc. Amer. J., 56: 1614–1620.

Brust, J., Claupein, W. and Gerhards, R. (2014). Growth and weed suppression ability of common and new cover crops in Germany. Crop Prot., 63: 1–8.

Camarguo, J.A. and Alonso, A. (2006). Ecological and toxicological effects of inorganic nitrogen pollution in aquatic ecosystems: A global assessment. Environ. Intl., 32: 831–849.

Carpenter, S.R., Caraco, N.F., Correll, D.L., Howarth, R.W., Sharpley, A.N. and Smith, V.H. (1998). Non-point pollution of surface waters with phosphorus and nitrogen. Ecol. Appl., 8: 559–568.

Castle, K., Arah, J.R.M. and Vinten, A.J.A. (1998). Denitrification in biology and fertility of soil, intact subsoil cores. Biol. Fert. Soils, 28: 12–18.

Chaillou, S. and Lamaze, T. (1997). Ammoniacal nutrition of plants. pp. 53–69. *In*: Morot-Gaudry, I.F. (ed.). Nitrogen Assimilation by Plants: Physiological, Biochemical and Molecular Aspects, Science Publishers, Enfield, NH.

Clark, A. (2012). Managing Cover Crops Profitably, third ed., Clark, A. (ed.). College Park, MD: Sustainable Agriculture Research and Education.

Clement, J.C., Pinay, G. and Marmonier, P. (2002). Seasonal dynamics of denitrification along topohydro sequences in three different riparian wetland. J. Environ. Qual., 31: 1025–1037.

Cole, L., Buckland, S.M. and Bardgett, R.D. (2008). Influence of disturbance and nitrogen addition on plant and soil animal diversity in grassland. Soil Biol. Biochem., 40: 505–514.

Conant, R., Berdanier, A. and Grace, P. (2013). Patterns and trends in nitrogen use and nitrogen recovery efficiency in world agriculture. Global Biogeochem. Cycles, 27: 558–566.

Cookson, W.R., Cornforth, I.S. and Rowarth, J.S. (2002). Winter soil temperature (2–1°C) effects on nitrogen transformations in clover green manure amended or unamended soils: a laboratory and field study. Soil Biol. Biochem., 34: 1401–1415.

Cordeau, S., Guillemin, J.P., Reibel, C. and Chauvel, B. (2015). Weed species differ in their ability to emerge in no-till systems that include cover crops. Ann. Appl. Biol., 166: 444–455.

Cosandey, A.C., Maitre, V. and Guenat, C. (2003). Temporal denitrification patterns in different horizons of two riparian soils. Eur. J. Soil Sci., 54: 25–37.

Dabney, S.M., Delgado, J.A. and Reeves, D.W. (2001). Using winter cover crops to improve soil and water quality. Comm. Soil Sci. Plant Anal., 32: 1221–1250.

Darwinkel, A. (1975). Aspects of Assimilation and Accumulation of Nitrate in Some cultivated Plants, Centre for Agricultural Publishing and Documentation, Wageningen.

De Datta, S.K., Buresh, R.J., Samaon, M.I., Obcemea, W.N. and Real, J.G. (1991). Direct measurement of ammonia and denitrification fluxes from urea applied to rice. Soil Sci. Soc. Amer. J., 55: 543–548.

Dinnes, D.L., Karlen, D.L., Jaynes, D.B., Kaspar, T.C., Hatfield, J.L., Colvin, T.S. and Cambardella, C.A. (2002). Nitrogen management strategies to reduce nitrate leaching in tile-drained Midwestern soils. Agron. J., 94: 153–171.

Dörsch, P., Palojärvi, A. and Mommertz, S. (2004). Overwinter greenhouse gas fluxes in two contrasting agricultural habitats. Nutr. Cycl. Agroecosyst., 70: 117–133.

Drury, C.F., McKeeney, D.J. and Findlay, W.I. (1991). Relationships between denitrification, microbial biomass and indigenous soil properties. Soil Biol. Biochem., 23: 751–755.

Ellert, B.H. and Janzen, H.H. (2008). Nitrous oxide, carbon dioxide and methane emissions from irrigated cropping systems as influenced by legumes, manure and fertiliser. Can. J. Soil Sci., 88: 207–217.

Elmi, A.A., Madramootoo, C., Hamel, C., Gordon, R. and Liu, A. (2003). Denitrification and nitrous to nitrous oxide plus nitrogen ratios in soil profile under three tillage systems. Biol. Fert. Soil, 38: 340–348.

Fageria, N.K., Moreira, A., Moraes, L.A.C. and Moraes, M.F. (2014). Root growth, nutrient uptake and nutrient use efficiency by roots of tropical legume cover crops as influenced by phosphorus fertilisers. Commun. Soil Sci. Plant Anal., 45: 555–569.

Flint, E. (2000). Comparisons of No-Tillage and Conventional Cotton (*Gossypium hirsutum* L.) with Evaluations of Mycorrhizal Associations, Ph.D. dissertation, Mississippi State University, Starkville, MS.

Galloway, J.N., Schlesinger, W.H., Levy, H., Michaels, A. and Schnoor, J.L. (1995). Nitrogen fixation: Anthropogenic enhancement-environmental response. Glob. Biogeochem. Cycles, 9: 235–252.

Groff, S. (2015). The past, present and future of the cover crop industry. J. Soil Water Conser., 70: 130.

Guardia, G., Abalos, D., García-Marco, S., Quemada, M., Alonso-Ayuso, M., Cárdenas, L.M., Dixon, E.R. and Vallejo, A. (2016). Effect of cover crops on greenhouse gas emissions in an irrigated field under integrated soil fertility management. Biogeosci., 13: 5245–5257.

Gul, S. and Whalen, J. (2013). Plant life history and residue chemistry influences emissions of CO_2 and N_2O from soil-perspectives for genetically modified cell wall mutants. Crit. Rev. Plant Sci., 32: 344–368

Hallama, M., Pekrun, C., Lambers, H. and Kandeler, E. (2019). Hidden miners—the roles of cover crops and soil microorganisms in phosphorus cycling through agro-ecosystems. Plant Soil, 434: 7.

Hao, X. (2015). Nitrous oxide and carbon dioxide emissions during the non-growing season from manured soils under rain-fed and irrigated conditions. Geomicrobiol. J., 32: 648–654.

Harter, T., Davis, H., Mathews, M.C. and Meyer, R.D. (2002). Shallow groundwater quality on dairy farms with irrigated forage crops. J. Contam. Hydrol., 55: 287–315.

Hu, B.L., Shen, L.D., Xu, X.Y. and Zheng, P. (2011). Anaerobic ammonium oxidation (anammox) in different natural ecosystems. Biochem. Soc. Trans., 39: 1811–1816.

Hubbard, R.K., Strickland, T.C. and Phatak, S. (2013). Effects of cover crop systems on soil physical properties and carbon/nitrogen relationships in the coastal plain of southeastern USA. Soil Till. Res., 126: 276–283.

Humbert, S., Tarnawski, S., Fromin, N., Mallet, M.P., Aragno, M. and Zopfi, J. (2010). Molecular detection of anammox bacteria in terrestrial ecosystems: distribution and diversity. Microbial Pop. Commun. Ecol., 4: 450–454.

IPCC. (2013). Climate Change 2013: The Physical Science Basis, Contribution of Working Group I to the Fifth Assessment Report of the Intergovernmental Panel on Climate Change. *In*: Stocker, T.F., Qin, D., Plattner, G.K., Tignor, M., Allen, S.K., Boschung, J., Nauels, A., Xia, Y., Bex, V. and Midggley, P.M. (eds.). Cambridge University Press, Cambridge.

Jackson, L.E. (2000). Fates and losses of nitrogen from a N^{15}-labelled cover crop in an intensively managed vegetable system. Soil Sci. Soc. Amer. J., 64: 1404–1412.

Jackson, L.E., Ramirez, I., Yokota, R., Fennimore, S.A., Koike, S.T., Henderson, D.M., Chaney, W.E., Calderón, F.J. and Klonsky, K. (2004). On-farm assessment of organic matter and tillage management on vegetable yield, soil, weeds, pests, and economics in California. Agric. Ecosys. Environ., 103: 443–463.

Jahangir, M.M.R., Khalil, M.I., Johnson, P., Cardenas, L.M., Hatch, D.J., Butler, M., Barrett, M., O'flaherty, V. and Richard, K.G. (2012). Denitrification potential in subsoil: A mechanism to reduce nitrate leaching to groundwater. Agric. Ecosys. Environ., 147: 13–23.

Jahangir, M.M.R., Minet, E.P., Johnson, P., Premrov, A., Coxon, C.E., Hackett, R. and Richard, K.G. (2014). Mustard catch crop enhances denitrification in shallow groundwater beneath a spring barley field. Chemosphere, 103: 234–239.

Jenkinson, D.S. (2001). The impact of humans on the nitrogen cycle, with focus on temperate arable agriculture. Plant Soil, 228: 3–15.

Jensen, L.S., Salo, T., Palmason, F. and Breland, T.A. (2005). Influence of biochemical quality on C and N mineralisation from a broad variety of plant materials in soil. Plant Soil, 273: 307–326.

Jensen, E.S., Peoples, M.B., Boddey, R.M., Gresshoff, P.M., Henrik, H.N., Alves, B.J.R. and Morrison, M.J. (2012). Legumes for mitigation of climate change and the provision of feedstock for biofuels and biorefineries. Agron. Sustain. Develop., 32: 329–364.

Jetten, M.S., Niftrik, L.V., Strous, M., Kartal, B., Keltjens, J.T. and Op den Camp, H.J. (2009). Biochemistry and molecular biology of anammox bacteria. Crit. Rev. Biochem. Molecular Biol., 44: 65–84.

Kallenbach, C.M., Rolston, D.E. and Horwath, W.R. (2010). Cover cropping affects soil N_2O and CO_2 emissions differently depending on type of irrigation. Agric. Ecosyst. Environ., 137: 251–260.

Kamh, M., Horst, W.J. and Chude, V.O. (1998). Mobilisation of soil and fertiliser phosphate by cover crops. pp. 167–17. *In*: Merbach, P.D.W. (ed.). Pflanzenernährung, Wurzelleistung und Exsudation, Vieweg+Teubner Verlag, Leipzig.

Karlen, R.L. (1990). Conservation tillage research needs. J. Soil Water Conser., 45: 365–369.

Kaspar, T.C., Jaynes, D.B., Parkin, T.B. and Moorman, T.B. (2007). Rye cover crop and gamagrass strip effects on NO_3^- concentration and load in tile drainage. J. Environ. Qual., 36: 1503–1511.

Kaye, J.P. and Quemada, M. (2017). Using cover crops to mitigate and adapt to climate change: A review. Agron. Sustain. Develop., 37: 4.

Khalil, M.I. and Baggs, E.M. (2005). Soil water-filled pore space affects the interaction between CH_4 oxidation, nitrification and N_2O emissions. Soil Biol. Biochem., 37: 1785–1794.

Khalil, M.I. and Richards, K.G. (2010). Denitrification enzyme activity and potential of subsoils under grazed grasslands assayed by membrane inlet mass spectrometer. Soil Biol. Biochem., 43: 1787–1797.

Kramberger, B., Gselman, A., Kapun, S. and Kaligaric, M. (2007). Effect of sowing rate of Italian ryegrass drilled into pea stubble on soil mineral nitrogen depletion and autumn nitrogen accumulation by herbage yield. Polish J. Environ. Stud., 16: 705–713.

Kuo, S. and Sainju, U.M. (1998). Nitrogen mineralisation and availability of mixed leguminous and non-leguminous cover crop residues in soil. Biol. Fert. Soils, 26: 346–353.

Kuo, S., Huang, B. and Bembenek, R. (2005). Effects of long-term phosphorus fertilization and winter cover cropping on soil phosphorus transformations in less weathered soil. Biol. Fertil. Soils, 41(2): 116–123.

Kustermann, B., Christen, O. and Hulsgergen, K. (2010). Modelling nitrogen cycles of farming systems as basis of site- and farm-specific nitrogen management. Agric. Ecosyst. Environ., 135: 70–80.

Lee, C., Fletcher, T.D. and Sun, G. (2009). Nitrogen removal in constructed wetland systems. Eng. Life Sci., 9: 11–22.

Lipson, D. and Monson, R. (1998). Plant-microbe competition for soil amino acids in the alpine tundra: effects of freeze-thaw and dry-rewet events. Oecologia, 3: 406–414.

Logsdon, S.D., Kaspar, T.C., Meek, D.W. and Preuger, J.H. (2002). Nitrate leaching as influenced by cover crops in large soil monoliths. Agron. J., 94: 807–814.

Macrae, R.J. and Mehuys, G.R. (1985). The effect of green manuring on the physical properties of temperate area soils. Adv. Soil Sci., 3: 71–93.

Marcillo, G.S. and Miguez, F.E. (2017). Corn yield response to winter cover crops: An updated meta-analysis. J. Soil Water Conserv., 72(3): 226–239.

McIsaac, G.F., David, M.B., Gertner, G.Z. and Goolsby, D.A. (2001). Nitrate flux in the Mississippi River. Nature, 414: 166–167.

Meisinger, J.J., Shipley, P.R. and Decker, A.M. (1990). Using winter cover crops to recycle nitrogen and reduce leaching. pp. 3–6. *In*: Mueller, J.P. and Wagger, M.G. (eds.). Proc. of the 1990 Southern Region Conservation Tillage Conference, NC State Univ., Raleigh.

Mitchell, D.C., Castellano, M.J., Sawyer, J.E. and Pantoja, J. (2013). Cover crop effects on nitrous oxide emissions: Role of mineralisable carbon. Soil Sci. Soc. Amer. J., 77: 1765–1773.

Muller, J.C., Denys, D., Morlet, G. and Mariotti, A. (1987). Influence of catch crops on mineral nitrogen leaching and its subsequent plant use. *In*: Jenkinson, D.S. and Smith, K.A. (eds.). Nitrogen Efficiency in Agricultural Soils, vol. 2, Elsevier, New York.

Parsons, L.L., Murray, R.E. and Smith, M.S. (1991). Soil denitrification dynamics-spatial and temporal variations of enzyme-activity, populations and nitrogen gas loss. Soil Sci. Soc. Amer. J., 55: 90–9.

Patrick, W.H., Jr., Haddon, C.B. and Hendrix, J.A. (1957). The effect of long-time use of winter cover crops on certain physical properties of commerce loam. Soil Sci. Soc. Amer. Proc., 21: 366–368.

Paul, E.A. and Clark, F.E. (1989). Soil Microbiology and Biochemistry, Academic Press, Inc. San Diego, pp. 273.

Paul, K.I., Black, A.S. and Conyers, M.K. (2001). Effect of plant residue return on the development of surface soil pH gradients. Biol. Fert. Soils, 33: 75–82.

Peoples, M.B., Landha, J.K. and Herridge, D.F. (1995). Enhancing legume N_2 fixation through plant and soil management. Plant Soil, 174: 83–101.

Petersen, S.O., Mutegi, J.K., Hansen, E.M. and Munkholm. L.J. (2011). Tillage effects on N_2O emissions as influenced by a winter cover crop. Soil Biol. Biochem., 43: 1509–1517.

Prasad, R. and Datta, S.K.D. (1979). Increasing efficiency of fertiliser nitrogen in wetland rice by manipulation of plant density and plant geometry. Field Crops Res., 2: 19–34.

Premrov, A., Coxon. C.E., Hackett, R., Kirwan, L. and Richrads, K.G. (2014). Effects of over-winter green cover on soil solution nitrate concentrations beneath tillage land. Sci. Total Environ., 470-471: 967–974.

Quemada, M., Baranski, M., Nobel-de Lange, M.N.J., Vallejo, A. and Cooper, J.M. (2013). Meta-analysis of strategies to control nitrate leaching in irrigated agricultural systems and their effects on crop yield. Agric. Ecosys. Environ., 174: 1–10.

Rijn, J.V., Tal, Y. and Schreier, H.J. (2006). Denitrification in re-circulating systems: Theory and applications. Aqua. Eng., 34: 364–376.

Risk, N., Snider, D. and Wagner-Riddle, C. (2013). Mechanisms leading to enhanced soil nitrous oxide fluxes induced by freeze-thaw cycles. Can. J. Soil Sci., 93: 401–414.

Rochette, P., Angers, D.A., Chantigny, M.H., MacDonald, J.D., Bissonnette, N. and Bertrand, N. (2009). Ammonia volatilisation following surface application of urea to tilled and no-till soils: a laboratory comparison. Soil Till. Res., 103: 310–315.

Romdhane, S., Spor, A., Busset, H., Falchetto, L., Martin, J., Bizouard, F., Bru, D., Breuil, M.C., Philippot, L. and Cordeau, S. (2019). Cover crop management practices rather than composition of cover crop mixtures affect bacterial communities in no-till agro-ecosystems. Front. Microbiol., 10: 1618.

Russo, T.A., Tully, K., Palm, C. and Neill, C. (2017). Leaching losses from Kenyan maize cropland receiving different rates of nitrogen fertiliser. Nutr. Cycl. Agroecosys., 108: 195–209.

Samson, M.I., Buresh, R.J. and De Datta, S.K. (1990). Evolution and soil entrapment of nitrogen gases formed by denitrification in flooded soil. Soil Sci. Plant Nutr., 36: 299–307.

Sanz-Cobena, A., García-Marco, S., Quemada, M., Gabriel, J.L., Almendros, P. and Vallejo, A. (2014). Do cover crops enhance N_2O, CO_2 or CH_4 emissions from soil in Mediterranean arable systems? Sci. Total Environ., 466-467: 164–174.

Sapkota, T.B., Mazzoncini, M., Bàrberi, P., Antichi, D. and Silvestri, N. (2012). Fifteen years of no till increase soil organic matter, microbial biomass and arthropod diversity in cover crop-based arable cropping systems. Agron. Sustain. Develop., 32: 853–863.

Schipanski, M.E., Barbercheck, M., Douglas, M.R., Finney, D.M., Haider, K., Kaye, J.P., Kemanian, K.R., Mortensen, D.A., Ryan, M.R., Tooker, J. and White, C. (2014). A framework for evaluating ecosystem services provided by cover crops in agro-ecosystems. Agric. Sys., 125: 12–22.

Schmidt, R., Gravuer, K., Bossange, A.V., Mitchell, J. and Scow, K. (2018). Long-term use of cover crops and no-till shift soil microbial community life strategies in agricultural soil. PLoS One, 13: e0192953.

Shipley, P.R., Meisinger, J.J. and Decker, A.M. (1992). Conserving residual corn fertiliser nitrogen with winter cover crops. Agron. J., 84: 869–876.

Sirivedhin, T. and Gray, K.A. (2006). Factors affecting denitrification rates in experimental wetlands: Field and laboratory studies. Ecol. Eng., 26: 167–81.

Skogland, T., Lomeland, S. and Goksøyr, J. (1988). Respiratory burst after freezing and thawing of soil: Experiments with soil bacteria. Soil Biol. Biochem., 20: 851–856.

Smith, M.S., Frye, W.W. and Varco, J.J. (1987). Advances in Soil Science, vol. 7, Stewart, B.A. (ed.). Springer-Verlag, New York, p. 96–139.

Snapp, S.S., Swinton, S.M., Labarta, R., Mutch, D., Black, J.R., Leep, R., Nyiraneza, J. and O'Neil, K. (2005). Evaluating benefits and costs of cover crops for cropping system niches. Agron. J., 97: 322–332.

Starr, R.C. and Gillham, R.W. (1993). Denitrification and organic carbon availability in two aquifers. Ground Water, 31: 934–947.

Steenwerth, K. and Belina, K.M. (2008a). Cover crops and cultivation: Impacts on soil N dynamics and microbiological function in a Mediterranean vineyard agro-ecosystem. Appl. Soil Ecol., 40: 370–380.

Steenwerth, K.L. and Belina, K.M. (2008b). Cover crops enhance soil organic matter, carbon dynamics and microbiological function in a Mediterranean vineyard agro-ecosystem. Appl. Soil Ecol., 40: 359–369.

Stehfest, E. and Bouwman, L. (2006). N_2O and NO emission from agricultural fields and soils under natural vegetation: Summarising available measurement data and modelling of global annual emissions. Nutr. Cycl. Agroecosys., 74: 207–228.

Stivers, L.J. and Shennan, C. (1991). Meeting the nitrogen needs of processing tomatoes through winter cover crops. J. Prod. Agr., 4: 330–334.

Stobart, R. and Morris, N.L. (2011). New Farming Systems Research (NFS) project: Long-term research seeking to improve the sustainability and resilience of conventional farming systems. Making Crop Rotations Fit for the Future, Aspects Appl. Biol., 113: 15–23.

Stobart, R., Morris, N.L., Fielding, H., Leake, A., Egan, J. and Burkinshaw, R. (2015). Developing the use of cover crops on farm through the Kellogg's Origins TM grower programme. Getting the most out of cover crops. Aspects Appl. Biol., 129: 27–34.

Stockdale, E.A., Gaunt, J.L. and Vos, J. (1996). Soil–plant nitrogen dynamics: What concepts are required? *In*: van Ittersum, M.K. and van de Geijn, C.S. (eds.). Persp. Agron., 25: 201–215.

Tesoriero, A.J., Liebscher, H. and Cox, S.E. (2000). Mechanism and rate of denitrification in an agricultural watershed: Electron and mass balance along groundwater flow paths. Water Resour. Res., 36: 1545–1559.

Thapa, R. and Chatterjee, A. (2017). Wheat production, nitrogen transformation and nitrogen losses as affected by nitrification and double inhibitors. Agron. J., 109: 1825–1835.

Thorup-Kristensen, K., Magid, J. and Jensen, L.S. (2003). Catch crops and green manures as biological tools in nitrogen management in temperate zones. Adv. Agron., 79: 227–302.

Tisdall, J.M. and Oades, J.M. (1982). Organic matter and water stable aggregates in soils. J. Soil Sci., 33: 141–163.

Tonitto, C., David, M.B. and Drinkwater, L.E. (2006). Replacing bare fallow with cover crops in fertiliser-intensive cropping systems: A meta-analysis of crop yields and N dynamics. Agric. Ecosys. Environ., 112: 58–72.

Tully, K. and Ryals, R. (2017). Nutrient cycling in agro-ecosystems: Balancing food and environmental objectives. Agroecol. Sustain. Food System, 41: 761–798.

USEPA. (2018). Global Mitigation of Non-CO_2 Greenhouse Gases: 2010–2030, Washington (DC): United States Environmental Protection Agency.

Ussiri, D. and Lal, R. (2013). Soil Emission of Nitrous Oxide and Its Mitigation, Springer, the Netherlands.

van der Linden, A.M.A., van Veen, J.A. and Frisel, M.J. (1987). Modelling soil organic matter levels after long-term applications of crop residue such as, farmyard manures and green manures. Plant Soil, 101: 21–28.

Vaughan, J.D., Hoyt, G.D. and Wollum, II, A.G. (2000). Cover crop nitrogen availability to conventional and no-till corn: Soil mineral nitrogen, corn nitrogen status and corn yield. Commun. Soil Sci. Plant Anal., 31: 1017–1041.

Wagner-Riddle, C. and Thurtell, G.W. (1998). Nitrous oxide emissions from agricultural fields during winter and spring thaw as affected by management practices. Nutr. Cycl. Agro-ecosys., 52: 151–163.

Wagner-Riddle, C., Hu, Q.C., van Bochove, E. and Jayasundara, S. (2008). Linking nitrous oxide flux during spring thaw to nitrate denitrification in the soil profile. Soil Sci. Soc. Amer. J., 72: 908–916.

Wendt, R.C. and Burwell, R.E. (1985). Runoff and soil losses for conventional, reduced and no-till corn. J. Soil and Water Conser., 40: 450–454.

White, C.A., Holmes, H.F., Morris, N.L. and Stobart, R.M. (2016). A review of the benefits, optimal crop management practices and knowledge gaps associated with different cover crop species. AHDB Res. Rev., 90: 1–93.

Whitehead, D.C. (1995). Grassland Nitrogen. CAB International. Wallingford. UK. World Meteorological Organization, 2015. WMO Greenhouse Gas Bulletin: the State of Greenhouse Gases in the Atmosphere Based on Observations through 2014. http://www.wmo.int/gaw/.

Williams, W.A. (1966). Management of non-leguminous green manures and crop residues to improve the infiltration rate of an irrigated soil. Soil Sci. Soc. Amer. Proc., 30: 631–634.

White, C.A., Holmes, H.F., Morris, N.L. and Stobart, R.M. (2016). A review of the benefits, optimal crop management practices and knowledge gaps associated with different cover crop species. AHDB Research Review, 90: 1–93.

Wittwer, R.A., Dorn, B., Jossi, W. and van der Heijden, M.G. (2017). Cover crops support ecological intensification of arable cropping systems. Sci. Rep., 7: 41911.

Wong, M.T.F., Nortcliff, S. and Swift, R.S. (1998). Method for determining the acid ameliorating capacity of plant residue compost, urban waste compost, farmyard manure, and peat applied to tropical soils. Commun. Soil Sci. Plant Anal., 29: 2927–2937.

Xing, G.X. and Zhu, Z.L. (2000). An assessment of N loss from agricultural fields to the environment in China. Nutr. Cycl. Agro-ecosys., 57: 67–73.

Xu, R.K. and Coventry, D.R. (2003). Soil pH changes associated with lupin and wheat plant materials incorporated in a red-brown earth soil. Plant Soil, 250: 113–119.

Zhao, X., Xie, Y., Xiong, Z., Yan, X., Xing, G. and Zhu, Z. (2009). Nitrogen fate and environmental consequence in paddy soil under rice-wheat rotation in the Taihu lake region, China. Plant Soil, 319: 225–234.

Zhao, X., Christianson, L.E., Harmel, D. and Pittelkow, C.M. (2016). Assessment of drainage nitrogen losses on a yield-scaled basis. Field Crops Res., 199: 156–166.

Zhao, Y.H., Deng, X.Z., Lu, Q. and Wei, H. (2010). Regional rural development, nitrogen input and output in farming-grazing system and its environmental impacts—a case study of the Wuliangsuhai catchment. Proc. Environ. Sci., 2: 542–556.

Zhu, G., Jetten, M.S.M., Kuschk, P., Ettwig, K. and Yin, C. (2010). Potential roles of anaerobic ammonia and methane oxidation in the nitrogen cycle of freshwater wetland ecosystems. Appl. Microbiol. Biotechnol., 86: 1043–1055.

13

Effect of Cover Crops on Soil Biology

Harit K Bal

1. Introduction

Importance of soil biology in delivering important ecosystem services, such as nutrient cycling, organic matter decomposition, pest and disease regulation, soil fertility maintenance and plant productivity enhancement forms the basis of agricultural sustainability and agro-ecosystem health (Brussaard et al., 2007; Scherr and McNeely, 2008). Soil food webs in agro-ecosystems comprise of the abundance and trophic relationship of soil organisms. These soil food webs are degraded, short and dominated by opportunists at the entry level (Ferris and Matute, 2003; Briar et al., 2007). Intensive production practices, such as tillage, seeding, harvesting and chemical applications contribute to the degradation of soil food webs in agro-ecosystems (Neher, 1999; Brussaard et al., 2007; Briar et al., 2007) and subsequently decline agricultural productivity (Millennium Ecosystem Assessment, 2005). Sustainable agricultural management practices including reduced tillage, cover cropping and crop rotations enhance soil food web diversity and associated ecosystem services, thereby acquiring popularity in the last three decades (Elliott and Stott, 1997; Kladivko, 2001; Baker et al., 2006). In this chapter, the effect of cover cropping practices on soil biology and overall soil food web is reviewed. Soil food-web health is discussed using soil nematodes and their community structure as bio-indicators of soil biological quality. Prospects to enhance soil biology using cover crops are also discussed.

2. Soil Biology: Major Players and Their Roles

This section briefly discusses the major groups of soil organisms and their roles within the belowground soil food web. Soil biota may be characterised on the basis of organism size, their function and structure of the soil food web (Cochran et al., 1994; Lavelle, 1997, 2000; Neher, 1999; Kladivko, 2001; Roger-Estrade et al., 2010). Based on the size, the major players of the soil food web are categorised as microflora, microfauna, mesofauna and macrofauna. Figure 1 shows the major players of the belowground soil food web in an agro-ecosystem and interactions between them and with the crop. These interactions are also discussed below. Soil organisms decrease in size and increase in number, from higher to lower trophic level. Soil organisms at each trophic level serve as predators of organisms present at a lower trophic level, or pathogens of the ones present at a higher trophic level, contributing to nutrient release and recycling and thereby impacting crop productivity.

Microflora, autotrophic and heterotrophic, comprises bacteria, fungi and green algae (Cochran et al., 1994). Autotrophic microflora constitutes a small proportion of the soil microbial biomass and

Bayer Crop Science, 700 Chesterfield Parkway West, Chesterfield, MO 63017.
Email: haritkaur.bal@bayer.com

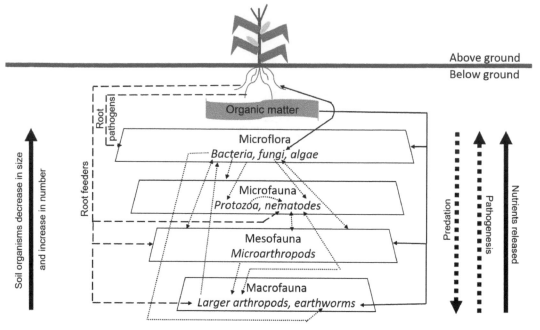

Fig. 1 Generalised diagram of belowground soil food web in an agro-ecosystem.

biological activity but plays a significant role in the oxidation of ammonium, nitrate and sulphur and in fixation of atmospheric nitrogen. On the other hand, heterotrophic microflora constitutes the largest soil microbial biomass and is primarily responsible for the decomposition of soil organic matter and nutrient cycling (Richardson et al., 2009). The soil microflora acts as both a sink by immobilisation and as a source by mineralisation of plant nutrients (Kumar and Goh, 2000). Bacteria and fungi serve as the principal resource base for soil food webs by forming key linkages between detritus and soil fauna (Ruess and Ferris, 2004). Members of the soil microflora exert beneficial or detrimental effects on the crop. Beneficial microflora includes N-fixing bacteria, endo- and ecto-mycorrhizal fungi and plant growth promoting rhizobacteria and fungi, all of which positively affects crop growth and nutrition, whereas microflora that has an adverse impact on crop productivity includes the pathogenic bacteria, fungi and oomycetes (Raaijmakers et al., 2009). Plant root exudates in the rhizosphere enhance the growth of bacterial populations, which utilise the carbon from the root exudates and other nutrients, particularly nitrogen from residual organic matter (Watt et al., 2006). Furthermore, soil microflora improves soil structure by binding soil particles together and enhancing soil aggregation, resulting in more water absorption, carbon retention in macroaggregates, reduction in erosion and maintenance of adequate pore space in soil (Kennedy and Papendick, 1995).

Microfauna comprises of soil organisms, such as protozoa and nematodes that are generally less than 0.2 mms long. Soil protozoa regulates bacterial and algal populations in soil (Clarholm, 1981) and release about one-third of the nitrogen from consumed microbial biomass, which then becomes available to the crop (Griffiths, 1994), enhancing crop productivity (Kuikman et al., 1990; Jentschke et al., 1995; Alphei et al., 1996; Bonkowski, 2004). Soil nematodes are functionally diverse members of the soil microfauna and are known to be bacteria or fungal feeders, plant parasitic, omnivorous and predatory (Yeates et al., 1993; Bongers and Bongers, 1998; Lee, 2002). They play a crucial role in nutrient cycling and microbial turnover (Neher, 2001; Bonkowski, 2004). In addition, soil protozoa and nematodes serve as prey for organisms at higher levels, or predators of bacteria and fungi, significantly affecting the population dynamics of the microflora and subsequent nutrient release from microbial biomass (Griffiths, 1994; Yeates and Wardle, 1996).

Mesofauna, including microarthropods (e.g., acarids, spiders, springtails, soil crustaceans) and enchytraeid worms (small Oligochaeta), ranges between 0.2–2 mm in length. Mesofauna has direct and indirect effects on primary production by root feeding, organic matter decomposition and nutrient mineralisation, respectively (Crossley et al., 1992). Mesofauna contributes to nutrient recycling by ingesting organic material and releasing fecal pellets, nutrients from which may subsequently be reabsorbed by the soil microflora (Lavelle, 1997). Microbial grazing by mesofauna on microbes excreting nitrogen-rich waste contributes up to 30 per cent of mineralised N (Griffiths, 1994; Lavelle et al., 2006), thus increasing plant uptake up to 50 per cent (Laakso et al., 2000). Soil microarthropods also interact with other organisms in the soil food web through their predatory behaviour. For instance, mites are one of the most abundant soil arthropods and play an important role in the belowground soil food web by serving as nematode predators and indicators of land use changes (Koehler, 1999). Furthermore, mesofauna, particularly enchytraeids, improves soil structure by creating burrows and enhances infiltration and aeration (van Vliet et al., 1995).

Macrofauna consists of individuals more than 2 mm in length, including earthworms and large arthropods. Macrofauna, particularly termites and earthworms, affects soil organic matter dynamics by building soil structures and by digesting soil organic matter, using their enzymes or their mutual association with the ingested soil microflora (Lavelle, 1997; Blouin et al., 2013). Soil organic matter digestion by earthworms' results in the release of high amounts of mineral nitrogen and phosphorus, thus regulating mineralisation and humification processes (Bertrand et al., 2015). A recent meta-analysis on the impact of earthworms on crop production has shown a 25 per cent increase in crop yield and 23 per cent increase in aboveground biomass in agro-ecosystems containing earthworms (van Groenigen et al., 2014). Earthworms also improve soil structure, infiltration and aeration by casting and burrowing in the soil (Lemtiri et al., 2014). Larger arthropods also contribute to nutrient recycling and stabilise soil structure and aggregation by releasing fecal pellets and casts, and by creating nests or digging burrows (Lavelle, 1997; Balesdent et al., 2000). Although larger arthropods may adversely impact crop productivity by feeding on plant roots, there are several predatory arthropods that provide natural biological control services in agro-ecosystems by regulating populations of agricultural insect and nematode pests (Tscharntke et al., 2005, *see* Chapter 13). The population of larger arthropods is also regulated by their bacterial and fungal pathogens in the soil and the interactions between microbes, arthropods and the crop (Biere and Tack, 2013).

3. Cover Crop Effects on Soil Biology

Cover cropping before or between the main crops, or along with the orchard crops, in no-till crop production systems improve soil's physical, chemical and biological properties by enhancing nutrient cycling and storage (primarily nitrogen) (Frye et al., 1988; Ruffo and Bollero, 2003); conserving soil moisture (Morse, 1993); reducing soil compaction and erosion by maintaining ground cover (Langdale et al., 1991); increasing soil fertility by building soil organic matter (Cavigelli and Thien, 2003); reducing weed, insect and disease incidence (Baldwin, 2006); and reducing greenhouse gas emissions (Robertson et al., 2000), thus improving crop productivity (Fageria et al., 2007). Soil biological properties have recently gained popularity as early and sensitive indicators of altered soil management practices in agro-ecosystems, in addition to soil physical and chemical properties that have been used widely (Kennedy and Papendick, 1995; Pompili et al., 2006; Babujia et al., 2010; Aziz et al., 2013). Specifically, cover crops have been adopted to improve soil biology and quality for developing sustainable and organic farming systems with improved crop productivity and soil health (Buyer et al., 2010). Available data on the impact of cover crops on the major players of soil biology has been summarised in Table 1.

3.1 Microflora

Agricultural practices have been shown to influence the size and diversity of microbial communities (Mummey et al., 2006), which could further have important implications for the functioning and

Table 1 Impact of cover crops on the major players of soil biology in different agro-ecosystems.

Major Player	Soil Organism or Parameter	Cover Crop	Main Crop	Impact on Soil Organism	Reference
Microflora	Total bacteria	Soybean, rye, red clover	Potato	+	Rouatt and Atkinson, 1950
		Oats-radish, oats-radish-vetch*	Soybean, corn	+	Chavarría et al., 2016
		Crimson clover	Corn	+	Kirchner et al., 1993
		Hairy vetch, rye	Soybean	+	Wagner et al., 1995
		Rye, crimson clover	Soybean	+	Reddy et al., 2003
		White clover	Maize	+	Nakamoto and Tsukamoto, 2006
		Hairy vetch	Tomato	ne[#]	Carrera et al., 2007
		Ryegrass	Cotton	+	Zablotowicz et al., 2007
		Rye	Cotton	+	Acosta-Martínez et al., 2010
		Rye	Potato	+	Larkin et al., 2010
		Oats, spring vetch	Scorzonera	+	Patkowska and Konopiński, 2013
		Ryegrass, Chinese milk vetch, rape	Rice	+	Hai-Ming et al., 2014
		Hairy vetch, winter wheat	Cotton	Ne	Mbuthia et al., 2015
	Gram positive bacteria	Rye	Cotton	+	Acosta-Martínez et al., 2010
		Rye, vetch	Tomato	-	Buyer et al., 2010
		Smooth vetch, pea	Tomato	+	Briar et al., 2011
		Hairy vetch	Cotton	+	Mbuthia et al., 2015
		Oats-radish, oats-radish-vetch	Soybean, corn	+	Chavarría et al., 2016
	Gram negative bacteria	Hairy vetch, rye	Soybean	+	Wagner et al., 1995
		Bushy rattle pod	Maize	+	Bünemann et al., 2004
		Ryegrass	Cotton	+	Zablotowicz et al., 2007
		Rye	Cotton	-	Acosta-Martínez et al., 2010
		Rye, vetch	Tomato	+	Buyer et al., 2010
		Winter pea	Wheat, sunflower, spelt, maize	+	Kuntz et al., 2013
		Winter wheat	Cotton	+	Mbuthia et al., 2015
	Actinomycetes	Soybean, red clover	Potato	+	Rouatt and Atkinson, 1950
		Rye	Cotton	-	Acosta-Martínez et al., 2010
		Rye, vetch	Tomato		Buyer et al., 2010

Table 1 Contd. ...

...Table 1 Contd.

Major Player	Soil Organism or Parameter	Cover Crop	Main Crop	Impact on Soil Organism	Reference
		Ryegrass, Chinese milk vetch, rape	Rice	+	Hai-Ming et al., 2014
	Fungi	Soybean, rye, red clover	Potato	+	Rouatt and Atkinson, 1950
		Rye	Tomato, safflower, corn	+	Lundquist et al., 1999
		Ryegrass-red clover-oats-buckwheat	Sweet corn, broccoli	+	Schutter et al., 2001
		Rye, mustard, oats	Sweet corn	+	Peachey et al., 2002
		Rye, crimson clover	Soybean	+	Reddy et al., 2003
		Bushy rattle pod	Maize	+	Bünemann et al., 2004
		Hairy vetch	Tomato	+	Carrera et al., 2007
		Ryegrass	Cotton	+	Zablotowicz et al., 2007
		Rye	Cotton	+	Simmons and Coleman, 2008
		Rye	Cotton	-	Acosta-Martínez et al., 2010
		Rye, vetch	Tomato	+	Buyer et al., 2010
		Red clover	Apple	+	Eo et al., 2010
		Rye	Potato	-	Larkin et al., 2010
		Oats, spring vetch, tansy phacelia	Salsify	-	Patkowska and Konopiński, 2011
		Winter pea	Wheat, sunflower, spelt, maize	+	Kuntz et al., 2013
		Oats	Scorzonera	-	Patkowska and Konopiński, 2013
		Ryegrass, Chinese milk vetch, rape	Rice	+	Hai-Ming et al., 2014
		Red, white clover	Wheat	-	Detheridge et al., 2016
	Vesicular-arbuscular mycorrhiza	Hairy vetch	Wheat, corn, soybean	+	Galvez et al., 1995
		Winter wheat	Corn	+	Boswell et al., 1998
		Winter wheat	Corn	+	Kabir and Koide, 2000
		Merced rye, Triticale	Grapes	+	Baumgartner et al., 2005
		Zorro fescue	Grapes	+	Cheng and Baumgartner, 2005
		Rye, vetch	Tomato	+	Buyer et al., 2010
		Forage radish-rye	Maize	ne	White and Weil, 2010

Table 1 Contd. ...

...Table 1 Contd.

Major Player	Soil Organism or Parameter	Cover Crop	Main Crop	Impact on Soil Organism	Reference
		Winter wheat	Cotton	+	Mbuthia et al., 2015
		Oats, clover	Maize, winter wheat, sunflower, spelt	+	Säle et al., 2015
	Microbial biomass	Red clover, hairy vetch	Corn, soybean	+	Doran, 1987
		Hairy vetch-oats	Tomato, safflower, corn, beans	+	Gunapala and Scow, 1998
		Bushy rattle pod	Maize	+	Bünemann et al., 2004
		Perennial grass mixture, green manure mixture, clover mixture, cereal mixture	Grapes	+	Ingels et al., 2005
		Hairy vetch	Tomato	+	Carrera et al., 2007
		Rye, vetch	Tomato	+	Buyer et al., 2010
	Microbial biomass C	Red clover	Barley	+	Angers et al., 1993
		Crimson clover	Corn	+	Kirchner et al., 1993
		Austrian winter pea	Wheat, pea	+	Bolton et al., 1995
		Red clover, triticale	Sweet corn, broccoli	+	Mendes et al., 1999
		Rye	Tomato, safflower, corn	+	Lundquist et al., 1999
		Ryegrass-red clover-oats-buckwheat	Sweet corn, broccoli	+	Ndiaye et al., 2000
		Gray oats, vetch	Green bean	+	Schutter and Dick, 2002
		Rye	Eggplant	+	Sainju et al., 2003
		Rye, Rye+Blend Balansa clover, hairy vetch, crimson clover	Cotton	+	Sainju et al., 2007
		Sorghum Sudangrass, cowpea, Sunn hemp, velvet bean	Tomato	+	Wang et al., 2007
		Red clover, and rapeseed	Maize	+	Tejada et al., 2008
		Barley, red clover	Potato	+	Carter et al., 2009
		Sunflower	Wheat, maize	+	Sapkota et al., 2012
		Ryegrass, milkvetch	Rice	+	Zhu et al., 2012
	Microbial biomass N	Hairy vetch, rye	Soybean	+	Wagner et al., 1995
		Hairy vetch, crimson clover	Eggplant	+	Sainju et al., 2003
		Sunn hemp, velvet bean	Tomato	+	Wang et al., 2007

Table 1 Contd. ...

...Table 1 Contd.

Major Player	Soil Organism or Parameter	Cover Crop	Main Crop	Impact on Soil Organism	Reference
		Ryegrass, milkvetch	Rice	+	Zhu et al., 2012
		Hairy vetch	Cotton	+	Mbuthia et al., 2015
Microfauna	Protozoa	Mustard	Beans, lupine	+	Schrader and Seibel, 2001
		Ryegrass-red clover-oats-buckwheat	Sweet corn, broccoli	+	Schutter et al., 2001
		White clover	Maize	+	Nakamoto and Tsukamoto, 2006
		Hairy vetch	Tomato	+	Carrera et al., 2007
		Rye, vetch	Tomato	+	Buyer et al., 2010
		Legume (not named)	Tomato	+	Briar et al., 2011
		Winter pea	Wheat, sunflower, spelt, maize	+	Kuntz et al., 2013
	Bacterivore nematodes	Rye	Tomato, safflower, corn	+	Lundquist et al., 1999
		Rye	Corn	+	Fu et al., 2000
		Rye-vetch	Tomato	+	Bulluck III et al., 2002
		Purple vetch, oats	Tomato, safflower, corn, wheat, dry beans	+	Berkelmans et al., 2003
		Oats-faba bean-field pea-common vetch	Tomato	+	Ferris et al., 2004
		Velvet bean	Corn	+	Blanchart et al., 2006
		Sudangrass	Wheat, corn, potato	+	Collins et al., 2006
		Red clover, hay	Corn, soybean	+	Briar et al., 2007
		Field pea-vetch-oats-triticale	Corn, tomato	+	DuPont et al., 2009
		Sunn hemp	Yellow squash	-	McSorley et al., 2009
		Cereal rye	Soybean	+	Takeda et al., 2009
		Sunn hemp	Bitter melon	+	Marahatta et al., 2010
		Sunn hemp	Cucumber, winter gourd	+	Wang et al., 2011
		Bahia grass	Banana	+	Djigal et al., 2012
		Cereal rye-hairy vetch	Tomato	+	Nair and Nguoajio, 2012

Table 1 Contd. ...

...Table 1 Contd.

Major Player	Soil Organism or Parameter	Cover Crop	Main Crop	Impact on Soil Organism	Reference
	Fungivore nematodes	Barley-white clover-red clover	Carrot	+	Bal et al., 2014
		Hairy vetch, rye	Soybean	+	Ito et al., 2015
		Ryegrass	Wheat	+	Crotty et al., 2016
		Rye	Tomato, safflower, corn	+	Lundquist et al., 1999
		Rye-vetch	Tomato	+	Bulluck III et al., 2002
		Oats-faba bean-field pea-common vetch	Tomato	+	Ferris et al., 2004
		Sudangrass	Wheat, corn, potato	+	Collins et al., 2006
		White clover	Maize	+	Nakamoto and Tsukamoto, 2006
		Sunn hemp	Bitter melon	+	Marahatta et al., 2010
		Legume (not named)	Tomato	+	Briar et al., 2011
		Sunn hemp	Cucumber, winter gourd	+	Wang et al., 2011
		Cereal rye-hairy vetch	Tomato	+	Nair and Nguoajio, 2012
		Rye	Soybean	+	Ito et al., 2015
		Clover	Wheat	+	Crotty et al., 2016
	Plant parasitic nematodes	Velvet bean	Tomato	-	Marban-Mendoza et al., 1992
		Rye, vetch, wheat, lupine, crimson clover	Corn, sorghum	-	McSorley and Gallaher, 1992
		Sudangrass	Potato	-	Davis et al., 1994
		Sudangrass	Lettuce	-	Viaene and Abawi, 1998
		Barley, oil radish	Potato	+	Hafez and Sundararaj, 2000
		Rye-vetch	Tomato	-	Bulluck III et al., 2002
		Vetch, oats	Tomato, safflower, corn, wheat, dry beans	-	Berkelmans et al., 2003
		Cowpea	Bell pepper	-	Wang et al., 2004
		Oats, rye, oilseed rape, sudangrass	Strawberry	ne	Forge et al., 2000
		Rye, vetch	Corn, peanut	ne	Gazaway et al., 2000

Table 1 Contd. ...

...*Table 1 Contd.*

Major Player	Soil Organism or Parameter	Cover Crop	Main Crop	Impact on Soil Organism	Reference
		Sudangrass, rye, flax	Potato	+	McKeown and Potter, 2001
		Oil radish	Potato, tulip, wheat, sugar beet, onion	-	Hartsema et al., 2005
		Italian ryegrass	Potato, tulip, wheat, sugar beet, onion	+	Hartsema et al., 2005
		Velvet bean	Corn	-	Blanchart et al., 2006
		Oats	Potato, cucumber	-	Everts et al., 2006
		Rye	Potato	-	Hoek et al., 2006
		Mustard	Potato	+	Hoek et al., 2006
		Sudangrass, oats	Potato	-	LaMondia, 2006
		Rye, oil radish, wheat	Potato	-	Runia et al., 2006
		Sunn hemp	Pepper	-	McSorley et al., 2008
		Sun hemp, cowpea	Turnip, lima bean	Ne	Wang et al., 2008
		Field pea-vetch-oats-triticale	Corn, tomato	+	DuPont et al., 2009
		Sunn hemp	Yellow squash	-	McSorley et al., 2009
		Marigold	Yellow squash	+	McSorley et al., 2009
		Marigold	Different crops	-	Hooks et al., 2010
		Marigold	Bitter melon	-	Marahatta et al., 2010
		Legume (not named)	Tomato	-	Briar et al., 2011
		Sunn hemp	Cucumber, winter gourd	-	Wang et al., 2011
		Bahia grass	Banana	-	Djigal et al., 2012
		Cereal rye-hairy vetch	Tomato	+	Nair and Nguoajio, 2012
		Rye	Soybean	-	Ito et al., 2015
		Hairy vetch	Soybean	+	Ito et al., 2015
		Ryegrass	Wheat	+	Crotty et al., 2016
	Predatory nematodes	Purple vetch, oats	Tomato, safflower, corn, wheat, dry beans	+	Berkelmans et al., 2003

Table 1 Contd. ...

...Table 1 Contd.

Major Player	Soil Organism or Parameter	Cover Crop	Main Crop	Impact on Soil Organism	Reference
		Velvet bean	Corn	+	Blanchart et al., 2006
		Perennial soybean, tropical kudzu, Brazilian Lucerne	Banana	+	Djigal et al., 2012
		Rye	Soybean	+	Ito et al., 2015
	Omnivore nematodes	Purple vetch, oats	Tomato, safflower, corn, wheat, dry beans	+	Berkelmans et al., 2003
		Chinese milk vetch	Apple	+	Eo et al., 2010
		Bahia grass	Banana	+	Djigal et al., 2012
Mesofauna	Collembola (springtails)	Rye, crimson clover	Sorghum, soybean	+	House and Parmelee, 1985
		White clover	Grass	+	Jággi et al., 1995
		Barley, red clover	Potato	+	Carter et al., 2009
		Alfalfa	Corn	+	Fox et al., 1999
		Mustard	Beans, lupine	+	Schrader and Seibel, 2001
		White clover	Maize	+	Nakamoto and Tsukamoto, 2006
		Barley	Corn	+	Rodríguez et al., 2006
		Rattle weed, velvet bean	Maize	+	Brévault et al., 2007
		Sunn hemp	Bitter melon	+	Marahatta et al., 2010
		Sunn hemp, marigold	Cucumber, winter gourd	+	Wang et al., 2011
	Acari (mites)	Rye, crimson clover	Sorghum, soybean	+	House and Parmelee, 1985
		Alfalfa	Corn	+	Fox et al., 1999
		Mustard	Beans, lupine	+	Schrader and Seibel, 2001
		Rye	Sweet corn	+	Peachey et al., 2002
		White clover	Maize	+	Nakamoto and Tsukamoto, 2006
		Rye	Corn	+	Reeleder et al., 2006
		Barley	Corn	+	Rodríguez et al., 2006
		Sunn hemp	Yellow squash	+	McSorley et al., 2009
		Rattail fescue, Chinese milk vetch	Apple	+	Eo et al., 2010
		Sunn hemp, marigold	Bitter melon	+	Marahatta et al., 2010

Table 1 Contd...

...Table 1 Contd.

Major Player	Soil Organism or Parameter	Cover Crop	Main Crop	Impact on Soil Organism	Reference
		Sunn hemp, marigold	Cucumber, winter gourd	+	Wang et al., 2011
		Sunflower	Wheat, maize	+	Sapkota et al., 2012
		Rye grass, red clover	Wheat, barley	+	Crotty et al., 2016
	Araneae (spiders)	Rye, crimson clover	Sorghum, soybean	+	House and Parmelee, 1985
		Barley	Corn	+	Rodríguez et al., 2006
		Chinese cabbage, rattle weed, velvet bean	Maize	+	Brévault et al., 2007
		Sunn hemp and marigold	Bitter melon	+	Marahatta et al., 2010
	Enchytraeid worms	Rye, crimson clover	Sorghum, soybean	-	House and Parmelee, 1985
		Mustard	Beans, lupine	+	Schrader and Seibel, 2001
		Rattle weed, velvet bean	Maize, sorghum	+	Brévault et al., 2007
		Sunn hemp	Yellow squash	+	McSorley et al., 2009
	Isopoda	Velvet bean	Corn	+	Blanchart et al., 2006
		Sunn hemp, marigold	Cucumber, winter gourd	+	Wang et al., 2011
		Sunflower	Wheat, maize	+	Sapkota et al., 2012
	Chilopoda	Velvet bean	Corn	+	Blanchart et al., 2006
		Rattle weed, velvet bean	Maize, sorghum	+	Brévault et al., 2007
	Diplopoda	Velvet bean	Corn	+	Blanchart et al., 2006
		Chinese cabbage, rattle weed, velvet bean	Maize	+	Brévault et al., 2007
		Sunflower	Wheat, maize	+	Sapkota et al., 2012
	Symphyla	Rye, oats	Sweet corn	-	Peachey et al., 2002
		Sunflower	Wheat, maize	+	Sapkota et al., 2012
Macrofauna	Coleoptera	Clover	Corn, tomato, cauliflower	+	Altieri et al., 1985
		Velvet bean	Corn	+	Blanchart et al., 2006

Table 1 Contd. ...

...Table 1 Contd.

Major Player	Soil Organism or Parameter	Cover Crop	Main Crop	Impact on Soil Organism	Reference
		White clover	Maize	+	Nakamoto and Tsukamoto, 2006
		Chinese cabbage	Maize	+	Brévault et al., 2007
	Diptera	Velvet bean	Corn	+	Blanchart et al., 2006
		White clover	Maize	+	Nakamoto and Tsukamoto, 2006
		Barley	Corn	+	Rodríguez et al., 2006
	Earthworms	Rye, crimson clover	Sorghum, soybean	+	House and Parmelee, 1985
		White clover	Grass	+	Jággi et al., 1995
		White clover	Wheat	+	Schmidt and Curry, 2001
		White clover	Wheat	+	Schmidt et al., 2001
		Red clover	Corn, wheat	+	Jordan et al., 2004
		Velvet bean	Maize	+	Ortiz-Ceballos and Fragoso, 2004
		Velvet bean	Corn	+	Blanchart et al., 2006
		Rye	Corn	+	Reeleder et al., 2006
		Alfalfa, oats	Maize, soybean, wheat	+	Peigné et al., 2009
		Pea	Wheat, maize	+	Pelosi et al., 2009
		Winter pea	Wheat, sunflower, spelt, maize	+	Kuntz et al., 2013
		Winter oilseed rape, winter rye, ryegrass	Pea, potato, winter wheat	+	Cima et al., 2016
		White clover	Wheat, barley	+	Crotty et al., 2016
		Alfalfa, oats, rye	Maize, soybean, wheat	+	Pelosi et al., 2016

* Dash (-) between two or more cover crops indicates cover crop mixture; # ne: no effect.

stability of soil food webs. Soil microbial biomass has been frequently used as an early and sensitive indicator of changes in microbial activity as well as soil quality (Brookes, 1995; Zagal and Córdova, 2005; Pompili et al., 2006; Babujia et al., 2010). Different methods to measure the soil microbial biomass and size of microbial populations have been reviewed in the past (Azam et al., 2003; Gonzalez-Quiñones et al., 2011). Crop management practices, such as cover cropping, significantly enhance soil microbial biomass (Angers et al., 1993; Mendes et al., 1999; Schutter and Dick, 2002;

Carrera et al., 2007; Sainju et al., 2007; Buyer et al., 2010; Briar et al., 2011; Zhu et al., 2012; Chavarría et al., 2016) as the nutrients available in cover crops get incorporated into the soil and recycled through the decomposition of organic matter, which is primarily regulated by microflora. This increase in the microbial biomass by cover cropping has been shown to occur both when the cover crops were incorporated into the soil and left untilled (Doran, 1987; Ndiaye et al., 2000; Ingels et al., 2005; Wang et al., 2007; Tejada et al., 2008; Buyer et al., 2010). Cover crops may provide favourable environmental conditions (moisture, temperature, nutrients) for the growth of microflora. However, the quality and quantity of cover crop residues in the soil, as well as cropping history of soils, have been found to significantly influence the structural and functional diversity of microflora communities and their associated microbial processes (Tian et al., 2011; Kumar and Goh, 2000; Govaerts et al., 2007; Vineela et al., 2008). For instance, greater bacterial and fungal populations were reported on legume residues than in grass or cereal residues (Kumar and Goh, 2000).

Recent studies have reported a significant increase in total bacterial and gram-positive bacterial populations after the incorporation of cover crops in continuously monocultured agricultural systems (Acosta-Martínez et al., 2010; Briar et al., 2011; Mbuthia et al., 2015; Chavarría et al., 2016). On the contrary, certain cover crops or cover crop mixtures have been found to lower the populations of gram-positive bacteria but increase the populations of gram-negative bacteria and actinomycetes (Wagner et al., 1995; Bünemann et al., 2004; Zablotowicz et al., 2007; Buyer et al., 2010; Mbuthia et al., 2015). Cover crops have also been shown to exert a significant effect on fungal communities in the soil (Lundquist et al., 1999; Schutter et al., 2001; Carrera et al., 2007; Zablotowicz et al., 2007; Buyer et al., 2010; Hai-Ming et al., 2014), particularly on vesicular-arbuscular mycorrhiza (VAM), which increase in field, vegetable, or fruit crops after cover cropping (Galvez et al., 1995; Boswell et al., 1998; Kabir and Koide, 2000; Dabney et al., 2001; Baumgartner et al., 2005; Cheng and Baumgartner, 2005; Buyer et al., 2010; Mbuthia et al., 2015), but is not affected or slightly decreased in some cases (Baltruschat and Dehne, 1989; White and Weil, 2010). The presence of VAM fungi significantly increases the populations of other soil microflora, such as total bacteria and P solubilising bacteria in the soil (Andrade et al., 1998).

Furthermore, previous studies have shown cover crops, such as soybean, red clover, alfalfa, rye, vetch and *Brassica* sp. to play an important role in reducing infection by disease-causing microorganisms in various crops (West and Hildebrand, 1941; Abawi and Widmer, 2000; Larkin et al., 2010; Patkowska and Konopiński, 2011, 2013). Reduction in plant pathogenic microorganisms could be due to competition or direct predation by other microflora whose populations are enhanced by cover cropping or due to toxicity caused by root exudates or other toxic compounds produced from the decomposition of cover crop residues (Patrick et al., 1965; Pal and McSpadden Gardener, 2006; Badri and Vivanco, 2009). However, cover crops may not always reduce soil-borne plant pathogens (Hansen et al., 1990; Pinkerton et al., 2000). To better understand the impact of cover cropping on crop productivity due to alterations in the soil microflora, more investigation is required to study the impact of different cover crops on different components of the soil microflora, including beneficial as well as pathogenic soil-borne bacterial and fungal populations and be examined collectively.

3.2 Microfauna

Cover crops significantly enhance the abundance of soil protozoa (Table 1), which play an important role in the soil ecosystem, functioning primarily by regulating soil bacteria via predation and by mineralising plant nutrients. Changes in soil bacterial populations resulting from cover cropping subsequently influence soil protozoa (Cochran et al., 1994). Therefore, cover crop practices that favour increase of soil bacterial populations or their predators, such as bacterivore nematodes, may tend to favour soil protozoa as well. With only a few studies reporting the impact of cover crops on soil protozoa, more research on the effect of cover crops and their interaction with other soil management practices on soil protozoa and their functional relationship with other soil organisms is warranted in future. Cover crops also have a distinct impact on different groups of soil nematodes, details of which are illustrated in Table 1 and discussed in the section on cover-crop effects on overall soil food web.

3.3 Mesofauna

This, and the next section on macrofauna, covers the impact of cover crops only on major components of mesofauna, including microarthropods and enchytraeid worms and macrofauna, including soil-dwelling larger arthropods and earthworms. Cover cropping has the potential to regulate the populations of insect-pests and their natural enemies belonging to soil mesofauna and macrofauna, the literature regarding which has been reviewed and discussed in detail in Chapter 7. Hence, it is only mentioned briefly in these two sections and not detailed in Table 1 except for a few examples on spiders, which act as general predators in different agro-ecosystems. Several studies have evaluated the impact of cover cropping on above- and below-ground insect-pest and predator populations in different cropping systems (Stinner and House, 1990; Bugg et al., 1991; Landis et al., 2000; Chaplin-Kramer et al., 2011; Peachey et al., 2002; Bianchi et al., 2006; Schmidt et al., 2007; Sirrine et al., 2008; Nyoike and Liburd, 2010; Roger-Estrade et al., 2010; Duyck et al., 2011). The impact of cover crops on insect-pest densities is either directly due to changes in the food sources to the pest, or indirectly due to changes in the overall soil food web influencing the prey-predator dynamics (Peachey et al., 2002).

Cover crops have been shown to positively influence populations of microarthropods, such as mites and collembola, in diverse agro-ecosystems (Badejo et al., 2002; Osler et al., 2008; more studies given in Table 1). Soil mesofauna utilise the organic matter supplied by the cover-crop roots and readily perform their ecosystem functions. Enhanced soil moisture, alleviated soil microclimate and increased soil pore space due to cover cropping are some of the factors that enhance soil mesofauna (Heisler and Kaiser, 1995; Nakomoto and Tsukamoto, 2006; Eo et al., 2010; Sapkota et al., 2012). In addition, microarthropod abundance, richness and diversity are enhanced in cover-cropped soils due to greater food supply from surface residues and protection from physical perturbations, soil desiccation and soil freezing (Sapkota et al., 2012). As discussed in the previous sections, the identity of the cover crop has a major impact on the soil mesofauna as well (St. John et al., 2006; Osler et al., 2008). Soil-dwelling crustaceans, such as isopods and amphipods, that act mainly as detritivores and play an important role in recycling nutrients from the soil organic matter that soil microflora is unable to digest, are also positively influenced by cover crops (Wang et al., 2011). Other soil microarthropods, such as centipedes and millipedes, are favoured by cover crops, particularly legumes (Blanchart et al., 2006). Due to limited information on population dynamics of these soil microarthropods in cover-cropped agro-ecosystems, future work on the impact of soil management with cover crops on these soil organisms and their functional interactions with other soil micro- and mesofauna is imperative.

3.4 Macrofauna

Significant positive impact of organic agriculture practices, including reduced tillage and cover cropping on the abundance, richness and diversity of soil macrofauna, have been reviewed by Bengtsson et al. (2005). Larger arthropods, such as ants and ground beetles, act as generalist predators and have been widely used as bioindicators to assess soil management and land-use changes in different agro-ecosystems (Kromp, 1999; Rainio and Niemelä, 2003). Their densities and species compositions are differentially influenced by cover crops (House and Parmelee, 1985; Carmona and Landis, 1999; Blanchart et al., 2006; Rodríguez et al., 2006; Diekötter et al., 2010; Rivers et al., 2017). Earthworms also serve as useful bioindicators for agro-ecosystem sustainability in different cropping systems (Paoletti, 1999; Fründ et al., 2011). Earthworm densities have also been shown to increase by cover cropping in different soil types and cropping systems (Jäggi et al., 1995; Schmidt et al., 2001; Reeleder et al., 2006; van Eekeren et al., 2009; Kuntz et al., 2013; Crotty et al., 2015, 2016; Cima et al., 2016). This is primarily attributed to enhanced soil infiltration and bio-porosity resulting from cover crop root architecture and biomass (Bronick and Lal, 2005; Kautz et al., 2014). Along with population density, earthworm species are also influenced by the quality and quantity of plant material (Flegel and Schrader, 2000; Schmidt et al., 2003). Certain cover crops, such as white clover, enhance earthworm abundance within a crop season and even beyond (Crotty et al., 2016).

Cover crops, such as cereal rye and hairy vetch, have also been shown to enhance the ability of earthworms to stabilise soil macro-aggregates, increase C and N storage in aggregates and facilitate decomposition of coarse organic matter accumulated on the soil surface (Ketterings et al., 1997). Other soil management practices, such as reduced tillage, work synergistically with cover crops in enhancing earthworm densities and biomass by decreasing soil perturbation and increasing food supply (Overstreet et al., 2010; van Capelle et al., 2012; Kuntz et al., 2013). However, there are some contrasting reports that show the positive impact of tillage on earthworm populations that have benefitted from crop residues being incorporated deep in the soil by intensive tillage (Wyss and Glasstetter, 1992; Chan, 2001; Pelosi et al., 2009). Therefore, crop-specific investigation on the combination of different tillage practices and cover crop quantity and identity would generate useful information on earthworm abundance, biomass and species composition in different agro-ecosystems.

4. Cover Crop Effects on Soil Food Web: Using Nematodes as Bio-indicators of Soil Health

In this section, the effect of cover crops on the overall soil food web will be discussed using nematodes as indicators of soil food web health. Nematodes belong to phylum Nematoda, which is the second largest phylum after arthropoda (Hugot et al., 2001). Soil nematodes are microscopic and occur in diverse habitats (Lee, 2002). Nematodes have been used to determine soil food web structural and functional diversity (Yeates, 1979; Ritz and Trudgill, 1999; Ferris et al., 2001; Berkelmans et al., 2003) because they are excellent indicators of changes in soil management (Neher and Olson, 1999; Yeates and Bongers, 1999; Forge and Simard, 2001; Briar et al., 2007) and represent all major trophic levels in the soil, including bacterivores, fungivores, plant parasites, predators and omnivores (Yeates et al., 1993; Bongers and Bongers, 1998; Lee, 2002).

Bacterivore and fungivore nematodes graze on decomposer microflora and contribute to nutrient mineralisation (Ingham et al., 1985; Ferris and Matatue, 2003; Ferris et al., 1996, 2004). Bacterivore and fungivore nematodes contribute as much as 27 per cent of the readily available N in the soil (Ekschmitt et al., 1999) and promote colonisation of rhizobacteria (Kimpinski and Sturz, 1996; Knox et al., 2003). Predatory nematodes contribute to pest suppression (Grewal et al., 2005). Nematode faunal analysis, based on the relative weighted abundance of coloniser-persister (c-p) classes (*r* versus *K* life strategists), provides a quantitative measure of the nematode community structure and the probable condition of the soil food web (Ferris et al., 2001). The analysis comprises food web indices including enrichment, structure, basal and decomposition channel (Ferris et al., 2001), which are indicative of status and function of the soil food web and provide critical information about soil processes, including N mineralisation and energy flow in ecosystems. Nematode indices may also determine environmental stress, major decomposition channels and crop susceptibility to pests (Bongers, 1990; Lenz and Eisenbeis, 2000; Ferris and Matute, 2003). The basal index (BI) indicates a soil food web deteriorated by environmental stress or limited nutrient availability, the structure index (SI) suggests presence of trophic linkages, the enrichment index (EI) represents resource availability and the channel index (CI) distinguishes bacteria-dominated from fungi-dominated food webs (Ferris et al., 2001). In addition, the maturity index (MI) monitors the evolution of complex soil nematode communities in agro-ecosystems and indicates differences among soil management practices, such as tillage systems and cover cropping (Berkelmans et al., 2003; Bongers and Bongers, 1998; Neher, 1999; Yeates and Bongers, 1999). Therefore, these nematode indices, indicating the condition of the soil food web, have been used to characterise the health status of agro-ecosystem soils (Neher, 1999; Neher and Olson, 1999; Porazinska et al., 1999; Urzelai et al., 2000; Yeates and Bongers, 1999; van Bruggen and Semenov, 2000). Furthermore, the metabolic activity of different nematode groups in an agro-ecosystem depicting their metabolic footprint has been shown to provide detailed explanation of the structure and function of the soil food web and soil health (Ferris, 2010; Zhang et al., 2012).

Soil management practices, such as cover cropping and crop rotations, affect the occurrence and abundance of different nematode feeding groups (Ettema and Bongers, 1993), thus reflecting changes in the soil food web structure. Cover-cropped soils are often associated with bacterial-dominated food webs with high enrichment, and low channel and basal indices (Ferris et al., 1996; Berkelmans et al., 2003; DuPont et al., 2009; Djigal et al., 2012). Plant community has been previously documented to have direct and indirect effects on the soil nematode community structure (Neher, 1999; De Deyn et al., 2004; Wardle et al., 2006; Duyck et al., 2009; Viketoft et al., 2009). Plant residue resulting from cover cropping has a significant positive impact on total nematode abundance, biomass and diversity as resources get abundantly supplied to the soil food web (Liebig et al., 2004; Liang et al., 2009; Sanchez-Moreno et al., 2010).

Residue quantity is also an important driver of nematodes belonging to different guilds, bacterivores, omnivores, predatory and plant parasitic, which further determine the nature and magnitude of ecosystem services provided by the soil food web (Ferris and Matute, 2003; DuPont et al., 2009; Zhang et al., 2012). The identity of the cover crop is another important driver regulating the abundance and diversity of nematodes and structuring of soil food webs (*see* Table 1). Some cover crops enhance the abundance of lower trophic nematode groups, such as bacterivores and fungivores, which support an increase in the abundance of higher trophic groups that further result in top-down control of plant parasitic nematodes. For instance, cover crops, such as rye, marigold, sudangrass, *Mucuna pruriens* and *Paspalum notatum* have been shown to negatively affect the plant parasitic nematode pest populations in different agro-ecosystems (McSorley and Gallaher, 1992; Viaene and Abawi, 1998; Wang et al., 2004; Blanchart et al., 2006; Hooks et al., 2010; Thoden et al., 2011; Djigal et al., 2012; Nair and Nguoajio, 2012). This could be a result of the allelopathic effect of the cover crops on plant parasitic nematodes (Patrick et al., 1965; Widmer and Abawi, 2002; Thoden et al., 2009; Hooks et al., 2010; Kruger et al., 2013) or the occurrence of top-down control performed by nematodes belonging to higher trophic groups due to their increased abundance in cover-cropped soils (Blanchart et al., 2006; Djigal et al., 2012).

On the other hand, the bottom-up effect of the new resources added by cover crops is evident by the abundance of bacterivore, omnivore and predatory nematodes (Neher, 2010). Furthermore, percent N in cover crops and the corresponding C:N ratio may also determine the diversity and abundance of nematode groups in the soil and influence crop productivity (Ettema and Bongers, 1993; Pimental et al., 1995; Porazinska et al., 1999; Ferris et al., 2004; Ilieva-Makulec et al., 2006; Wang et al., 2006; Hoy et al., 2008). While cover crops with high C:N ratio result in fungal-dominated soil food webs, low C:N ratio cover crops favour bacterial-based soil food webs, which further have an impact on nutrient cycling and nitrogen availability to the crop (Porazinska et al., 1999; Ferris and Matute, 2003). Therefore, cover crop mixtures maintaining medium range C:N ratio may regulate steady nutrient cycling by soil fauna and synchronise nitrogen release with plant requirements (Ferris et al., 2004). A combination of cover crops with organic amendments, such as compost, also enhances soil health by increasing the population of decomposers and their activity in soil. Specifically, there is a significant positive impact of such combinations on the population of bacterivore and fungivore nematodes (Ferris et al., 1996; Gunapala and Scow, 1998; Wang et al., 2011; Bulluck III et al., 2002; Nair and Nguoajio, 2012).

Based on the literature reviewed above, cover crops appear to have a significant impact on the soil food web as indicated by the changes in the nematode community structure. Due to the ease of handling and identifying nematodes and their presence at several levels of the soil food-web (Bongers and Bongers, 1998), all measures of nematode community structure and diversity utilised in these studies provide information about belowground soil food web processes in agro-ecosystems. While considerable progress has been made, more efforts are needed to identify cover crops and their compatibility with other soil amendments and soil-management practices that will provide enriched, structured and matured soil food webs of agro-ecosystems with sufficiently large populations of bacterivore and fungivore nematodes important for nutrient cycling and predatory and omnivore nematodes for more soil food web stability, together ensuring increased soil health.

5. Conclusion and Future Considerations

Cover crops appear to have a significant impact on soil biology and soil food web, which play an important role in regulating soil quality, crop productivity and agro-ecosystem health. Several studies have shown that cover cropping alone, or in combination with other soil-management practices, result in changes in soil microbial structure and faunal diversity in different agro-ecosystems. However, there is still lack of information on the effects of these management practices and their interactions on soil biology and overall soil food web over the long term. The soil food web shows observable differences after long-term ($\geq$ 5 years) transitioning of conventional to an organic or alternative soil-management system (Ferris et al., 1996; Neher, 1999; Yeates et al., 1999; Fu et al., 2000; Berkelmans et al., 2003). Therefore, long-term agro-ecosystem studies are imperative to evaluate the impact of soil management on soil fauna, quality and productivity (Rasmussen et al., 1998). Although several studies have shown the impact of cover crops on microbial biomass, more investigation is required on the composition of bacterial and fungal communities and their interaction using modern tools, such as metagenomics, in order to elucidate the effect of cover cropping on soil microflora and the ecosystem services associated with them (Carbonetto et al., 2014; Dias et al., 2015; Detheridge et al., 2016). Furthermore, a comprehensive approach to study the impact of sustainable farming systems, including cover crops on all or most of the major players of the soil biota by assessing simultaneous changes in their abundance and diversity, is an essential step for understanding the use of cover crops that are intended to enhance soil biology and sustain soil quality and soil health.

The quality or identity of the cover crops also significantly impacts the structural and functional diversity of soil biology and their functional relationships. Cover crops generally serve as nutrient-management tools for the main crops (Ruffo and Bollero, 2003). While leguminous cover crops provide nitrogen to the succeeding cash crop and reduce the need for nitrogen fertilisers for the cash crop, non-leguminous cover crops reduce nitrate leaching and erosion. A combination of a legume and a grass cover crop may provide both benefits simultaneously (Ranells and Wagger, 1996; Sainju et al., 2007). Therefore, improved information on different cover crops or cover crop mixtures that do not aggressively compete with the main crop for resources (Tixier et al., 2011; Keene et al., 2017; Rivers et al., 2017), enhance the abundance of beneficial soil organisms, reduce soil-borne plant pathogenic microflora and fauna, yet maintain the soil C and N pools, is essential to make informed decisions about species choice, residue management and timing of operations (Dabney et al., 2001; Snapp et al., 2005). It is also important to evaluate the interactions of cover crops with different tillage practices that may influence residue decomposition and soil nutrient mineralisation (Cambardella and Elliott, 1993). Furthermore, quantification of soil biological properties influenced by cover cropping and their economic analyses are the key areas of future research that need attention for enhanced adoption of these alternative soil-management practices.

References

Abawi, G.S. and Widmer, T.L. (2000). Impact of soil health management practices on soil-borne pathogens, nematodes and root diseases of vegetable crops. Appl. Soil Ecol., 15: 37–47.

Acosta-Martínez, V., Bell, C.W., Morris, B.E.L., Zak, J. and Allen, V.G. (2010). Long-term soil microbial community and enzyme activity responses to an integrated cropping-livestock system in a semi-arid region. Agric. Ecosyst. Environ., 137: 231–240.

Alphei, J., Bonkowski, M. and Scheu, S. (1996). Protozoa, nematoda and lumbricidae in the rhizosphere of *Hordelymus europeaus* (Poaceae): Faunal interactions, response of microorganisms and effects on plant growth. Oecologia, 106: 111–126.

Altieri, M.A., Wilson, R.C. and Schmidt, L.L. (1985). The effects of living mulches and weed cover on the dynamics of foliage- and soil-arthropod communities in three crop systems. Crop Prot., 4: 201–213.

Andrade, G., Mihara, K.L., Linderman, R.G. and Bethlenfalvay, G.J. (1998). Soil aggregation status and rhizobacteria in the mycorrhizosphere. Plant Soil, 202: 89–96.

Angers, D.A., Bissonnette, N., Légère, A. and Samson, N. (1993). Microbial and biochemical changes induced by rotation and tillage in a soil under barley production. Can. J. Soil. Sci., 73: 39–50.

Azam, F., Farooq, S. and Lodhi, A. (2003). Microbial biomass in agricultural soils-determination, synthesis, dynamics and role in plant nutrition. Pak. J. Biol. Sci., 6: 629–639.

Aziz, I., Mahmood, T. and Islam, K.R. (2013). Effect of long term no-till and conventional tillage practices on soil quality. Soil Till. Res., 131: 28–35.

Babujia, L.C., Hungria, M., Franchini, J.C. and Brookes, P.C. (2010). Microbial biomass and activity at various soil depths in a Brazilian oxisol after two decades of no-tillage and conventional tillage. Soil Biol. Biochem., 42: 2174–2181.

Badejo, M.A., Espindola, J.A.A., Guerra, J.G.M., de Aquino, A.M. and Correa, M.E.F. (2002). Soil oribatid mite communities under three species of legumes in an ultisol in Brazil. Exp. Appl. Acarol., 27: 283–296.

Badri, D.V. and Vivanco, J.M. (2009). Regulation and function of root exudates. Plant Cell Environ., 32: 666–681.

Baker, C.J., Saxton, K.E., Ritchie, W.R., Chamen, W.C.T., Reicosky, D.C., Ribeiro, M.F.S., Justice, S.E. and Hobbs, P.R. (2006). No-tillage Seeding in Conservation Agriculture, second ed., CAB International/FAO, Oxford, UK.

Bal, H.K., Acosta, N., Cheng, Z., Whitehead, H., Grewal, P.S. and Hoy, C.W. (2014). Effect of soil management on *Heterorhabditis bacteriophora* GPS11 persistence and biological control in a vegetable production system. Biol. Control, 79: 75–83.

Baldwin, K.R. (2006). Conservation Tillage on Organic Farms, North Carolina Cooperative Extension Service, 1–13.

Balesdent, J., Chenu, C. and Balabane, M. (2000). Relationship of soil organic matter dynamics to physical protection and tillage. Soil Till. Res., 53: 215–230.

Baltruschat, H. and Dehne, H.W. (1989). The occurrence of vesicular-arbuscular mycorrhiza in agro-ecosystems. Plant Soil, 113: 251–256.

Baumgartner, K., Smith, R.F. and Bettiga, L. (2005). Weed control and cover crop management affect mycorrhizal colonisation of grapevine roots and arbuscular mycorrhizal fungal spore populations in a California vineyard. Mycorrhiza, 15: 111–119.

Bengtsson, J., Ahnström, J. and Weibull, A.-C. (2005). The effects of organic agriculture on biodiversity and abundance: a meta-analysis. J. Appl. Ecol., 42: 261–269.

Berkelmans, R., Ferris, H., Tenuta, M. and van Bruggen, A.H.C. (2003). Effects of long-term crop management on nematode trophic levels other than plant feeders disappear after 1 year of disruptive soil management. Appl. Soil Ecol., 23: 223–235.

Bertrand, M., Barot, S., Blouin, M., Whalen, J., de Oliveira, T. and Roger-Estrade, J. (2015). Earthworm services for cropping systems: A review. Agron. Sustain. Dev., 35: 553–567.

Bianchi, F.J.J.A., Booij, C.J.H. and Tscharntke, T. (2006). Sustainable pest regulation in agricultural landscapes: a review on landscape composition, biodiversity and natural pest control. Proc. R. Soc. Lond. B Biol. Sci., 273: 1715–1727.

Biere, A. and Tack, A.J.M. (2013). Evolutionary adaptation in three-way interactions between plants, microbes and arthropods. Funct. Ecol., 27: 646–660.

Blanchart, E., Villenave, C., Viallatoux, A., Barthès, B., Girardin, C., Azontonde, A. and Feller, C. (2006). Long-term effect of a legume cover crop (*Mucuna pruriens* var. *utilis*) on the communities of soil macrofauna and nematofauna, under maize cultivation, in southern Benin. Eur. J. Soil Biol., 42: S136–S144.

Blouin, M., Hodson, M.E., Delgado, E.A., Baker, G., Brussaard, L., Butt, K.R., Dai, J., Dendooven, L., Peres, G., Tondoh, J.E., Cluzeau, D. and Brun, J.-J. (2013). A review of earthworm impact on soil function and ecosystem services. Eur. J. Soil Sci., 64: 161–182.

Bolton, H., Elliott, L.F., Papendick, R.I. and Bezdicek, D.F. (1995). Soil microbial biomass and selected soil enzyme activities: Effect of fertilization and cropping practices. Soil Biol. Biochem., 17: 297–302.

Bongers, T. (1990). The maturity index: An ecological measure of environmental disturbance based on nematode species composition. Oecologia, 83: 14–19.

Bongers, T. and Bongers, M. (1998). Functional diversity of nematodes. Appl. Soil Ecol., 10: 239–251.

Bonkowski, M., Cheng, W., Griffiths, B.S., Alphei, J. and Scheu, S. (2000). Microbial-faunal interactions in the rhizosphere and effects on plant growth. Eur. J. Soil Biol., 36: 135–147.

Bonkowski, M. (2004). Protozoa and plant growth: The microbial loop in soil revisited. New Phytol., 162: 617–631.

Boswell, E.P., Koide, R.T., Shumway, D.L. and Addy, H.D. (1998). Winter wheat cover cropping, VA mycorrhizal fungi and maize growth and yield. Agric. Ecosyst. Environ., 67: 55–65.

Brévault, T., Bikay, S., Maldès, J.M. and Naudin, K. (2007). Impact of a no-till with mulch soil management strategy on soil macrofauna communities in a cotton cropping system. Soil Till. Res., 97: 140–149.

Briar, S.S., Grewal, P.S., Somasekhar, N., Stinner, D. and Miller, S.A. (2007). Soil nematode community, organic matter, microbial biomass and nitrogen dynamics in field plots transitioning from conventional to organic management. Appl. Soil Ecol., 37: 256–266.

Briar, S.S., Fonte, S.J., Park, I., Six, J., Scow, K. and Ferris, H. (2011). The distribution of nematodes and soil microbial communities across soil aggregate fractions and farm management systems. Soil Biol. Biochem., 43: 905–914.

Bronick, C.J. and Lal, R. (2005). Soil structure and management: A review. Geoderma, 124: 3–22.

Brookes, P.C. (1995). The use of microbial parameters in monitoring soil pollution by heavy metals. Biol. Fert. Soils, 19: 269–279.

Brussaard, L., de Ruiter, P.C. and Brown, G.G. (2007). Soil biodiversity for agricultural sustainability. Agric. Ecosyst. Environ., 121: 233–244.

Bugg, R.L., Sarrantonio, M., Dutcher, J.D. and Phatak, S.C. (1991). Understory cover crops in pecan orchards: Possible management systems. Am. J. Alternative Agr., 6: 50–62.

Bulluck III, L.R., Barker, K.R. and Ristaino, J.B. (2002). Influences of organic and synthetic soil fertility amendments on nematode trophic groups and community dynamics under tomatoes. Appl. Soil Ecol., 21: 233–250.

Bünemann, E.K., Bossio, D.A., Smithson, P.C., Frossard, E. and Oberson, A. (2004). Microbial community composition and substrate use in a highly weathered soil as affected by crop rotation and P fertilisation. Soil Biol. Biochem., 36: 889–901.

Buyer, J.S., Teasdale, J.R., Roberts, D.P., Zasada, I.A. and Maul, J.E. (2010). Factors affecting soil microbial community structure in tomato cropping systems. Soil Biol. Biochem., 42: 831–841.

Cambardella, C.A. and Elliott, E.T. (1993). Carbon and nitrogen distribution in aggregates from cultivated and native grassland soils. Soil Sci. Soc. Am. J., 57: 1071–1076.

Carbonetto, B., Rascovan, N., Álvarez, R., Mentaberry, A. and Vázquez, M.P. (2014). Structure, composition and metagenomic profile of soil microbiomes associated to agricultural land use and tillage systems in Argentine Pampas. PLoS One, 9: e99949.

Carmona, D.M. and Landis, D.A. (1999). Influence of refuge habitats and cover crops on seasonal activity-density of ground beetles (Coleoptera: Carabidae) in field crops. Environ. Entomol., 28: 1145–1153.

Carrera, L.M., Buyer, J.S., Vinyard, B., Abdul-Baki, A.A., Sikora, L.J. and Teasdale, J.R. (2007). Effects of cover crops, compost, and manure amendments on soil microbial community structure in tomato production systems. Appl. Soil Ecol., 37: 247–255.

Carter, M.R., Noronha, C., Peters, R.D. and Kimpinski, J. (2009). Influence of conservation tillage and crop rotation on the resilience of an intensive long-term potato cropping system: Restoration of soil biological properties after the potato phase. Agric. Ecosys. Environ., 133: 32–39.

Cavigelli, M.A. and Thien, S.J. (2003). Phosphorus bioavailability following incorporation of green manure crops. Soil Sci. Soc. Am. J., 67: 1186–1194.

Chan, K.Y. (2001). An overview of some tillage impacts on earthworm population abundance and diversity— Implications for functioning in soils. Soil Till. Res., 57: 179–191.

Chaplin-Kramer, R., O'Rourke, M.E., Blitzer, E.J. and Kremen, C. (2011). A meta-analysis of crop pest and natural enemy response to landscape complexity. Ecol. Lett., 14: 922–932.

Chavarría, D.N., Verdenelli, R.A., Serri, D.L., Restovich, S.B., Andriulo, A.E., Meriles, J.M. and Vargas-Gil, S. (2016). Effect of cover crops on microbial community structure and related enzyme activities and macronutrient availability. Eur. J. Soil Biol., 76: 74–82.

Cheng, X. and Baumgartner, K. (2005). Overlap of grapevine and cover-crop roots enhances interactions among grapevines, cover crops, and arbuscular mycorrhizal fungi. pp. 171–174. In: Soil Environment and Vine Mineral Nutrition: Symposium Proc. and Related Papers, San Diego, CA: Am. J. Enol. Vitic.

Cima, D.S. de, Tein, B., Eremeev, V., Luik, A., Kauer, K., Reintam, E. and Kahu, G. (2016). Winter cover crop effects on soil structural stability and microbiological activity in organic farming. Biol. Agric. Hortic., 32: 170–181.

Clarholm, M. (1981). Protozoan grazing of bacteria in soil—Impact and importance. Microb. Ecol., 7: 343–350.

Cochran, V.L., Sparrow, S.D. and Sparrow, E.B. (1994). Residues effects on soil micro- and macroorganisms. pp. 163–184. In: Unger, P.W. (ed.). Managing Agricultural Residues, CRC Press, Boca Raton, FL, USA.

Collins, H.P., Alva, A., Boydston, R.A., Cochran, R.L., Hamm, P.B., McGuire, A. and Riga, E. (2006). Soil microbial, fungal, and nematode responses to soil fumigation and cover crops under potato production. Biol. Fert. Soils, 42: 247–257.

Crossley, D.A., Mueller, B.R. and Perdue, J.C. (1992). Biodiversity of microarthropods in agricultural soils: relations to processes. Agric. Ecosyst. Environ., 40: 37–46.

Crotty, F.V., Fychan, R., Scullion, J., Sanderson, R. and Marley, C.L. (2015). Assessing the impact of agricultural forage crops on soil biodiversity and abundance. Soil Biol. Biochem., 91: 119–126.

Crotty, F.V., Fychan, R., Sanderson, R., Rhymes, J.R., Bourdin, F., Scullion, J. and Marley, C.L. (2016). Understanding the legacy effect of previous forage crop and tillage management on soil biology, after conversion to an arable crop rotation. Soil Biol. Biochem., 103: 241–252.

Dabney, S.M., Delgado, J.A. and Reeves, D.W. (2001). Using winter cover crops to improve soil and water quality. Commun. Soil Sci. Plant Anal., 32: 1221–1250.

Davis, J.R., Huisman, O.C., Westermann, D.T., Sorensen, L.H., Schneider, A.T. and Stark, J.C. (1994). The influence of cover crops on the suppression of Verticillium wilt of potato. Advances in Potato Pest Biology, Management, The Amer. Phytopathological Soc., St. Paul, MN.

De Deyn, G.B., Raaijmakers, C.E., Van Ruijven, J., Berendse, F. and Van Der Putten, W.H. (2004). Plant species identity and diversity effects on different trophic levels of nematodes in the soil food web. Oikos, 106: 576–586.

Detheridge, A.P., Brand, G., Fychan, R., Crotty, F.V., Sanderson, R., Griffith, G.W. and Marley, C.L. (2016). The legacy effect of cover crops on soil fungal populations in a cereal rotation. Agric. Ecosyst. Environ., 228: 49–61.

Dias, T., Dukes, A. and Antunes, P.M. (2015). Accounting for soil biotic effects on soil health and crop productivity in the design of crop rotations. J. Sci. Food Agric., 95: 447–454.

Diekötter, T., Wamser, S., Wolters, V. and Birkhofer, K. (2010). Landscape and management effects on structure and function of soil arthropod communities in winter wheat. Agric. Ecosyst. Environ., 137: 108–112.

Djigal, D., Chabrier, C., Duyck, P.-F., Achard, R., Quénéhervé, P. and Tixier, P. (2012). Cover crops alter the soil nematode food web in banana agro-ecosystems. Soil Biol. Biochem., 48: 142–150.

Doran, J.W. (1987). Microbial biomass and mineralisable nitrogen distributions in no-tillage and plowed soils. Biol. Fert. Soils, 5: 68–75.

DuPont, S.T., Ferris, H. and Van Horn, M. (2009). Effects of cover crop quality and quantity on nematode-based soil food webs and nutrient cycling. Appl. Soil Ecol., 41: 157–167.

Duyck, P.-F., Pavoine, S., Tixier, P., Chabrier, C. and Quénéhervé, P. (2009). Host range as an axis of niche partitioning in the plant-feeding nematode community of banana agro-ecosystems. Soil Biol. Biochem., 41: 1139–1145.

Duyck, P.-F., Lavigne, A., Vinatier, F., Achard, R., Okolle, J.N. and Tixier, P. (2011). Addition of a new resource in agroecosystems: Do cover crops alter the trophic positions of generalist predators? Basic Appl. Ecol., 12: 47–55.

Ekschmitt, K., Bakonyi, G., Bongers, M., Bongers, T., Boström, S., Dogan, H., Harrison, A., Kallimanis, A., Nagy, P., O'Donnell, A.G., Sohlenius, B., Stamou, G.P. and Wolters, V. (1999). Effects of the nematofauna on microbial energy and matter transformation rates in European grassland soils. Plant Soil, 212: 45–61.

Elliott, L.F. and Stott, D.E. (1997). Influence of no-till cropping systems on microbial relationships. pp. 121–147. *In*: Sparks, D.L. (ed.). Adv. Agron., Academic Press.

Eo, J.-U., Kang, S.-B., Park, K.-C., Han, K.-S. and Yi, Y.-K. (2010). Effects of cover plants on soil biota: A study in an apple orchard. Korean J. Environ. Agric., 29: 287–292.

Ettema, C.H. and Bongers, T. (1993). Characterisation of nematode colonisation and succession in disturbed soil using the Maturity Index. Biol. Fert. Soils, 16: 79–85.

Everts, K.L., Sardanelli, S., Kratochvil, R.J., Armentrout, D.K. and Gallagher, L.E. (2006). Root-knot and root-lesion nematode suppression by cover crops, poultry litter and poultry litter compost. Plant Dis., 90: 487–492.

Fageria, N.K., Baligar, V.C. and Bailey, B.A. (2005). Role of cover crops in improving soil and row crop productivity. Commun. Soil Sci. Plant Anal., 36: 2733–2757.

Ferris, H., Venette, R.C. and Lau, S.S. (1996). Dynamics of nematode communities in tomatoes grown in conventional and organic farming systems, and their impact on soil fertility. Appl. Soil Ecol., 3: 161–175.

Ferris, H., Bongers, T. and de Goede, R.G.M. (2001). A framework for soil food web diagnostics: extension of the nematode faunal analysis concept. Appl. Soil Ecol., 18: 13–29.

Ferris, H. and Matute, M.M. (2003). Structural and functional succession in the nematode fauna of a soil food web. Appl. Soil Ecol., 23: 93–110.

Ferris, H., Venette, R.C. and Scow, K.M. (2004). Soil management to enhance bacterivore and fungivore nematode populations and their nitrogen mineralisation function. Appl. Soil Ecol., 25: 19–35.

Ferris, H. (2010). Form and function: Metabolic footprints of nematodes in the soil food web. Eur. J. Soil Biol., 46: 97–104.

Flegel, M. and Schrader, S. (2000). Importance of food quality on selected enzyme activities in earthworm casts (Dendrobaena octaedra, Lumbricidae). Soil Biol. Biochem., 32: 1191–1196.

Forge, T.A., Ingham, R.E., Kaufman, D. and Pinkerton, J.N. (2000). Population growth of *Pratylenchus penetrans* on winter cover crops grown in the pacific northwest. J. Nematol., 32: 42–51.

Forge, T. and Simard, S. (2001). Structure of nematode communities in forest soils of southern British Columbia: Relationships to nitrogen mineralization and effects of clear-cut harvesting and fertilisation. Biol. Fertil. Soils, 34: 170–178.

Fox, C.A., Fonseca, E.J.A., Miller, J.J. and Tomlin, A.D. (1999). The influence of row position and selected soil attributes on Acarina and Collembola in no-till and conventional continuous corn on a clay loam soil. Appl. Soil Ecol., 13: 1–8.

Fründ, H.-C., Graefe, U. and Tischer, S. (2011). Earthworms as bioindicators of soil quality. pp. 261–278. *In*: Karaca, A. (ed.). Biology of Earthworms, Springer Berlin, Heidelberg.

Frye, W.W., Blevins, R.L., Smith, M.S., Corak, S.J. and Varco, J.J. (1988). Role of annual legume cover crops in efficient use of water and nitrogen cropping strategies for efficient use of water and nitrogen. Cropping Strategies for Efficient Use of Water and Nitrogen, Special Publication No. 51: 129–154.

Fu, S., Coleman, D.C., Hendrix, P.F. and Crossley Jr., D.A. (2000). Responses of trophic groups of soil nematodes to residue application under conventional tillage and no-till regimes. Soil Biol. Biochem., 32: 1731–1741.

Galvez, L., Douds, D.D., Wagoner, P., Longnecker, L.R., Drinkwater, L.E. and Janke, R.R. (1995). An overwintering cover crop increases inoculum of VAM fungi in agricultural soil. Am. J. Alternative Agr., 10: 152–156.

Gazaway, W.S., Akridge, J.R. and McLean, K. (2000). Impact of various crop rotations and various winter cover crops on reniform nematode in cotton. *In*: 2000 Proceedings Beltwide Cotton Conferences, San Antonio, USA, 4–8 January 2000, vol. 1, National Cotton Council, 2000.

Gonzalez-Quiñones, V., Stockdale, E.A., Banning, N.C., Hoyle, F.C., Sawada, Y., Wherrett, A.D., Jones, D.L. and Murphy, D.V. (2011). Soil microbial biomass—Interpretation and consideration for soil monitoring. Soil Res., 49: 287–304.

Govaerts, B., Mezzalama, M., Unno, Y., Sayre, K.D., Luna-Guido, M., Vanherck, K., Dendooven, L. and Deckers, J. (2007). Influence of tillage, residue management and crop rotation on soil microbial biomass and catabolic diversity. Appl. Soil Ecol., 37: 18–30.

Grewal, P.S., Ehlers, R.-U. and Shapiro-Ilan, D.I. (2005). Nematodes as Biocontrol Agents. CABI Publishing, Wallingford, UK.

Griffiths, B.S. (1994). Microbial-feeding nematodes and protozoa in soil: Their effects on microbial activity and nitrogen mineralisation in decomposition hotspots and the rhizosphere. Plant Soil, 164: 25–33.

Gunapala, N. and Scow, K.M. (1998). Dynamics of soil microbial biomass and activity in conventional and organic farming systems. Soil Biol. Biochem., 30: 805–816.

Hafez, S.L. and Sundararaj, P. (2000). Evaluation of chemical strategies along with cultural practices for the management of *Meloidogyne chitwoodi* on potato. Int. J. Nematol., 10: 89–93.

Hai-Ming, T., Xiao-Ping, X., Wen-Guang, T., Ye-Chun, L., Ke, W. and Guang-Li, Y. (2014). Effects of winter cover crops residue returning on soil enzyme activities and soil microbial community in double-cropping rice fields. PLoS One, 9: e100443.

Hansen, E.M., Myrold, D.D. and Hamm, P.B. (1990). Effects of soil fumigation and cover crops on potential pathogens, microbial activity, nitrogen availability and seedling quality in conifer nurseries. Phytopathology, 80: 698–704.

Hartsema, O.H., Koot, P., Molendijk, L.P.G., Van Den Berg, W., Plentinger, M.C. and Hoek, J. (2005). Rotatie onderzoek Paratrichodorus teres, Praktijkonderzoek Plant en Omgeving, Wageningen UR, The Netherlands.

Heisler, C. and Kaiser, E.-A. (1995). Influence of agricultural traffic and crop management on collembola and microbial biomass in arable soil. Biol. Fert. Soils, 19: 159–165.

Hoek, J., Brommer, E. and Molendijk, L.P.G. (2006). Groenbemesters als voorvrucht van zetmeelaardappelen, PPO - AGV, Lelystad, p. 23.

Hooks, C.R.R., Wang, K.-H., Ploeg, A. and McSorley, R. (2010). Using marigold (*Tagetes* spp.) as a cover crop to protect crops from plant-parasitic nematodes. Appl. Soil Ecol., 46: 307–320.

House, G.J. and Parmelee, R.W. (1985). Comparison of soil arthropods and earthworms from conventional and no-tillage agro-ecosystems. Soil Till. Res., 5: 351–360.

Hoy, C.W., Grewal, P.S., Lawrence, J.L., Jagdale, G. and Acosta, N. (2008). Canonical correspondence analysis demonstrates unique soil conditions for entomopathogenic nematode species compared with other free-living nematode species. Biol. Control, 46: 371–379.

Hugot, J.-P., Baujard, P. and Morand, S. (2001). Biodiversity in helminths and nematodes as a field of study: An overview. Nematol., 3: 199–208.

Ilieva-Makulec, K., Olejniczak, I. and Szanser, M. (2006). Response of soil micro- and mesofauna to diversity and quality of plant litter. Eur. J. Soil Biol., 42, supplement 1: S244–S249.

Ingels, C.A., Scow, K.M., Whisson, D.A. and Drenovsky, R.E. (2005). Effects of cover crops on grapevines, yield, juice composition, soil microbial ecology, and gopher activity. Am. J. Enol. Vitic., 56: 19–29.

Ingham, R.E., Trofymow, J.A., Ingham, E.R. and Coleman, D.C. (1985). Interactions of bacteria, fungi, and their nematode grazers: effects on nutrient cycling and plant growth. Ecol. Monogr., 55: 119–140.

Ito, T., Araki, M., Komatsuzaki, M., Kaneko, N. and Ohta, H. (2015). Soil nematode community structure affected by tillage systems and cover crop managements in organic soybean production. Appl. Soil Ecol., 86: 137–147.

Jäggi, W., Oberholzer, H.R. and Waldburger, M. (1995). Vier Maisanbauverfahren 1990 bis 1993: Auswirkungen auf das Bodenleben. Agrarforschung, 2: 361–364.

Jentschke, G., Bonkowski, M., Godbold, D.L. and Scheu, S. (1995). Soil protozoa and forest tree growth: non-nutritional effects and interaction with mycorrhizae. Biol. Fert. Soils, 20: 263–269.

Jordan, D., Miles, R.J., Hubbard, V.C. and Lorenz, T. (2004). Effect of management practices and cropping systems on earthworm abundance and microbial activity in Sanborn Field: A 115-year-old agricultural field. Pedobiologia, 48: 99–110.

Kabir, Z. and Koide, R.T. (2000). The effect of dandelion or a cover crop on mycorrhiza inoculum potential, soil aggregation and yield of maize. Agric. Ecosyst. Environ., 78: 167–174.

Kautz, T., Lüsebrink, M., Pätzold, S., Vetterlein, D., Pude, R., Athmann, M., Küpper, P.M., Perkons, U. and Köpke, U. (2014). Contribution of anecic earthworms to biopore formation during cultivation of perennial ley crops. Pedobiologia, 57: 47–52.

Keene, C.L., Curran, W.S., Wallace, J.M., Ryan, M.R., Mirsky, S.B., VanGessel, M.J. and Barbercheck, M.E. (2017). Cover crop termination timing is critical in organic rotational no-till systems. Agron. J., 109: 272–282.

Kennedy, A.C. and Papendick, R.I. (1995). Microbial characteristics of soil quality. J. Soil Water Conser., 50: 243–248.

Ketterings, Q.M., Blair, J.M. and Marinissen, J.C.Y. (1997). Effects of earthworms on soil aggregate stability and carbon and nitrogen storage in a legume cover crop agro-ecosystem. Soil Biol. Biochem., 29: 401–408.

Kimpinski, J. and Sturz, A.V. (1996). Population growth of a rhabditid nematode on plant growth promoting bacteria from potato tubers and rhizosphere. Soil J. Nematol., 28: 682–686.

Kirchner, M.J., Wollum, A.G. and King, L.D. (1993). Soil microbial populations and activities in reduced chemical input agro-ecosystems. Soil Sci. Soc. Am. J., 57: 1289–1295.

Kladivko, E.J. (2001). Tillage systems and soil ecology. Soil Till. Res., 61: 61–76.

Knox, O.G.G., Killham, K., Mullins, C.E. and Wilson, M.J. (2003). Nematode-enhanced microbial colonisation of the wheat rhizosphere FEMS. Microbiol. Lett., 225: 227–233.

Koehler, H.H. (1999). Predatory mites (Gamasina, Mesostigmata). Agric. Ecosys. Environ., 74: 395–410.

Kromp, B. (1999). Carabid beetles in sustainable agriculture: A review on pest control efficacy, cultivation impacts and enhancement. Agric. Ecosyst. Environ., 74: 187–228.

Kruger, D.H.M., Fourie, J.C. and Malan, A.P. (2013). Cover crops with biofumigation properties for the suppression of plant-parasitic nematodes: A review. S. Afr. J. Enol. Vitic., 34: 287–295.

Kuikman, P.J., Jansen, A.G., van Veen, J.A. and Zehnder, A.J.B. (1990). Protozoan predation and the turnover of soil organic carbon and nitrogen in the presence of plants. Biol. Fert. Soils, 10: 22–28.

Kumar, K. and Goh, K.M. (2000). Biological nitrogen fixation, accumulation of soil nitrogen and nitrogen balance for white clover (*Trifolium repens* L.) and field pea (*Pisum sativum* L.) grown for seed. Field Crops Res., 68: 49–59.

Kuntz, M., Berner, A., Gattinger, A., Scholberg, J.M., Mäder, P. and Pfiffner, L. (2013). Influence of reduced tillage on earthworm and microbial communities under organic arable farming. Pedobiologia, 56: 251–260.

Laakso, J., Setälä, H. and Palojärvi, A. (2000). Influence of decomposer food web structure and nitrogen availability on plant growth. Plant Soil, 225: 153–165.

LaMondia, J.A. (2006). Management of lesion nematodes and potato early dying with rotation crops. J. Nematol., 38: 442–448.

Landis, D.A., Wratten, S.D. and Gurr, G.M. (2000). Habitat management to conserve natural enemies of arthropod pests in agriculture. Ann. Rev. Entomol., 45: 175–201.

Langdale, G.W., Blevins, R.L., Karlen, D.L., McCool, D.K., Nearing, M.A., Skidmore, E.L., Thomas, A.W., Tyler, D.D. and Williams, J.R. (1991). Cover crop effects on soil erosion by wind and water. Cover Crops for Clean Water, 15–22.

Larkin, R.P., Griffin, T.S. and Honeycutt, C.W. (2010). Rotation and cover crop effects on soil-borne potato diseases, tuber yield, and soil microbial communities. Plant Dis., 94: 1491–1502.

Lavelle, P. (1997). Faunal activities and soil processes: Adaptive strategies that determine ecosystem function. pp. 93–132. *In*: Fitter, A.H. (ed.). Adv. Ecol. Res., Academic Press.

Lavelle, P. (2000). Ecological challenges for soil science. Soil Sci., 165: 73–86.

Lavelle, P., Decaëns, T., Aubert, M., Barot, S., Blouin, M., Bureau, F., Margerie, P., Mora, P. and Rossi, J.-P. (2006). Soil invertebrates and ecosystem services. Eur. J. Soil Biol., 42: S3–S15.

Lee, D.L. (2002). The Biology of Nematodes. CRC Press, Boca Raton, FL.

Lemtiri, A., Colinet, G., Alabi, T., Cluzeau, D., Zirbes, L., Haubruge, É. and Francis, F. (2014). Impacts of earthworms on soil components and dynamics. A review. Biotechnol. Agron. Soc. Environ, Gembloux, 18: 121–133.

Lenz, R. and Eisenbeis, G. (2000). Short-term effects of different tillage in a sustainable farming system on nematode community structure. Biol. Fert. Soils, 31: 237–244.

Liang, W., Lou, Y., Li, Q., Zhong, S., Zhang, X. and Wang, J. (2009). Nematode faunal response to long-term application of nitrogen fertiliser and organic manure in Northeast China. Soil Biol. Biochem., 41: 883–890.

Liebig, M.A., Tanaka, D.L. and Wienhold, B.J. (2004). Tillage and cropping effects on soil quality indicators in the northern Great Plains. Soil Till. Res., 78: 131–141.

Lundquist, E.J., Jackson, L.E., Scow, K.M. and Hsu, C. (1999). Changes in microbial biomass and community composition, and soil carbon and nitrogen pools after incorporation of rye into three California agricultural soils. Soil Biol. Biochem., 31: 221–236.

Marahatta, S.P., Wang, K.-H., Sipes, B.S. and Hooks, C.R.R. (2010). Strip-tilled cover cropping for managing nematodes, soil mesoarthropods and weeds in a bitter melon agro-ecosystem. J. Nematol., 42: 111–119.

Marban-Mendoza, N., Bess Dicklow, M. and Zuckerman, B.M. (1992). Control of *Meloidogyne incognita* on tomato by two leguminous plants. Fundam. Appl. Nematol., 15: 97–100.

Mbuthia, L.W., Acosta-Martínez, V., DeBruyn, J., Schaeffer, S., Tyler, D., Odoi, E., Mpheshea, M., Walker, F. and Eash, N. (2015). Long-term tillage, cover crop and fertilisation effects on microbial community structure, activity: Implications for soil quality. Soil Biol. Biochem., 89: 24–34.

McKeown, A.W. and Potter, J.W. (2001). Yield of 'Superior' potatoes (*Solanum tuberosum*) and dynamics of root-lesion nematode (*Pratylenchus penetrans*) populations following 'nematode suppressive' cover crops and fumigation. Phyto., 82: 13–23.

McSorley, R. and Gallaher, R.N. (1992). Comparison of nematode population densities on six summer crops at seven sites in North Florida. J. Nematol., 24: 699–706.

McSorley, R., Wang, K.H. and Frederick, J.J. (2008). Integrated effects of solarisation, Sunn hemp cover crop, and amendment on nematodes, weeds and pepper yields. Nematropica, 38: 115–126.

McSorley, R., Seal, D.R., Klassen, W., Wang, K.H. and Hooks, C.R.R. (2009). Non-target effects of Sunn hemp and marigold cover crops on the soil invertebrate community. Nematropica, 39: 235–245.

Mendes, I.C., Bandick, A.K., Dick, R.P. and Bottomley, P.J. (1999). Microbial biomass and activities in soil aggregates affected by winter cover crops. Soil Sci. Soc. Am. J., 63: 873–881.

Millennium Ecosystem Assessment. (2005). Ecosystems and Human Well Being: Synthesis. Island Press, Washington DC.

Morse, R.D. (1993). Components of sustainable production systems for vegetables-conserving soil moisture. Hort. Tech., 3: 211–214.

Mummey, D., Holben, W., Six, J. and Stahl, P. (2006). Spatial stratification of soil bacterial populations in aggregates of diverse soils. Microb. Ecol., 51: 404–411.

Nair, A. and Ngouajio, M. (2012). Soil microbial biomass, functional microbial diversity, and nematode community structure as affected by cover crops and compost in an organic vegetable production system. Appl. Soil Ecol., 58: 45–55.

Nakamoto, T. and Tsukamoto, M. (2006). Abundance and activity of soil organisms in fields of maize grown with a white clover living mulch. Agric. Ecosyst. Environ., 115: 34–42.

Ndiaye, E.L., Sandeno, J.M., McGrath, D. and Dick, R.P. (2000). Integrative biological indicators for detecting change in soil quality. Am. J. Alternative Agric., 15: 26–36.

Neher, D.A. (1999). Nematode communities in organically and conventionally managed agricultural soils. J. Nematol., 31: 142–154.

Neher, D.A. and Olson, R.K. (1999). Nematode communities in organically and conventionally managed agricultural soils. J. Nematol., 31: 142–154.

Neher, D.A. (2001). Role of nematodes in soil health and their use as indicators. J. Nematol., 33: 161–168.

Neher, D.A. (2010). Ecology of plant and free-living nematodes in natural and agricultural soil. Annual Review of Phytopathology, 48: 371–394.

Nyoike, T.W. and Liburd, O.E. (2010). Effect of living (buckwheat) and UV reflective mulches with and without imidacloprid on whiteflies, aphids and marketable yields of zucchini squash. Int. J. Pest Manage., 56: 31–39.

Ortiz-Ceballos, A.I. and Fragoso, C. (2004). Earthworm populations under tropical maize cultivation: the effect of mulching with velvet bean. Biol. Fert. Soils, 39: 438–445.

Osler, G.H.R., Harrison, L., Kanashiro, D.K. and Clapperton, M.J. (2008). Soil microarthropod assemblages under different arable crop rotations in Alberta, Canada. Appl. Soil Ecol., 38: 71–78.

Overstreet, L.F., Hoyt, G.D. and Imbriani, J. (2010). Comparing nematode and earthworm communities under combinations of conventional and conservation vegetable production practices. Soil Till. Res., 110: 42–50.

Pal, K.K. and McSpadden Gardener, B. (2006). Biological control of plant pathogens. The Plant Health Instructor, 2: 1117–1142.

Paoletti, M.G. (1999). The role of earthworms for assessment of sustainability and as bioindicators. Agric. Ecosyst. Environ., 74: 137–155.

Patkowska, E. and Konopinski, M. (2011). Cover crops and soil-borne fungi dangerous towards the cultivation of salsify (*Tragopogon porrifolius* var. *sativus* (Gaterau) Br.). Acta Sci. Pol. Hortorum Cultus, 2: 167–181.

Patkowska, E. and Konopiński, M. (2013). Effect of cover crops on the microorganism's communities in the soil under Scorzonera cultivation. Plant Soil Environ., 59: 460–464.

Patrick, Z.A., Sayre, R.M. and Thorpe, H.J. (1965). Nematocidal substances selective for plant-parasitic nematodes in extracts of decomposing rye. Phytopathology, 55: 702–704.

Peachey, R.E., Moldenke, A., William, R.D., Berry, R., Ingham, E. and Groth, E. (2002). Effect of cover crops and tillage system on symphylan (Symphlya: *Scutigerella immaculata*, Newport) and *Pergamasus quisquiliarum* Canestrini (Acari: Mesostigmata) populations, and other soil organisms in agricultural soils. Appl. Soil Ecol., 21: 59–70.

Peigné, J., Cannavacuolo, M., Gautronneau, Y., Aveline, A., Giteau, J.L. and Cluzeau, D. (2009). Earthworm populations under different tillage systems in organic farming. Soil Till. Res., 104: 207–214.

Pelosi, C., Bertrand, M. and Roger-Estrade, J. (2009). Earthworm community in conventional, organic and direct seeding with living mulch cropping systems. Agron. Sustain. Dev., 29: 287–295.

Pelosi, C., Pey, B., Caro, G., Cluzeau, D., Peigné, J., Bertrand, M. and Hedde, M. (2016). Dynamics of earthworm taxonomic and functional diversity in ploughed and no-tilled cropping systems. Soil Till. Res., 156: 25–32.

Pimentel, D., Harvey, C., Resosudarmo, P. and Sinclair, K. (1995). Environmental and economic costs of soil erosion and conservation benefits. Science, 267: 1117.

Pinkerton, J.N., Ivors, K.L., Miller, M.L. and Moore, L.W. (2000). Effect of soil solarisation and cover crops on populations of selected soil-borne plant pathogens in western Oregon. Plant Dis., 84: 952–960.

Pompili, L., Mellina, A.S. and Benedtti, A. (2006). Microbial indicators for evaluating soil quality in differently managed soils. Geophysic. Res. Abst., 8: 06991.

Porazinska, D.L., Duncan, L.W., McSorley, R. and Graham, J.H. (1999). Nematode communities as indicators of status and processes of a soil ecosystem influenced by agricultural management practices. Appl. Soil Ecol., 13: 69–86.

Raaijmakers, J.M., Paulitz, T.C., Steinberg, C., Alabouvette, C. and Moënne-Loccoz, Y. (2009). The rhizosphere: A playground and battlefield for soil-borne pathogens and beneficial microorganisms. Plant Soil, 321: 341–361.

Rainio, J. and Niemelä, J. (2003). Ground beetles (Coleoptera: Carabidae) as bioindicators. Biodivers. Conserv., 12: 487–506.

Ranells, N.N. and Wagger, M.G. (1996). Nitrogen release from grass and legume cover crop monocultures and bicultures. Agron. J., 88: 777–882.

Rasmussen, P.E., Goulding, K.W.T., Brown, J.R., Grace, P.R., Janzen, H.H. and Körschens, M. (1998). Long-term agro-ecosystem experiments: Assessing agricultural sustainability and global change. Science, 282: 893–896.

Reddy, K.N., Zablotowicz, R.M., Locke, M.A. and Koger, C.H. (2003). Cover crop, tillage, and herbicide effects on weeds, soil properties, microbial populations and soybean yield. Weed Sci., 51: 987–994.

Reeleder, R.D., Miller, J.J., Ball Coelho, B.R. and Roy, R.C. (2006). Impacts of tillage, cover crop, and nitrogen on populations of earthworms, microarthropods and soil fungi in a cultivated fragile soil. Appl. Soil Ecol., 33: 243–257.

Richardson, A.E., Barea, J.-M., McNeill, A.M. and Prigent-Combaret, C. (2009). Acquisition of phosphorus and nitrogen in the rhizosphere and plant growth promotion by microorganisms. Plant Soil, 321: 305–339.

Ritz, K. and Trudgill, D.L. (1999). Utility of nematode community analysis as an integrated measure of the functional state of soils: Perspectives and challenges. Plant Soil, 212: 1–11.

Rivers, A., Mullen, C., Wallace, J. and Barbercheck, M. (2017). Cover crop-based reduced tillage system influences Carabidae (Coleoptera) activity, diversity and trophic group during transition to organic production. Renew. Agr. Food Syst., 1–14.

Robertson, G.P., Paul, E.A. and Harwood, R.R. (2000). Greenhouse gases in intensive agriculture: Contributions of individual gases to the radiative forcing of the atmosphere. Science, 289: 1922–1925.

Rodríguez, E., Fernández-Anero, F.J., Ruiz, P. and Campos, M. (2006). Soil arthropod abundance under conventional and no tillage in a Mediterranean climate. Soil Till. Res., 85: 229–233.

Roger-Estrade, J., Anger, C., Bertrand, M. and Richard, G. (2010). Tillage and soil ecology: Partners for sustainable agriculture. Soil Till. Res., 111: 33–40.

Rouatt, J.W. and Atkinson, R.G. (1950). The effect of the incorporation of certain cover crops on the microbiological balance of potato scab infested soil. Can. J. Res., 28: 140–152.

Ruess, L. and Ferris, H. (2004). Decomposition pathways and successional changes. Nematol. Monogr. Persp., 2: 547–556.

Ruffo, M.L. and Bollero, G.A. (2003). Modelling rye and hairy vetch residue decomposition as a function of degree-days and decomposition-days. Agron. J., 95: 900–907.

Runia, W.T., van Gastel, W. and Korthals, G.W. (2006). Inventarisatie en beheersing van het quarantaine aaltje *Meloidogyne chitwoodi* binnen de pootgoedteelt in de Wieringermeer. PPO-AGV Rapport Project, 520117: 19.

Sainju, U.M., Whitehead, W.F. and Singh, B.P. (2003). Cover crops and nitrogen fertilisation effects on soil aggregation and carbon and nitrogen pools. Can. J. Soil. Sci., 83: 155–165.

Sainju, U.M., Schomberg, H.H., Singh, B.P., Whitehead, W.F., Tillman, P.G. and Lachnicht-Weyers, S.L. (2007). Cover crop effect on soil carbon fractions under conservation tillage cotton. Soil Till. Res., 96: 205–218.

Säle, V., Aguilera, P., Laczko, E., Mäder, P., Berner, A., Zihlmann, U., van der Heijden, M.G.A. and Oehl, F. (2015). Impact of conservation tillage and organic farming on the diversity of arbuscular mycorrhizal fungi. Soil Biol. Biochem., 84: 38–52.

Sánchez-Moreno, S., Nicola, N.L., Ferris, H. and Zalom, F.G. (2009). Effects of agricultural management on nematode–mite assemblages: Soil food web indices as predictors of mite community composition. Appl. Soil Ecol., 41: 107–117.

Sapkota, T.B., Mazzoncini, M., Bàrberi, P., Antichi, D. and Silvestri, N. (2012). Fifteen years of no till increase soil organic matter, microbial biomass and arthropod diversity in cover crop-based arable cropping systems. Agron. Sustain. Dev., 32: 853–863.

Scherr, S.J. and McNeely, J.A. (2008). Biodiversity conservation and agricultural sustainability: Towards a new paradigm of 'ecoagriculture' landscapes. Philos. Trans. R. Soc., Lond., B, Biol. Sci., 363: 477–494.

Schmidt, N.P., O'Neal, M.E. and Singer, J.W. (2007). Alfalfa living mulch advances biological control of soybean aphid. Environ. Entomol., 36: 416–424.

Schmidt, O. and Curry, J.P. (2001). Population dynamics of earthworms (Lumbricidae) and their role in nitrogen turnover in wheat and wheat clover cropping systems. Pedobiologia, 45: 174–187.

Schmidt, O., Curry, J.P., Hackett, R.A., Purvis, G. and Clements, R.O. (2001). Earthworm communities in conventional wheat monocropping and low-input wheat-clover intercropping systems. Ann. Appl. Biol., 138: 377–388.

Schmidt, O., Clements, R.O. and Donaldson, G. (2003). Why do cereal-legume intercrops support large earthworm populations? Appl. Soil Ecol., 22: 181–190.

Schrader, S. and Seibel, C. (2001). Impact of cultivation management in an agro-ecosystem on hot spot effects of earthworm middens. Eur. J. Soil Biol., 37: 309–313.

Schutter, M.E. and Dick, R.P. (2002). Microbial community profiles and activities among aggregates of winter fallow and cover-cropped soil. Soil Sci. Soc. Am. J., 66: 142–153.

Schutter, M., Sandeno, J. and Dick, R. (2001). Seasonal, soil type, and alternative management influences on microbial communities of vegetable cropping systems. Biol. Fert. Soils, 34: 397–410.

Simmons, B.L. and Coleman, D.C. (2008). Microbial community response to transition from conventional to conservation tillage in cotton fields. Appl. Soil Ecol., 40: 518–528.

Sirrine, J., Letourneau, D.K., Shennan, C., Sirrine, D., Fouch, R., Jackson, L. and Mages, A. (2008). Impacts of groundcover management systems on yield, leaf nutrients, weeds, and arthropods of tart cherry in Michigan, USA. Agric. Ecosyst. Environ., 125: 239–245.

Snapp, S.S., Swinton, S.M., Labarta, R., Mutch, D., Black, J.R., Leep, R., Nyiraneza, J. and O'Neil, K. (2005). Evaluating cover crops for benefits, costs and performance within cropping system niches. Agron. J., 97: 322–332.

St. John, M.G., Wall, D.H. and Behan-Pelletier, V.M. (2006). Does plant species co-occurrence influence soil mite diversity? Ecol., 87: 625–633.

Stinner, B.R. and House, G.J. (1990). Arthropods and other invertebrates in conservation-tillage agriculture. Ann. Rev. Entomol., 35: 299–318.

Takeda, M., Nakamoto, T., Miyazawa, K., Murayama, T. and Okada, H. (2009). Phosphorus availability and soil biological activity in an Andosol under compost application and winter cover cropping. Appl. Soil Ecol., 42: 86–95.

Tejada, M., Gonzalez, J.L., García-Martínez, A.M. and Parrado, J. (2008). Effects of different green manures on soil biological properties and maize yield. Bioresour. Tech., 99: 1758–1767.

Thoden, T.C., Hallmann, J. and Boppré, M. (2009). Effects of plants containing pyrrolizidine alkaloids on the northern root-knot nematode *Meloidogyne hapla*. Eur. J. Plant Pathol., 123: 27.

Thoden, T.C., Korthals, G.W. and Termorshuizen, A.J. (2011). Organic amendments and their influences on plant-parasitic and free-living nematodes: A promising method for nematode management? Nematology, 13: 133–153.

Tian, Y., Zhang, X., Liu, J. and Gao, L. (2011). Effects of summer cover crop and residue management on cucumber growth in intensive Chinese production systems: Soil nutrients, microbial properties and nematodes. Plant Soil, 339: 299–315.

Tixier, P., Lavigne, C., Alvarez, S., Gauquier, A., Blanchard, M., Ripoche, A. and Achard, R. (2011). Model evaluation of cover crops, application to eleven species for banana cropping systems. Eur. J. Agron., 34: 53–61.

Tscharntke, T., Klein, A.M., Kruess, A., Steffan-Dewenter, I. and Thies, C. (2005). Landscape perspectives on agricultural intensification and biodiversity—Ecosystem service management. Ecol. Lett., 8: 857–874.

Urzelai, A., Hernández, A.J. and Pastor, J. (2000). Biotic indices based on soil nematode communities for assessing soil quality in terrestrial ecosystems. Sci. Total Environ., 247: 253–261.

van Bruggen, A.H.C. and Semenov, A.M. (2000). In search of biological indicators for soil health and disease suppression. Appl. Soil Ecol., 15: 13–24.

van Capelle, C., Schrader, S. and Brunotte, J. (2012). Tillage-induced changes in the functional diversity of soil biota—A review with a focus on German data. Eur. J. Soil Biol., 50: 165–181.

van Eekeren, N., van Liere, D., de Vries, F., Rutgers, M., de Goede, R. and Brussaard, L. (2009). A mixture of grass and clover combines the positive effects of both plant species on selected soil biota. Appl. Soil Ecol., 42: 254–263.

van Groenigen, J.W., Lubbers, I.M., Vos, H.M.J., Brown, G.G., Deyn, G.B.D. and van Groenigen, K.J. (2014). Earthworms increase plant production: A meta-analysis. Sci. Rep., 4: 6365.

Van Vliet, P.C.J., Beare, M.H. and Coleman, D.C. (1995). Population dynamics and functional roles of Enchytraeidae (Oligochaeta) in hardwood forest and agricultural ecosystems. Plant and Soil, 170: 199–207.

Viaene, N.M. and Abawi, G.S. (1998). Management of *Meloidogyne hapla* on lettuce in organic soil with sudangrass as a cover crop. Plant Dis., 82: 945–952.

Viketoft, M., Bengtsson, J., Sohlenius, B., Berg, M.P., Petchey, O., Palmborg, C. and Huss-Danell, K. (2009). Long-term effects of plant diversity and composition on soil nematode communities in model grasslands. Ecol., 90: 90–99.

Vineela, C., Wani, S.P., Srinivasarao, C., Padmaja, B. and Vittal, K.P.R. (2008). Microbial properties of soils as affected by cropping and nutrient management practices in several long-term manurial experiments in the semi-arid tropics of India. Appl. Soil Ecol., 40: 165–173.

Vliet, P.C.J. van, Beare, M.H. and Coleman, D.C. (1995). Population dynamics and functional roles of *Enchytraeidae* (Oligochaeta) in hardwood forest and agricultural ecosystems. Plant Soil, 170: 199–207.

Wagner, S.C., Zablotowicz, R.M., Locke, M.A., Smeda, R.J. and Bryson, C.T. (1995). Influence of herbicide-desiccated cover crops on biological soil quality in the Mississippi Delta conservation farming: A focus on water quality. MAFES Spec. Bull., 88: 86–89.

Wang, K.-H., McSorley, R. and Gallaher, R.N. (2004). Effect of winter cover crops on nematode population levels in North Florida. J. Nematol., 36: 517–523.

Wang, K.-H., McSorley, R. and Kokalis-Burelle, N. (2006). Effects of cover cropping, solarisation, and soil fumigation on nematode communities. Plant Soil, 286: 229–243.

Wang, K.-H., McSorley, R., Gallaher, R.N. and Kokalis-Burelle, N. (2008). Cover crops and organic mulches for nematode, weed and plant health management. Nematology, 10: 231–242.

Wang, K.-H., Hooks, C.R.R. and Marahatta, S.P. (2011). Can using a strip-tilled cover cropping system followed by surface mulch practice enhance organisms higher up in the soil food web hierarchy? Appl. Soil Ecol., 49: 107–117.

Wang, Q.R., Li, Y.C. and Klassen, W. (2007). Changes of soil microbial biomass carbon and nitrogen with cover crops and irrigation in a tomato field. J. Plant Nutr., 30: 623–639.

Wardle, D.A., Yeates, G.W., Barker, G.M. and Bonner, K.I. (2006). The influence of plant litter diversity on decomposer abundance and diversity. Soil Biol. Biochem., 38: 1052–1062.

Watt, M., Kirkegaard, J.A. and Passioura, J.B. (2006). Rhizosphere biology and crop productivity—A review. Soil Res., 44: 299–317.

West, P.M. and Hildebrand, A.A. (1941). The microbiological balance of strawberry root rot soil as related to the rhizosphere and decomposition effects of certain cover crops. Can. J. Res., 19c: 199–210.

White, C.M. and Weil, R.R. (2010). Forage radish and cereal rye cover crop effects on mycorrhizal fungus colonization of maize roots. Plant Soil, 328: 507–521.

Widmer, T.L. and Abawi, G.S. (2002). Relationship between levels of cyanide in sudangrass hybrids incorporated into soil and suppression of *Meloidogyne hapla*. J. Nematol., 34: 16–22.

Wyss, E. and Glasstetter, M. (1992). Tillage treatments and earthworm distribution in a Swiss experimental corn field. Soil Biol. Biochem., 24: 1635–1639.

Yeates, G.W. (1979). Soil nematodes in terrestrial ecosystems. J. Nematol., 11: 213–229.

Yeates, G.W., Bongers, T., De Goede, R.G.M., Freckman, D.W. and Georgieva, S.S. (1993). Feeding habits in soil nematode families and genera—An outline for soil ecologists. J. Nematol., 25: 315–331.

Yeates, G.W. and Wardle, D.A. (1996). Nematodes as predators and prey: Relationships to biological control and soil processes. Pedobiologia, 40: 43–50.

Yeates, G.W. and Bongers, T. (1999). Nematode diversity in agro-ecosystems. Agric. Ecosyst. Environ., 74: 113–135.

Yeates, G.W., Wardle, D.A. and Watson, R.N. (1999). Responses of soil nematode populations, community structure, diversity and temporal variability to agricultural intensification over a seven-year period. Soil Biol. Biochem., 31: 1721–1733.

Zablotowicz, R.M., Locke, M.A. and Gaston, L.A. (2007). Tillage and cover effects on soil microbial properties and fluometuron degradation. Biol. Fer. Soils, 44: 27–35.

Zagal, E. and Córdova, C. (2005). Soil organic matter quality indicators in a cultivated Andisol. Agricultura Técnica, 65: 186–197.

Zhang, X., Li, Q., Zhu, A., Liang, W., Zhang, J. and Steinberger, Y. (2012). Effects of tillage and residue management on soil nematode communities in North China. Ecol. Indic., 13: 75–81.

Zhu, B., Yi, L., Guo, L., Chen, G., Hu, Y., Tang, H., Xiao, C., Xiao, X., Yang, G., Acharya, S.N. and Zeng, Z. (2012). Performance of two winter cover crops and their impacts on soil properties and two subsequent rice crops in Dongting Lake Plain, Hunan, China. Soil Till. Res., 124: 95–101.

14

Cover Cropping Improves Soil Quality and Physical Properties

Yilmaz Bayhan

1. Introduction

Recent environmental concerns and ecological awareness have focused on a resurgence of cover crop use in sustainable agriculture. Cover crops improve soil physical, chemical and biological properties, supply nutrients to the succeeding crop, suppress weeds, increase soil structural stability, prevent erosion and break pest cycles. Some cover crops are able to break into compacted soil layers, making it easier for the succedding cash crops' roots to develop more fully. However, the actual benefits from cover crops depend on the species and productivity of the crop to grow and the duration it is allowed to grow before the soil is prepared for the next crop (Anonymous, 2009).

Cover crops are planted between two main crops and are known to provide various ecological services in agro-ecosystems, such as protection against soil erosion, reduction of nutrient losses, improvement of soil and water quality, increase soil fertility and to some extent, the reduction of weeds and pests (Dabney et al., 2001; Dorn et al., 2015; Wittwer et al., 2017). Furthermore, adding a legume species as a cover crop can improve N nutrition of the succeeding main crop and increase the soil N organic pool. Thus, cover crops can contribute to sustainable agriculture and alleviate weed and crop-nutrition issues related to both organic and conservation agriculture. Despite these advantages, cover crops are generally not widely used by producers, mainly due to seed availability, additional costs and labour requirements. Moreover, cover crop effects on productivity, crop nutrition and weed control are variable and depend on the cover-crop species, soil type and climate (Thorup-Kristensen et al., 2003; Wittwer et al., 2017).

The use of cover-crop cultivars and blends enhance soil quality by increasing biodiversity and efficiency (by providing food and energy resources) for soil-borne beneficial organisms, which in turn creates a balanced soil that resists degradation and responds to management practices in a predictable manner (Sylvester and Bird, 2017). An illustration of how cover crops work is shown in Fig. 1.

Grass cover crops, such as rye, generally reduce nitrate-N loss, but the magnitude of the leaching reduction effect also varies widely across years, locations and management (Dabney et al., 2010; Dinnes et al., 2002; Kaspar and Singer, 2011; Thorup-Kristensen et al., 2003; Martinez-Feria et al., 2016). This indicates that rye effects on the maize system depend on specific combinations of management choices and environmental conditions. Most studies have focused on quantifying rye effects on final maize yields and/or annual nitrate-N losses but a knowledge gap still exists regarding the mechanisms by which rye affects these systems.

Tekirdag Namik Kemal University, Tekirdag, Turkey.

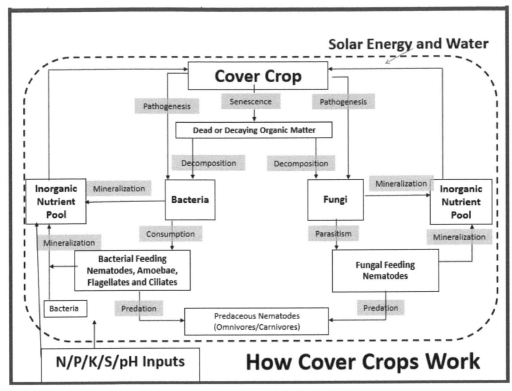

Fig. 1 How cover crops work? (*Source*: Sylvester and Bird, 2017).

A larger body of evidence exists for abiotic factors, which allow us to develop a generalised framework of the abiotic effects of rye on the maize system (Fig. 2; Martinez-Feria et al., 2016). Research findings have shown maize yield reductions following a rye cover crop, to be related to the depletion of soil moisture or suspected allelopathic effects. The collection of data during years

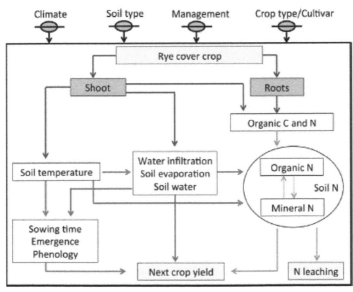

Fig. 2 A generalised diagram showing the abiotic mechanisms by which rye cover crop can affect crop yield and N losses in maize-based systems (Martinez-Feria et al., 2016).

in which crops experienced drought, flood and historically average weather, as well as included measurements of many system variables, are shown in Fig. 2 (Martinez-Feria et al., 2016).

Each cover crop has a niche or special purpose and offers many benefits to producers that increase farm profitability and environmental sustainability. Legumes, as cover crops, are typically used to fix atmospheric N (produce homegrown nitrogen) biologically and increase soil fertility. Grass cover crops are used to increase soil organic matter (SOM), recycle excess nutrients and reduce soil compaction. *Brassica* crops are grown to loosen the soil, recycle nutrients and suppress weeds. Some other cover crops are grown to suppress insects, disease, weeds, or attract beneficial insects. Therefore, cover crops should be considered an integral part of any farming system in order to efficiently utilise water, nutrients, improve soil biological, chemical and physical properties associated with soil quality to improve farm stability and profitability (Hoorman et al., 2009). In this chapter, we will focus more on soil physical properties as core indicators of soil quality as directly or indirectly influenced by cover crops.

2. Cover Crops and Soil Compaction (Bulk Density)

Soil compaction is a global concern in agriculture as farm equipment such as tractors, combines and grain carts become larger and heavier. For example, tractor weights have increased from 4 Mg in the 1940s to 20–45 Mg in the 2000s (Sidhu and Duiker, 2006). Moreover, farmers sometimes need to access their fields while the soil is wet and susceptible to compaction in order to achieve a timely crop harvest through planting, fertilisation, weed control and other field operations. It is documented that compaction reduces water, heat and gas flow; nutrient and water uptake; root growth and crop yields (Schafer-Landefeld et al., 2004).

Historically, tillage was employed to reduce soil compaction and improve field workability. While tillage may temporarily reduce soil compaction (bulk density), but rain, gravity and equipment traffic compact the soil. In recent years, cover crops were used to alleviate soil compaction. Long-term studies have shown that cover crops can alleviate and reduce the susceptibility of the soil to compaction; however, the extent of this benefit will depend on the cover-crop species, duration of growth and the amount and characteristics of the belowground biomass input (Chen and Weil, 2010).

Cover crops with deep taproots, such as *Brassicas* (radish) can alleviate soil compaction by penetrating compacted layers and acting like tillage tools, biological plow or bio-drills. Across different soils in Maryland, Chen and Weil (2010) reported that the number of roots that penetrated compacted layers under no-till soils in the 0–50 cm depth by species were in the following order: forage radish > rapeseed > rye. Taproots have more biological drilling potential than fibrous roots because the latter are often concentrated near the soil surface (Cresswell and Kirkegaard, 1995). Under favourable growing conditions, radish roots can extend more than 3 feet deep in 60 days, with the thickened storage portion of the root (commonly referred to as the tuber, though not botanically correct) extending more than 12 inches. After radishes winter-killed and their large fleshy roots desiccate, the channels created by the roots tend to remain open at the soil surface, improving infiltration, surface drainage and soil warming (Fig. 3). Radish rooting effects on soil porosity also extend into the subsoil, which can improve root growth by subsequent crops and provide better access to subsoil moisture, resulting in greater resilience under drought conditions (Chen and Weil, 2010; Gruver et al., 2017).

Penetration resistance is another measure of soil compaction, or strength. Root growth is restricted when resistance is greater than 2–3 kPa which affects water and nutrient uptake and consequently, decrease crop yield (Hakansson and Lipiec, 2000; Hamza and Anderson, 2005), Several studies show that cover crops reduce penetration resistance. Folorunso et al. (1992) reported that cover crops reduce near-surface (< 1 cm depth) penetration resistance values by about 65% at two sites in California. Abdollahi et al. (2014) reported that Brassicaceae cover crops reduce penetration resistance at 32 38 cm depth in tilled soils in Denmark after 10 years. Islam (2009) reported that cover crop (radish) roots, as a biological plow, significantly reduced soil penetration

Fig. 3 Radish growth in subsoil and effect on soil compaction (*Photo*: Rafiq Islam, Alan Sundermeier and Dave Bradnt).

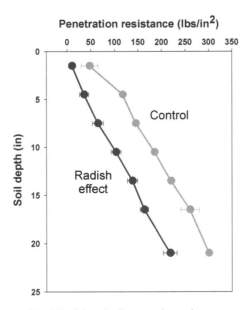

Fig. 4 Radish and soil penetration resistance.

resistance by 40 per cent, when compared between cropped vs non-cover cropped plots. The effect was more pronounced at surface depth than at sub-surface depth (Fig. 4).

Figure 4 shows that cover crops can reduce soil compaction by improving soil aggregation with an associated increase in SOM (Blanco-Canquia et al., 2015). Microaggregate and soil primary particles (clay, silt, and sand) are held together by humus which is resistant to decomposition (Hoorman et al., 2011). After termination, cover crop taproots when decompose create large bio-channels, or macropores, to increase water and air flow along with root proliferation of the main crop to deeper layers (Chen and Weil, 2010).

Alvarez et al. (2017) have shown that soil physical properties generally improve when replacing a fallow period with the inclusion of cover crops. Nascente et al. (2014) reported that the deployment of cover crop biomass on the soil surface provides a greater accumulation of SOM, which positively correlated with the soil aggregate stability and the SOM content inversely correlated with the bulk

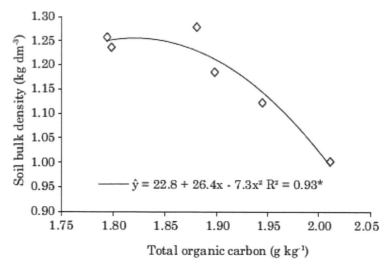

Fig. 5 Relationship between soil bulk density and the concentration of TOC throughout the soil at the depth of 0.00–0.10 m. The diamond symbol represents the average of soil bulk density for the respective TOC in each cover crop treatment (*Source*: Nascente et al., 2015).

density (Fig. 5). However, it is reported that bulk density is minimally impacted by cover crops (Hubbard et al., 2013; Zhu et al., 2012), or it does not differ significantly in comparison to fallow soils (Chen et al., 2014; Hubbard et al., 2013). Accepted thresholds, beyond which root growth is affected by high bulk density, usually vary from 1.5 to 1.8 g cm^{-3} depending on soil texture (Arshad et al., 1996; USDA, 2014; Alvarez et al., 2017).

Overall, soil compaction has a biological component and its root cause is a lack of actively growing plants all the year around and active roots in the soil. A continuous living cover, plus long-term continuous no-till, reduce soil compaction in five ways: (1) organic residues on the soil surface cushion the soil from heavy equipment and rainfall splashs; (2) plant roots create voids and macropores in the soil for air and water movement; (3) plant roots act like a biological conduit to regulate oxygen diffusion in the soil to sequester SOM; (4) fine roots and root exudates supply food and nutrients for soil microbes and fauna; and (5) residual organic residues (plants, roots, microbes) are lighter and less dense than soil particles. All of these factors reduce soil compaction by forming soil macroaggregation (Hoorman et al., 2011).

3. Cover Crops and Soil Porosity Characteristics

Cover cropping in agriculture is expected to enhance many agricultural and ecosystem functions and services. The positive effects of cover crops include an increase in the soil micro- and macro-porosity, which in turn improves the soil structure and hydraulic characteristics. The effect of cover crops can be explained in terms of rootsize, architecture and distribution on the development of micropore size, structure and distribution. The root effect on micropore-characteristics development has been reported as being due to the micro-fissuring produced by the wetting-drying process, enhanced by the presence of roots and also by the radial pressures exerted by the roots themselves (Scanlan, 2009; Bodner et al., 2014; Gabriel et al., 2017).

Gabriel et al. (2017) suggested an increase in macroporosity produced by biopores from dead roots and by improvements in soil structure. This result is consistent with observations of other researchers, who reported that macroporosity could increase in cover cropping systems (Cresswell and Kirkegaard, 1995; Bodner et al., 2014; Yu et al., 2016). Moreover, several studies report that cover crops increase SOM and soil aggregate stability (Six et al., 2006; Peregrina et al., 2010), providing more stability to both macro- and microporosity improvement. The difference is most pronounced in deeper soil layers because roots tend to homogenise the macroporosity of both treatments in the upper soil layers (Gabriel et al., 2017).

Results on soil pore size distribution two weeks after spring tillage and cover crop termination at four depths was reported by Haruna et al. (2017), which shows that macropores and coarse mesopores are affected by a significant tillage x cover crop (Fig. 6a,b,c,d). They reported that cover crops improved macropores by 24 per cent compared with no cover crop; this can potentially increase water infiltration and reduce surface runoff. As a result of higher macroporosity, saturated hydraulic conductivity is higher in the cover crop when compared to no-cover crop management.

Intense rainfall over a short period of time may cause the soil to 'settle', thus diminishing the proportion of large pores quickly. This may lead to accelerated loss of fertile soil, water and nutrient with possible low crop yields over time. In such conditions, the inclusion of cover crops may help transpire excess water from the soil and their roots can hold the soil better in place and consequently, help to mitigate against soil, water and nutrient loss. Auler et al. (2014) reported that annual ryegrass (*Lolium multflorum*, L.) when used as a cover crop significantly reduces soil bulk density and microporosity with an associated increase in macroporosity and total porosity, which can lead to better water flow in the soil. They also reported that water retention is higher in the top 10 cm of the soil when ryegrass is planted in combination with different tillage systems. In their study of tillage systems and cover crops on soil quality, Abdollahi et al. (2014) reported that the use of a cover crop increases air-filled porosity at −10 kPa, air permeability, pore organisation and reduces the value of blocked air porosity at all depths for all tillage treatments (Haruna et al., 2017).

Raut et al. (2015) found that mean values of total porosity increased by 47 per cent with radish, compared to the fields without radish (Table 1). This increase ranged from 71 per cent in the upper soil depth (0–13 cm) to 25 per cent at the 56–64 cm depth (Table 1). As a result, cover crops bio-drilling (using radish) is an inexpensive and ecologically friendly way to alleviate compaction

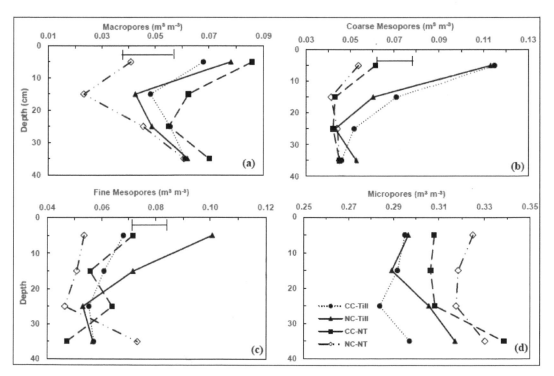

Fig. 6 Pore size distributions at various depths; (a–d) two weeks after cover crop termination as influenced by cover crop with tillage (CC-Till), no cover crop with tillage (NC-Till), cover crop with no till (CC-NT) and no cover crop with no till (NC-NT) treatments (Haruna et al., 2017).

Table 1 Radish effects on compaction and porosity at different soil depths (Raut et al., 2015).

Soil Depth (cm)	Soil Compaction Without Radish (kPa)	Soil Compaction with Radish (kPa)	Increase in Porosity (%)
56–64	2069	1448	25
46–56	1724	1034	32
38–46	1552	862	34
30–38	1241	690	30
18–30	1034	414	43
13–18	862	207	62
00–13	345	34	71
Mean	**1261**	**670**	**47**

and has the potential to improve the overall soil quality by increasing porosity, the air and waterholding capacity of soil.

4. Cover Crops and Soil Aggregate Stability

Soil aggregate stability is one of the important factors for plant growth and directly affects soil quality functions, such as water infiltration, porosity and oxygen availability to the roots, microbial activity, SOM accumulation and protection and soil erosion. Cover crop roots contribute as much N as cover crop shoots to the total N pool for use by succeeding crops from the soil structure's most easily accessible zones, which are the surfaces of larger soil aggregates. Therefore, maintaining active plant roots and well-aggregated soil structure enhances C:N stoichiometry and maximises N availability to plants (Kavdir and Smucker, 2005).

Cover crops improve soil aggregate stability and increase water infiltration; legume cover crops also fix atmospheric N and scavenge nutrients that are subject to leaching. They also promote soil aggregation and greater porosity, which indirectly reduces erosion and increases water infiltration capacity and provides support for root growth (Kemper and Rosenau, 1986; McVay et al., 1989; Drury et al., 1991; Robertson et al., 1995; Sainju et al., 2001). Cover crop root systems are valuable because they make the soil more resistant to abrasive forces caused by rainfall impacts and wind and/or water. In response to rainfall splash and soil erosion, soil primary particles detachment is reduced because roots penetrate into the soil and physically modify soil pore architecture, thereby augmenting water and air retention across soil layers (Bronick and Lal, 2005). Moreover, complex polysaccharides found in cover crop root exudates and decomposed organic matter (OM) act as binding agents to hold soil aggregates together (Angers and Caron, 1998; Kabir and Koide, 2002; Bronick and Lal, 2005; Jokela et al., 2009).

Sainju et al. (2003) suggested that cover crops may improve soil aggregation and accumulate C and N in whole soil and aggregates. Non-legume cover crops, such as deep rooted grass species, may be more effective in increasing soil aggregation and C sequestration in whole soil and aggregates, thereby improving soil quality by reducing erosion, increasing microbial activities and SOM level, and helping to reduce global warming. In contrast, legume cover crops (such as soybeans, winter peas, cowpea, clover, etc.) may be more effective in increasing labile N pools, thereby increasing soil productivity by sequestering atmospheric N and increasing soil N mineralisation and producing agronomic crop yields similar to those produced by the recommended rates of N fertilisation. A mixture of legume and non-legume cover crops may be needed to improve soil aggregation and other soil quality properties.

Blanco-Canquia et al. (2015) investigated cover crops and ecosystem services and reported that cover crops appear to rapidly improve soil aggregation. They suggested that soil aggregate stability is one of the most responsive parameters to cover crop management. The rapid improvement in soil

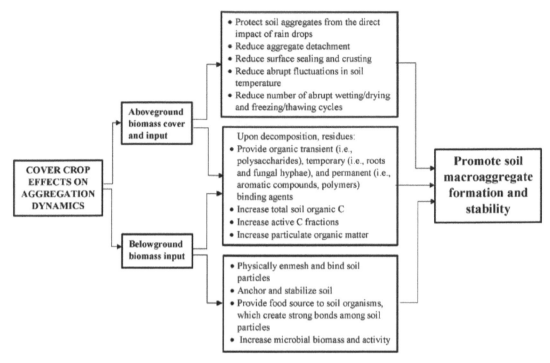

Fig. 7 Mechanisms by which cover crops influence soil physical, chemical, and biological processes contributing to the formation of stable macroaggregates (*Source*: Blanco-Canquia et al., 2015).

aggregate stability under cover crops can enhance water, nutrient and C storage; soil macroporosity; and root growth while reducing soil erodibility (Blanco-Canqui et al., 2013). Cover crops increase aggregate stability by protecting the soil surface from raindrop impact, providing additional biomass input (i.e., roots) and increasing SOM content and efficient microbial activity (Fig. 7). An increase in SOM content, in return, is positively correlated with an increase in soil aggregate stability (Blanco-Canqui et al., 2013). Upon decomposition, cover crop biomass often produces transient, temporary and permanent organic binding agents to promote soil aggregation (Tisdall and Oades, 1982).

5. Cover Crops and Soil Moisture Dynamics

There are two major factors responsible for the increase in water infiltration by cover crops. The decomposed roots of cover crops and enhanced earthworm activities create channels in the soil that help water to infiltrate faster through these holes and channels (Lal et al., 1991). On fallow land where the ground is bare, a seal or crust is formed on the soil surface by rainfall splashes which may lead to a decrease in water infiltration with an associated increase in surface runoff from the edge of the field (Louw et al., 1991; Morin et al., 1981). Cover crops, alive or dead, provide blanket the soil surface and protect it from the direct impact of rainfall and reduce wind and water erosion (Dabney et al., 2001). The addition of organic matter by cover crops improves the water-holding capacity of soil, which in turn significantly increases the water permeability and infiltration and lowers soil erosion (Fageria et al., 2005).

Stobart and Morris (2014) demonstrated improvement in water infiltration rates associated with cover crops when clover was used as a clover bi-crop approach (Fig. 8). The study indicates an increase in water infiltrate from 0.78 mm min^{-1} in conventional standard practices to 2.19 mm min^{-1} in clover bi-crop system. The increased moisture infiltration rates recorded with the clover bi-crop systems are analogous to those recorded in other studies (Stobart and Morris, 2014). It is reported that these changes in water infiltration relate to the development of a more open soil structure associated with the use of the clover bi-crop.

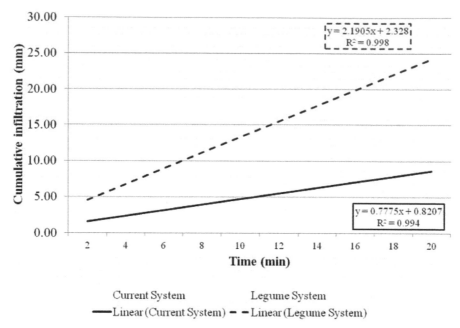

Fig. 8 The effect of cover crop treatment on water infiltration rates comparing a standard practice (no cover crop) to clover bi-crop cover crop system (mm min^{-1}) in 2012 (*Source*: Stobart and Morris, 2014).

It is reported that cover-crop biomass serves as a food and shelter to support earthworms which, in turn, create natural biochannels and facilitates drainage. The soil aggregation effect of fine roots seems be the most probable cause. The greater water infiltration under cover crops does not necessarily increase soil water content (Lal et al., 1978) but water consumption by the cover crops could offset infiltration increases, and in some experiments, the impact of cover crops on infiltration is reported to be time-dependent (Steele et al., 2012). Several other studies report that cover crops increase soil moisture retention (at water potentials associated with field capacity [–33 kPa]) by 10–11 per cent, as well as increase water availability to the plant by 21–22 per cent. It is further reported that winter rye, when used as a cover crop, can improve soil water dynamics (Basche et al., 2016).

6. Cover Crops and Soil Erosion

Soil erosion is a naturally-occurring degradative process that affects soil quality under diverse agricultural management practices. In agriculture, soil erosion refers to the wearing and tearing away of topsoils by the natural forces of water and wind or through accelerated forces associated with conventional agricultural practices, such as tillage, monocropping and fallow (Ritter et al., 2012). Soil erosion, either by water, wind or tillage operations, involves three distinct steps— soil detachment, movement and deposition which ultimately affects soil quality, reduces crop productivity and contributes to water and air pollution. With sustainable management practices, soil erosion can be minimised to improve soil quality for supporting economic crop production (Fig. 9).

Cover crops, as one of the components of 21st century agricultural practices, are essential to minimise soil erosion by providing living or dead ground cover that reduces raindrop impact, reduces runoff, slows down air and water velocities and increases water infiltration into the soil. With several benefits, cover crops control both water and wind erosion.

Water Erosion – Water erosion comes in several different forms due to various natural and anthropogenic causes. While the causes of erosion by water are generally natural or tillage-induced effects, cover crops play a prominent role in reducing both water and wind erosion. Water erosion is usually caused by rainfall and runoff on a slope and the magnitude of water erosion depends on the type of soil and management practices.

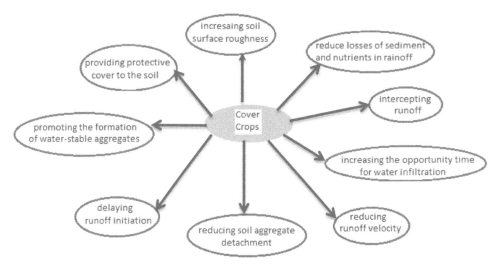

Fig. 9 Cover crops have multiple benefits to control soil erosion.

While the benefits of cover crops for reducing water erosion are widely recognised, actual runoff and sediment loss data are rather limited (Table 2). Runoff loss can be decreased by up to 80 per cent and sediment loss between 40–96 per cent, with the use of cover crops (Table 2). The magnitude by which cover crops reduce water erosion is a function of root distribution and characteristics, biomass production and cover crop species. During a three-year study in Iowa, Kaspar et al. (2001) observed that rye reduced runoff by 10 per cent in one of the three years, but oat (*Avena sativa* L.) did not reduce runoff. Across the three years, rye and oat reduced rill erosion by 54 and 89 per cent, respectively, but interrill erosion was reduced in two of the three years by rye, and in only one of the years by oat. In western Kansas, winter triticale reduced water erosion more than winter lentil or spring pea; this is attributed to the greater biomass production of winter triticale (Blanco-Canqui et al., 2015).

Results reported from several other studies have shown decrease in soil loss from fields planted with different types of cover crops (Clark, 2019). While non-legumes, such as cereal rye, ryegrass, teff, triticale, barley, oat and wheat reduced soil loss by 31–100 per cent%, legumes, such as red clover, crimson clover, lentil and pea, in contrast, reduced soil loss by 38–69 per cent when compared without any cover crops. *Brassicas*, such as mutard, reduced soil loss by upto 82 per cent with respect to control plots. Averaged across, cover crops have shown to reduce sediment losses from erosion by 20.8 tons/ac. on conventional-till fields, 6.5 tons/ac. on reduced-till fields and 1.2 tons/ac. on no-till fields.

Cover crops also reduce the loss of dissolved and particle-bound nutrients in runoff, particularly total P and NO_3-N, which can result in improved water quality, soil fertility and crop productivity (Kaspar et al., 2001). For example, in Missouri, winter cover crops reduce dissolved nutrient losses by 7–77 per cent (Zhu et al., 1989). Likewise, in New York, Kleinman et al. (2005) reported that total P loads in runoff were 74 per cent lower in plots with cover crops than in plots without any cover crops. A substantial reduction in water erosion by cover crops subsequently reduce nutrient and sediment pollution in streams, rivers, lakes and other water sources with an associated improvement in water quality (Blanco-Canqui et al., 2015). It is reported that planting a mixture or blends cover crop species can provide more canopy cover, more total biomass yield, extensive rooting effects and more uniform surface cover than a single species alone, resulting in greater water erosion control (Wortman et al., 2012; Blanco-Canqui et al., 2015).

Wind Erosion – Cover crops reduce wind erosion risks by physically protecting the soil surface, improving soil structural properties, increasing the SOM content and anchoring the soil with their roots when primary crops are not in place, thereby reducing potential soil erodibility. An increase in

Table 2 Data on Cover Crop (CC) effects on runoff and soil loss from select studies. Studies that did not report soil loss in Megagrams per hectare were not included (*Source*: Blanco-Canqui et al., 2015).

Study Site	Precipitation mm yr^{-1}	Soil and Slope	Tillage	Crop	Cover Crop Treatments	Runoff mm	Soil Loss Mg ha^{-1}	Runoff Reduction %	Soil Loss Reduction %	References
Kingdom City, MI	996	silt loam, 3–3.5%	no-till	corn	no CC	245a†‡	22a‡			Wendt and Burwell (1985)
					rye	122b	0.9b	50	96	
Kingdom City, MI	996	silt loam, 3%	no-till	soybean	no CC	179a‡	1.517a‡			Zhu et al. (1989)
					chickweed (*Stelidria media* L.)	100b	0.197b	44	87	
					Canada bluegrass (*Poa compressa* L.)	99b	0.062b	45	95	
					downy brome (*Bromus teetorum* L.)	85b	0.078b	53	96	
Reidsville, NC	1129	sandy loam, 4%	moldboard plow	corn	no CC	23.6a‡	41.3‡			Martin and Cassel (1992)
					rye and hairy vetch	20.4a	3.3	13	92	
Garden City, KS	462§	silt loam, 1–3%	no-till	wheat	no CC	45a	1.59a			Blanco-Canqui et al. (2013a)
					winter lentil	26ab	0.97ab	42	39	
					spring triticale	17b	0.61ab	62	61	
					spring pea	12b	0.51b	73	68	
					winter triticale	10b	0.34b	78	79	

† Means followed by different lowercase letters in a column within the same study are different at $P < 0.05$.

‡ Data averaged across study years.

§ This study was conducted under simulated rainfall at 63.5 mm h^{-1}.

SOM content with cover crops is one of the main factors contributing to increased aggregate stability and reduced wind-erodible fraction, especially silt, because SOM can physically, chemically and biologically bind primary soil particles and form stable macroaggregates (Colazo and Buschiazzo, 2010).

Reduced wind erosion by cover crops has many beneficial impacts to agricultural viability, environmental compatibility and society. In addition to conserving the soil resources, it can improve air quality and public health. Particulate or dust emissions from agricultural lands by conventional agrocultural can be transported long distances, posing a threat to human and animal health (i.e., shortness of breath, respiratory disorders) (USDA–ARS, 2000). The 'dust bowl' is a reminder of the consequences of severe wind erosion on agriculture and society. The inclusion of cover crops in current cropping systems offers promise to manage dust emissions from agricultural lands, thereby reducing air pollution (Bilbro, 1991; Blanco-Canqui et al., 2013a; Blanco-Canqui et al., 2015).

References

Abdollahi, L., Munkholm, L.J. and Garbout, A. (2014). Tillage system and cover crop effects on soil quality: II. Pore characteristics. Soil Sci. Soc. Am. J., 78: 271–279.

Alvarez, R., Steinbach, H.S. and De Paepe, J.L. (2017). Cover crop effects on soils and subsequent crops in the pampas: A meta-analysis. Soil Tillage Res., 170: 53–65.

Angers, D.A. and Caron, J. (1998). Plant-induced changes in soil structure: processes and feedbacks. Biogeochem., 42: 55–72.

Anonymous. (2009). Building Soil for Better Crops. Sustainable Soil Management, third ed. (Ed.). Fred Magdoff and Harold Van, Handbook Series Book 10. Published in 2009 by the Sustainable Agriculture Research and Education (SARE) programme with funding form the National Institute of Food and Agriculture, U.S. Department of Agriculture.

Arshad, M.A., Lowery, B. and Grossman, B. (1996). Physical tests for monitoring soil quality. pp. 123–142. *In*: Doran, J.W. and Jones, J. (eds.). Methods for Assessing Soil Quality, SSSA, Madison, WI. Soil Sci. Soc. Am. Spec. Publ. 49.

Auler, A.C., Miara, S., Pires, L.F., da Fonseca, A.F. and Barth, G. (2014). Soil physico-hydrical properties resulting from the management in integrated production systems. Rev Ciênc Agronômica, 45: 976–989.

Basche, A.D., Archontoulis, S.V., Kaspar, T.C., Jaynes, D.B., Parkin, T.B. and Miguez, F.E. (2016). Simulating long-term impacts of cover crops and climate change on crop production and environmental outcomes in the midwestern United States. Agric. Ecosyst. Environ., 218: 95–106.

Bilbro, J.D. (1991). Cover crops for wind erosion control in semiarid regions. pp. 36–38. *In*: Hargrove, W.L. (ed.). Cover Crops for Clean Water: Proc. International Conference, Jackson, TN. 9–11 Apr. 1991, Soil Water Conserv. Soc., Ankeny, IA.

Blanco-Canqui, H., Holman, J.D., Schlegel, A.J., Tatarko, J. and Shaver, T. (2013a). Replacing fallow with cover crops in a semiarid soil: Effects on soil properties. Soil Sci. Soc. Am. J., 77: 1026–1034.

Blanco-Canqui, H., Shapiro, C.A., Wortmann, C.S., Drijber, R.A., Mamo, M., Shaver, T.M. and Ferguson, R.B. (2013b). Soil organic carbon: The value to soil properties. J. Soil Water Conserv., 68: 129A–134A.

Blanco-Canqui, H., Shaver, T.M., Lindquis, J.L., Shapiro, C.A., Elmore, R.W., Francis, C.A. and Hergert, G.W. (2015). Cover crops and ecosystem services: insights from studies in temperate soils. Agron. J., 107: 2049–2074.

Bodner, G., Leitner, D. and Kaul, H.P. (2014). Coarse and fine root plants affect pore size distributions differently. Plant Soil, 380: 133–151.

Bronick, C.J. and Lal, R. (2005). Soil structure and management: A review. Geoderma, 124: 3–22.

Chen, G. and Weil, R.R. (2010). Penetration of cover crop roots through compacted soils. Plant Soil, 331: 31–43.

Chen, G., Weil, R.R. and Hill, R.L. (2014). Effects of compaction and cover crops on soil least limiting water range and air permeability. Soil Till. Res., 136: 61–69.

Clark, A. (2019). Cover Crops for Sustainable Crop Rotations, Topic Room Series. www.SARE.org/Cover-Crops.

Colazo, J.C. and Buschiazzo, D.E. (2010). Soil dry aggregate stability and wind erodible fraction in a semiarid environment of Argentina. Geoderma, 159: 228–236.

Cresswell, H.P. and Kirkegaard, J.A. (1995). Subsoil amelioration by plant roots: The process and the evidence. Aust. J. Soil Res., 33: 221–239.

Dabney, S.M., Delgado, J.A. and Reeves, D.W. (2001). Using winter cover crops to improve soil and water quality. Commun. Soil Sci. Plant Anal., 32: 1221–1250.

Dabney, S.M., Delgado, J.A., Collins, F., Meisinger, J.J., Schomberg, H.H., Liebig, M.A., Kaspar, T. and Mitchell, J. (2010). Using cover crops and cropping systems for nitrogen management. pp. 230–281. *In*: Delgado, J.A. and

Follett, R.F. (eds.). Advances in Nitrogen Management for Water Quality, Soil and Water Conservation Society, Ankeny, Iowa.

Dinnes, D.L., Karlen, D.L., Jaynes, D.B., Kaspar, T.C., Hatfield, J.L., Colvin, T.S. and Cambardella, C. (2002). Nitrogen management strategies to reduce nitrate leaching in tile-drained midwestern soils. Agron. J., 94: 153–171.

Dorn, B., Jossi, W. and van der Heijden, M.G.A. (2015). Weed suppression by cover crops: Comparative on-farm experiments under integrated and organic conservation tillage. Weed Res., 55: 586–597.

Drury, C.F., Stone, J.A. and Findlay, W.I. (1991). Microbial biomass and soil structure associated with corn, grass and legumes. Soil Sci. Soc. Am. J., 55: 805–811.

Fageria, N.K., Baligar, V.C. and Bailey, B.A. (2005). Role of cover crops in improving soil and row crop productivity. Commun. Soil Sci. Plant Anal., 36: 2733–2757.

Folorunso, O.A., Rolston, D.E., Prichard, T. and Louie, D.T. (1992). Soil surface strength and infiltration rate as affected by winter cover crops. Soil Technol., 5: 189–197.

Gabriel, J.L., Quemada, M., Martin-Lammerding, D. and Vanclooster, M. (2017). Assessing the cover crop effect on soil hydraulic properties by inverse modelling in a 10-year field trial. Hydrol. Earth Syst. Sci. Discuss. https://doi.org/10.5194/hess-2017-643.

Gruver, J. (2010). Cover Crops, Tillage and Soil Quality. http://mccc.msu.edu/wp-content/uploads/2016/09/IL_2010_Cover-Crops-Tillage-and-Soil-Quality.pdf.

Gruver, J., Weil, R.R., White, C. and Lawley, T. (2017). Radishes—A New Cover Crop for Organic Farming Systems. http://articles.extension.org/pages/64400/radishes-a-new-cover-crop-for-organic-farming-systems.

Islam, K.R. (2009). Plant Roots: The Biological Plow, Professional presentation at annual meetings of the Conservation Tillage and Technology Conference (CTTC), Ada, Ohio.

Hakansson, I. and Lipiec, J. (2000). A review of the usefulness of relative bulk density values in studies of soil structure and compaction. Soil Till. Res., 53: 71–85.

Hamza, M.A. and Anderson, W.K. (2005). Soil compaction in cropping systems. A review of the nature: causes and possible solutions. Soil Till. Res., 82: 121–145.

Haruna, S.I., Anderson, S.H., Nkongola, N.V. and Zabion, S. (2017). Soil Hydraulic Properties: Influence of Tillage and Cover Crops. Pedosphere ISSN 1002-0160/CN 32-1315/P doi:10.1016/S1002-0160(17)60387-4.

Hoorman, J.J., Islam, R. and Sundermeier, A. (2009). Sustainable Crop Rotations with Cover Crops. http://mccc.msu.edu/wp-content/uploads/2016/10/OH_2009_Sustainable-crop-rotations-with-cover-crops.pdf.   SAG-9, Agriculture and Natural Resources.

Hoorman, J.J., Moraes Sá, J.C. and Reeder, R. (2011). The Biology of Soil Compaction. https://www.exapta.com/wp-content/uploads/2015/07/Biology-Soil-Compaction.pdf. Science 583–587.

Hubbard, R.K., Strickland, T.C. and Phatak, S. (2013). Effects of cover crop systems on soil physical properties and carbon/nitrogen relationships in the plain of southeastern USA. Soil Till. Res., 126: 276–283.

Jokela, W.E., Grabber, J.H., Karlen, D.L., Balser, T.C. and Palmquist, D.E. (2009). Cover crop and liquid manure effects on soil quality indicators in a corn silage system. Agron. J., 101: 727–737.

Kabir, Z. and Koide, R.T. (2002). Effect of autumn and winter mycorrhizal cover crops on soil properties, nutrient uptake and yield of sweet corn in Pennsylvania, USA. Plant Soil, 238: 205–215.

Kaspar, T.C., Radke, J.K. and Laflen, J.M. (2001). Small grain cover crops and wheel traffic effects on infiltration, runoff, and erosion. J. Soil Water Conserv., 56: 160–164.

Kaspar, T.C. and Singer, J.W. (2011). The use of cover crops to manage soil. pp. 321–338. *In*: Hatfield, J. and Sauer, T. (eds.). Soil Management: Building a Stable Base for Agriculture, Amer. Soc. Agronomy, Madison, WI.

Kavdır, Y. and Smucker, A.J.M. (2005). Soil aggregate sequestration of cover crop root and shoot-derived nitrogen. Plant Soil, 272: 263–276.

Kemper, W.D. and Rosenau, R.C. (1986). Aggregate stability and size distribution. pp. 425–442. *In*: Klute, A. (ed.). Methods of Soil Analysis, Part 1: Physical and Mineralogical Methods. ASA and SSSA, Madison, WI.

Kleinman, P.J.A., Salon, P., Sharpley, A.N. and Saporito, L.S. (2005). Effect of cover crops established at time of corn planting on phosphorus runoff from soils before and after dairy manure application. J. Soil Water Conserv., 60: 311–322.

Krutz, L.J., Locke, M.A. and Steinriede, R.W. (2009). Interactions of tillage and cover crop on water, sediment, and pre-emergence herbicide loss in glyphosate-resistant cotton: Implications for the control of glyphosate-resistant weed biotypes. J. Environ. Qual., 38: 1240–1247.

Lal, R., Wilson, G.F. and Okigbo, B.N. (1978). No-till framing alter various grasses and leguminous cover crops in tropical Alfisol, I: Crop performance. Field Crops Res., 1: 71–84.

Lal, R., Regnier, E., Eckert, D.J., Edwards, W.M. and Hammond, R. (1991). Expectations of cover crops for sustainable agriculture. pp. 1–11. *In*: Hargrove, W.L. (ed.). Cover Crops for Clean Water, Soil and Water Society, Ankeny, Iowa.

Louw, P.J.E. and Bennie, A.T.P. (1991). Soil surface effects on runoff and erosion on selected vineyard soils. pp. 25–26. *In*: Hargrove, W.L. (ed.). Cover Crops for Clean Water, Soil and Water Society, Ankeny, Iowa.

Martinez-Feria, R.A., Dietzela, R., Liebmana, M., Helmers, M.J. and Archontoulis, S.V. (2016). Rye cover crop effects on maize: A system-level analysis. Field Crops Res., 196: 145–159.

McVay, K.A., Radcliffe, D.E. and Hargrove, W.L. (1989). Winter legume effects on soil properties and nitrogen fertiliser requirements. Soil Sci. Soc. Am. J., 53: 1856–1862.

Morin, J., Benyamini, Y. and Michaeli, A. (1981). The effect of raindrop impact on the dynamics of soil surface crusting and water movement in the profile. Journal of Hydrology, 52: 321–33.

Nascente, A.S., Yuncong Li, Y. and Crusciol, C.A. (2014). Cover Crops Species as Affecting Soil Aggregation, Aggregate Stability, Organic Carbon Concentration and Soil Bulk Density in Different Soil Aggregate Fractions (Abstract), The 20th World Congress of Soil Science, June 8–13, 2014 Jeju, Korea.

Nascente, A.S., Li, C.Y. and Crusciol, C.A. (2015). Soil aggregation, organic carbon concentration, and soil bulk density as affected by cover crop species in a no-tillage system. R. Bras. Ci. Solo., 39: 871–879.

Peregrina, F., Larrieta, C., Ibáñez, S. and García-Escudero, E. (2010). Labile organic matter, aggregates and stratification ratios in 40 semiarid vineyard with cover crops. Soil Sci. Soc. Am. J., 74: 2120–2130.

Raut, Y., Warren Dick, W., Weaks, E., Jahan, H. and Islam, K.R. (2015). Bio-drilling, Compaction Alleviation and Fate of Storm-water Management, presented at the Environmental Science, Annual Symposium on March 6, 2015 (Spring Semester) at the Ohio State University, Ohio Union OH, USA.

Ritter, J. (2012). Soil Erosion—Causes and Effects. Factsheet Order No. 12-053 AGDEX 572/751.www.ontario.ca/omafra.

Robertson, E.B., Sarig, S. and Firestone, M.K. (1991). Cover crop management of polysaccharide-mediated aggregation in an orchard soil. Soil Sci. Soc. Am. J., 55: 734–739.

Robertson, E.B., Sarig, S., Shennan, C. and Firestone, M.K. (1995). Nutritional management of microbial polysaccharide production and aggregation in an agricultural soil. Soil Sci. Soc. Am. J., 59: 1587–1594.

Sainju, U.M., Singh, B.P. and Whitehead, W.F. (2001). Comparison of the effects of cover crops and nitrogen fertilization on tomato yield, root growth and soil properties. Sci. Hortic., 91: 201–214.

Sainju, U.M., Whitehead, W.F. and Singh, B.P. (2003). Cover crops and nitrogen fertilisation effects on soil aggregation and carbon and nitrogen pools. Can. J. Soil Sci., 83: 155–165.

Scanlan, C.A. (2009). Processes and Effects of Root-induced Changes to Soil Hydraulic Properties. University of Western Australia.

Schafer-Landefeld, L., Brandhuber, R., Fenner, S., Koch, H.J. and Stockfisch, N. (2004). Effects of agricultural machinery with high axle load on soil properties of normally managed fields. Soil Tillage Res., 75: 75–86.

Sidhu, D. and Duiker, S.W. (2006). Soil compaction in conservation tillage: Crop impacts. Agron. J., 98: 1257–1264.

Six, J., Frey, S.D., Thiet, R.K. and Batten, K.M. (2006). Bacterial and fungal contributions to carbon sequestration in agroecosystems. Soil Sci. Soc. Am. J., 70: 555–569.

Steele, M.K., Coale, F.J. and Hill, R.L. (2012). Winter annual cover crop impacts on no-till soil physical properties and organic matter. Soil Sci. Soc. Am. J., 76: 2164–2173.

Stobart, R.M. and Morris, N.L. (2014). The impact of cover crops on yield and soils in the New Farming Systems programme: Crop production in Southern Britain: Precision decisions for profitable cropping. Aspects of Appl. Biol., 127: 223–231.

Sylvester, M. and Bird, G. (2017). Role of Cover Crops in Field Crop Nematode Management and Soil Health Enhancement. http://mccc.msu.edu/wp-content/uploads/2017/03/2017MCCC-FC-Bird-Sylvester-Warner.pdf.

Thorup-Kristensen, K., Magid, J. and Jensen, L.S. (2003). Catch crops and green manures as biological tools in nitrogen management in temperate zones. Adv. Agron., 79: 227–302.

Tisdall, J.M. and Oades, J.M. (1982). Organic matter and water-stable aggregates in soils. J. Soil Sci., 33: 141–163.

USDA. (2014). Soil Survey Field and Laboratory Methods Manual, Soil Survey Investigations Report No. 51, version 2, pp. 457.

Wittwer, R.A., Dorn, B., Jossi1, W. and van der Heijden, M.G.A. (2017). Cover crops support ecological intensification of arable cropping systems. Scientific Reports, 7: 41911.

Wortman, S.E., Francis, C.A. and Lindquist, J.L. (2012). Cover crop mixtures for the western Corn Belt: Opportunities for increased productivity and stability. Agron. J., 104: 699–705.

Yu, Y., Loiskandl, W., Kaul, H.P., Himmelbauer, M., Wei, W., Chen, L.D. and Bodner, G. (2016). Estimation of runoff mitigation by morphologically different cover crop root systems. J. Hydrol., 538: 667–676.

Zhu, B., Yi, L., Gou, L., Chen, G., Hu, Y., Tang, H., Xiao, C., Xiao, X., Yang, G., Acharya, S.A. and Zeng, Z. (2012). Performance of two winter cover crops and their impacts on soil properties and two subsequent rice crops in Dongting Lake Plain Hunan, China. Soil Till. Res., 124: 95–101.

Zhu, J.C., Gantzer, C.J., Anderson, S.H., Alberts, E.E. and Beuselinck, P.R. (1989). Runoff, soil, and dissolved nutrient losses from no-till soybean with winter cover crops. Soil Sci. Soc. Am. J., 53: 1210–1214.

15

Cover Crops Effects on Soil Erosion and Water Quality

Beenish Saba[1,2,]* *and Ann D Christy*[1]

1. Introduction

The term 'cover crops' is used to describe the crops as one of the components of the sustainable agricultural management practices, that cover the soil to protect from erosion, improve soil quality and enhance agro-ecosystem services. These crops are used during fallow periods in annual grain systems. Both grasses and legumes are widely used as cover crops. Legumes are usually used as cover crops for green manuring, as they fix the nitrogen (N) from the atmosphere via biological N fixation in soil to provide it to the grain crops. In winter, small grain crops, such as winter rye, wheat, barley, oats and triticale are used as cover crops. These crops are generally managed by mechanical methods or herbicide application before the planting of main seasonal crops, like corn, soybeans, sunflower, etc. Among the legumes, some excellent cover crops include alfalfa, cowpeas, winter peas, Sunn hemp and red, white and sweet clover. Grasses and *Brassicas*, such as ryegrass, mustard and radish can also potentially be used as cover crops. The benefits of cover crops include N fertility and organic matter, protection from soil erosion, increased soil moisture, capture and recycling of nutrients, decreased soil compaction and suppression of weeds, nematodes and other soil-borne diseases (Fig. 1).

Fig. 1 Cover crops reduce soil erosion (*Source*: Uttech, 2010). Cover crops reduce erosion, runoff (public release: 18 May 2010. http://www.agronomy.org).

[1] Department of Food, Agricultural and Biological Engineering, The Ohio State University, 590 Woody Hayes Drive, Columbus OH 43210, USA.

[2] Department of Environmental Sciences, PMAS Arid Agriculture University Rawalpindi, Murree Road Rawalpindi, Pakistan, 46300.

* Corresponding author: beenishsabaosu@gmail.com

In a natural ecosystem, the soil is covered most of the time by growing plants that maintain soil moisture through the transpiration of water, increasing rainfall infiltration, fixing carbon, keeping reactive nutrients like N and phosphorus (P) in place, preventing runoff and supporting soil flora and fauna. In an annual cropping system, management practices affect residues distribution and placement, leaving the soil partially exposed and unprotected to the erosive forces of wind and water, loss of nutrient and organic matter, increased runoff and stressed soil organisms with low soil productivity during the fall, winter, and early spring. Thus, planting cover crops during fallow periods helps to maintain soil health and protect water quality.

Duran and Pleguezuelo (2008) have studied the effects of cover crops on soil and observed that an increased duration of plant cover is an effective strategy to reduce N and P losses from annual crops, like corn silage, but the degree of improvement depends on management. Cereal rye (*Secale cereal* L.) planting after corn silage harvest provides an over-the-winter soil cover, but post-harvest planting does not address the risk of soil and nutrient losses that occur during silage production (Wendt and Burwell, 1985). To prevent nutrient losses during and after corn silage production, Siller et al. (2016) used living mulch systems with Kura clover (*Trifolium ambiguum* L.) alone and in combination with a cereal rye cover crop.

Inter-seeding of cover crops with corn have reportedly reduced nutrient losses following the harvest or during corn production (Kleinman et al., 2005; Grabber and Jokela, 2013; Grabber, 2014). It is reported that cover crops are only helpful when managed properly, otherwise they can lead to yield decline of main crops, though contributing little to farm profitability due to market constraints (Sarrantonio and Gallandt, 2003; Roesch-McNally et al., 2017). However, current technologies and selection of economically valuable and suitable cover crops can overcome these barriers and improve agro-ecosystem services. Inter-seeding of alfalfa with corn silage within a few days of corn planting provides a valuable forage crop, as well as a good cover crop as seen in a research experiment conducted in Wisconsin (Osterholz et al., 2019). Corn silage-alfalfa rotations achieve double the alfalfa production than that through solo seeding of alfalfa in the first year of its application (Grabber, 2016; Osterholz et al., 2017).

2. Effectiveness of Cover Crop in Protecting Ecosystem Resources

In recent decades, agricultural production has been intensified to meet global food security and this is inconsistent with environmental impact and ecological resources (soil, water, or air quality) deterioration (Tilman, 2002; Lal, 2008). Mono-cropping, over-fertilisation, pesticide application, overgrazing, nutrient leaching, fallow periods without cover crops and excessive tillage are some of the conventional management practices that convert once-fertile agricultural lands into degraded and marginal lands (Scherr and Yadav, 1996; Maharjan et al., 2016). Agricultural land degradation is one of the great concerns to farm income, ecosystem resources and decreased yield as has been most frequently observed (and still increasing) in Asia (206 M ha) (Oldeman, 1994; Scherr and Yadav, 1996). The United Nations Convention to Combat Desertification has postulated global zero net land degradation by 2030 (Stavi and Lal, 2015; UNCCD, 2012).

Soil erosion and water quality not only depend on agricultural management practices, but also natural factors, such as natural vegetation, climate and topography. Generally, for agricultural lands, soil-loss rates are tenfold higher than soil-formation rates (Pimentel et al., 1987). Bare soils on slopes are at higher risk of erosion due to the intensity of rainfall and surface runoff (Morgan, 2005). Over-fertilisation also contributes to higher nutrient runoff through surface flow and percolation into aquatic ecosystems, thus deteriorating the freshwater resources. A study by Tilman et al. (2002) reported that only 30–50 per cent of N and 45 per cent of phosphorus is taken in by crops and the rest of the nutrients are lost from the fields.

To reduce pollution load, improve water quality and prevent soil erosion, best management practices (BMPs) for sustainable agriculture have been adapted worldwide. In addition to policies, incentives for farmers to implement BMPs, awareness and education about suitable cover crops in the area to maintain productive farmlands need to be provided. Sustainable agriculture ensures an economically-viable yield, environmental benefits and watershed and water quality protection. In a study, Maharjan et al. (2016) recommended that split fertiliser application and cover crop integration in cropping systems reduce the sediment load and nitrate and soluble reactive P levels significantly when compared to conventional agricultural management practices.

3. Cover Crops and Soil Erosion

3.1 Soil Loss

Soil loss affects fertility, long-term productivity and nutrient availability, structural stability, the soil organic matter (SOM) content and increases compaction in agricultural lands (Lal, 1995; Morgan, 2005; Ramos and Martinez Casasnovas, 2006). Soil erosion also becomes a non-point source of water pollution and deteriorates surface water quality (Fisher et al., 2000; Verstraeten et al., 2003; Chu et al., 2004). Agricultural practices are a major source of soil erosion (Montgomery, 2007). A study by Verheijen et al. (2009) showed that 3–40 ton/ha of fine soil material is eroded each year in Europe. Several other studies (Six et al., 2000; Zhang et al., 2007; Erhart and Hartl, 2009) reported that any tillage system that maintains 30 per cent of cover on the soil surface decreases erosion by improving soil structural porosity and water infiltration. A greater soil surface cover provides better protection to the soil surface, soil structure and soil organisms such as earthworms (Mikha and Rice, 2004; Blanco-Canqui and Lal, 2008).

Like sustainable agriculture, organic farming (cropping intensity and rotation) is one alternative strategy to reduce edge-of-the-field soil loss (Gomiero et al., 2011a; Gomiero et al., 2011b). It can lead to a reduction of crop yield, but increases soil fertility, biological diversity and enhances soil surface cover over time (Wittwer et al., 2017; Knaapp and van der Heijden, 2018). In a study by Seitz et al. (2019), sedimentary loss decreased in organic farming as compared to conventionally-tilled farming by 30 per cent and reduced tillage in organic farming compared to intensive tillage by 61 per cent. Similarly, erosion rates were higher in maize growth during June when compared to fallow land after winter wheat (2.92 to 0.23 ton/ha).

Seitz et al. (2019) observed that when surface cover is 30 per cent, it had the greatest influence on soil-erosion reduction. Results presented in Fig. 2 show a comparison of different farming systems (conventional-intensive tillage [red], no-till [blue], organic intensive tillage [green] and organic reduce tillage [purple]) during two different years (2014 in triangle and 2017 in dot) and how each affect surface cover and sediment delivery (ton/ha). This data shows that organic farming

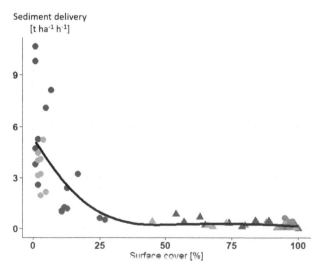

Fig. 2 Surface cover affects sediment delivery (ton/ha) in 2014 (triangles) and 2017 (dots) (adapted from Seitz et al. (2019) © Agron. Sustain. Dev.).

and conservation agriculture are the best strategies to reduce soil losses and reduced tillage leads to significant improvement in organic farming in terms of soil erosion control.

3.2 Soil Nutrients

Cover crops help to improve carbon (C) input into the soil. Both N and C are needed to make SOM. No tillage and C input make SOM. While legumes help to fix atmospheric N, grasses add C into the soil while maintaining the average C:N (10–12:1) in soil (Hoorman, 2017). Cover crop helps to accelerate soil microbial activity and microbial biomass, which in turn increases SOM and other properties, like soil permeability, microporosity, aggregate size, etc. It is reported that bare soil holds 1.7 inches of water, but with cover crops, its water-holding capacity is 4.2 inches (USDA-NRCS engineering book). Korucu et al. (2018) studied the effect of rye as a cover crop with corn-soybean rotation in reducing runoff, as a nutrient recycler and enhancer of SOM. The results showed that runoff decreased by 65 per cent, sediment loss by 68 per cent, while biomass increased by 1.4–2.5 times with a cover crop as compared to no-cover crop. An increase (1.2–3.2 times higher) in the earthworm population was also observed.

Carbon sequestration in agricultural soils is a non-expensive climate-change mitigation and adaptation tool (Lal, 2010; Vicente-Vicente et al., 2016) and cover crops accelerate C sequestration (Pardo et al., 2017; Paustian et al., 2016). Sastre et al. (2018) studied a trial of a rain-fed olive grove with legume and barley as two annuals and *Brachypodium distachyon* as one permanent cover crop in Spain. Their results suggested that after three years, *B. distachyon* improved SOM and stoichiometrically maintained C:N. Cover crops facilitated higher SOM content in deeper layers by improving C sequestration. Therefore, cover crops should be planted continuously over longer periods to improve SOM.

Cover crops and SOM are correlated with slope gradient of the agricultural land. Novara et al. (2019) compared the effect of slope and flat agricultural land on SOM content. The experiment was conducted in the Mediterranean vineyards of Sicily, Italy for a period of five years. The results showed an SOM increase of 6 per cent in the flat area and 9 per cent in the sloping vineyards when *Vicia fava* was the cover crop. The higher C sequestration rates on the slopes could be attributed to role of cover crops in the prevention of sediment and nutrient erosion.

Nutrient losses from the edge of the agricultural lands due to runoff are often associated with the growth of freshwater algal blooms; specifically the reason behind seasonal algal blooms in the Gulf of Mexico and Lake Erie is nutrient leaching and/or runoff from agricultural farms (Rabalais, 2002; Van Meter et al., 2018). The Midwest United States contributes > 50 per cent of the N entering into the Gulf as a result of the N fertiliser applied (Alexander et al., 2008; Robertson and Saad, 2013). Sub-surface tile drainage also facilitates the runoff of nitrate (NO^-_3-N) into water streams (Jaynes et al., 1999; Sugg, 2007; Blann et al., 2009). Drainage discharge bypasses the biogeochemical system, which can help retain the nitrate removed by the process of denitrification and directly discharged into ditches and streams (Kladivko et al., 2004; William et al., 2015). Several studies have compared the yield of nitrate from tile drainage (15–60 Kg NO^-_3-N ha$^-$) and watersheds dominated by tall-grass prairie (0.16 Kg NO^-_3-N ha$^-$) or hardwood forest (< 0.25 Kg NO^-_3-N ha$^-$) and found orders of magnitude difference (Dodds et al., 1996; Royer et al., 2006; Ikenberry et al., 2014; Swonk and Vose, 1997; Williams et al., 2015). According to the US Geological Survey, agricultural sources in the watersheds of the Mississippi River basin contribute more than 70 per cent of the nitrogen and phosphorus, versus about 9–12 per cent from urban sources (Alexander et al., 2008).

Cover crops have been used historically in the United States to improve soil quality associated with soil fertility (Dabney et al., 2001; Blanco-Canqui et al., 2015). Watershed conservation practices that can retain NO^-_3-N in the field and sustain productive agriculture are suggested as potential solutions (Magdoff et al., 1997; Dinnes et al., 2002; Schilling et al., 2012). Cover crops planted at the end of the growing season for cash crops (corn or soybean) and then terminated before planting of the next cash crop is one of the sustainable conservation practices. A number of studies have reported that cover crops immobilise N in their biomass and prevent NO^-_3-N leaching from

Table 1 Average annual flow-weighted NO₃-N concentrations for each treatment for the 2014–2016 hydrologic years (adopted from Ruffati et al., 2019; ©Agric. Water Manag.).

Treatment	2014 Hydrologic Year	2015 Hydrologic Year	2016 Hydrologic Year
	mg L^{-1}		
Fall nitrogen (N)	10.48^{Bb}φ	5.44^{AB}	13.98^{A}
Fall N with cover crop	9.52^{AB}	3.53^{A}	5.48^{B}
Spring N	8.88^{AC}	5.97^{A}	11.80^{A}
Spring N cover crop	7.93^{AC}	3.64^{B}	4.65^{B}
Zero control	7.18^{C}	4.29^{AB}	8.92^{C}

φ Different capital letters within a column indicate significant differences between experimental. Treatments for that given hydrologic year at an alpha level of 0.05.

tile drains (Kaspar et al., 2007, 2012; Krueger et al., 2012). Cover crops also reduce the volume of water from saturated soil by increased evapo-transpiration (Strock et al., 2004; Qi and Helmers, 2010). As evidenced in Table 1, the average NO₃-N loss from cover cropped land is reducing each subsequent year.

Hanrahan et al. (2018) studied the NO⁻₃-N export after planting a cover crop (> 60 per cent of cropped acres) in a tile drainage agricultural field. The results revealed a 69–90 per cent reduction in NO₃⁻-N export in the cover crop-planted field over the non-planted field during winter-spring growing seasons. In recent years, the state of Indiana (US) has increased cover crop planting by 446 per cent, according to data from 2015–2016 (ISDA, 2017). Cover crop planting reduces nutrient loss but assessment of the efficiency of conservation practices has some challenges.

3.3 Soil Moisture

Cover crops help improve the soil's physical, chemical and biological quality. Overall, cover crops have both positive as well as negative impacts on soil moisture. They can help infiltrate water into the root zone and increase soil moisture, as well as water availability, for a cash crop. They also transpire soil water and dry out fields faster, affecting main crop yield. In the areas where precipitation is higher, transpiration is effective to improve field workability. A study by Hoorman (2017) suggested that in the state of Ohio, fields (with clay content) are wet seven out of 10 years during the spring season, so early transpiration with the help of a cover crop is better to dry the soils for planting succeeding crops. Deep-rooted crops help to attain subsoil moisture and conserve for dry periods. If prolonged drought occurs, cover crops may hurt the yield of the cash crop.

While grassy cover crops (such as cereal rye) increase water infiltration by 8–42 per cent, legume cover crops (crimson clover, strawberry clover, hairy vetch) increase water infiltration by 39–528 per cent and crop residue increases water infiltration by 180 per cent (SARE, 2017). A study by Snapp and Surpu (2018) did not find any positive or negative effects of cover crops on soil moisture and corn yields in the same season. However, the use of cover crops depleted the soil moisture for the following crop (Dabney et al., 2001). In contrast, Sij et al. (2004) documented a higher soil moisture availability under conservation tillage system with a cover crop as compared to one without a cover crop. Likewise, annual winter cover crops do not reduce soil moisture for the subsequent crop, according to a study in North Texas by Teague et al. (2019). However, as these results have been obtained in years when precipitation was above average, it may vary in subsequent dry years.

3.4 Soil Temperature

Cover crops affect soil temperature and significantly decrease day and night temperatures. They protect against cold nights and slow down the cooling. Temperature drop is beneficial only in hot regions; in cold regions, it can slow down the growth. Temperature and precipitation are the two

Soil temperature differences

Conventional tillage

No-till + cover crops

Fig. 3 Comparative effectives of conventional tillage vs. no-till cover crops on soil temperature (*Courtesy*: Jim Hoorman).

variables that affect cover-crop selection. Hoorman (2017) revealed that standing cover crops can maintain a higher temperature than flat crops. In no-till fields, row cleaners can be used to manage residues and improve soil temperature. Figure 3 shows that soil temperatures are significantly lower under no-till cover-cropped plots (87.6°F or 30.9°C) when compared to conventionally-tilled plots without any cover crops (107.4°F or 41.9°C) in hot sunny days.

4. Cover Crops and Water Quality

The efficacy of cover crops in surface runoff reduction and water quality improvement has been well established (Dabney et al., 2001; Kaspar and Singer, 2011; Blanco-Canqui et al., 2015). It is well known that surface freshwater quality is under continuous deterioration and an increase in the pollution load can be observed at conventional agricultural fields. Soluble and sediment-bound reactive nutrients flow (especially soluble reactive P) in the runoff cause algal blooms in surface water bodies and both soluble reactive P and nitrate cause algal blooms in estuarine and ocean waters. These algal blooms in surface freshwater release toxins and their decomposition depletes the concentration of dissolved oxygen levels in water, which causes death of aquatic species. Aesthetically, water also looks unpleasant and not drinkable and is not suitable for recreational sports. To improve water quality, the planting of cover crops in the fallow periods is one of the BMPs. It increases water infiltration in the soil, raises groundwater level, keeps moisture locked in the soil for dry periods and decreases soil erosion. Cover crops can increase infiltration six-fold in some systems (SARE, 2017).

4.1 Reduction in Sediment Loads and Water Quality

Sediments are suspended soil particles (silt and clay) that move via runoff after intense precipitation. Korucu et al. (2018) studied the effect of sediment loss and runoff reduction from the edge of the fields after fertiliser application. The results showed that rye cover crop significantly delayed runoff by 5.7 min, which was a 65 per cent improvement when compared to non-cover crop fields. Overall reported reduction in sediment loss was 68 per cent, including a reduction in the nutrient losses of NH_4^- N (86 per cent), TP (83 per cent) and K (91 per cent). Simulated rainfall on an experimental field by Siller et al. (2016) has proven that planting cereal rye as a cover crop after corn harvest reduces runoff by 67 per cent, sediment losses by 81 per cent, SRP losses by 94 per cent and N losses by 83 per cent as compared to a non-cover crop field. Cover crops reduce sediments from erosion by 20.8 ton/ac in conventional fields, 6.58 ton/ac in reduced till-fields, and 1.28 ton/ac in no-till fields (SARE, 2017).

Osterholz et al. (2019) studied total suspended solids (TSS) loss under cover crops in two experiments and accounted 89 per cent variation in TSS loss (Fig. 4). Cover crops significantly reduce sediment loss and improve water quality. Large reductions in dissolved solids, soluble and

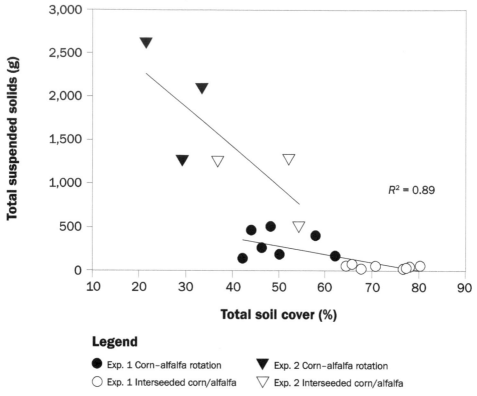

Fig. 4 Relationship between total soil cover under different cover crops and total suspended solids from run off (adopted from Osterholz et al., 2019).

sediment-bound P, soluble nitrogen (organic N such as amino acids, amino sugars, etc.) and NO_3^- were observed as well, which are clear indications of improved water quality. Cover crops impact water and soil quality gradually and are not a one-time phenomena. Cercioglu et al. (2019) studied the effect on soil hydraulic conductivity, moisture retention, bulk density and pore size distribution in a corn-soybean rotation with radish, cereal rye, cowpeas, buckwheat, barley, red clover, turnips, hairy vetch, triticale and winter peas as cover crops. The results showed that a significant and maximum effect on soil-water retention (K_{sat}) values was observed at five years (64.4 per cent) at 0–10 cm and 10–20 cm depths. The study proves that in clay-pan soils, the changes take several years to improve hydraulic properties.

4.2 Scavenging of Nutrients and Pesticides

Nutrients loss from agricultural fields in runoff and leaching goes into surface water and groundwater. These nutrients in the waterway imbalance the sensitivity of the ecosystems. Surface runoff and leaching and nutrient loss are directly linked to each other. As the nutrients are dissolved in water or adsorb on soil, particles move in runoff, while the cover crops trap the soil particles, reducing erosion and capturing nutrients as well. Nutrients also move into groundwater with infiltration followed by percolation. Studies showed, on an average, a 48–89 per cent reduction in N and 15–92 per cent reduction in P, with the presence of a cover crop in diverse agro-ecosystems (SARE, 2017).

The introduction of cover crops between two cash crops not only improves water and soil quality but also reduces edge-of-the-field pesticide losses in runoff by influencing water dynamics and solute movement in soil (Alleto et al., 2012). With extensive root distribution and labile organic matter, cover crops reduce the transfer of pesticides in soils (Potter et al., 2007). Moreover, with increased biodiversity and efficient biological activity, cover crops help to inactivate or degrade pesticides much faster in ecosystems (Reeves, 1994). Upon termination or winter-kill, cover crop residues form vegetative surface mulch that intercepts pesticides and their metabolites by slowing down water movement and reducing runoff (Ghadiri et al., 1984).

Cassigneul et al. (2015) studied the nature and decomposition of byproducts of cover crops influencing pesticide sorption by studying three different pesticides (epoxiconazole [EPX], S-metolachlor [SMOC] and glyphosate [GLY]). Results suggested that cover crops increased pesticide decomposition and sorption 1.6–4.7-folds. The decomposition and sorption followed the EPX > SMOC > GLY pattern. The conclusion is that the sorption and decomposition of pesticides depends on the biochemical characteristics of cover crops. Sorption of pesticides increases with the degree of decomposition.

4.3 Prevention of Harmful Algal Blooms

Harmful algal blooms (HABs), especially cyanobacteria, occur when densities of micro- and macroalgaes are so high that they can cause damage to environment, specially to aquatic life and water quality (Anderson, 1995; Erdner et al., 2008). Many of the largest and most productive water bodies in the world are under the threat of habitat loss and decrease in biodiversity, leading to public-health risk and economic loss due to eutrophication (Brooks et al., 2017). The chief contributing factor to eutrophication and algal growth is nutrient over-enrichment, such as SRP in fresh water bodies and both nitrate and SRP in estuarine and sea-water systems. It creates a favourable environment for producing vertically-stratified low-oxygen systems (hypoxia) for the growth of cyanobacteria (Heisler et al., 2008).

Lake Erie is one of the examples of freshwater eutrophication, HABs and hypoxia in the United States and Canada. In the mid-20th century, an increase in the SRP and N loading from agricultural, industrial and urban runoff led to the degradation of water quality in the lake (Makarewicz et al., 1989; Bertram, 1993). Figure 5 shows bloom sensitivity to SRP from 2002–2016. Nutrient runoff from midwest US states stimulates algal blooms in the Gulf of Mexico (Van Meter et al., 2018).

The transport of N as nitrate and a small amount of SRP is facilitated by subsurface tile drainage that carries excess water from agricultural fields to adjacent water bodies (Blann et al., 2009). In this process, biogeochemical interception of nutrients is bypassed and results in a higher concentration of nutrients in the waterways (Williams et al., 2015). Nitrate loads from tile-drained water in the midwest United States ranges from 15–60 Kg NO_3^--N/ha as compared to hardwood forest (< 0.25 Kg NO_3^--N/ha) and grass prairie (0.16 Kg NO_3^--N/ha) (Williams et al., 2015; Swank and Vose, 1997).

An in-depth study conducted by Hoorman et al. (2008) showed that the water quality index (WQ_{Index}) was influenced by both season and feeding streams in St. Mary's Lake in Ohio (Fig. 6). The WQ_{Index} of streams showed peak degradation in the summer at 30 per cent above the annual

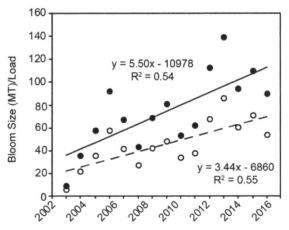

Fig. 5 Bloom sensitivity from total phosphorus and (open circles) and bioavailable phosphorus (filled circles) for the year 2002–2016 (adapted from Manning et al., 2019).

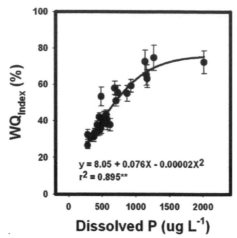

Fig. 6 Relationship between dissolved phosphorus with water quality index (WQ_{Index}) in St. Mary's Lake, Ohio, USA (adapted from Hoorman et al., 2008).

mean and improvement during the winter at 26 per cent below the annual mean. Among the streams, the Beaver and Coldwater had a significantly greater WQ_{Index} degradation (15–30 per cent higher) as compared to other creeks. Likewise, the WQ_{Index} of the lake had peak degradation in the summer at 26 per cent above the annual mean. A significant quadratic relationship between WQ_{Index} and dissolved P suggested that 90 per cent of the variations in stream and lake quality can be credited to dissolved P concentration (Fig. 6).

Agricultural lands covered under forest, grasses, or winter cover crops are a solution to intercept reactive nutrients and prevent algal blooms. Planting of watershed-scale cover crops is encouraged as one of the components of the BMPs to reduce soil erosion and nutrient loads for controlling seasonal algal blooms and water quality degradation.

References

Alexander, R.B., Smith, R.A., Schwarz, G.E., Boyer, E.W., Nolan, J.V. and Brakebill, J.W. (2008). Differences in phosphorus and nitrogen delivery to the Gulf of Mexico from the Mississippi River Basin. Environ. Sci. Technol., 42: 8220830.

Alletto, L., Benoit, P., Justes, E. and Coquet, Y. (2012). Tillage and fallow period management effects on the fate of the herbicide isoxaflutole in an irrigated contiguous maize field. Agric. Ecosyst. Environ., 153: 40–49.

Anderson, D.M. (1995). Toxic red tides and harmful algal blooms: A practical challenge in coastal oceanography. Rev. Geophys., 33: 1189–1200.

Bertram, P.E. (1993). Total phosphorus and dissolved oxygen trends in the central basin of Lake Erie, 1970–1991. J. Great Lakes Res., 19: 224–236.

Blanco-Canqui, H. and Lal, R. (2008). Principles of Soil Conservation and Management, Springer, Dordrecht.

Blanco-Canqui, H., Shaver, T.M., Lindquist, J.L., Shapiro, C.A., Elmore, R.W., Francis, C.A. and Hergert, G.W. (2015). Cover crops and ecosystem services: Insights from studies in temperate soils. Agron. J., 107: 2449–2474.

Blann, K.L., Anderson, J.L., Sands, G.R. and Vondracek, B. (2009). Effects of agricultural drainage on aquatic ecosystems: A review. Environ. Sci. Technol., 39: 909–1001.

Brooks, B.W., Lazorchak, J.M., Howard, M.D., Johnson, M.V.V., Morton, S.L., Perkins, D.A., Reavie, E.D., Scott, G.I., Smith, S.A. and Steevens, J.A. (2017). In some places, in some cases and at sometimes, harmful algal blooms are the greatest threat to inland water quality. Environ. Toxicol. Chem., 36: 1125–1127.

Cassigneul, A., Alletto, L., Benoit, P., Bergheaud, V., Etievant, V., Dumeny, V., Le Gac, A.L., Chuette, D., Rumpel, C. and Justes, E. (2015). Nature and decomposition degree of cover crops influence pesticide sorption: Quantification and modelling. Chemosphere, 119: 1007–1014.

Cercioglu, M., Anderson, S.H., Udawatta, R.P. and Algele, S. (2019). Effect of cover crop management on soil hydraulic properties. Geoderma, 343: 247–253.

Chu, T.W., Shirmohammadi, A., Montas, H. and Sadeghi, A. (2004). Evaluation of the SWAT model's sediment and nutrient components in the Piedmont physiographic region of Maryland. Trans. ASAE, 47: 1523–1538.

Dabney, S.M., Delgado, J.A. and Reeves, D.W. (2001). Using winter cover crops to improve soil and water quality. Comm. Soil Sci. Plant Anal., 32: 1221–1250.

Dinnes, D.L., Karlen, D.L., Jaynes, D.B., Kaspar, T.C., Hatfield, J.L., Colvin, T.S. and Cambardella, C.A. (2002). Nitrogen management strategies to reduce nitrate leaching in tile-drained midwestern soils. Agron. J., 94: 153–171.

Dodds, W.K., Blair, J.M., Henebry, G.M., Koelliker, J.K., Ramundo, R. and Tate, C.M. (1996). Nitrogen transport from tall grass prairie watersheds. J. Environ. Qual., 25: 973–981.

Durán Zuazo, V.H. and Rodríguez Pleguezuelo, C.R. (2008). Soil erosion and runoff prevention by plant covers: A review. Agron. Sustain. Dev., 28: 65–86.

Erdner, D., Dyble, J., Parsons, M.L., Stevens, R.C., Hubbard, K.A., Wrabel, M.L., Moore, S.K., Lefebvre, K.A., Anderson, D.M., Bienfang, P., Bidigare, R.R., Parker, M.S., Moeller, P., Brand, L.E. and Trainer, V.L. (2008). Centres for oceans and human health: A unified approach to the challenge of harmful algal blooms. Environ. Health, 7: (Suppl. 2).

Erhart, E. and Hartl, W. (2009). Soil protection through organic farming: A review. pp. 203–226. *In*: Lichtfouse E. (ed.). Organic Farming, Pest Control and Remediation of Soil Pollutants. Springer, Dordrecht.

Fisher, D.S., Steiner, J.L., Endale, D.M., Stuedemann, J.A., Schomberg, H.H., Franzluebbers, A.J. and Wilkinson, S.R. (2000). The relationship of land use practices to surface water quality in the Upper Oconee watershed of Georgia. For. Ecol. Manage., 128: 39–48.

Ghadiri, H., Shea, P.J., Wicks, G.A. and Haderlie, L.C. (1984). Atrazine dissipation in conventional-till and no-till sorghum. J. Environ. Qual., 13: 549–552.

Gomiero, T., Pimentel, D. and Paoletti, M.G. (2011a). Environmental impact of different agricultural management practices: Conventional vs. organic agriculture. Crit. Rev. Plant Sci., 30: 95–124.

Gomiero, T., Pimentel, D. and Paoletti, M.G. (2011b). Is there a need for a more sustainable agriculture? Crit. Rev. Plant Sci., 30: 6–23.

Grabber, J.H. and Jokela, W.E. (2013). Off-season groundcover and runoff characteristics of perennial clover and annual grass companion crops for no-till corn fertilised with manure. J. Soil Water Conser., 68: 411–418.

Grabber, J.H., Jokela, W.E. and Lauer, J.G. (2014). Soil nitrogen and forage yields of corn grown with clover or grass companion crops and manure. Agron. J., 106: 952–961.

Grabber, J.H. (2016). Prohexadione-calcium improves stand density and yield of alfalfa inter-seeded into silage corn. Agron. J., 108: 726–735.

Hanrahan, B.R., Tank, J.L., Christopher, S.F., Mahl, U.H., Trentman, M.T. and Royer, T.V. (2018). Winter cover crops reduce nitrate loss in an agricultural watershed in the central US. J. Agri. Ecosyst. Environ., 265: 513–523.

Heisler, J., Glibert, P.M., Burkholder, J.M., Anderson, D.M., Cochlan, W., Dennison, W.C., Dortch, Q., Gobler, C.J., Heil, C.A., Humphries, E. and Lewitus, A. (2008). Eutrophication and harmful algal blooms: a scientific consensus. Harmful Algae, 8: 3–13.

Hoorman, J., Hone, T., Sudman, T., Dirksen, T., Iles, J. and Islam, K.R. (2008). Agricultural impacts on lake and stream water quality in Grand Lake St. Mary's, Western Ohio. Water Air Soil Pollut., 193: 309–322.

Ikenberry, C.D., Soupir, M.L., Schilling, K.E., Jones, C.S. and Seeman, A. (2014). Nitrate-nitrogen export: Magnitude and patterns from drainage districts to downstream river basins. J. Environ. Qual., 43: 2024–2033.

ISDA. (2017). Cover Crop Trends 2011–2016 Statewide. (Accessed 11 August 2017). http://www.in.gov/isda/files/Cover%20Crop%20Trends%202011-2016%20Statewide.pdf.

Jaynes, D.B., Hatfield, J.L. and Meek, D.W. (1999). Water quality in Walnut Creek watershed: Herbicides and nitrate in surface waters. J. Environ. Qual., 28: 45–59.

Kaspar, T.C., Jaynes, D.B., Parkin, T.B. and Moorman, T.B. (2007). Rye cover crop and Gama grass strip effects on NO_3 concentration and load in tile drainage. J. Environ. Qual., 36: 1503–1511.

Kaspar, T.C., Jaynes, D.B., Parkin, T.B., Moorman, T.B. and Singer, J.W. (2012). Effectiveness of oat and rye cover crops in reducing nitrate losses in drainage water. Agric. Water Manage., 110: 25–33.

Kaspar, T. and Singer, J.W. (2011). The use of cover crops to manage soil. Published in Soil Management: Building a stable base for agriculture. 321–337. American Society of Agronomy and Soil Science Society of America. DOI:10.2136/2011.soilmanagement.c21.

Kladivko, E.J., Frankenberger, J.R., Jaynes, D.B., Meek, D.W., Jenkinson, B.J. and Fausey, N.R. (2004). Nitrate leaching to subsurface drains as affected by drain spacing and changes in crop production system. J. Environ. Qual., 33: 1803–1813.

Kleinman, P.J., Salon, P., Sharpley, A.N. and Saporito, L.S. (2005). Effect of cover crops established at time of corn planting on phosphorus runoff from soils before and after dairy manure application. J. Soil Water Conser., 60: 311–322.

Knapp, S. and Van der Heijden, M.G.A. (2018). A global meta-analysis of yield stability in organic and conservation agriculture. Nat Commun., 9: 3632.

Korucu, T., Shipitalo, M.J. and Kasper, T.C. (2018). Rye cover crop increases earthworm populations and reduces losses of broadcast, fall-applied fertilisers in surface runoff. Soil Till. Res., 180: 99–106.

Krueger, E.S., Ochsner, T.E., Baker, J.M., Porter, P.M. and Reicosky, D.C. (2012). Rye-corn silage double-cropping reduces corn yield but improves environmental impacts. Agron. J., 104: 888–896.

Lal, R. (1995). Erosion crop productivity relationships for soils of Africa. Soil Sci. Soc. Amer. J., 59: 661–667.

Lal, R. (2008). Soils and sustainable agriculture: A review. Agron. Sustain. Dev., 28: 57–64.

Lal, R. (2010). Beyond Copenhagen: Mitigating climate change and achieving food security through soil carbon sequestration. Food Sec., 2: 169–177.

Magdoff, F., Lanyon, L. and Liebhardt, B. (1997). Nutrient cycling, transformation and flows: Implications for a more sustainable agriculture. Adv. Agron., 60: 2–73.

Maharjan, G.R., Ruidisch, M., Shope, C.L., Choi, K., Huwe, B., Kim, S.J., Tenhunen, J. and Arnhold, S. (2016). Assessing the effectiveness of split fertilisation and cover crop cultivation in order to conserve soil and water resources and improve crop productivity. Agric. Water Manage., 163: 305–318.

Makarewicz, J.C., Lewis, T. and Bertram, P. (1989). Phytoplankton and Zooplankton Composition, Abundance and Distribution and Trophic Interactions: Offshore Regions of Lake Erie, Lake Huron and Lake Michigan, 15 (1985 CIESIN).

Mikha, M.M. and Rice, C.W. (2004). Tillage and manure effects on soil and aggregate-associated carbon and nitrogen. Soil Sci. Soc. Am. J., 68: 809–816.

Montgomery. D.R. (2007). Soil erosion and agricultural sustainability. PNAS, 104: 13268–13272.

Morgan, R.P.C. (2005). Third ed. *In*: Soil Erosion and Conservation, Blackwell, Malden.

Novara, A., Minacapilli, M., Santoro, A., Rodrigo-Comino, J., Carrubba, A., Sarno, M. Venezia, G. and Gristina, L. (2019). Real cover crops contribution to soil organic carbon sequestration in sloping vineyard. Sci. Total Environ., 652: 300–306.

Oldeman, L.R. (1994). The global extent of soil degradation. pp. 99–118. *In*: Greenland, D.J. and Szabolcs, I. (eds.). Soil Resilience and Sustainable Land Use. CAB International, Wallingford.

Osterholz, W.R., Renz, M.J., Lauer, J.G. and Grabber. J.H. (2017). Prohexadione-calcium rate and timing effects on alfalfa inter-seeded into silage corn. Agron. J., 110: 1–10.

Osterholz, W.R., Renz, M.J., Jokela, W.E. and Grabber. J.H. (2019). Inter-seeded alfalfa reduces soil and nutrient losses during and after corn silage production. J. Soil Water Conserv., 74: 85–90.

Pardo, G., Prado, A., del Martinez-Mena, M., Bustamante, M.A., Martin, J.A.R., Alvaro-Fuentes, A. and Moral, R. (2017). Orchard and horticulture systems in Spanish Mediterranean coastal areas: Is there a real possibility to contribute to C sequestration? Agric. Ecosyst. Environ., 238: 153–167.

Paustian, K., Lehmann, J., Ogle, S., Reay, D., Roberston, G.P. and Smith, P. (2016). Climate-smart soils. Nature, 532: 49–57.

Pimentel, D., Allen, J., Beers, A., Guinand, L., Linder, R., McLaughlin, P., Meer, B., Musonda, D., Perdue, D., Poisson, S., Siebert, S., Stoner, K., Salazar, R. and Hawkins, A. (1987). World agriculture and soil erosion. Bioscience, 37: 277–283.

Potter, T.L., Bosch, D.D., Joo, H., Schaffer, B. and Munoz-Carpena, R. (2007). Summer cover crops reduce atrazine leaching to shallow groundwater in southern Florida. J. Environ. Qual., 36: 1301–1309.

Qi, Z. and Helmers, M.J. (2010). Soil water dynamics under winter rye cover crop in central Iowa. Vadose Zone J., 9: 53–60.

Rabalais, N.N. (2002). Nitrogen in aquatic ecosystems. AMBIO, 31: 102–112.

Ramos, M.C. and Martínez-Casasnovas, J.A. (2006). Impact of land levelling on soil moisture and runoff variability in vineyards under different rainfall distributions in Mediterranean climate and its influence on crop productivity. J. Hydrol., 321: 131–146.

Reeves, D.W. (1994). Cover crops and rotations. *In*: Hatfield, J.L. and Stewart, B.A. (eds.). Advances in Soil Science: Crops Residue Management. Lewis, Boca Raton, FL.

Robertson, D.M. and Saad, D.A. (2013). SPARROW models used to understand nutrient sources in the Mississippi/Atchafalaya River Basin. J. Environ. Qual., 42: 1422–1440.

Roesch-McNally, G.E., Basche, A.D., Arbuckle, J.G., Tyndall, J.C., Miguez, F.E., Bowman, T. and Clay. R. (2017). The trouble with cover crops: Farmers' experiences with overcoming barriers to adoption. Renew. Agr. Food Syst., 33: 322–333.

Royer, T.V., David, M.B. and Gentry, L.E. (2006). Timing of riverine export of nitrate and phosphorus from agricultural watersheds in Illinois: Implications for reducing nutrient loading to the Mississippi River. Environ. Sci. Technol., 40: 4126–4131.

Ruffatti, M.D., Roth, R.T., Lacey, C.G. and Armstrong, S.D. (2019). Impacts of nitrogen application timing and cover crop inclusion on subsurface drainage water quality. Agric. Water Manag., 211: 81–88.

SARE. (2017). Cover crops at work: Increasing infiltration. *In*: Sustainable Agriculture and Research Education. https://sare.org/wp-content/uploads/Cover-Crops-at-Work-Increasing-Infiltration.pdf.

Sarrantonio, M. and Gallandt. E. (2003). The role of cover crops in North American cropping systems. J. Crop Prod., 8: 53–74.

Sastre, B., Marques, M.J., Diaz, A.G. and Bienes, R. (2018). Three years of management with cover crops protecting sloping olive groves soils, carbon and water effects on gypsiferous soil. Catena, 171: 115–124.

Scherr, S.J. and Yadav, S.N. (1996). Land degradation in the developing world: Implications for food, agriculture, and the environment to 2020, Food Agric. Environ., discussion paper 14, International Food Policy Research Institute, Washington, D.C., 20036–3006, USA.

Schilling, K.E., Jones, C.S., Seeman, A., Bader, E. and Filipiak, J. (2012). Nitrate-nitrogen patterns in engineered catchments in the upper Mississippi River basin. Ecol. Eng., 42: 1–9.

Seitz, S., Goebes, P., Puerta, V.L., Pereira, E.I.P., Wittwer, R., Six, J., Van der Heijden M.G.A. and Scholten, T. (2019). Conservation tillage and organic farming reduces soil erosion. Agron. Sustain. Dev., 39: 4.

Sij, J., Ott, J., Olson, B., Baughman, T. and Bordovsky, D. (2004). Dryland cropping systems to enhance soil moisture capture and water-use efficiency in cotton. Proc. 2003 Beltwide Cotton Conf.

Siller, A.R.S., Albrecht, K.A. and Jokela, W.E. (2016). Soil erosion and nutrient runoff in corn silage production with Kura clover living mulch and winter rye. Agron. J., 108: 989–999.

Six, J., Paustian, K., Elliott, E.T. and Combrink, C. (2000). Soil structure and organic matter I. distribution of aggregate-size classes and aggregate-associated carbon. Soil Sci. Soc. Amer. J., 64: 681–689.

Snapp, S. and Surapur, S. (2018). Rye cover crop retains nitrogen and doesn't reduce corn yields. Soil Till. Res., 180: 107–115.

Stavi, I. and Lal, R. (2015). Achieving zero net land degradation: Challenges and opportunities. J. Arid. Environ., 112: 44–51.

Strock, J.S., Porter, P.M. and Russelle, M.P. (2004). Cover cropping to reduce nitrate loss through subsurface drainage in the northern U.S. corn belt. J. Environ. Qual., 33: 1010–1016.

Sugg, Z. (2007). Assessing U.S. Farm Drainage: Can GIS Lead to Better Estimates of Subsurface Drainage Extent? World Resources Institute.

Swank, W.T. and Vose, J.M. (1997). Long-term nitrogen dynamics of Coweeta forested watersheds in the southeastern United States of America. Global Biochem. Cycles, 11: 657–671.

Teague, R., Delaune, P.B. and Dowhower, S.T. (2019). Impacts of over-seeding Bermuda grass pasture with multispecies cover crops on soil water availability, microbiology, and nutrient status in North Texas. Agric. Ecosyst. Environ., 273: 117–129.

Tilman, D., Cassman, K.G., Matson, P.A., Naylor, R. and Polasky, S. (2002). Agricultural sustainability and intensive production practices. Nature, 418: 671–677.

UNCCD. (2012). Zero net land degradation, A Sustainable Development Goal for Rio +20, UNCCD Secretariat Policy Brief.

Uttech, S. (2010). Cover Crops Reduce Erosion, Runoff. Public release: 18 May, ww.agronomy.org.

Van Meter, K.J., Van Cappellen, P. and Basu, N.B. (2018). Legacy nitrogen may prevent achievement of water quality goals in the Gulf of Mexico. Science, 360: 427–430.

Verheijen, F.G.A., Jones, R.J.A., Rickson, R.J. and Smith, C.J. (2009). Tolerable versus actual soil erosion rates in Europe. Earth Sci. Rev., 94: 23–38.

Verstraeten, G., Poesen, J., Govers, G., Gillijns, K., Van Rompaey, A. and Van Oost, K. (2003). Integrating science, policy, and farmers to reduce soil loss and sediment delivery in Flanders, Belgium. Environ. Sci. Policy, 6: 95–103.

Vicente-Vicente, J.L., García-Ruiz, R., Francaviglia, R., Aguilera, E. and Smith, P. (2016). Soil carbon sequestration rates under Mediterranean woody crops using recommended management practices: a meta-analysis. Agric. Ecosyst. Environ., 235: 204–214.

Wendt, R.C. and Burwell, R.E. (1985). Runoff and soil losses for conventional, reduced and no-till corn. J. Soil Water Conserv., 40: 450–454.

Williams, M.R., King, K.W. and Fausey, N.R. (2015). Contribution of tile drains to basin discharge and nitrogen export in a headwater agricultural watershed. Agric. Water Manage., 158: 2–50.

Wittwer, R., Dorn, B., Jossi, W. and Van der Heijden, M.G.A. (2017). Cover crops support ecological intensification of arable cropping systems. Sci Rep., 7: 41911.

Zhang, G., Chan, K., Oates, A., Heenan, D. and Huang, G. (2007). Relationship between soil structure and runoff/soil loss after 24 years of conservation tillage. Soil Till. Res., 92: 122–128.

16

Effects of Cover Crops on Greenhouse Gas Emissions

Somayyeh Razzaghi

1. Introduction

Sustainable agriculture is an integration of economically viable, environmentally compatible and socially acceptable approaches with enhanced ecosystem services (Linquist et al., 2012; Negassa et al., 2015). While conservation tillage has been promoted as a climate-mitigation tool, incorporating cover crops into no-till agriculture means that there is another valuable tool for agricultural climate mitigation. Cover crops, in different names or forms, have long been used for their ability to provide organic matter, reduce soil erosion and nutrient leaching, fix atmospheric nitrogen (N) and improve soil health. Moreover, they play an important role in mitigating the effects of climate change on agriculture.

Cover crops are valuable for conservation agriculture (Jaffuel et al., 2017). They improve soil physical properties (Fageria et al., 2005), using bio-physical actions through their roots, which are able to increase the porosity, provide the pathways (bio-channels) for water infiltration (Folorunso et al., 1992) and decrease soil compaction (Hubbard et al., 2013). Cover-crop residues can increase soil organic matter (SOM) content (Masilionyte and Maikšténiené, 2016; Poeplau and Don, 2015; Sainju et al., 2007; Smith et al., 2014; Tribouillois et al., 2015), thereby improving the soil structure (Poeplau and Don, 2015), chemical and biological properties and overall soil quality (Cuello et al., 2015; Gómez et al., 2009; Jaffuel et al., 2017; Schipanski et al., 2014; Tribouillois et al., 2015; Verzeaux et al., 2016). In addition, cover crops are a biological, a valuable option to control weeds and pests in sustainable agriculture (Hartwig and Ammon, 2002; Cherr et al., 2006; Kassam and Brammer, 2013; Thomson and Hoffmann, 2013; Veres et al., 2013; Webster et al., 2013; Masilionyte et al., 2017; Muscas et al., 2017; Seufert et al., 2012). Therefore, using cover crops as organic agro-technical methods is important for the reduction of soil and environmental pollution (Keating et al., 2010). Consequently, according to Balota et al. (2014) and Calderon et al. (2016), using cover crops the correct way can also modify aboveground biodiversity. In general, utilising cover crops within crop rotation has a positive impact on enhancing crop yield in sustainable agriculture (Kim et al., 2017; Nygaard Sorensen and Thorup-Kristensen, 2011). One of the major benefits of cover crops discussed in this section is their influence on mitigating global climate change and decreasing greenhouse gas emissions (GHGs).

Dept. of Soil Science and Plant Nutrition, Faculty of Agriculture, Çukurova University, Adana, Turkey.

2. Sources of Greenhouse Gases (GHGs)

Several of the gases, such as carbon dioxide (CO_2), nitrous oxide (N_2O), methane (CH_4) and others (Fig. 1), accumulated in the troposphere in response to natural processes, accelerate industrial and household activities and conventional agriculture. These are the primary reasons for the greenhouse effects in the form of global warming and climate change worldwide (Le Treut et al., 2007).

The GHGs have a long life-time (Ramaswamy et al., 2001). These gases can trap long-wave radiation and permit shortwave radiation to pass through the troposphere (Fig. 2). Some other portions of these shortwave radiations are absorbed by the surface of the earth and the remaining portions return to the troposphere as heat (Allison, 2015). The GHGs with specific molecular structures can absorb the escaping heat and then re-emit it to the earth. This is how GHGs cause global warming and climate change (Luis et al., 2012; Timmermann et al., 1999). These occurrences have remarkable negative effects on human life and ecosystems (Anomohanran, 2012; Cuellar-Bermudez et al., 2015; Dhillon and von Wuehlisch, 2013; Flessa et al., 2002; Joshi et al., 2017; Tang et al., 2012). It is worth noting that the existence of GHGs in the atmosphere is important because without them, the mean of surface temperatures would be $-32.5°C$ colder than the present average and life development and initial support would have been impossible on Earth (Karl and Trenberth, 2003). It is important to know the sources and volume of GHGs emissions associated with global warming and climate-change effects and develop proactive approaches to decrease their growing concentration in the troposphere to mitigate climate change effects (Meinshausen et al., 2009).

The GHGs are generally released from accelerated human activities, like deforestation, fossil fuel use, industrial production, traditional agriculture and natural processes, like animal and plant

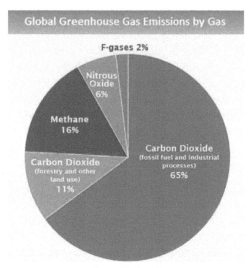

Fig. 1 Global greenhouse gas emission by gas from sectors (*Source*: IPCC, 2015).

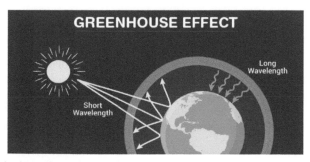

Fig. 2 Mechanisms of greenhouse effect (*Source*: http://blog.nigurha.com/greenhouse-effect).

respiration. Despite the numerous climate change mitigation studies, GHG emissions from human activities reached their highest range (49 ± 4.5 GtCO$_2$-e) in history in 2013 (Mikulčić et al., 2016). It is expected that climate-smart agriculture can mitigate GHG emissions resulting from human activities (Liebig et al., 2012; Tilman et al., 2011).

2.1 Carbon Dioxide

The C is the main building block of life on the earth but as CO$_2$, it is the key greenhouse gas in the atmosphere (Laws and Berning, 1991). The CO$_2$, by definition, has a global warming potential (GWP) of 1 regardless of the time period used, because it is the gas being used as the reference. The natural sinks and sources of CO$_2$ are in balance with one other and the concentration of this gas is in the safe level and quite steady in the atmosphere (Menon et al., 2007). Natural sources of CO$_2$ emissions include oceans (42.8 per cent), plant and animal respiration (28.6 per cent), soil respiration (28.6 per cent) and volcanic eruptions (0.03 per cent) (Gerlach, 2011; Graber et al., 2008; Menon et al., 2007). The CO$_2$ emissions from the soil are associated with temperature (Negassa et al., 2015). Frank et al. (2002) reported that soil temperature, moisture and air temperature influence 65 per cent, 5 per cent and 3 per cent of CO$_2$ emissions, respectively. The CO$_2$ emission in the atmosphere, associated with human activities, is smaller than natural sources, but this content has increased significantly and reached the 'dangerous' range in the recent years (Van de Wal et al., 2011; Le Quéré et al., 2012). It is reported that if we all stepped up our efforts to adopt climate-smart agriculture, especially increasing cover crops adoption, they could offset 8 per cent of C emissions released from agriculture worldwide (Poeplau and Don, 2015).

2.2 Nitrous Oxides

The N$_2$O is one of the most important GHGs in the earth's atmosphere (IPCC, 2007; Pachauri et al., 2014). The N$_2$O has a GWP 265–298 times that of CO$_2$ for a 100-year time-scale. Natural (soils and the oceans) and human sources (agriculture, fossil fuels and industrial activities) produce 62 per cent and 38 per cent of total emissions of N$_2$O to the atmosphere, respectively (Menon et al., 2007). Like CO$_2$, N$_2$O concentration from the human activities has increased rapidly during the last 800,000 years (Schilt et al., 2010). Higher levels of this gas could alter the balance of the N cycle and earth's temperature (Butterbach-Bahl et al., 2013; Ravishankara et al., 2009).

2.3 Methane

Like CO$_2$ and N$_2$O, the accumulation of CH$_4$ in the atmosphere is associated with increasing the global temperature and is responsible for climate change (Liu et al., 2015). The CH$_4$ is estimated to have a GWP of 28–36 over 100 years. The anaerobic decomposition of organic matter in the wastewater is believed to be responsible for CH$_4$ emissions into the atmosphere (Hwang et al., 2016). The concentration of CH$_4$ has quickly increased since the industrial revolution (Griggs and Noguer, 2002; Yuping et al., 2008). Therefore, finding the best way to decrease the concentration of this gas in the atmosphere should be considered, to mitigate climate change and save human life in our planet.

2.4 Fluorinated Gases

Fluorinated gases ('F-gases') are a family of man-made gases used in a range of industrial applications. While the F-gases (hydrofluorocarbons [HFCs], perfluorocarbons [PFCs] and sulphur hexafluoride [SF$_6$]) constitute a very minor part of GHGs in the atmosphere (Lucas et al., 2007; Meinshausen et al., 2011) but they have a global warming effect up to 23,000 times greater than CO$_2$. Their emissions are rising strongly and expected to pose a dangerous threat for human life (Schaefer et al., 2006).

3. Effect of Cover Crops on CO_2 Emissions

Agricultural land use (Houghton, 2010) and plant and animal respiration (Graber et al., 2008) are fundamental sources of CO_2 emissions in the atmosphere (Houghton, 2010). According to Le Quéré et al. (2012), about 3.3 billion tons of CO_2 emissions in 2011 came from land use changes caused by human activities. It has been suggested that GHG emissions in agricultural soils is highly dependent on cover-crop management (Mosier et al., 1998). According to Tang et al. (2013), winter cover crops have a strong influence on SOC sequestration and CO_2 emission from soils in southern China.

While there are published reports on the effect of cover crops on CO_2 emissions from the soil, but conflicting results concerning whether cover crops can increase or decrease CO_2 emissions. Negassa et al. (2015) indicated that CO_2 emissions from the soil were higher in rye cover crop treatments in southwest and central Michigan. These higher CO_2 emissions with cover crop applications can be associated with the decomposition of labile organic matter derived from the presence of cover crops (Negassa et al., 2015). In contrast, cover crops are expected to increase C sequestration and play an important role in mitigating climate change (Duiker and Curran, 2005; Miguez and Bollero, 2005; Scholberg et al., 2010; Shipley et al., 1992). Sanz-Cobena et al. (2014) reported that soil CO_2 emissions increased by 21–28 per cent under barley (*Hordeum vulgare* L.) and vetch (*Vicia villosa* L.) cover crops in the central Tajo River basin, near Aranjuez (Madrid, Spain). Likewise, Kallenbach et al. (2010) indicated that CO_2 emissions under winter legume cover cropping were 40 per cent higher than no-cover crop under furrow irrigation in central California. Hendrix et al. (1988) indicated that CO_2 emissions from the soil under the cover of clover was significantly higher than the soil covered by rye. In contrast, Jarecki and Lal (2003) stated that winter legume cover crops can add a considerable amount of C to the soil and therefore can decrease soil CO_2 emissions. Atmojo et al. (2015) demonstrated that planting leguminous cover crops on peat land decreased CO_2 emissions from the soil. Rigon et al. (2018) reported that Sunnhemp in crop rotation can be used to mitigate CO_2 emissions and increase soybean yield (Table 1).

According to Hatfield and Parkin (2012) and Morell et al. (2012), root respiration plays an important role in CO_2 emissions in soil. They suggested that up to 50 per cent of total CO_2 emissions come from soils, with even higher levels observed during growing and flowering stages of plants. Results from several studies (Fig. 3) have also reported that root respiration at different growth stages of crops could increase CO_2 emissions in soil (Fu et al., 2002; Omonode et al., 2007; Wilson and Al-Kaisi, 2008; Sainju and Singh, 2008; Feng et al., 2013).

Sustainable soil management and the use of cover crops can reduce the amount of CO_2 in the atmosphere, according to Delgado et al. (2011). It has been demonstrated that soil management and no-till with increasing crop residue content can help accumulate SOM and mitigate CO_2 emissions (ECO_2) (Stewart et al., 2009). In addition, according to Brito et al. (2015), cover-crop residues can help retain water in the soil and therefore, decrease the soil temperature and reduce CO_2

Table 1 Accumulated C-CO_2 emissions, soybean grain yield and relative C-CO_2 emissions in Typic Rhodudalf soil (*Source*: Rigon et al., 2018).

Crop Sequence before Soybean	Accumulated Emission C-CO_2	Soybean Yield	Yield Scaled C-CO_2
	kgha[-1]		
Fall-Winter crops			
Sunflower	5727a[≠]	2838a	2.01a
Triticale	4810a	2915a	1.65b
Sunn hemp	5062a	3023a	1.67b
Pearl millet	5728a	2756ab	2.07a
Forage soyghum	5312a	2627b	2.02a
Fallow; chisel	4974a	3100a	1.60b

≠ Mean values followed by different letters in the column differ between themselves by the t test (LSD = least significant difference) at the 5 per cent probability level.

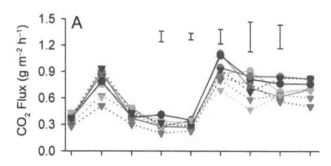

Fig. 3 Carbon dioxide (CO_2) flux in g m^{-2} h^{-1}, at 1, 2, 3, 8, 15, 30, 60, 90, and 120 days after sowing of soybean in accordance with different crop sequences in Typic Rhodudalf soil. Vertical bars correspond to the LSD (least significant difference) at the 5 per cent probability level (Rigon et al., 2018).

emissions (Brito et al., 2015). Low N content of cover crop residues lead to a low mineralisation rate (immobilisation effect) and, consequently, low levels of CO_2 emissions (Sainju et al., 2012). Even though these results differ from the studies of Zhou et al. (2016), who reported that crop residues with high N content are decomposed faster by microbes with subsequent increase CO_2 emissions from the soil.

Kaye and Quemada (2017) studied the variations in climate mitigation in units of CO_2 equivalents (CO_2-e) that come from the albedo change in two experiment sites (Aranjuez, Spain and Pennsylvania, USA) (Table 2, Fig. 4). They reported that cover cropping can decrease 12 to 46 g CO_2-e. m^{-2} per year at both experiment sites.

Table 2 Changes in climate mitigation (CO_2 Equivalents, CO_2-e) Result from the albedo change when shifting from bare soils to winter cover cropping (*Source*: Kaye and Quemada, 2017).

Scenario Name		Scenario Parameters		Changes in g CO_2-e $m^{-2}yr^{-1}$ with Cover Crop			
				Pennsylvania USA			Aranjuez Spain
Plant albedo	Soil albedo	Plant albedo	Soil albedo	No snow	Full snow	Partial snow	No snow
Typical	Typical	0.26	0.17	45	25	12	46
High	Low	0.30	0.10	111	87	67	101
High	High	0.30	0.24	33	26	6	30
Low	Low	0.21	0.10	61	48	22	56
Low	High	0.21	0.24	−17	−13	−39	−15

Positive values reflect increased albedo, which mitigates warming. The full snow scenario buried all cover crops with snow (albedo = 0.65) from mid-December to mid-March. The partial snow scenario buried all the soil, but only half of the cover crop canopy for the same time period.

Fig. 4 A cover cropping experiment in Pennsylvania (USA) within a maize-soybean wheat rotation in late summer (A). Different coloured plots reflect different species of cover crop (A) and Two cover cropping experiments in Aranjuez, Spain, in spring (B). All cover crops were planted after harvesting irrigated maize for grain (B) (*Source*: Kaye and Quemada, 2017).

4. Effect of Cover Crops on N₂O Emission

Agriculture is one of the important anthropogenic factors in emitting N_2O to the atmosphere. According to Robertson et al. (2000), due to high radiative force per mass unit (300 times) of N_2O in comparison to CO_2, small reducing of N_2O emissions from agricultural soil can have a large effect on the global warming potential.

It is reported that agriculture creates 4.5 million tons of N_2O (about 60 per cent of the anthropogenic N_2O) that is released annually into the atmosphere. The production of N_2O mostly occurs during the microbial processes (i.e., nitrification and denitrification) in both cultivated and natural soils (Barnard et al., 2005; Butterbach-Bahl et al., 2013; Lam et al., 2016; Mannina et al., 2016; Meixner, 2006; Menon et al., 2007; Ostrom et al., 2010; Van Groenigen et al., 2015). This means that during organic C oxidation in the denitrification process under anaerobic conditions, NO_3 is reduced to N_2O (Senbayram et al., 2012). Previous research studies (Haider et al., 1987; Smith and Tiedje, 1979) have established that living and growing crops preferably absorb available N (NO_3) and reduce N_2O loss by denitrification. Han et al. (2017) found that the spelt-clover phase had the lowest N_2O emissions because of low availability of NO_3^- and consequently, low denitrification.

Data from other studies (Khalil et al., 2002; Mosier et al., 1998; Stott et al., 1986) suggested that denitrification depends on soil pH, NO_3^- concentration, moisture and temperatures. The N_2O and CO_2 emissions from the soil contribute to weather, moisture and temperature conditions (Izaurralde et al., 2004; Liu et al., 2008; Philibert et al., 2012; Schindlbacher et al., 2004); NO_3^- content and fertiliser application (Abdalla et al., 2014; Hoben et al., 2011); surface topography (Vilain et al., 2010); landscape characteristics (Ladoni et al., 2016) and soil microbial activity (Wickings et al., 2016). In this regard, Corre et al. (1996), Negassa et al. (2015) and Nishina et al. (2009) indicated that higher soil moisture is responsible for higher N_2O emissions. According to Hansen et al. (2014), N_2O emissions in a warmer year were 10 times higher than a colder year. Agricultural management practices with cover crops can influence soil physical, chemical and biological properties, which exert positive effects on N_2O emissions (Delon et al., 2007; Kim et al., 2017; Pilegaard, 2013).

As mentioned by Kim et al. (2013), N_2O emissions can vary depending on cover crop types. Others (Basche et al., 2014; Hwang et al., 2015) have highlighted that using legumes in combination with non-legumes can contribute N to the soil due to their physiological characteristics, increasing transpiration, crop production and effect on N_2O emissions. Literatures on cover crops have revealed the emergence of several contrasting results on whether cover crops can decrease or increase N_2O emissions. This subject is still under debate. According to Basche et al. (2014), the ratio between a N_2O flux with a cover crop treatment to a N_2O flux without a cover crop (Response ratio, RR in Eq. 1) and natural log of RR (in Eq. 2) (Hedges et al., 1999) is an important factor in determining the cover crop effect on N_2O emissions.

$$RR = \frac{N_2O \; emission \; cover \; crop \; treatment}{N_2O \; emission \; noncover \; crop \; treatment} \tag{1}$$

$$LRR = ln \; RR \; Natural \; logarithm \; of \; response \; ratio \; (RR) \tag{2}$$

Basche et al. (2014) stated that 60 per cent of the results associated with cover crops can increase N_2O emissions (positive Logarithm response ratio, LRR) and 40 per cent of reported results indicated decrease in N_2O emissions (negative LRR) due to cover crops (Fig. 5). A number of studies (Berger et al., 2013; Liu et al., 2014) have reported that although the overall estimation of N_2O emissions due to agricultural management practices have spatial variability, using cover crops is one of the best agricultural management practices for controlling the amount of N_2O emissions to the atmosphere among agricultural management practices.

Kaspar et al. (2007) noted that rye winter cover crops grown after corn and soybean can decrease NO_3 concentrations. From these studies, it is inferred that rye, as a winter cover crop, is able to reduce N_2O outputs from the soil. Likewise, Tonitto et al. (2006) stated that rye, as a non-legume cover crop, can decrease NO_3 leaching from the soils. Baggs et al. (2000) also outlined

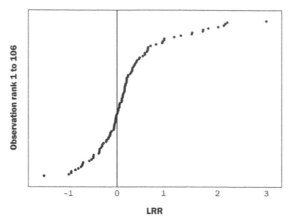

Fig. 5 Natural logarithm of response ratio (LLR) for 106 observations in the data set (*Source*: Basche et al., 2014).

that using cover crops can reduce the amount of soil NO_3 and lead to a decrease of N_2O emissions. Similarly, Snyder et al. (2009) indicated that cover crops can decrease NO_3 leaching from the soil and total N_2O emissions.

Reeves (1994) and Alvarez et al. (2017) concluded that using cover crops is a sustainable management approach for improving soil quality, reducing erosion and decreasing N_2O emission from the soil. Similarly, according to Parkin and Kaspar (2006), N_2O emissions from soil under corn were significantly higher than from the soil under soybean. It has been suggested (Peoples et al., 1995; Rochette et al., 2004; Fageria et al., 2005; Miguez and Bollero, 2005; Carter and Ambus, 2006; Kramberger et al., 2014; Ketterings et al., 2015) that cover crops, specially legumes, can fix atmospheric N symbiotically with rhizobia bacteria, provide available N for plants and increase the crop grain yield. Jeuffroy et al. (2013) indicated that pea, as a legume cover crop, can decrease N_2O emissions by 20–25 per cent.

As noted by Rochette and Janzen (2005), the mean of N_2O emissions from legumes for annual, pure forage and grass + legume mixtures were 1, 1.8, and 0.4 kg N ha^{-1}, respectively. Schwenke et al. (2015) also reported higher N_2O emissions from N-fertilized canola compared to N-fixing legumes from Vertosols. According to Hénault and Revellin (2011), N_2O can be consumed by the nodules of legume root. In addition, according to Pappa et al. (2011) and Thorup-Kristensen et al. (2003), several cover crops spp. are referred to as 'catch crops'; in that, they re-consume the N that comes from the previously harvested plants, prevent both nitrification and denitrification processes and help to reduce N_2O emissions. About 50–60 per cent of the N of the catch crops' tissues are derived from N uptake and recycling (Gabriel and Quemada, 2011).

The effect of cover crops on soil N_2O emissions is dependent on the C:N of crop residue. Low and high C:N of the cover crop residue can increase and decrease N_2O emissions, respectively (Huang et al., 2004; Gentile et al., 2008). Several other studies (Delgado et al., 2010a,b) suggested that N_2O emissions, NO_3^- leaching and fertiliser inputs are reduced by using cover-crop residues with high C:N.

Much of the literatures (McCracken et al., 1994; Novoa and Tejeda, 2006; Rosecrance et al., 2000) emphasised that non-N fixing cover crops (like oats) or those with a higher C:N or deeper roots (like rye) provide lower content of N to the soil susceptible to denitrification and therefore reduce soil N_2O emissions. Data from several sources (Kirchmann and Witter, 1992; Rosecrance et al., 2000; Kim et al., 2013) have identified that a higher C:N cover crop can decrease N loss by leaching or volatilisation as N_2O. These results seem to be inconsistent with Abdalla et al. (2014), who reported that a mustard cover crop can significantly increase N_2O emissions from the soils.

Several studies (Jensen et al., 2012; Basche et al., 2014) indicated that the decomposition of labile portion of cover crop residues that remain after harvest can increase N_2O emissions. Schwenke et al. (2015) also concluded that 75 per cent of annual N_2O emissions are attributed to legumes, especially after harvest. In this context, Rochette and Janzen (2005) concluded that legume crops can increase N_2O emissions from the soil, either by N diffusion from the root exudates, or after harvest through the decomposition of labile crop residues, more than biological N fixation itself. The results of Rochette and Janzen (2005) are in line with those of previous studies (Carter and Ambus, 2006; Zhong et al., 2009; Ingram et al., 2015).

Sarkodie-Addo et al. (2003) and Abalos et al. (2013) reported that labile organic C, which released from cover-crop residue, may be utilised by denitrifiers as an energy and food source. They concluded that a combination of N fertiliser and cover-crop residue can increase N_2O emissions compared to a solely N fertilisation. Likewise, Jarecki et al. (2009) concluded that rate of N fertilisation is one of the controlling factors affecting soil N_2O emissions. While they indicated that increasing the N fertilisation rates can lead to an increase of N_2O emissions in the fine, loamy soil of central Iowa; in contrast, Mitchell et al. (2013) reported that rye as a cover crop decreased N_2O emissions when N fertilisers were not applied. They showed that N_2O emissions increased with N fertilisation. McSwiney et al. (2010), on the other hand, indicated that N fertilisation can lead to N immobilisation by cover crops, which can decrease N_2O emissions from the soil. Similarly, Basche et al. (2014) indicated that the type of cover crop is an important factor for N_2O emissions from the soil in the presence of N fertilisation. Their study reported that N_2O emissions from the soil with legumes at low N rates measured high, but non-legumes measured low.

There is a positive correlation between the labile C and N_2O emissions (Azam et al., 2002; Mitchell et al., 2013). Cover crops add considerable amounts of labile C and N to soil, and have the potential to increase heterotrophic microbial activities, thus N_2O emissions increase (Kallenbach et al., 2010). According to Steenwerth and Belina (2008), when cover crop residues decompose, the soil N immobilises and pools of soil labile C and inorganic N increase, which are expected to affect the N_2O emission dynamics. Basche et al. (2014) summarised a list of environmental and management factors that associated with N_2O emissions (Table 3).

Table 3 Description of database factors included to analyse variability in the cover crops' effect on nitrous oxide (N_2O) (*Source*: Basche et al., 2014).

Factor	Management Factors and Soil Properties	Number of Observations
Tillage	No-till, conventional tillage	74
Residue C: N	4 to 48	57
Soil bulk density	1.2 to 2.65	67
Soil pH	5.5 to 8.1	89
Type of cover crop	Legume, non-legume and biculture	106
N rate (kgha^{-1})	0 to 303	103
Soil incorporated residues	Yes, no	84
Cover crop kill date	Days between cover crop termination and cash crop planting (1 to 25)	71
Sand (%)	8 to 80	106
Silt (%)	11 to 73	106
Clay (%)	5 to 45	106
Organic carbon (0–30 cm, %)	0.38 to 2.1	97
Cover crop biomass (kg ha^{-1})	280 to 14,400	65
Total precipitation (mm)	11 to 906	77
Std. deviation rainfall (mm)	0.5 to 40	77
Drainage	Well-drained, poorly drained	69
Period of measurement	Full year, cover crop growth, cover crop decomposition, and cash crop growth	80
Experiment type	Field, model, growth chamber	106

Nevertheless, the results of above studies differ slightly to the results of other (Liebig et al., 2010; Sanz-Cobena et al., 2014; Negassa et al., 2015; Peyrard et al., 2016), who reported a minor impact of cover crops on N_2O emissions. Likewise, Smith et al. (2011) and Jarecki et al. (2009) reported that cover crops have no significant effect on N_2O emissions. In contrast, Mitchell et al. (2013) reported that the cover crop can have positive effects on C and NO_3 availability, and by this way, can indirectly control soil N_2O emissions. Therefore, finding the relationship between cover crops and N_2O emissions is important to mitigate GHG emissions and climate change effects (Eagle and Olander, 2012).

The type of cover crops and N fertilisation rates also have important effects on N_2O emissions from the soil (Fig. 6). Without any N fertilisations, legumes showed higher LRRs (natural log of the ratio between a N_2O flux with a cover crop treatment to a N_2O flux without a cover crop) than non-legume cover crops (Basche et al., 2014). Unlike legumes, winter cereals as non-legume cover crops, can extract soil N by their deep roots and lead to a reduction in N_2O emissions (Kallenbach et al., 2010). In this sense, Ussiri and Lal (2012) estimated N_2O mitigation potential for winter cover crops from 0.2 to 1.1 kg N_2O ha^{-1} yr^{-1}. Sarkodie-Addo et al. (2003) also indicated that N_2O emissions decreased with cover crops in the no-fertilised plots when incorporating a wheat and winter rye cover.

Cover-crop residues with high C:N's generally decreased N_2O emissions (Bouwman et al., 2002; Parkin et al., 2006). De Gryze et al. (2010) and Basche et al. (2014) reported that the decrease in global warming potential was a result of increasing SOC content in cover-cropped systems. In general, Basche et al. (2014) reported that cover crops increased N_2O emissions from the soil in 60 per cent of published researches, while in 40 per cent of observations, cover crops were the reason for decreasing of N_2O emissions.

According to Sanz-Cobena et al. (2014), in comparison to barley or rape, inclusion of vetch after four years of a cover crop-maize rotation, significantly increased N_2O emissions due to its ability to fix atmospheric N_2. Basche et al. (2014) showed the magnitude of changes of N_2O with and without cover crops under N fertilisation (Table 4).

Guardia et al. (2016) studied the effect of vetch and barley on GHG emissions during the intercrop and the maize cropping period. They reported that N_2O emissions in barley plots were

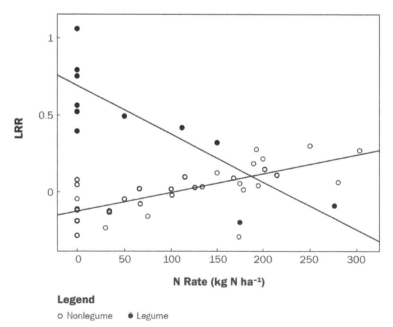

Fig. 6 Response ratios (LLR) of legume versus grass cover crop species as a function of fertiliser nitrogen (N) rate (*Source*: Basche et al., 2014).

Table 4 Magnitude of nitrous oxide (N$_2$O) changes with- and without cover crops (*Source*: Basche et al., 2014).

Study No.	Control-N$_2$O Emission (kg N ha^{-1})	Cover crop-N$_2$O emission (kg N ha^{-1})	Crop Rotation and Cover Crop spp.	Study Duration	Nitrogen Rates (kg ha^{-1})	Reference
1	7.5	5.3	Corn - corn-soybean, 70 per cent rye and 30 per cent oat	Full year	175	(Jarecki et al., 2009)
2	3.7	2.3	Soybean, winter rye	Winter cover crop time	195	(Parkin et al., 2006)
3	1.5	1.4	Soybean - corn-soybean, annual ryegrass	Full year	0N	(Smith et al., 2011)
4	11.	15.4	Corn - corn-soybean, winter rye	Full year	215	(Parkin and Kaspar, 2006)
5	3.8	5.1	Corn - corn-soybean, annual ryegrass	Full year	193	(Smith et al., 2011)
6	9.3	50.2	Rice-wheat, Sesbania	Spring cover crop	0-N (176 kg from legumes)	(Aulakh et al., 2001)

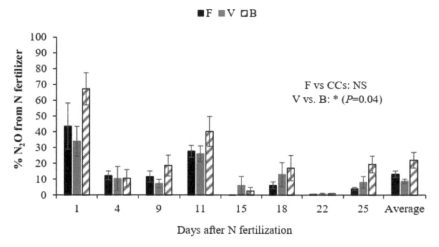

Fig. 7 Proportion of N$_2$O losses (%) that come from N synthetic fertilizer, for the three CC treatments (fallow, F; vetch, V; and barley, B). Vertical lines indicate standard errors. 'NS' and * denote not significant and significant at P < 0.05, respectively (*Source*: Guardia et al., 2016).

higher than fallow or vetch residue plots (Fig. 7). Accordingly, higher N$_2$O emissions in response to non-legume barley residue can be indicated by the higher C:N (20.7 ± 0.7), compared to vetch (11.1 ± 0.1). Shan and Yan (2013) indicated that low C:N in leguminous vetch residue leads to release of N for soil microorganisms, therefore N$_2$O derived from the synthetic fertiliser decrease.

5. Effect of Cover Crops on CH$_4$ Emission

The effect of cover crops on CH4 fluxes is very low and often negative (Abdalla et al., 2014). According to Lee et al. (2006), the sandy loam soils, even with cover crops, are a sink for atmospheric CH$_4$. Ruis et al. (2017) indicated that cover crops have no effect on CH$_4$ fluxes. Szerencsits et al. (2016) reported that from 4.5 tons of cover crop dry matter, about 1,300 million CH$_4$ ha^{-1} can be produced. Dubey (2005) reported that vetch and rye cover crops supply labile C to methanogens, which increase CH$_4$ emission from the soil. However, the increase of CH$_4$ emissions in rice with winter cover crop amendments highly depends on the rice growing stage (Kim et al., 2013). In this regard, Sinha (1995) found that after flowering, the rate of photosynthesis and the supply of

Table 5 Summary of literatures on the effect of different types of cover types on greenhouse gas emissions (GHGs).

Cover Crop Type	GHGs Emissions	References
Rye	$N_2O\downarrow$	Kaspar et al. (2007); Mitchell et al. (2013); Parkin et al. (2006); Tonitto et al. (2006)
Rye	NE on N_2O	Negassa et al. (2015)
Pea	$N_2O\downarrow$	Jeuffroy et al. (2013)
Oat	$N_2O\downarrow$	McCracken et al. (1994); Novoa and Tejeda (2006); Rosecrance et al. (2000)
Winter cover crops (radish, mustard, Phacelia, rye)	Sustained mineralisable N = Denitrification$\downarrow$ = Indirectly $N_2O\downarrow$	Jackson et al. (1993)
Grass species	Sustained mineralisable N = Denitrification$\downarrow$ = Indirectly $N_2O\downarrow$	(Poudel et al. (2001)
Grass or cereal species	Sustained mineralisable N = Denitrification$\downarrow$ = Indirectly $N_2O\downarrow$	Bergstrom (1986); Meisinger et al. (1991); Scott et al. (1987)
Winter cover crops	Sustained mineralisable N = Denitrification$\downarrow$ = Indirectly $N_2O\downarrow$	(King and Buchanan, 1993)
Vetch + high C/N organic matter	$N_2O\downarrow$	(Kim et al., 2013)
Hairy vetch/barley biomass	$N_2O\downarrow$	(Kim et al., 2017)
Mustard	Less effect (LE) on CH_4	Abdalla et al. (2014)
Mustard	$N_2O\uparrow$	Abdalla et al. (2012)
Rye and Trios	$N_2O\uparrow$	Steenwerth and Belina (2008)
Rye	Non-legume but can increase available N more that vetch/clover = Denitrification$\uparrow$ = Indirectly $N_2O\uparrow$	(Sainju et al., 2002)
Hairy vetch	Increased available N = Denitrification$\uparrow$ = Indirectly $N_2O\uparrow$	Kim et al. (2013); Brown et al. (1993)
Grass + N fertilizer	$N_2O\uparrow$	Abdalla et al. (2009)
Grass (lowland soil)	$N_2O\uparrow$	Vilain et al. (2010)
Canola + N fertilizer	$N_2O\uparrow$	Schwenke et al. (2015)
Rye	$CO_2\uparrow$	Negassa et al. (2015)
Clover	$CO_2\uparrow$	Hendrix et al. (1988)
Vetch	$CO_2\uparrow$	Sanz-Cobena et al. (2014); Kallenbach et al. (2010)
Barley	$CO_2\uparrow$	Sanz-Cobena et al. (2014)
Pea	$CO_2\uparrow$	Kallenbach et al. (2010)
Vetch and Rye	$CH_4\uparrow$	Dubey (2005)

$\uparrow$= Increase, $\downarrow$= Decrease, NE= No effect, LE= Less effect.

CH_4 construction assimilates reduce. Because of this reason, CH_4 emissions decreased after the flowering growing stage.

6. Conclusion

Cover crops are one of the important components of climate-smart agriculture. Based on current research results, understanding whether cover cropping reduces or increases GHG emissions is still unclear. Cropping diversity, type of cover crop (legumes vs. non-legumes), climatic conditions, soil properties (especially soil moisture) and N fertilisation are important factors that control the

GHG emissions from agricultural soils. Research has shown that if we all stepped up our efforts to increase cover crop adoption, it may be possible to offset about 8 per cent of C emissions currently produced by agricultural activities worldwide.

References

Abalos, D., Sanz-Cobena, A., Garcia-Torres, L., Van Groenigen, J.W. and Vallejo, A. (2013). Role of maize stover incorporation on nitrogen oxide emissions in a non-irrigated Mediterranean barley field. Plant Soil, 364: 357–371.

Abdalla, M., Jones, M., Smith, P. and Williams. M. (2009). Nitrous oxide fluxes and denitrification sensitivity to temperature in Irish pasture soils. Soil Use Manage., 25: 376–388.

Abdalla, M., Rueangritsarakul, K., Jones, M., Osborne, B., Helmy, M., Roth, B., Burke, J., Nolan, P., Smith, P. and Williams, M. (2012). How effective is reduced tillage–cover crop management in reducing N_2O fluxes from arable crop soils? Water, Air, Soil Pollut., 223: 5155–5174.

Abdalla, M., Hastings, A., Helmy, M., Prescher, A., Osborne, B., Lanigane, G., Forristal, D., Killi, D., Maratha, P., Williams, M., Rueangritsarakul, K., Smith, P., Nolan, P. and Jones, M.B. (2014). Assessing the combined use of reduced tillage and cover crops for mitigating greenhouse gas emissions from arable ecosystem. Geoderma, 223: 9–20.

Allison, I. (2015). The science of climate change: Questions and answers. Aust. Acad. Sci., 1–44.

Alvarez, R., Steinbach, H.S. and De Paepe, J.L. (2017). Cover crop effects on soils and subsequent crops in the pampas: A meta-analysis. Soil Till. Res., 170: 53–65.

Anomohanran, O. (2012). Determination of greenhouse gas emission resulting from gas flaring activities in Nigeria. Energy Policy, 45: 666–670.

Atmojo, S., Setyono, P. and Dewi, W. (2015). Temperature effect investigation toward peat surface CO_2 emissions by planting leguminous cover crops in oil palm plantations in West Kalimantan. J. Agric. Sci. Tech. B., 5: 170–183.

Aulakh, M.S., Khera, T.S., Doran, J.W. and Bronson, K.F. (2001). Denitrification, N_2O and CO_2 fluxes in rice-wheat cropping system as affected by crop residues, fertiliser N and legume green manure. Biol. Fert. Soils, 34: 375–389.

Azam, F., Müller, C., Weiske, A., Benckiser, G. and Ottow, J. (2002). Nitrification and denitrification as sources of atmospheric nitrous oxide-role of oxidisable carbon and applied nitrogen. Biol. Fert. Soils, 35: 54–61.

Baggs, E., Watson, C. and Rees, R. (2000). The fate of nitrogen from incorporated cover crop and green manure residues. Nutr. Cycl. Agroecosys., 56: 153–163.

Balota, E.L., Calegari, A., Nakatani, A.S. and Coyne, M.S. (2014). Benefits of winter cover crops and no-tillage for microbial parameters in a Brazilian Oxisol: A long-term study. Agr. Ecosyst. Environ., 197: 31–40.

Barnard, R., Leadley, P.W. and Hungate, B.A. (2005). Global change, nitrification, and denitrification: A review. Global Biogeochem. Cy., 19.

Basche, A.D., Miguez, F.E., Kaspar, T.C. and Castellano, M.J. (2014). Do cover crops increase or decrease nitrous oxide emissions? A meta-analysis. J. Soil Water Conserv., 69: 471–482.

Berger, S., Kim, Y., Kettering, J. and Gebauer, G. (2013). Plastic mulching in agriculture – Friend or foe of N_2O emissions? Agr. Ecosyst. Environ., 167: 43–51.

Bergstrom, L. (1986). Distribution and temporal changes of mineral nitrogen in soils supporting annual and perennial crops. Swed. J Agr. Res., 16: 105–112.

Bouwman, A.F., Boumans, L.J.M. and Batjes, N.H. (2002). Emissions of N_2O and NO from fertilized fields: Summary of available measurement data. Global Biogeo. Cyc., 16: 6–13.

Brito, L.F. et al. (2015). Seasonal fluctuation of soil carbon dioxide emission in differently managed pastures. Agron. J., 107: 957–962.

Brown, R., Varvel, G. and Shapiro, C. (1993). Residual effects of inter-seeded hairy vetch on soil nitrate-nitrogen levels. Soil Sci. Soc. Am. J., 57: 121–124.

Butterbach-Bahl, K., Baggs, E.M., Dannenmann, M., Kiese, R. and Zechmeister-Boltenstern, S. (2013). Nitrous oxide emissions from soils: How well do we understand the processes and their controls? Phil. Trans. R. Soc. B, 368: 20130122.

Calderon, F.J., Nielsen, D., Acosta-Martinez, V., Vigil, M.F. and Drew, L. (2016). Cover crop and irrigation effects on soil microbial communities and enzymes in semi-arid agro-ecosystems of the central Great Plains of North America. Pedosphere, 26: 192–205.

Carter, M.S. and Ambus, P. (2006). Biologically fixed N_2 as a source for N_2O production in a grass-clover mixture, measured by $_{15}N_2$. Nutr. Cycl. Agro-ecosys., 74: 13–26.

Cherr, C., Scholberg, J. and McSorley, R. (2006). Green manure approaches to crop production. Agron. J., 98: 302–319.

Corre, M., Van Kessel, C. and Pennock, D. (1996). Landscape and seasonal patterns of nitrous oxide emissions in a semiarid region. Soil Sci. Soc. Am. J., 60: 1806–1815.

Cuellar-Bermudez, S.P., Garcia-Perez, J.S., Rittmann, B.E. and Parra-Saldivar, R. (2015). Photosynthetic bioenergy utilising CO_2: An approach on flue gases utilisation for third generation biofuels. J. Clean. Product., 98: 53–65.

Cuello, J.P., Hwang, H.Y., Gutierrez, J., Kim, S.Y. and Kim, P.J. (2015). Impact of plastic film mulching on increasing greenhouse gas emissions in temperate upland soil during maize cultivation. Appl. Soil Ecol., 91: 48–57.

De Gryze, S., Wolf, A., Kaffka, S.R., Mitchell, J., Rolston, D.E., Temple, S.R., Lee, J. and Six, J. (2010). Simulating greenhouse gas budgets of four California cropping systems under conventional and alternative management. Ecol. Appl., 20: 1805–1819.

Delgado, J., Del Grosso, S. and Ogle, S. (2010a). 15[N] Isotopic crop residue cycling studies suggest that IPCC methodologies to assess N_2O-N emissions should be reevaluated. Nutr. Cycl. Agro-ecosyst., 86: 383–390.

Delgado, J.A. (2010b). Crop residue is a key for sustaining maximum food production and for conservation of our biosphere. J. Soil Water Conserv., 65: 111A–116A.

Delgado, J.A., Groffman, P.M., Nearing, M.A., Goddard, T., Reicosky, D., Lal, R., Kitchen, N.R., Rice, C.W., Towery, D. and Salon, P. (2011). Conservation practices to mitigate and adapt to climate change. J. Soil Water Conserv., 66: 118A–129A.

Delon, C., SerçA, D., Boissard, C., Dupont, R., Dutot, A., Laville, P., De Rosnay, P. and Delmas, R. (2007). Soil NO emissions modelling using artificial neural network. Tellus B., 59: 502–513.

Dhillon, R. and von Wuehlisch, G. (2013). Mitigation of global warming through renewable biomass. Biomass Bioener., 48: 75–89.

Dubey, S. (2005). Microbial ecology of methane emission in rice agro-ecosystem: A review. Appl. Ecol. Environ. Res., 3: 1–27.

Duiker, S.W. and Curran, W.S. (2005). Rye cover crop management for corn production in the northern Mid-Atlantic region. Agron. J., 97: 1413–1418.

Eagle, A.J. and Olander, L.P. (2012). 3 Greenhouse gas mitigation with agricultural land management activities in the United States—A side-by-side comparison of biophysical potential. Adv. Agron., 115: 79.

Fageria, N., Baligar, V. and Bailey, B. (2005). Role of cover crops in improving soil and row crop productivity. Commun. Soil Sci. Plan., 36: 2733–2757.

Feng, J., Chen, C., Zhang, Y., Song, Z., Deng, A., Zheng, C. and Zhang, W. (2013). Impacts of cropping practices on yield-scaled greenhouse gas emissions from rice fields in China: A meta-analysis. Agr. Ecosyst. Environ., 164: 220–228.

Flessa, H., Ruser, R., Dörsch, P., Kamp, T,, Jimenez, M.A., Munch, J.C. and Beese, F. (2002). Integrated evaluation of greenhouse gas emissions (CO_2, CH_4, N_2O) from two farming systems in southern Germany. Agr. Ecosyst. Environ., 91: 175–189.

Folorunso, O., Rolston, D., Prichard, T. and Loui, D. (1992). Soil surface strength and infiltration rate as affected by winter cover crops. Soil Technol., 5: 189–197.

Frank, A., Liebig, M. and Hanson, J. (2002). Soil carbon dioxide fluxes in northern semiarid grasslands. Soil Biol. Biochem., 34: 1235–1241.

Fu, S., Cheng, W. and Susfalk, R. (2002). Rhizosphere respiration varies with plant species and phenology: A greenhouse pot experiment. Plant Soil, 239: 133–140.

Gabriel, J. and Quemada, M. (2011). Replacing bare fallow with cover crops in a maize cropping system: Yield, N uptake and fertiliser fate. Eur. J. Agron., 34: 133–143.

Gentile, R., Vanlauwe, B., Chivenge, P. and Six, J. (2008). Interactive effects from combining fertilizer and organic residue inputs on nitrogen transformations. Soil Biol. Biochem., 40: 2375–2384.

Gerlach, T. (2011). Volcanic versus anthropogenic carbon dioxide. Ecos., 92: 201–208.

Gómez, J.A., Guzmán, M.G., Giráldez, J.V. and Fereres, E. (2009). The influence of cover crops and tillage on water and sediment yield, and on nutrient, and organic matter losses in an olive orchard on a sandy loam soil. Soil Till. Res., 106: 137–144.

Graber, J., Amthor, J., Dahlman, R., Drell, D. and Weatherwax, S. (2008). Carbon Cycling and Biosequestration Integrating Biology and Climate Through Systems Science Report from the March 2008 Workshop, DOESC (USDOE Office of Science (SC)).

Griggs, D.J. and Noguer, M. (2002). Climate change 2001: The scientific basis, Contribution of working group I to the third assessment report of the intergovernmental panel on climate change. Weather, 57: 267–269.

Guardia, G., Abalos, D., García-Marco1, S., Quemada, M., Alonso-Ayuso1, M., Cárdenas, L.M., Dixon, E.R. and Vallejo, A. (2016). Effect of cover crops on greenhouse gas emissions in an irrigated field under integrated soil fertility management. Biogeosciences, 13: 5245–5257.

Haider, K., Mosier, A. and Heinemeyer, O. (1987). The effect of growing plants on denitrification at high soil nitrate concentrations. Soil Sci. Soc. Am. J., 51: 97–102.

Han, Z., Walter, M.T. and Drinkwater, L.E. (2017). Impact of cover cropping and landscape positions on nitrous oxide emissions in northeastern US agro-ecosystems. Agr. Ecosyst. Environ., 245: 124–134.

Hansen, S., Bernard, M.E., Rochette, P., Whalen, J.K. and Dörsch, P. (2014). Nitrous oxide emissions from a fertile grassland in western Norway following the application of inorganic and organic fertilisers. Nutrient Cycl. Agroecosys., 98: 71–85.

Hartwig, N.L. and Ammon, H.U. (2002). Cover crops and living mulches. Weed Sci., 50: 688–699.

Hatfield, J.L. and Parkin, T.B. (2012). Spatial variation of carbon dioxide fluxes in corn and soybean fields. Agr. Sci., 3: 986.

Hedges, L.V., Gurevitch, J. and Curtis, P.S.J.E. (1999). The meta-analysis of response ratios in experimental ecology. Ecol. Soc. Am., 80: 1150–1156.

Hénault, C. and Revellin, C. (2011). Inoculants of leguminous crops for mitigating soil emissions of the greenhouse gas nitrous oxide. Plant Soil, 346: 289–296.

Hendrix, P., Han, C.R. and Groffman, P. (1988). Soil respiration in conventional and no-tillage agro-ecosystems under different winter cover crop rotations. Soil Till. Res., 12: 135–148.

Hoben, J., Gehl, R., Millar, N., Grace, P. and Robertson, G. (2011). Non-linear nitrous oxide (N_2O) response to nitrogen fertilizer in on-farm corn crops of the US midwest. Global Change Biol., 17: 1140–1152.

Houghton, R.A. (2010). How well do we know the flux of CO_2 from land-use change? Tellus B., 62: 337–351.

Huang, Y., Zou, J., Zheng, X., Wang, Y. and Xu, X. (2004). Nitrous oxide emissions as influenced by amendment of plant residues with different C:N ratios. Soil Biol. Biochem., 36: 973–981.

Hubbard, R.K., Strickland, T.C. and Phatak, S. (2013). Effects of cover crop systems on soil physical properties and carbon/nitrogen relationships in the coastal plain of southeastern USA. Soil Till. Res., 126: 276–283.

Hwang, H.Y., Kim, G.W., Lee, Y.B., Kim, P.J. and Kim, S.Y. (2015). Improvement of the value of green manure via mixed hairy vetch and barley cultivation in temperate paddy soil. Field Crop Res., 183: 138–146.

Hwang, K.L., Bang, C.H. and Zoh, K.D. (2016). Characteristics of methane and nitrous oxide emissions from the wastewater treatment plant. Bioresource Tech., 214: 881–884.

Ingram, L., Schuman, G., Parkin, T. and Mortenson, M. (2015). Trace gas fluxes from a Northern mixed-grass prairie inter-seeded with alfalfa. Plant Soil, 386: 285–301.

IPCC. (2007). AR4 Climate Change 2007: The Physical Science Basis. https://www.ipcc.ch/report/ar4/wg1.

IPCC. (2015). Climate Change, 2014: Mitigation of Climate Change. Cambridge Univ. Press., London, U.K.

Izaurralde, R., Lemke, R.L., Goddard, T.W., McConkey, B. and Zhang, Z. (2004). Nitrous oxide emissions from agricultural topo-sequences in Alberta and Saskatchewan. Soil Sci. Soc. Am. J., 68: 1285–1294.

Jackson, L., Wyland, L. and Stivers, L. (1993). Winter cover crops to minimise nitrate losses in intensive lettuce production. J. Agr. Sci., 121: 55–62.

Jaffuel, G., Blanco-Pérez, R., Büchi, L., Mäder, P., Fliessbach, A., Charles, R., Degen, T., Turlings, T. and Campos-Herrera, R. (2017). Effects of cover crops on the overwintering success of entomopathogenic nematodes and their antagonists. Appl. Soil Ecol., 114: 62–73.

Jarecki, M.K. and Lal, R. (2003). Crop management for soil carbon sequestration. Crit. Rev. Plant Sci., 22: 471–502.

Jarecki, M.K. et al. (2009). Cover crop effects on nitrous oxide emission from a manure-treated. Mollisol., Agr. Ecosyst. Environ., 134: 29–35.

Jensen, E.S., Peoples, M.B., Boddey, R.M., Gresshoff, P.M., Hauggaard-Nielsen, H., Alves, B.J.R. and Morrison, M.J. (2012). Legumes for mitigation of climate change and the provision of feedstock for biofuels and biorefineries. A review. Agron. Sustain. Dev., 32: 329–364.

Jeuffroy, M.H., Baranger, E., Carrouee´, B., de Chezelles, E., Gosme, M., Henault´, C., Schneider, A. and Cellier, P. (2013). Nitrous oxide emissions from crop rotations including wheat, oilseed rape and dry peas. Biogeosciences, 10: 1787–1797.

Joshi, G., Pandey, J.K., Rana, S. and Rawat, D.S. (2017). Challenges and opportunities for the application of biofuel. Renew Sust. Ener. Rev., 79: 850–866.

Kallenbach, C.M., Rolston, D.E. and Horwath, W.R. (2010). Cover cropping affects soil N_2O and CO_2 emissions differently depending on type of irrigation. Agr. Ecosyst. Environ., 137: 251–260.

Karl, T.R. and Trenberth, K.E. (2003). Modern global climate change. Science, 302: 1719–1723.

Kaspar, T., Jaynes, D., Parkin, T. and Moorman, T. (2007). Rye cover crop and gamagrass strip effects on NO concentration and load in tile drainage. J. Environ. Qual., 36: 1503–1511.

Kassam, A. and Brammer, H. (2013). Combining sustainable agricultural production with economic and environmental benefits. Geographical J., 179: 11–18.

Kaye, J.P. and Quemada, M. (2017). Using cover crops to mitigate and adapt to climate change. A review. Agron. Sust. Dev., 37: 4.

Keating, B.A., Carberry, P.S., Bindraban, P.S., Asseng, S., Meinke, H. and Dixon, J. (2010). Eco-efficient agriculture: Concepts, challenges, and opportunities. Crop Sci., 50: S-109–S-119.

Ketterings, Q.M., Swink, S.N., Duiker, S.W., Czymmek, K.J., Beegle, D.B. and Cox, W.J. (2015). Integrating cover crops for nitrogen management in corn systems on northeastern US dairies. Agron. J., 107: 1365–1376.

Khalil, M., Rosenani, A., Van Cleemput, O., Boeckx, P., Shamshuddin, J. and Fauziah, C. (2002). Nitrous oxide production from an ultisol of the humid tropics treated with different nitrogen sources and moisture regimes. Biol. Fert. Soils, 36: 59–65.

Kim, G.W., Das, S., Hwang, H.Y. and Kim, P.J. (2017). Nitrous oxide emissions from soils amended by cover-crops and under plastic film mulching: Fluxes, emission factors and yield-scaled emissions. Atmos. Environ., 152: 377–388.

Kim, S.Y., Lee, C.H., Gutierrez, J. and Kim, P.J. (2013). Contribution of winter cover crop amendments on global warming potential in rice paddy soil during cultivation. Plant Soil, 366: 273–286.

King, L.D. and Buchanan, M. (1993). Reduced chemical input cropping systems in the southeastern United States. I. Effect of rotations, green manure crops and nitrogen fertilizer on crop yields. Am. J. Alt. Agr., 8: 58–77.

Kirchmann, H. and Witter, E. (1992). Composition of fresh, aerobic and anaerobic farm animal dungs. Bioresource Tech., 40: 137–142.

Kramberger, B., Gselman, A., Kristl, J., Lesnik, M., Sem, V., Mursec, M. and Podvršnik, M. (2014). Winter cover crop: The effects of grass–clover mixture proportion and biomass management on maize and the apparent residual N in the soil. Eur. J. Agron., 55: 63–71.

Ladoni, M., Basir, A., Robertson, P.G. and Kravchenko, A.N. (2016). Scaling-up: Cover crops differentially influence soil carbon in agricultural fields with diverse topography. Agric. Ecosyst. Environ., 225: 93–103.

Lam, S.K., Suter, H., Mosier, A.R. and Chen, D. (2016). Using nitrification inhibitors to mitigate agricultural N_2O emission: a double-edged sword? Global Change Biol., 23: 485–489.

Laws, E. and Berning, J. (1991). Photosynthetic efficiency optimisation studies with the macroalga *Gracilaria tikvihae*: Implications for CO_2 emission control from power plants. Bioresource Tech., 37: 25–33.

Le Quéré, C., Andres, R.J., Boden, T., Conway, T., Houghton, R.A., House, J.I., Marland, G., Peters, G.P., van der Werf, G., Ahlstrom, A., Andrew, R.M., Bopp, L., Canadell, J.G., Ciais, P., Doney, S.C., Enright, C., Friedlingstein, P., Huntingford, C., Jain, A.K., Jourdain, C., Kato, E., Keeling, R.F., Klein Goldewijk, K., Levis, S., Levy, P., Lomas, M., Poulter, B., Raupach, M.R., Schwinger, J., Sitch, S., Stocker, B.D., Viovy, N., Zaehle, S. and Zeng, N. (2012). The global carbon budget 1959–2011. Earth Syst. Sci. Data Discussions, 5: 1107–1157.

Le Treut, H., Somerville, R., Cubasch, U., Ding, Y., Mauritzen, C., Mokssit, A., Peterson, T. and Prather, M. (2007). Historical Overview of Climate Change. *In*: Solomon, S., Qin, D., Manning, M., Chen, Z., Marquis, M., Averyt, K.B., Tignor, M. and Miller, H.L. (eds.). Climate Change 2007: The Physical Science Basis, Contribution of Working Group I to the Fourth Assessment Report of the Intergovernmental Panel on Climate Change. Cambridge University Press, Cambridge, United Kingdom and New York, NY, USA.

Lee, J., Six, J., King, A.P., Van Kessel, C. and Rolston, D.E. (2006). Tillage and field scale controls on greenhouse gas emissions. J. Environ. Qual., 35: 714–725.

Liebig, M., Tanaka, D. and Gross, J. (2010). Fallow effects on soil carbon and greenhouse gas flux in central North Dakota. Soil Sci. Soc. Am. J., 74: 358–365.

Liebig, M., Franzluebbers, A. and Follett, R. (2012). Agriculture and climate change: Mitigation opportunities and adaptation imperatives, Managing gricultural Greenhouse Gases: Coordinated Agricultural Research through GRACEnet to Address our Changing Climate. Acad. Press, San Diego CA, 3–11.

Linquist, B., Groenigen, K.J., Adviento-Borbe, M.A., Pittelkow, C. and Kessel, C. (2012). An agronomic assessment of greenhouse gas emissions from major cereal crops. Global Change Biol., 18: 194–209.

Liu, H., Zhao, P., Lu, P., Wang, Y., Lin, Y. and Rao, X. (2008). Greenhouse gas fluxes from soils of different land-use types in a hilly area of South China. Agr. Ecosys. Environ., 124: 125–135.

Liu, J., Zhu, L., Luo, S., Bu, L., Chena, X., Yue, S. and Lia, S. (2014). Response of nitrous oxide emission to soil mulching and nitrogen fertilisation in semi-arid farmland. Agr. Ecosys. Environ., 188: 20–28.

Liu, J., Hu, C., Yang, P., Ju, Z., Olesen, J.E. and Tang, J. (2015). Effects of experimental warming and nitrogen addition on soil respiration and CH_4 fluxes from crop rotations of winter wheat–soybean/fallow. Agr. Forest Meteor., 207: 38–47.

Lucas, P.L., van Vuuren, D.P., Olivier, J.G. and Den Elzen, M.G. (2007). Long-term reduction potential of non-CO_2 greenhouse gases. Environ. Sci. Policy, 10: 85–103.

Luis, P., Van Gerven, T. and Van der Bruggen, B. (2012). Recent developments in membrane-based technologies for CO_2 capture. Prog. Ener. Combustion Sci., 38: 419–448.

Mannina, G., Morici, C., Cosenza, A., Di Trapani, D. and Degaard, H. (2016). Greenhouse gases from sequential batch membrane bioreactors: A pilot plant case study. Biochem. Eng. J., 112: 114–122.

Masilionyte, L. and Maikšténiené, S. (2016). The effect of alternative cropping systems on the changes of the main nutritional elements in the soil. Zemdirbyste Agri., 103: 3–10.

Masilionyte, L., Maiksteniene, S., Kriauciuniene, Z., Jablonskyte-Rasce, D., Zou, L. and Sarauskis, E. (2017). Effect of cover crops in smothering weeds and volunteer plants in alternative farming systems. Crop Prot., 91: 74–81.

McCracken, D.V., Smith, M.S., Grove, J.H., Blevins, R.L. and MacKown, C.T. (1994). Nitrate leaching as influenced by cover cropping and nitrogen source. Soil Sci. Soc. Am. J., 58: 1476–1483.

McSwiney, C.P., Snapp, S.S. and Gentry, L.E. (2010). Use of N immobilisation to tighten the N cycle in conventional agro-ecosystems. Ecol. Appl., 20: 648–662.

Meinshausen, M., Meinshausen, N., Hare, W., Raper, S.C.B., Frieler, K., Knutti, R. Frame, D.J. and Allen, M.R. (2009). Greenhouse-gas emission targets for limiting global warming to 2°C. Nature, 458: 1158–1162.

Meinshausen, M., Smith, S.J., Calvin, K., Daniel, J.S., Kainuma, M.L.T., Lamarque, J.-F., Matsumoto, K., Montzka, S.A., Raper, S.C.B., Riahi, K., Thomson, A., Velders, G.J.M. and van Vuuren, D.P. P. (2011). The RCP greenhouse gas concentrations and their extensions from 1765 to 2300. Climatic Change, 109: 213.

Meisinger, J., Hargrove, W., Mikkelsen, R., Williams, J. and Benson, V. (1991). Effects of cover crops on groundwater quality. Cover Crops for Clean Water, Soil Water Conserv. Soc., Ankeny, Iowa, 266: 793–799.

Meixner, F.X. (2006). Biogenic emissions of nitric oxide and nitrous oxide from arid and semi-arid land, Dryland Ecohydrology. Springer, 233–255.

Menon, S., Denman, K.L., Brasseur, G., Chidthaisong, A., Ciais, P., Cox, P.M., Dickinson, R.E., Hauglustaine, D.; Heinze, C., Holland, E., Jacob, D., Lohmann, U., Ramachandran, S., Wofsy, S.C. and Zhang, X. (2007). Couplings Between Changes in the Climate System and Biogeochemistry, Lawrence Berkeley National Lab., Berkeley, CA, United States.

Miguez, F.E. and Bollero, G.A. (2005). Review of corn yield response under winter cover cropping systems using meta-analytic methods. Crop Sci., 45: 2318–2329.

Mikulčić, H., Klemeš, J.J., Vujanović, M., Urbaniec, K. and Duić, N. (2016). Reducing greenhouse gasses emissions by fostering the deployment of alternative raw materials and energy sources in the cleaner cement manufacturing process. J. Clean Produc., 136: 119–132.

Mitchell, D.C., Castellano, M.J., Sawyer, J.E. and Pantoja, J. (2013). Cover crop effects on nitrous oxide emissions: role of mineralisable carbon. Soil Sci. Soc. Am. J., 77: 1765–1773.

Morell, F.J., Whitmore, A., Álvaro-Fuentes, J., Lampurlanés, J. and Cantero-Martínez, C. (2012). Root respiration of barley in a semiarid Mediterranean agro-ecosystem: Field and modelling approaches. Plant Soil, 351: 135–147.

Mosier, A., Duxbury, J., Freney, J., Heinemeyer, O. and Minami, K. (1998). Assessing and mitigating N_2O emissions from agricultural soils. Climatic Change, 40: 7–38.

Muscas, E., Cocco, A., Mercenaro, L., Cabras, M., Lentini, A., Porqueddu, C. and Nieddu, G. (2017). Effects of vineyard floor cover crops on grapevine vigor, yield, and fruit quality, and the development of the vine mealybug under a Mediterranean climate. Agr. Ecosyst. Environ., 237: 203–212.

Negassa, W., Price, R.F., Basir, A., Snapp, S.S. and Kravchenko, A. (2015). Cover crop and tillage systems effect on soil CO_2 and N_2O fluxes in contrasting topographic positions. Soil Till. Res., 154: 64–74.

Nishina, K., Takenaka, C. and Ishizuka, S. (2009). Spatiotemporal variation in N_2O flux within a slope in a Japanese cedar (*Cryptomeria japonica*) forest. Biogeochem., 96: 163–175.

Novoa, R.S. and Tejeda, H.R. (2006). Evaluation of the N_2O emissions from N in plant residues as affected by environmental and management factors. Nutr. Cycl. Agroecosys., 75: 29–46.

Nygaard Sorensen, J. and Thorup-Kristensen, K. (2011). Plant-based fertilisers for organic vegetable production. J. Plant Nutr. Soil Sci., 174: 321–332.

Omonode, R.A., Vyn, T.J., Smith, D.R., Hegymegi, P. and Gál, A. (2007). Soil carbon dioxide and methane fluxes from long-term tillage systems in continuous corn and corn–soybean rotations. Soil Till. Res., 95:182–195.

Ostrom, N.E., Sutka, R., Ostrom, P., Grandy, S., Huizinga, K.M., Gandhi, H., von Fischer, J. and Robertson, G.P. (2010). Isotopologue data reveal bacterial denitrification as the primary source of N_2O during a high flux event following cultivation of a native temperate grassland. Soil Biol. Biochem., 42: 499–506.

Pachauri, R.K. (2014). Climate Change 2014: Synthesis Report, Contribution of Working Groups I, II and III to the fifth assessment report of the IPCC. https://www.ipcc.ch/site/assets/uploads/2018/05/SYR_AR5_FINAL_full_wcover.pdf.

Pappa, V.A., Rees, R.M., Walker, R.L., Baddeley, J.A. and Watson, C.A. (2011). Nitrous oxide emissions and nitrate leaching in an arable rotation resulting from the presence of an intercrop. Agr. Ecosyst. Environ., 141: 153–161.

Parkin, T., Kaspar, T. and Singer, J. (2006). Cover crop effects on the fate of N following soil application of swine manure. Plant Soil, 289: 141–152.

Parkin, T.B. and Kaspar, T.C. (2006). Nitrous oxide emissions from corn-soybean systems in the midwest. J. Environ. Qual., 35: 1496–1506.

Peoples, M., Herridge, D. and Ladha, J. (1995). Biological nitrogen fixation: An efficient source of nitrogen for sustainable agricultural production? Plant Soil, 174: 3–28.

Peyrard, C., Mary, B., Perrin, P., Véricel, G., Gréhan, E., Justes, E. and Léonard, J. (2016). N₂O emissions of low input cropping systems as affected by legume and cover crops use. Agr. Ecosyst. Environ., 224: 145–156.

Philibert, A., Loyce, C. and Makowski, D. (2012). Assessment of the quality of meta-analysis in agronomy. Agr. Ecosyst. Environ., 148: 72–82.

Pilegaard, K. (2013). Processes regulating nitric oxide emissions from soils. Phil. Trans. R. Soc. B., 368: 20130126.

Poeplau, C. and Don, A. (2015). Carbon sequestration in agricultural soils via cultivation of cover crops—A meta-analysis. Agr. Ecosyst. Environ., 200: 33–41.

Poudel, D., Horwath, W., Mitchell, J. and Temple, S. (2001). Impacts of cropping systems on soil nitrogen storage and loss. Agric. Sys., 68: 253–268.

Ramaswamy, V., Boucher, O., Haigh, J., Hauglustaine, D., Haywood, J., Myhre, G., Nakajima, T., Shi, G., Solomon, S., Betts, R., Charlson, R., Chuang, C.C., Daniel, J.S., Del Genio, A.D., Feichter, J., Fuglestvedt, J., Forster, P.M., Ghan, S.J., Jones, A., Kiehl, J.T., Koch, D., Land, C., Lean, J., Lohmann, U., Minschwaner, K., Penner, J.E., Roberts, D.L., Rodhe, H., Roelofs, G-J., Rotstayn, L.D., Schneider, T.L., Schumann, U., Schwartz, S.E., Schwartzkopf, M.D., Shine, K.P., Smith, S.J., Stevenson, D.S., Stordal, F., Tegen, I., van Dorland, R., Zhang, Y., Srinivasan, J. and Joos, F. (2001). Radiative Forcing of Climate Change, Pacific Northwest National Laboratory (PNNL), Richland, WA (US).

Ravishankara, A., Daniel, J.S. and Portmann, R.W. (2009). Nitrous oxide (N₂O): The dominant ozone-depleting substance emitted in the 21st century. Science, 326: 123–125.

Reeves, D. (1994). Cover crops and rotations. Advances in Soil Science: Crops Residue Manag., 125–172.

Rigon, J.P.G., Calonego, J.C., Rosolem, C.A. and Scala Jr, N.L. (2018). Cover crop rotations in no-till system: short-term CO₂ emissions and soybean yield. Sci. Agr., 75: 18–26.

Robertson, G.P., Paul, E.A. and Harwood, R.R.J.S. (2000). Greenhouse gases in intensive agriculture: contributions of individual gases to the radiative forcing of the atmosphere. Sci., 289: 1922–1925.

Rochette, P., Angers, D., Bélanger, G., Chantigny, M.H., Prévost, D. and Lévesque, G. (2004). Emissions of NO from alfalfa and soybean crops in Eastern Canada. Soil Sci. Soc. Am. J., 68: 493–506.

Rochette, P. and Janzen, H.H. (2005). Towards a revised coefficient for estimating N₂O emissions from legumes. Nutr. Cycl. Agr., 73: 171–189.

Rosecrance, R., McCarty, G., Shelton, D. and Teasdale, J. (2000). Denitrification and N mineralization from hairy vetch (*Vicia villosa* Roth) and rye (*Secale cereale* L.) cover crop monocultures and bicultures. Plant Soil, 227: 283–290.

Ruis, S., Blanco-Canqui, H., Jasa, P. and Ferguson, R.B. (2017). Cover Crop Termination Date Effects on Greenhouse Gas Fluxes in Rain-fed and Irrigated No-till Corn, Managing Global Resources for a Secure Future, Tampa, FL, Annual Meeting, SSSA.

Sainju, U., Singh, B. and Whitehead, W. (2002). Long-term effects of tillage, cover crops, and nitrogen fertilisation on organic carbon and nitrogen concentrations in sandy loam soils in Georgia, USA. Soil Till. Res., 63: 167–179.

Sainju, U.M., Schomberg, H., Singh, B.P., Whitehead, W., Tillman, P.G. and Weyers, S.L. (2007). Cover crop effect on soil carbon fractions under conservation tillage cotton. Soil Till. Res., 96: 205–218.

Sainju, U.M. and Singh, B.P. (2008). Nitrogen storage with cover crops and nitrogen fertilisation in tilled and non-tilled soils. Agr. J., 100: 619–627.

Sainju, U.M., Stevens, W.B., Caesar-TonThat, T. and Liebig, M.A. (2012). Soil greenhouse gas emissions affected by irrigation, tillage, crop rotation, and nitrogen fertilisation. J. Environ. Qual., 41: 1774–1786.

Sanz-Cobena, A. et al. (2014). Do cover crops enhance N₂O, CO₂ or CH₄ emissions from soil in Mediterranean arable systems? Sci. Total Environ., 466: 164–174.

Sarkodie-Addo, J., Lee, H. and Baggs, E. (2003). Nitrous oxide emissions after application of inorganic fertiliser and incorporation of green manure residues. Soil Use Manag., 19: 331–339.

Schaefer, D.O., Godwin, D. and Harnisch, J. (2006). Estimating future emissions and potential reductions of HFCs, PFCs, and SFCs. Energ. J., 63–88.

Schilt, A., Baumgartner, M., Blunier, T., Schwander, J., Spahni, R., Fischer, H. and Stocker, T. (2010). Glacial-interglacial and millennial-scale variations in the atmospheric nitrous oxide concentration during the last 800,000 years. Quaternary Sci. Rev., 29: 182–192.

Schindlbacher, A., Zechmeister-Boltenstern, S. and Butterbach-Bahl, K. (2004). Effects of soil moisture and temperature on NO, NO₂, and N₂O emissions from European forest soils. Journal of Geophysical Research: Atmospheres, 109.

Schipanski, M.E., Barbercheck, M., Douglas, M.R., Finney, D., Haider, K., Kaye, J.P., Kemanian, A.R., Mortensen, D.A., Ryan, M.R., Tooker, J.F. and White, C. (2014). A framework for evaluating ecosystem services provided by cover crops in agro-ecosystems. Agr. Syst., 125: 12–22.

Scholberg, J.M.S., Dogliotti, S., Leoni, C., Cherr, C., Zotarelli, L. and Rossing, W.A.H. (2010). Cover crops for sustainable agrosystems in the Americas. In: Genetic engineering, biofertilisation, soil quality and organic farming (Lichtfouse, L.). Sustainable Agriculture Review vol. 4., Springer, pp. 23–58.

Schwenke, G.D., Haigh, B.M., Scheer, C., Herridge, D.F., Rowlings, D.W. and McMullen, K.G. (2015). Soil N$_2$O emissions under N$_2$-fixing legumes and N-fertilised canola: A reappraisal of emissions factor calculations. Agr. Ecosys. Environ., 202: 232–242.

Scott, T., Burt, R. and Otis, D. (1987). Contributions of ground cover, dry matter, and nitrogen from intercrops and cover crops in a corn polyculture system. Agron. J., 79: 792–798.

Senbayram, M., Chen, R., Budai, A., Bakken, L. and Dittert, K. (2012). N$_2$O emission and the N$_2$O/(N$_2$O + N$_2$) product ratio of denitrification as controlled by available carbon substrates and nitrate concentrations. Agr. Ecosyst. Environ., 147: 4–12.

Seufert, V., Ramankutty, N. and Foley, J.A. (2012). Comparing the yields of organic and conventional agriculture. Nature, 485: 229–232.

Shan, J. and Yan, X.J.A.E. (2013). Effects of crop residue returning on nitrous oxide emissions in agricultural soils. Atmos. Environ., 71: 170–175.

Shipley, P.R., Messinger, J. and Decker, A. (1992). Conserving residual corn fertiliser nitrogen with winter cover crops. Agron J., 84: 869–876.

Sinha, S.K. (1995). Global methane emission from rice paddies: Excellent methodology but poor extrapolation. Curr. Sci., 643–646.

Smith, D., Hernandez-Ramirez, G., Armstrong, S., Bucholtz, D. and Stott, D. (2011). Fertiliser and tillage management impacts on non-carbon-dioxide greenhouse gas emissions. Soil Sci. Soc. Am. J., 75: 1070–1082.

Smith, M.S. and Tiedje, J.M. (1979). The effect of roots on soil denitrification. Soil Sci. Soc. Am. J., 43: 951–955.

Smith, R.G., Atwood, L.W. and Warren, N.D. (2014). Increased productivity of a cover crop mixture is not associated with enhanced agro-ecosystem services. PLoS ONE, 9: e97351.

Snyder, C., Bruulsema, T., Jensen, T. and Fixen, P. (2009). Review of greenhouse gas emissions from crop production systems and fertiliser management effects. Agr. Ecosyst. Environ., 133: 247–266.

Steenwerth, K. and Belina, K. (2008). Cover crops enhance soil organic matter, carbon dynamics and microbiological function in a vineyard agro-ecosystem. Appl. Soil Ecol., 40: 359–369.

Stewart, C.E., Paustian, K., Conant, R.T., Plante, A.F. and Six, J. (2009). Soil carbon saturation: Implications for measurable carbon pool dynamics in long-term incubations. Soil Biol. Biochem., 41: 357–366.

Stott, D., Elliott, L., Papendick, R. and Campbell, G. (1986). Low temperature or low water potential effects on the microbial decomposition of wheat residue. Soil Biol. Biochem., 18: 577–582.

Szerencsits, M., Weinberger, C., Kuderna, M., Feichtinger, F., Erhart, E. and Maier, S. (2016). Biogas from cover crops and field residues: effects on soil, water, climate and ecological footprint. World Academy of Science, Engineering and Technology, Int. J. Environ. Chem. Ecol. Geolog. Geophysic Eng., 9: 413–416.

Tang, H., Chen, M., Simon Ng, K. and Salley, S.O. (2012). Continuous microalgae cultivation in a photobioreactor. Biotech. Bioenergy, 109: 2468–2474.

Tang, H., Xiao, X., Tang, W. and Yang, G. (2013). Effect of different winter cover crops on carbon dioxide emission in paddy field of double cropping rice area in Southern China. J. Environ. Sci. Eng. B., 2.

Thomson, L.J. and Hoffmann, A.A. (2013). Spatial scale of benefits from adjacent woody vegetation on natural enemies within vineyards. Biol. Control, 64: 57–65.

Thorup-Kristensen, K., Magid, J. and Jensen, L.S. (2003). Catch crops and green manures as biological tools in nitrogen management in temperate zones. Adv. Agron., 79: 227–302.

Tilman, D., Balzer, C., Hill, J. and Befort, B.L. (2011). Global food demand and the sustainable intensification of agriculture. Proc. Nt. Acad. Sci., 108: 20260–20264.

Timmermann, A., Oberhuber, J., Bacher, A., Esch, M., Latif, M. and Roeckner, E. (1999). Increased El Niño frequency in a climate model forced by future greenhouse warming. Nature, 398: 694–697.

Tonitto, C., David, M. and Drinkwater, L. (2006). Replacing bare fallows with cover crops in fertilizer-intensive cropping systems: A meta-analysis of crop yield and N dynamics. Agr. Ecosyst. Environ., 112: 58–72.

Tribouillois, H., Cruz, P., Cohan, J.-P. and Justes, É. (2015). Modelling agro-ecosystem nitrogen functions provided by cover crop species in bispecific mixtures using functional traits and environmental factors. Agr. Ecosys. Environ., 207: 218–228.

Ussiri, D. and Lal, R. (2012). Soil Emission of Nitrous Oxide and Its Mitigation, Springer, Sci. Business Media.

Van de Wal, R., Boer, B.D., Lourens, L., Köhler, P. and Bintanja, R. (2011). Reconstruction of a continuous high-resolution CO$_2$ record over the past 20 million years. Clim. Past, 7: 1459–1469.

Van Groenigen, J. et al. (2015). The soil N cycle: New insights and key challenges. Soil, 1: 235.

Veres, A., Petit, S., Conord, C. and Lavigne, C. (2013). Does landscape composition affect pest abundance and their control by natural enemies? A review. Agr. Ecosys. Environ., 166: 110–117.

Verzeaux, J., Alahmad, A., Habbib, H., Nivelle, E., Roger, D., Lacoux, J., Decocq, G., Hirel, B., Catterou, M., Spicher, F., Dubois, F., Duclercq, J. and Tétu, T. (2016). Cover crops prevent the deleterious effect of nitrogen fertilisation on bacterial diversity by maintaining the carbon content of ploughed soil. Geoderma, 281: 49–57.

Vilain, G., Garnier, J., Tallec, G. and Cellier, P. (2010). Effect of slope position and land use on nitrous oxide (N$_2$O) emissions (Seine Basin, France). Agr. Forest Meteo., 150: 1192–1202.

Webster, T.M., Scully, B.T., Grey, T.L. and Culpepper, A.S. (2013). Winter cover crops influence *Amaranthus palmeri* establishment. Crop Prot., 52: 130–135.

Wickings, K., Grandy, A.S. and Kravchenko, A.N. (2016). Going with the flow: Landscape position drives differences in microbial biomass and activity in conventional, low input, and organic agricultural systems in the midwestern US. Agr. Ecosys. Environ., 218: 1–10.

Wilson, H. and Al-Kaisi, M. (2008). Crop rotation and nitrogen fertilisation effect on soil CO$_2$ emissions in central Iowa. Appl. Soil Ecol., 39: 264–270.

Yuping, Y., Sha, L., Cao, M., Zheng, Z., Tang, J., Wang, Y., Zhang, Y., Wang, R., Liu, G., Wang, Y. and Sun, Y. (2008). Fluxes of CH$_4$ and N$_2$O from soil under a tropical seasonal rain forest in Xishuangbanna, southwest China. J. Environ. Sci., 20: 207–215.

Zhong, Z., Lemke, R.L. and Nelson, L.M. (2009). Nitrous oxide emissions associated with nitrogen fixation by grain legumes. Soil Biol. Biochem., 41: 2283–2291.

Zhou, G., Zhang, J., Zhang, C., Feng, Y., Chen, L., Yu, Z., Xin, X. and Zhao, B. (2016). Effects of changes in straw chemical properties and alkaline soils on bacterial communities engaged in straw decomposition at different temperatures. Sci. Rep., 6: 22186.

17

Cover Crops Influence Soil Microbial and Biochemical Properties

Amoakwah, E

1. Introduction

In the wake of rapid population growth and climate change effects, there is an urgent need for transition to sustainable agricultural-production systems to achieve food security. Such production systems have the propensity to address numerous challenges, militating against mainstream conventional agriculture, namely soil degradation, loss of soil fertility and quality. One of the core philosophies of sustainable agricultural-production systems is the development of healthy and productive soils that promote crop production and soil quality with special reference to biodiversity and biotic communities (Maeder et al., 2002) and improve ecosystem functions. Soil biological and biochemical properties, such as microbial biomass (SMB) and diversity, respiratory activities, enzymatic functions, residue decomposition, carbon dynamics and mycorrhizal associations are directly linked to agricultural sustainability (Table 1). Soil biology is reported to be the driving force that facilitates decomposition processes that break down complex organic molecules and convert them into forms that are readily available to growing plants (Islam and Weil, 2000; Friedel and Gabel, 2001). Soil biological properties play a pivotal role in defining soil quality, which is a critical and effective indicator of soil's ability to perform ecological functions. Also it reflects changes in soil properties that are both inherent and anthropogenic (Nair and Ngouajio, 2012). A large, diverse, more stable and active soil microbial community and composition are crucial to maintain soil productivity (Nair and Ngouajio, 2012).

Cover cropping is one of the promising sustainable agricultural methods with the potential to enhance soil health and mitigate consequences of soil degradation. Because cover cropping can form a balanced agro-ecosystem distinct from that of bare fallow or conventionally managed agro-ecosystems (Kim et al., 2020), cover cropping has proven to effectively develop such sustainable production systems for enhanced soil fertility and crop productivity by significantly increasing soil organic matter (SOM) turnover and improving soil biology and quality (Buyer et al., 2010). The effects of cover crops on soil biodiversity have been largely explored relative to nutrient recycling and soil biology. However, there is paucity of information on their related effects on the soil microclimate. Understanding the effects of cover crops on soil's biological and biochemical properties is, therefore, important for developing cover-crop management methods.

CSIR - Soil Research Institute, Kumasi, Ghana.

Table 1 Influence of different cover crop types on soil microbes involved in plant-soil feedback (*Source*: Vukicevich et al., 2016).

Cover Crop	Microbial Effects	Species Studied	System Type	Reference
Legumes	Increases diversity of AM fungi	*Lotus corniculatus, Trifolium repens, Ononis repens*	Annual; natural	Klabi et al. (2014)
	Increases entomopathogenic fungi persistence	*T. repens*	Perennial	Shapiro-Ilan et al. (Wood, 2012)
	Decreases abundance of DAPG and PRN-producing bacteria	*Lathyrus pratensis, Lotus corniculatus, Medicago lupulina, Medicago varia, Onobrychis viciifolia, Trifolium* spp., *Vicia cracca*	Experimental	Latz et al. (2012)
C_3 grasses	Increases abundance of DAPG and PRN-producing bacteria	*Lolium perenne*	Experimental	Latz et al. (2015)
	Cultivar-specific disease-suppressive bacterial community	*Triticum* (different cultivars)	Perennial	Mazzola et al. (2004)
	Low mycorrhizal response (less AM fungi?)	*Koeleria cristata, Bromus inermis, Festuca arundinacea, Lolium perenne, Agropyron smithii, Elymus cinereus*	Natural	Hetrick et al. (1988)
C_4 grasses	High mycorrhizal response (more AM fungi?)	*Andropogon gerardi, Panicum virgatum, Sorghastrum nutans, Bouteloua curtipendula*	Natural	Hetrick et al. (1988)
Brassicas	Decreases fungal pathogens	*Brassica napus*/mustard green manure; *B. napus* seed meal	Annual; perennial	Larkin et al. (2010)
	Increases disease-suppressive bacteria	*B. napus* (living plant or seed meal)	Annual	Hollister et al. (2013); Mazzola et al. (2015)
	Favors *Trichoderma* and other disease-protective fungi	*B. juncea, B. napus* (crop rotation); seed meal	Annual	Galletti et al. (2008)
	Alters microbial community	*B. napus* seed meal	Perennial	Mazzola et al. (2015)
	Inhibits AM spore germination; decreases AM fungal diversity	*B. kaber, B. nigra; B. napus* seed meal	Experimental; perennial	Mazzola et al. (2015)
	Increases disease-suppressive bacteria	*N/A*	Experimental; annual/natural	Latz et al. (2012)
	Increases AM fungal diversity	*N/A*	Perennial/ natural	Holland et al. (2016)
	Decreases overall negative feedback	*N/A*	Experimental	Maron et al. (2011)
	Increases diversity and abundance of entomopathogenic fungi	*N/A*	Annual/ natural	Meyling et al. (2009)

2. Cover Crops' Effect on Soil Microbial Activity and Diversity

The effects of cover crops on soil microbial biomass (SMB) and activity can influence N transformations (Shelton et al., 2018). However, the impacts of cover crops on soil microbial biomass and enzyme activity are still poorly understood (Brennan and Acosta-Martinez, 2018), particularly beyond initial, short-term laboratory studies. Cover cropping systems have been reported to have induced large changes in SMB composition and activity, with beneficial effects on soil and/or plant productivity (Martínez-García et al., 2018) (Table 1). Cover crop-induced changes in pH-value through carboxylate exudation or phosphatase release (Hallama et al., 2019), generation of carbon-nutrient agglomerates in soil and provision of an additional C-source (Malik et al., 2018) may result in these positive effects on soil microbial biomass in cover-copping systems. The afore-mentioned conditions create a conducive environment with greater resource diversity and a more consistent nutrient supply to induce changes in microbial abundance and diversity.

The myriads of organic C provided by cover-crop-root exudates and plant residues have been reported to increase biomass and change the composition of soil microbial communities (de Graaff et al., 2010; Ramirez-Villanueva et al., 2015). Wang et al. (2019) reported the impact of cover crops on the abundance of colonies for three major microbial groups showing the order of bacteria > actinomycetes > fungi in soil (Table 2). The number of bacteria colonies under alfalfa cover (AC) and fungi colonies under sweet clover (SC) was 0.02–0.49 and 0.02–0.24 log10 CFU g^{-1} higher than that under other treatments, respectively; and winter wheat cover (WC) showed the maximum number of actinomycete colonies (5.72 and 5.82 log10 CFU g^{-1}).

Considering the wide influence of SMB on the decomposition of SOM, nutrient cycling and on plant growth, it is imperative to identify the effects of cover crops on soil microorganisms more succinctly. However, there is limited scientific knowledge on the long-term effects of cover crops on SMB to precisely predict SMB responses to cover-cropping systems. Cover cropping promotes soil's biological activity by enhancing the build-up of the SOM ecosystem. It has been shown that temporary release of labile C by cover crops (Zhou et al., 2012) can considerably promote soil biological activity. Crop residue mulches have been reported to reduce soil water evaporation, increase soil water content and decrease daily soil temperature (Dahiya et al., 2007)—conditions favourable for enhancing soil microbial activity. Soil temperature and moisture content are reported to be the key determinants of soil biological activity (Gallo et al., 2006); therefore, cover crop-management methods should be designed to encompass this knowledge.

Table 2 Changes in the number of microbial colonies in soil (log10 CFU g^{-1}) under different treatments (*Source*: Wang et al., 2019).

Trts.	Bacteria Number		Fungi Number		Actinomycetes Number	
	2018	2019	2018	2019	2018	2019
AC	6.86 ± 0.01aB	6.95 ± 0.02aA	4.91 ± 0.04abA	4.86 ± 0.01aA	5.67 ± 0.05abA	5.70 ± 0.06bcA
SC	6.81 ± 0.05aA	6.93 ± 0.03aA	4.94 ± 0.05aA	4.88 ± 0.04aA	5.65 ± 0.05abA	5.77 ± 0.02abA
WC	6.68 ± 0.07bA	6.66 ± 0.04cA	4.83 ± 0.03bA	4.70 ± 0.01bA	5.59 ± 0.04bB	5.68 ± 0.05cA
RC	6.72 ± 0.03bA	6.79 ± 0.07bA	4.88 ± 0.07abA	4.83 ± 0.04aA	5.72 ± 0.05aB	5.82 ± 0.03aA
CK	6.37 ± 0.06cA	6.56 ± 0.07dA	4.70 ± 0.06cA	4.67 ± 0.03bA	5.63 ± 0.03bA	5.57 ± 0.02dA

AC: alfalfa cover; SC: sweet clover cover; WC: winter wheat cover; RC: ryegrass cover; CK: bare land.
Values are means of three replications ± standard error. Means in each column followed by a different lowercase letter denote significant differences among the treatments within the same year (Duncan's new multiple range test; $p < 0.05$); means in each row followed by a different uppercase letter denote significant difference among the years within the same treatment (t-test; $p < 0.05$).

3. Cover Crops' Effect on Soil Biological and Physico-chemical Interactions

The observed effects of cover crops on soil microbiological activity are reported to result from at least two effects: (1) alteration of phyico-chemical interactions, such as increased water and nutrient retention, and (2) provision of micro-habitat that protects microorganisms from predation.

There is also a body of evidence that supports the potential of cover crops to influence soil biology and, consequently, increase SOM to improve the chemical and physical properties that enhance soil water dynamics (Daigh et al., 2014). Further, there is a complex interaction of soil physical and chemical properties (such as SOM content, aggregation and porosity) that contributes to soil-water retention capacity (Emerson, 1995). Cover crops enhance the water-holding capacity of the soil (Basche et al., 2016), nutrient adsorption capabilities (Williams et al., 2018) and improve the availability of soil nutrients through a reduction in soil loss (via erosion and leaching), by supporting the growth of diverse and beneficial microbes in the soil, and in some cases, through the fixation of atmospheric N. These conditions provide a conducive environment for microbes to thrive and flourish soil ecosystem's functionality.

4. Cover Crops and Habitat Provision

Cover crops can enhance SOM build-up in soils. Organic materials are directly responsible for the formation of macro aggregates (Six et al., 2004) through the actions of fungal hyphae and microbial extracellular polysaccharide gums provided by the root exudates of cover crops in the soil matrix. Complexation processes resulting from organic functional groups and/or microbial activity are the most likely mechanisms that result in the formation of macro aggregates in cover cropping systems. The pores in the aggregates, together with the organic matter in cover cropping systems, can provide support surfaces for microbial colonisation which, together with improved water-holding capacity, provide a suitable habitat for microorganisms (Dai et al., 2017).

5. Effects of Cover Crops on Enzyme Activities

Enzymes are biological catalysts crucial for SOM decomposition, cycling of nutrients in nature and the activities of enzymes can be used as an indicator of microbial activity and soil quality (Benítez et al., 2000). Soil enzymes are reported to be involved in the transfer of energy (Gu et al., 2019) and, consequently, affect soil quality and crop productivity. The production, activity and the stability of free and adsorbed enzymes are basically controlled by environmental conditions and ecological interactions.

Soil enzymes have the propensity to respond quickly to changes in soil management interventions than other soil variables. Therefore, soil enzyme activities might be useful as early indicators of biological changes in response to management practices. Profiles of soil enzyme activities depict soil functional diversity, which is influenced by the genetic diversity of soil microorganisms, plants and animals and is closely linked to environmental factors and ecological interactions (Nannipieri et al., 2002). Soil enzyme activities and microbial community play a critical role in ecosystem functions and cycling of soil nutrients. Moreover, the ability of the soil to support plant growth and development or regulate nutrient cycling may largely depend on the composition of soil microbial communities and soil enzyme activities (Singh et al., 2010).

Wang et al. (2019) studied the effects of different cover crops on soil enzymes with special reference to urease, dehydrogenase, protease and sucrase activities (Fig. 1). Results showed that in topsoil (0–20 cm), the urease (Fig. 1a) and dehydrogenase activities (Fig. 1b) under AC were 10.3–62.7 and 1.1–31.3 per cent higher, respectively, than that under other cover crop treatments and significantly higher than the control (CK). However, the protease activity (Fig. 1c) under SC was significantly higher (32.6–56.8 per cent) than that under WC and CK treatments. The sucrase activity (Fig. 1d) in RC was the largest (13.5 and 15.2 mg g^{-1}) among the cover crop treatments.

Among the different soil enzymes, Hai-Ming et al. (2014) opined that arylamidase, alkaline phosphatase, β-glucosidase, and arylsulfatase are vital for the transformation of plant nutrients. In a study conducted by Hai-Ming et al. (2014), the authors reported that winter cover crop residue significantly increased the activities of all four enzymes, primarily due to the addition of organic matter to soils.

Herencia (2015) equally reported an increase in dehydrogenase enzyme activity in a study to elucidate the long-term effects of cover crops on organic carbon and soil enzyme activities at three different soil depths in a Mediterranean olive orchard, in southeast Spain. The increase in dehydrogenase activity in the cover cropping systems may be attributed to a potential increase in soil water content due to the propensity of cover crops to conserve soil moisture and enhance water retention. Low water availability has been reported to inhibit soil dehydrogenases by lowering intracellular water potential and by reducing hydration and dehydrogenase enzymes activity (Wall

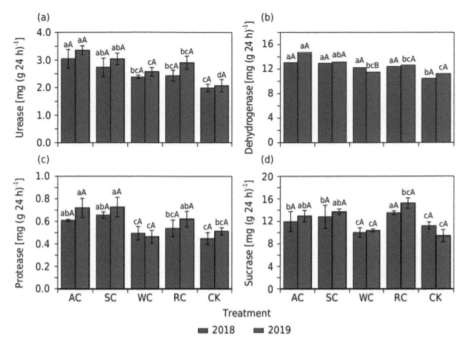

Fig. 1 Changes of soil urease activity (a), dehydrogenase activity (b), protease activity (c), and sucrase activity (d) under different treatments. AC: alfalfa cover; SC: sweet clover cover; WC: winter wheat cover; RC: ryegrass cover; CK: bare land. Different lowercase letters denote significant differences among the treatments within the same year (Duncan's new multiple range test; $p < 0.05$), different uppercase letters indicate significant differences among the years within the same treatment (t-test; $p < 0.05$). Bars represent the standard error (*Source*: Hai-Ming et al., 2014).

and Heiskanen, 2003). Moreover, the increased dehydrogenase activity under cover cropping systems may also be ascribed to increased organic matter content under cover cropping systems. In a study to determine the effect of a biculture of oat (*Avena sativa* L.) and grazing vetch (*Vicia dasycarpa* L.) cover crops on carbon pools and activities of selected enzymes in a loam soil under warm temperate conditions, Mukumbareza et al. (2016) reported an increase in urease enzyme activity. The increase in the urease enzyme activity may be ascribed to the fact that the cover crops potentially increase the activity of specific enzymes related to N utilisation (Huang et al., 2017) in the soil medium.

Effects of cover crops on soil enzyme activities are dependent on the cover crop species. Liang (2013) reported on the short-term effects of winter legumes on soil microbial activity and particulate organic matter (POM) at two separate experimental sites (Kinston vs. Goldboro). The effect of hairy vetch was found to be different from Austrian winter pea and crimson clover as shown by non-overlapping PCA scores (Fig. 2).

In this study, it was apparent that comparatively, Austrian winter pea and crimson clover positively affected the activities of β-glucosidase and β-glucosaminidase enzymes more than hairy vetch at the Kinston site. At the Goldsboro site, Austrian winter pea also affected more positively the activities of β-glucosidase and β-glucosaminidase, notwithstanding the fact that three cover crops were not separated completely (Fig. 2). Furthermore, Austrian winter pea tended to stimulate exoglycanase activity more positively than hairy vetch.

6. Cover Crops and Biological Carbon Cycling

Soil management stimulates bacterial biomass and activity and including cover crops in rotation increases the biologically active labile fractions of SOM. A significant increase in the labile C pool has been reported by several authors (Weil et al., 2003; Rosolem et al., 2016; Martínez-García et al., 2018) following the growing of cover crops. The more biologically active pools of SOM (microbial

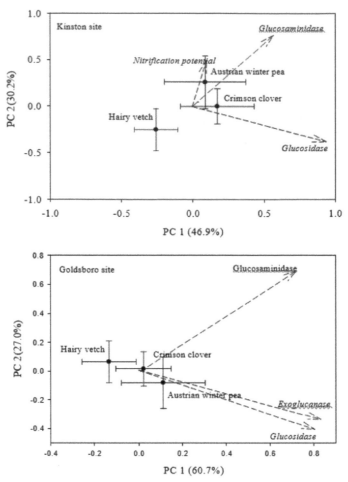

Fig. 2 Principal component analysis (PCA) showing the separation of cover crop species and the contribution of microbial properties. The loading scores along the first and second principal components were the mean values over sampling times, termination method, and field replicates. Bars represent standard errors (n = 18 for Kinston site and n = 45 for Goldsboro site) (*Source*: Liang, 2013).

biomass carbon) have been reported to be more sensitive to changes in soil quality by cover cropping (Schutter and Dick, 2002). Soil microbial biomass is the living component of total organic carbon (TOC) and plays a key role in nutrient cycling and the decomposition and transformation of TOC in the soil medium (Liang et al., 2011). The SMB also serves as a useful indicator of changes in soil C stabilisation and nutrient dynamics following soil management practices, such as cover cropping.

By increasing plant C inputs to the soil, cover crop root biomass increases the content and quality of TOC with a resultant increase in the active carbon fraction of the TOC. This indicates that a higher rate of SOM and nutrient cycling would accelerate the conversion of nutrients from organic to inorganic forms through mineralisation (Weil et al., 2003; Wang et al., 2015), possibly due to high biological activity in cover cropping systems.

7. Cover Crops and Mycorrhizal Associations

Cover crops are reported to provide a habitat for diverse microorganisms, such as bacteria and mycorrhizal fungi (Schmidt et al., 2018). Several authors have reported the positive effects of cover crops on growth, root colonisation and spore germination of mycorrhizal fungi (Iligo et al., 2018; Morimoto et al., 2018) as well as on the SMB activity and abundance (Hallama et al., 2019).

Hallama et al. (2018) determined several biological parameters – the abundance of arbuscular mycorrhizal fungus (AMF), microbial biomass P (P_{mic}) and on extracellular P-cycling enzymes (phosphatases) under diverse cover cropping systems (Fig. 3). Results showed that the abundance of AMF spores and root colonisation increased after mycorrhizal covercrop mixtures of *Fabaceae* and *Poaceae* but did not change or increase slightly under non-mycorrhizal cover crops of several species of *Brassicaceae* and Lupinus. Cover cropping significantly increased the P_{mic}; with *Poaceae*, *Fabaceae* and Lupinus spp. resulting in the highest increase (~ 25 per cent); however, the effect of *Poaceae* on P_{mic} was significant. Moreover, extracellular phosphatase enzyme activity increased by 20 per cent with cover cropping, with *Brassicaceae* having the lowest and the *Fabaceae*, lupins, and *Poaceae* having the highest effect when compared to the control (Fig. 3).

Further analysis of the results showed that cover crop effects on soil microbial parameters were influenced by soil P (Pa) status and availability, reflecting in a much higher increase in AMF abundance in soils with low Pa compared to soils with high Pa (Fig. 4). In general, under low availability of Pa, the cover crop benefit on mycorrhizal root colonisation was highest.

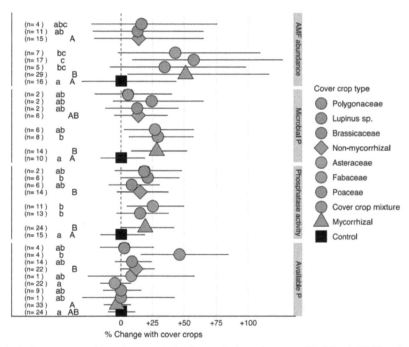

Fig. 3 Soil biological parameters: change in percent abundance of arbuscular mycorrhizal fungi (AMF), microbial biomass phosphorus (P) content, and phosphatase activity, as well as available P after different cover crops, relative to the respective controls. On the left are displayed the number of observations. The lower-case letters indicate, for a single main crop type with a Tukey post-hoc test ($p < 0.05$), significant differences among cover crop types (including the control), and the upper-case letters between mycorrhizal cover crops, nonmycorrhizal cover crops and the controls.

8. Conclusion

The effects of cover crops on soil biological properties cannot be overemphasised. Soil biological property is a critical component of the overall soil quality index and it is consistent, sensitive and considered as an early indicator of changes in soil quality long before changes are detected in the other soil quality indicator properties. The soil biological quality, among other properties, is greatly responsible for decomposition of organic materials, facilitating nutrient cycling in the soil, metabolising labile carbon pools and improving macro aggregation and structural stability. Therefore, an improvement in the soil biological properties using cover crops relates to enhancement

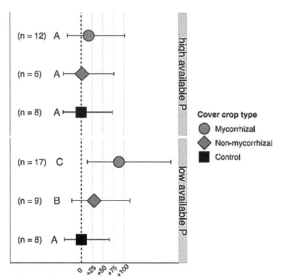

% Change of mycorrhizal abundance with cover crops

Fig. 4 Effect of cover crops on arbuscular mycorrhizal fungi (AMF) in soils with different P availability. The points represent the percentage change of the modeled median (+/– 95 per cent CI) of the cover crop treatments relative to the controls without cover crops. On the left are displayed the number of observations. The letters indicate significant differences among groups with a Tukey post-hoc test (p < 0.05).

in the soil's chemical and physical properties and subsequently, enhancement in the overall soil quality.

References

Basche, A.D., Kaspar, T.C., Archontoulis, S.V., Jaynes, D.B., Sauer, T.J., Parkin, T.B. and Miguez, F.E. (2016). Soil water improvements with the long-term use of a winter rye cover crop. Agric. Water Manage., 172: 40–50.

Benítez, E., Melgar, R., Sainz, H., Gómez, M. and Nogales, R. (2000). Enzyme activities in the rhizosphere of pepper (*Capsicum annuum*, L.) grown with olive cake mulches. Soil Biol. Biochem., 32: 1829–1835.

Brennan, E.B. and Acosta-Martinez, V. (2018). Soil microbial biomass and enzyme data after six years of cover crop and compost treatments in organic vegetable production. Data in Brief, 21: 212–227.

Buyer, J.S., Teasdale, J.R., Roberts, D.P., Zasada, I.A. and Maul, J.E. (2010). Factors affecting soil microbial community structure in tomato cropping systems. Soil Biol. Biochem., 42: 831–841.

Dahiya, R., Ingwersen, J. and Streck, T. (2007). The effect of mulching and tillage on the water and temperature regimes of a loess soil: Experimental findings and modelling. Soil Tillage Res., 96: 52–63.

Dai, Z., Barberán, A., Li, Y., Brookes, P.C. and Xu, J. (2017). Bacterial community composition associated with pyrogenic organic matter (biochar) varies with pyrolysis temperature and colonisation environment. MSphere, 2(2): e00085-17.

Daigh, A.L., Helmers, M.J., Kladivko, E., Zhou, X., Goeken, R., Cavdini, J. and Sawyer, J. (2014). Soil water during the drought of 2012 as affected by rye cover crops in fields in Iowa and Indiana. J. Soil Water Conser., 69: 564–573.

de Graaff, M.A., Classen, A.T., Castro, H.F. and Schadt, C.W. (2010). Labile soil carbon inputs mediate the soil microbial community composition and plant residue decomposition rates. New Phytologist, 188: 1055–1064.

Emerson, W. (1995). Water-retention, organic-C and soil texture. Soil Res., 33: 241–251.

Friedel, J.K. and Gabel, D. (2001). Nitrogen pools and turnover in arable soils under different durations of organic farming: I: Pool sizes of total soil nitrogen, microbial biomass nitrogen, and potentially mineralizable nitrogen. J. Plant Nutr. Soil Sci., 164: 415–419.

Galletti, S., Sala, E., Leoni, O., Burzi, P.L. and Cerato, C. (2008). Trichoderma spp. tolerance to *Brassica carinata* seed meal for a combined use in biofumigation. Biol. Control, 45: 319–327. doi:10.1016/j.biocontrol.2008.01.014.

Gallo, M.E., Sinsabaugh, R.L. and Cabaniss, S.E. (2006). The role of ultraviolet radiation in litter decomposition in arid ecosystems. Appl. Soil Ecol., 34: 82–91.

Gu, C., Zhang, S., Han, P., Hu, X., Xie, L., Li, Y. and Qin, L. (2019). Soil enzyme activity in soils subjected to flooding and the effect on nitrogen and phosphorus uptake by oilseed rape. Frontiers in Plant Sci., 10: 368.

Hai-Ming, T., Xiao-Ping, X., Wen-Guang, T., Ye-Chun, L., Ke, W. and Guang-Li, Y. (2014). Effects of winter cover crops residue returning on soil enzyme activities and soil microbial community in double-cropping rice fields. PLoS ONE, 9: e100443–e100443.

Hallama, M., Pekrun, C., Lambers, H. and Kandeler, E. (2019). Hidden miners—The roles of cover crops and soil microorganisms in phosphorus cycling through agro-ecosystems. Plant Soil, 434: 7–45.

Herencia, J.F. (2015). Enzymatic activities under different cover crop management in a Mediterranean olive orchard. Biol. Agric. Hort., 31: 45–52.

Hetrick, B.A.D., Kitt, D.G. and Wilson, G.T. (1988). Mycorrhizal dependence and growth habit of warm-season and cool-season tallgrass prairie plants. Can. J. Bot., 66: 1376–1380.

Higo, M., Sato, R., Serizawa, A., Takahashi, Y., Gunji, K., Tatewaki, Y. and Isobe, K. (2018). Can phosphorus application and cover cropping alter arbuscular mycorrhizal fungal communities and soybean performance after a five-year phosphorus-unfertilized crop rotational system? Peer J., 6: e4606–e4606.

Holland, T.C., Bowen, P.A., Bogdanoff, C.P., Lowery, T.D., Shaposhnikova, O., Smith, S. and Hart, M.M. (2016). Evaluating the diversity of soil microbial communities in vineyards relative to adjacent native ecosystems. Appl. Soil Ecol., 100: 91–103. doi:10.1016/j.apsoil.2015.12.001.

Huang, D., Liu, L., Zeng, G., Xu, P., Huang, C., Deng, L. and Wan, J. (2017). The effects of rice straw biochar on indigenous microbial community and enzymes activity in heavy metal-contaminated sediment. Chemosphere, 174: 545–553.

Kim, N., Zabaloy, M.C., Guan, K. and Villamila, M.B. (2020). Do cover crops benefit soil microbiome? A meta-analysis of current research. Soil Biol. Biochem., 142: 107701.

Klabi, R., Bell, T.H., Hamel, C., Iwaasa, A., Schellenberg, M., Raies, A. and St-Arnaud, M. (2015). Plant assemblage composition and soil P concentration differentially affect communities of AM and total fungi in a semi-arid grassland. FEMS Microbiol. Ecol., 91: 1–13. doi:10.1093/femsec/fiu015.

Larkin, R.P., Griffin, T.S. and Honeycutt, C.W. (2010). Rotation and cover crop effects on soilborne potato diseases, tuber yield, and soil microbial communities. Plant Dis., 94: 1491–1502. doi:10.1094/pdis-03-10-0172.

Liang, B., Yang, X., He, X. and Zhou, J. (2011). Effects of 17-year fertilisation on soil microbial biomass C and N and soluble organic C and N in loessial soil during maize growth. Biol. Fert. Soils, 47: 121–128.

Maeder, P., Fliessbach, A., Dubois, D., Gunst, L., Fried, P. and Niggli, U. (2002). Soil fertility and biodiversity in organic farming. Science, 296(5573): 1694–1697.

Malik, A.A., Puissant, J., Buckeridge, K.M., Goodall, T., Jehmlich, N., Chowdhury, S. and Griffiths, R.I. (2018). Land use driven change in soil pH affects microbial carbon cycling processes. Nature Communications, 9: 3591–3591.

Martínez-García, L.B., Korthals, G., Brussaard, L., Jørgensen, H.B. and Deyn, G.B.D. (2018). Organic management and cover crop species steer soil microbial community structure and functionality along with soil organic matter properties. Agric. Ecosys. Environ., 263: 7–17.

Maron, J.L., Marler, M., Klironomos, J.N. and Cleveland, C.C. (2011). Soil fungal pathogens and the relationship between plant diversity and productivity. Ecol. Lett., 14: 36–41. doi:10.1111/j.1461-0248.2010.01547.x.

Mazzola, M., Hewavitharana, S.S. and Strauss, S.L. (2015). *Brassica* seed meal soil amendments transform the rhizosphere microbiome and improve apple production through resistance to pathogen reinfestation. Phytopathology, 105: 460–469. doi:10.1094/phyto-09-14-0247-r.

Meyling, N.V., Lubeck, M., Buckley, E.P., Eilenberg, J. and Rehner, S.A. (2009). Community composition, host range and genetic structure of the fungal entomopathogen Beauveria in adjoining agricultural and seminatural habitats. Mol. Ecol., 18: 1282–1293. doi:10.1111/j.1365-294X.2009.04095.x.

Morimoto, S., Uchida, T., Matsunami, H. and Kobayashi, H. (2018). Effect of winter wheat cover cropping with no-till cultivation on the community structure of arbuscular mycorrhizal fungi colonising the subsequent soybean. Soil Sci. Plant Nutr., 64: 545–553.

Mukumbareza, C., Muchaonyerwa, P. and Chiduza, C. (2016). Bicultures of oat (*Avena sativa* L.) and grazing vetch (*Vicia dasycarpa* L.) cover crops increase contents of carbon pools and activities of selected enzymes in a loam soil under warm temperate conditions. Soil Sci. Plant Nutr., 62: 447–455.

Nair, A. and Ngouajio, M. (2012). Soil microbial biomass, functional microbial diversity, and nematode community structure as affected by cover crops and compost in an organic vegetable production system. Appl. Soil Ecol., 58: 45–55.

Nannipieri, P., Kandeler, E. and Ruggiero, P. (2002). Enzyme activities and microbiological and biochemical processes in soil. *In*: Burns, R.G. and Dick, R.P. (eds.). Enzymes in the Environment. Activity, Ecology and Applications. Dekker, New York.

Ramirez-Villanueva, D.A., Bello-López, J.M., Navarro-Noya, Y.E., Luna-Guido, M., Verhulst, N., Govaerts, B. and Dendooven, L. (2015). Bacterial community structure in maize residue amended soil with contrasting management practices. Appl. Soil Ecol., 90: 49–59.

Rosolem, C.A., Li, Y. and Garcia, R.A. (2016). Soil carbon as affected by cover crops under no-till under tropical climate. Soil Use Manage., 32: 495–503.

Schmidt, R., Gravuer, K., Bossange, A.V., Mitchell, J. and Scow, K. (2018). Long-term use of cover crops and no-till shift soil microbial community life strategies in agricultural soil. PLoS ONE, 13: e0192953–e0192953.

Schutter, M.E. and Dick, R.P. (2002). Microbial Community Profiles and Activities among Aggregates of Winter Fallow and Cover-Cropped Soil, published as Paper No. 11590 of the Oregon Agric. Exp. Station., Oregon State Univ., Corvallis, OR.

Shapiro-Ilan, D.I., Gardner, W.A., Wells, L. and Wood, B.W. (2012). Cumulative impact of a clover cover crop on the persistence and efficacy of Beauveria bassiana in suppressing the pecan weevil (Coleoptera: Curculionidae). Environ. Entomol., 41: 298–307. doi:10.1603/en11229.

Shelton, R.E., Jacobsen, K.L. and McCulley, R.L. (2018). Cover crops and fertilisation alter nitrogen loss in organic and conventional conservation agriculture systems. Frontiers in Plant Sci., 8: 2260–2260.

Singh, B.K., Bardgett, R.D., Smith, P. and Reay, D.S. (2010). Microorganisms and climate change: terrestrial feedbacks and mitigation options. Nature Reviews Microbiol., 8: 779.

Six, J., Bossuyt, H., Degryze, S. and Denef, K. (2004). A history of research on the link between (micro)aggregates, soil biota and soil organic matter dynamics. Soil Till. Res., 79: 7–31.

Wall, A. and Heiskanen, J. (2003). Water-retention characteristics and related physical properties of soil on afforested agricultural land in Finland. Forest Ecol. Manage., 186: 21–32.

Wang, X., Song, D., Liang, G., Zhang, Q., Ai, C. and Zhou, W. (2015). Maize biochar addition rate influences soil enzyme activity and microbial community composition in a fluvo-aquic soil. Appl. Soil Ecol., 96: 265–272.

Wang, W., Han, L. and Zhang, X. (2018). Winter cover crops effects on soil microbial characteristics in sandy areas of northern Shaanxi, China. Rev Bras Cienc Solo, 2020: 44: e0190173.

Weil, R.R., Islam, K.R., Stine, M.A., Gruver, J.B. and Sampson-Liebig, S.E. (2003). Estimating active carbon for soil quality assessment: A simplified method for laboratory and field use. Amer. J. Altern. Agric., 18: 3–17.

Williams, A., Jordan, N.R., Smith, R.G., Hunter, M.C., Kammerer, M., Kane, D.A. and Davis, A.S. (2018). A regionally-adapted implementation of conservation agriculture delivers rapid improvements to soil properties associated with crop yield stability. Scientific Reports, 8: 8467.

Zhou, X., Chen, C., Lu, S., Rui, Y., Wu, H. and Xu, Z. (2012). The short-term cover crops increase soil labile organic carbon in southeastern Australia. Biol. Fert. Soils, 48: 239–244.

18

Economics of Cover Crops

Mohammad S Rahman[1] and *James J Hoorman*[2,]*

1. Introduction

A cover crop is a secondary crop grown for the purpose of soil protection and enrichment. Generally, a cover crop is grown in the off-season before the field is needed for growing the main crop. In essence, a cover crop readies the land for an incoming cash crop. Cover crops are commonly used to provide organic matter and recycle nutrients for plants, suppress weeds, control diseases and pests, act as surface mulch, control soil erosion, help to develop and/or restore soil quality, promote biodiversity and bio-efficiency and improve crop yield. Cover crops are typically grass or legume species, but may comprise of other plants, such as *brassicas*, buckwheat, etc.

Cover crops began being referred to as 'green manures' within the last 50 years or so. In fact, it was under research conducted as late as the 1990s that researchers really started to pinpoint what cover crops are and what they can do for producers (Clark, 2007). Although cover crops are generally used for multiple purposes, as discussed above, overall they improve soil quality and crop productivity and consequently, improve farm economy. Most importantly, the effect of cover crops indirectly influences farm economics by decreasing inputs, improving soil quality and increasing crop productivity over time.

2. Cover Cropping and Benefits

Many farmers ask, "What is the value of planting cover crops on my farm?" This is a common question with many answers. Cover crops have many benefits and uses, so the answer varies according to farm field and farming operation. They have value, even if main crop yields do not immediately increase because they may reduce some pests (weeds, insects, diseases), decrease fertility costs, improve soil structure, decrease soil erosion and nutrient loss, and can be used for grazing or forage. Many cover crop benefits accrue over time, so changes may be difficult to see at first or measure numerically.

One of the biggest benefits to cover crops is the accumulation of soil organic matter (SOM). Each 1% of SOM addition is worth between $500–600 in terms of soil fertility value. A typical cover crop adds 0.1–0.15% SOM annually valued at $50 to $90/acre. Cover crops add roots and surface cover that greatly increase SOM and improve soil fertility. Calculations done by Myers et al. (2019) showed that cover crop seed costs range between $10–50/acre, planting from $5–18, and termination from $0–10, with a cover crop cost range between $15–78 and an average of

[1] Dept. of Agricultural Economics, Bangladesh Agricultural University, Mymensingh, Bangladesh.
[2] Former Ohio State University Extension Educator and Assistant Professor. Former USDA-NRCS Regional Soil Health Specialist. Private Consultant for Hoorman Soil Health Services, 22133 TR 60, Jenera, Ohio 45841.
* Corresponding author: hoormansoilhealthservices@gmail.com

Table 1 Cost of seedling cover crops (*Source*: Myers et al., 2019).

Item	Cost/acre
Cover crop seed	$10–$50
Seeding of cover crops	$5–$18
Termination	$0–$50
Subtotal range	$15–$78
Median cost from survey	$37.

$37/acre (Table 1). The yearly increase in SOM is enough to pay for the cover crop seed, planting and termination, especially on no-till farms (Hoorman, 2020).

Some of the reduced costs include out-competing hardy weeds, like mare's tail, water hemp, giant ragweed and pigweed species, like Palmer amaranth, saving $27/acre in reduced herbicide costs. In no-till systems, yearly reductions average $24/acre in reduced tillage cost and $15/acre in compaction cost. For farmers with livestock, cover cropping can be grazed or used as forage, adding $50/acre on an average. It is worthwhile noting that federal and state cover crop payments may range between $50–60/acre, depending on the programme. Cover crops can greatly improve the bottom line, depending on the farming operation (Hoorman, 2020).

One of the biggest savings from cover crops comes in the form of reduced soil erosion and edge-of-the-field nutrient runoff. The average Ohio soil erosion rate is 2.62 tons/acre/year; with cover crops, this loss could be down to hundreds of pounds versus tons of topsoil lost/acre/year. Assuming 2.5 tons saved at $10 per ton, farmers save $25 per acre in topsoil. Levelling soil ruts and fixing erosion problems average at $24/acre. Cover crops reduce nitrogen (N) losses from fertilizer or manure by an average of 48%. The fertility savings in keeping nutrients on the land and out of surface water is at least another $15 per acre. Legumes, as cover crops, while more expensive, may add 75–200 lbs of additional slow-release N at $0.40/lbs of N worth $30 to $80/acre. Thus, cover crops improve farm economy by increasing water infiltration, reducing nutrient loss, adding SOM and fertility and improving water quality (Hoorman, 2020).

The biggest increase in crop yields impacted by cover crops occurs in dry years due to improved water infiltration and moisture storage, averaging 9.6% on corn and 11.6% on soybeans, with added income ranging between $59–93/acre/farm. The benefits accumulate and increase as soil health is restored and becomes more resilient over time. In wet years, the crop yield increase may only average 1.9% for corn and 2.8% for soybeans, which may only be enough to pay for the cover crop seed. The average yield increase from 2012–2016 was 3.6% for corn and 5.3% for soybeans (Myers et al., 2019; Hoorman, 2020).

If cover crops are used consistently, on an average, they start paying off after around three years. It generally takes three years as a transition period to begin seeing soil quality improvements from stable soil structure, improved water infiltration and moisture retention, higher SOM content, greater biodiversity and bio-efficiency and improved soil fertility. Why would a farmer invest in cover crops if it takes three years to see a significant difference? The same question can be asked about tile. Systematic tile costs about $1,000/acre and may take 10 years to pay off. Like tile, cover crop usage is a long-term investment in your soil. Some farmers have found they can even reduce splitting or adding more tile by planting cover crops. The added roots increase water infiltration and drainage to make the existing tile work more efficient, so that new tile may not be needed (Hoorman, 2020).

Depending upon farm-soil problems, cover crops can start paying off immediately or the returns may be slower (two to five years) but cover crops do pay off in the long-run. The USDA Census of Agriculture says cover crop acreage increased by 50% nationally between 2012–2017. If cover crops did not pay or return, farmers would not be planting cover crops. Several examples of using cover crops and their benefits on crop growth and farm economics are discussed in the following sections.

2.1 Cover Crop Benefits on Potato Yield, Soil Quality and Farm Economics

Cover crop use as green manure is now quite popular in diverse crop production systems. Cover crop usage in commercial potato production is helpful in alleviating the severe soil erosion recorded in Mediterranean-climate agriculture. In a study, Blanco-Canqui et al. (2015) oats, triticale, oats, purple vetch, clover and rapeseed were used as cover crop blends and their effect on potato production in Mediterranean area, particularly in Israel, was examined.

Soil loss reduced by 97% while potato yield and quality were retained (Blanco-Canqui et al., 2015). Crop infection by pests also decreased, thus improving potato yield and quality. When the soil was completely covered with cover crops, runoff and soil erosion were not observed. The more abundant the ground cover, the more efficient the control of runoff and erosion. The incorporated cover crops (oat, mixture of oat and purple vetch, rapeseed) decreased soil erosion by 95%. It minimized runoff by more than 60% and saved 13% of variable cost incurred in potato production (Blanco-Canqui et al., 2015). The highest benefit from using cover crops to control runoff was observed with oats and the oat-hairy vetch blends. The number of weed species and their biomass were lower in the plots of potato where oats were harvested before planting potato as compared to the conventional method (bare-soil). Oat was the most efficient economic cover crop with respect to weed suppression in the potato field.

Some financial investment is required when using cover crops—these include seeds, labor for planting and growing and additional irrigation for cover crops seed germination. However, the average net cost for growing potatoes under cover crop practices was 1.3% lower than the total cost required for growing potatoes under conventional agriculture, whereas water for cooling the soil and uncontrolled soil erosion imposed relatively higher costs. Additional savings come in the form of irrigation costs due to the ability of the cover crop to conserve soil moisture and in direct infrastructure, as cover crops can prevent soil-erosion damage, such as road and waterway canal destruction and eliminate the cost of returning soil to the fields. However, it is reported that cover crops have no significant impact on crop yield. The only benefits to the grower from the adoption of cover crop cultivation is an estimated 1.3% cost savings.

2.2 Cover Crops (Wheat-Barley-Canola) Benefits on Alfalfa (Lucerne) Forage

Alfalfa (Lucerne) is cultivated worldwide as a cover crop and green manure, as well as an important forage crop. It is used for grazing, hay and silage. There are two methods of planting alfalfa with other species of cover crops: (1) planting alfalfa with wheat, barley and canola during the cropping phase and (2) planting alfalfa alone, following the cover crop. Cultural practices' expense is less to plant alfalfa with the final crop as in that, it can provide full income from the sale of grain, but this practice can reduce pasture quantity and quality in very dry or wet years. Field research shows that planting alfalfa alone is the most reliable method to establish a pasture in farm areas, and it increases the financial benefit of the farm.

The cost of planting alfalfa alone is $204/ha, whereas with the final crop in rotation with cover crops, it adds only $55/ha costs. Nordblom et al. (2017) reported that crop establishment costs are already committed by cover crops (wheat/barley/canola), so by planting alfalfa seeds, farmers can establish a 'lowest cost' pasture. Supplementary feeding costs for sheep enterprise were minimal at low stocking rates because the pasture produces sufficient energy. Expense is less to plant alfalfa with the final crop, but the yield is less because both are competing for space, water, light and nutrients. In response to the competition with the cover crops, the performance of alfalfa to produce biomass for forage is generally reduced because alfalfa and cover crops (wheat/barley/canola) are cash crops and these require much more water and nutrients.

2.3 Cover Crop Impacts on Cotton Production Economics

Varco et al. (1999) evaluated the long-term effects of winter cover crops (cereal rye vs. hairy vetch) on net return for upland cotton production. Results showed that cover crops increased the cost of production, but hairy vetch, as one of the legume cover crops, minimized the N fertilization cost by complementing N for cotton through biological N-fixation. Hairy vetch improves profit by maximizing N rate relative to no-cover crop. Both cover crop systems were found to be more profitable at all fertilizer N rates and lint prices, with rye averaging $24–44/acre and hairy vetch $33–51/acre greater returns than for winter fallow. Marginal profitability of rye, compared to winter fallow, was greatest at a fertilizer N-to-lint price ratio of 0.125, while for vetch it was at 0.375. Results also showed that average net returns were most sensitive to changes in lint prices and were greater for the winter cover crop treatments when compared to the winter fallow (Table 2). Cotton, after rye and hairy vetch, resulted in similar net returns at low fertilizer and lint prices. However, as the price of fertilizer N increased, cotton after hairy vetch gained an advantage as compared to cotton following rye, especially at high lint prices.

Table 2 Effects of winter cover management on average per-acre returns above specified costs at the profit-maximising fertilizer N rate for no-tillage cotton at selected net lint and fertilizer N prices, Lowndes County, Mississippi, 1989–1992.

	N price, $/lb				
Net Lint Price	**0.10**	**0.15**	**0.20**	**0.25**	**0.30**
$/lb			**$/acre**		
Fallow					
0.6	207	203	198	194	190
0.7	284	279	275	270	266
0.8	360	355	351	346	342
Rye					
0.6	237	231	225	220	214
0.7	320	314	309	303	298
0.8	404	398	392	387	381
Hairy vetch					
0.6	235	232	229	226	223
0.7	320	317	314	311	308
0.8	406	402	399	396	393

Source: Varco et al. (1999)

2.4 Cover Crops' Effect on Potato Production and Farm Economics

Several studies conducted by Hamzaev et al. (2007) aimed to evaluate the impact of cover crops such as barley, pea, radish, rape, pea/barley, and pea/radish compared to chemical fertilization (as a control) on irrigated potato production in Samarkand, Uzbekistan. Results showed that potato yields were significantly higher under cover crop amendments compared to the control. Pea had the highest beneficial effects on potato yields followed by pea/radish and pea/barley in mixed cropping (Table 3). Compared to the control, the greatest potato yields (31 Mg/ha) were obtained when using pea as a cover crop and the lowest yield difference (23.8 Mg/ha) was obtained when barley was used as a cover crop. Pea,when mixed with radish or barley, significantly increased (29 to 30 Mg/ha) potato yields; however, the use of radish as a cover crop resulted in a small increase in yields compared to the control.

The total and average costs of potato production did not vary significantly among the cover crop treatments (Hamzaev et al., 2007); however, the total market price of potatoes was higher when pea alone was used as a cover crop or when pea was mixed with radish or barley, as opposed to the control (Table 3). The higher returns were generated by successive use of pea as a cover crop in pure stands or in blends with radish and barley. Economic efficiencies were 20 to 30% higher

Table 3 Economic efficiency of cover crops as soil amendments for irrigated potato production in Uzbekistan (*Source*: Hamzaev et al., 2007).

Cover Crops	Potato Yield (Mg/ha)	Production Cost Average ($/Mg)	Production Cost Total ($/ha)	Market Price Average ($/Mg)	Market Price Total ($/ha)	Profit ($/ha)	Economic Efficiency (%)*	Net Efficiency (%)**
Control	21.8	38.7	842.7	63.0	1382.9	540.1	64.1	0
Barley	23.8	39.2	932.7	63.0	1502.6	569.9	61.1	−3
Pea	31.1	32.4	1007.9	63.0	1959.3	951.4	94.3	30.2
Pea + barley	28.8	34.2	984.3	63.0	1814.4	830.1	84.3	20.2
Pea + radish	30.1	33.2	998.8	63.0	1896.3	897.5	89.8	25.7
Radish	25.8	36.9	953.2	63.0	1628.6	675.3	70.8	6.7
Rape	27.1	37.3	1011.1	63.0	1710.5	699.4	69.4	5.3
$LSD_{p \leq 0.05}$***	2.62	Ns	ns	63.0	ns	353.9	18.4	24.1
Orthogonal linear contrast								
Crop vs. control	6.1	−3.2	138.7	-----	36.9	230.5	14.2	14.2
T-test (p)	0.013	ns	0.004	-----	0.013	0.045	0.017	0.015

*   Efficiency = (total market price − total production cost) × 100/total production cost.
** Net efficiency = (efficiency of individual cover crops − efficiency of control treatment).
*** LSD = least significant difference.

when potatoes were grown as a cash crop only after pea was used, or when used in combination with radish or barley to improve soil quality. A linear contrast between the effects of chemical fertilization and cover crops shows that cover crops, as amendments, significantly increased potato yields (about 28%), profit levels (about 43%) and economic efficiency levels (about 22%) when compared to the control (Table 4). Significant positive effects of cover crops on potato yield and farm economics were possibly due to the incorporation of large amounts of labile and nutrient-enriched residues via cover crops, which resulted in soil quality improvement.

2.5 SARE/CTIC Studies on Economics of Cover Crops on Corn-Soybean Systems

With cover crops as one of the core components of the sustainable agricultural management practices that are becoming popular throughout the United States, the economic benefits are not generally obvious immediately (Myers et al., 2019). As expected, a short-term budget analysis of cover crops on crop yields may indeed show an economic loss when compared to the results of the long-term studies (Tables 4 and 5).

To better understand how the economics of cover crops change over time with reductions of input costs and improvements in soil health and crop yields, economic data on corn and soybean production were compiled from a variety of data sources (Myers et al., 2019). The numbers used in the tables are based on a combination of Sustainable Agricultural Research and Extension (SARE)/ Conservation Tillage Information Centre (CTIC) survey data, published input prices and research data (Myers et al., 2019). However, prices shown are from spring 2019, unless otherwise mentioned. Much of the baseline economic information that underlies the financial analysis of these situations is derived from five years of data from the US National Cover Crop Survey conducted by SARE/CTIC for the 2012–2016 growing seasons. In this survey, farmer profiles share real-world examples of how the diverse and cumulative benefits of cover crops translate into economics of farm profitability. Data presented in Tables 4 and 5 show the impact of cover crops on corn-and-soybean-based farm economics under each of the seven situations.

Results presented in Table 4 show that, when used, cover crops in a corn system saved fertilizer cost by $14.10 per acre after three years as compared to an outstanding cumulative savings of

Table 4 Impact of cover crops on costs, returns and net profit for corn following one, three and five years of cover crops use and with various management scenarios.

Budget Item	Years of Cover Cropping		
All figures are per acre	One	Three	Five
Estimated input savings when using cover crops			
Fertilizer[1]	$0	$14.10	$21.90
Weed control[2]	$0–$15	$10–$25	$10–$25
Erosion repair[3]	$2–$4	$2–$4	$2–$4
Subtotal	**$2–$19**	**$26.10–$43.10**	**$33.90–$50.90**
a. Savings on inputs (the low end of the subtotal range from above)	$2	$26.10	$33.90
b. Income from extra yield in normal weather year (survey data)[4]	$3.64	$12.32	$21
c. Cost of seed and seeding (survey data)[5]	$37	$37	$37
Net return in a normal weather year (a + b − c)	**−$31.36**	**$1.42**	**$17.90**
Special situations where cover crops can pay off faster			
I. When facing severe herbicide-resistant weeds[6]	$27	$27	$27
Adjusted net return	**−$4.36**	**$28.42**	**$44.90**
II. Potential grazing income[7]	$49.23	$49.23	$49.23
Adjusted net return	**$17.87**	**$50.65**	**$67.13**
III. Compaction addresses by cover crops[8]	$15.30	$15.30	$15.30
Adjusted net return	**−$16.06**	**$16.72**	**$33.20**
IV. Assisting the conversion to no-till from conventional[9]	$23.96	$23.96	$23.96
Adjusted net return	**−$7.40**	**$25.38**	**$41.86**
V. Income from extra yield in a drought year (survey data)[10]	$58.70	$75.73	$92.55
Adjusted net return	**$27.34**	**$77.15**	**$110.45**
VI. Extra fertilizer savings from improved fertility[11]	$15.20	$15.20	$15.20
Adjusted net return	**−$16.16**	**$16.62**	**$33.10**
VII. Federal or state incentive payments received[12]	$50	$50	$50
Adjusted net return	**$18.64**	**$51.42**	**$67.90**

[1] Assumes no fertilizer savings in year one, then a savings of 15 pounds of nitrogen per acre in year three and 30 pounds per acre in year five, at $0.38 per pound. Also assumes a phosphorus saving of 20 pounds per acre in year three and 25 pounds per acre in year five, at $0.42 per pound.

[2] The first year assumes a reduction of one herbicide pass if sufficient cover crop biomass is achieved. Savings are higher in later years due to reducing by two passes or by using less-expensive herbicide products.

[3] Based on the cost of machinery operations and labor to repair gullies and clean ditches (assumes average cost, but fields will vary).

[4] Assumes a corn price of $3.50 per bushel and a 200-bushel yield times the percent yield increases shown in Table 2.

[5] Costs for seed, seeding and termination can vary from a low of about $10 to over $50 per acre; most farms estimated to be $25–$40 per acre.

[6] In a field with a severe herbicide-resistant weed infestation, this figure assumes that a thick-biomass cover crop will reduce herbicide and labor costs and will reduce dockage for weed seed at harvest.

[7] Assumes that grazing a cover crop (cereal rye in this example) results in a reduction of 1,093 pounds of jay fed per acre of cover crops. This is based on 1,500 pounds per acre of dry matter generated by rye, then reduced effective use of the rye by 50% due to hoof action and selective grazing. Assumes average feedlot waste of 22% for hay fed (88% dry matter). The hay is valued at $80 per ton. Additional savings of approximately $5.50 per acre generated due to lower labor, fuel and machinery depreciation from reduced hay fed. Assumes grazer already has water access for their grazing area and an electric fencing system.

[8] This is based on a University of Minnesota machinery cost estimate for subsoiling at $15.30 per acre (2017 data used for machinery costs).

[9] No-till savings versus conventional: No fall chisel plow ($11.22 per acre) and savings on two field cultivator passes in the spring (2 × $6.37 per acre).

Table 4 Contd. ...

...Table 4 Contd.

[10] Assumes a corn price in drought of $6.89 per bushel and reduced base yield of 142 bushels per acre × percent yield increase for drought. Numbers are based on actual national average corn yield for 2012 and national average corn price in the 2012–13 marketing year (USDA-NASS).

[11] Assumes using legumes as a cover crop and that overall improved soil health allow nitrogen to be cut by an extra 40 pounds per acre over basic fertilizer savings.

[12] The basic NRCS EQIP rate in the majority of Corn Belt states starts at $50 per acre or higher, some states have lower rates.

Table 5 Impact of cover crops on costs, returns and net profit for soybean following one, three and five years of cover crop use and with various management scenarios.

Budget Item	Years of Cover Cropping		
All figures are per acre	One	Three	Five
Estimated input savings when using cover crops			
Fertilizer[1]	$0	$6.30	$8.40
Weed control[2]	$0–$15	$10–$25	$10–$25
Erosion repair[3]	$2–$4	$2–$4	$2–$4
Subtotal	**$2–$19**	**$18.30–$35.30**	**$20.40–$37.40**
a. Savings on inputs (the low end of the subtotal range from above)	$2	$18.30	$20.40
b. Income from extra yield in normal weather year (survey data)[4]	$11.45	$19.12	$26.73
c. Cost of seed and seeding (survey data)[5]	$37	$37	$37
Net return in a normal weather year (a + b – c)	**–$23.55**	**$0.42**	**$10.13**
Special situations where cover crops can pay off faster			
I. When facing severe herbicide-resistant weeds[6]	$27	$27	$27
Adjusted net return	**$3.45**	**$27.42**	**$37.13**
II. Potential grazing income[7]	$49.23	$49.23	$49.23
Adjusted net return	**$25.68**	**$49.65**	**$59.41**
III. Compaction addresses by cover crops[8]	$15.30	$15.30	$15.30
Adjusted net return	**–$8.25**	**$15.72**	**$25.43**
IV. Assisting the conversion to no-till from conventional[9]	$23.96	$23.96	$23.96
Adjusted net return	**$0.41**	**$24.38**	**$34.34**
V. Income from extra yield in a drought year (survey data)[10]	$65.24	$69.80	$74.35
Adjusted net return	**$41.69**	**$70.22**	**$84.54**
VI. Extra fertilizer savings from improved fertility[11]	$7	$7	$7
Adjusted net return	**–$16.55**	**$7.42**	**$17.13**
VII. Federal or state incentive payments received[12]	$50	$50	$50
Adjusted net return	**$26.45**	**$50.42**	**$60.13**

[1] Assumes no fertilizer savings in year one, then a savings of 15 pounds of phosphorus per acre in year three and 20 pounds per acre in year five, at $0.42 per pound.

[2] The first year assumes either no herbicide savings or a possible saving of $15 per acre by avoiding a fall herbicide pass ($7.50 per acre for the chemical and $7.50 per acre for application). The third and fifth years assume using a less expensive residual chemistry that costs $10 per acre, with the possibility of saving $15 per acre in the fall.

[3] Based on the cost of machinery operations and labor to repair gullies and clean ditches (assumes average cost, but fields will vary).

[4] Assumes a soybean price of $9 per bushel and a 60-bushel yield times the percent yield increases shown in Table 2.

[5] Costs for seed, seeding and termination can vary from a low of about $10 to over $50 per acre; most farms estimated to be $25–$40 per acre.

[6] In a field with a severe herbicide-resistant weed infestation, this figure assumes that a thick-biomass cover crop will reduce herbicide and labor costs and will reduce dockage for weed seed at harvest.

Table 5 Contd. ...

...Table 5 Contd.

[7] Assumes that grazing a cover crop (cereal rye in this example) results in a reduction of 1,093 pounds of jay fed per acre of cover crops. This is based on 1,500 pounds per acre of dry matter generated by rye, then reduced effective use of the rye by 50% due to hoof action and selective grazing. Assumes average feedlot waste of 22% for hay fed (88% dry matter). The hay is valued at $80 per ton. Additional savings of approximately $5.50 per acre generated due to lower labor, fuel and machinery depreciation from reduced hayfed. Assumes grazer already has water access for their grazing area and an electric fencing system.

[8] This is based on a University of Minnesota machinery cost estimate for subsoiling at $15.30 per acre (2017 data used for machinery costs).

[9] No-till savings versus conventional: No fall chisel plow ($11.22 per acre) and savings on two field cultivator passes in the spring (2 × $6.37 per acre).

[10] Assumes a soybean price in drought of $14.40 per bushel and reduced yield of 39.6 bushels per acre × percent yield increase for drought. Numbers are based on actual national average soybean yield for 2012 and national average price in the 2012–13 marketing year (USDA-NASS).

[11] Assumes that overall improved soil health allows an additional reduction in phosphorus of 10 pounds per acre ($0.42 per pound) and 10 pounds per acre of potassium ($0.28 per pound) over basic fertilizer savings.

[12] The basic NRCS EQIP rate in the majority of Corn Belt states starts at $50 per acre or higher, some states have lower rates.

$21.90 per acre after five years. Weed control savings was $0–15 per acre in the first year than that of $10–25 per acre over a period of three to five years. Net return to a normal year was estimated at $17.90 per acre after five years; however, under variable situations, the estimated net economic return varied from $18.64 per acre after one year, $51.42 per acre after three years and $67.90 after five years of cover cropping. Data presented in Table 5 showed that cover cropping in a soybean system produced a slightly less than estimated net return of $0.42–10.13 per acre after three and five years, respectively. Under diverse situations, the estimated net economic returns ranged from $26.45, $50.42, and $60.13 after one, three and five years of cover cropping, respectively.

3. Rulon Farm Enterprises LLC Evaluation on Cover Crop Economic Benefits

Looker (2018), based on his in-depth discussion with an innovative and experienced cover crop farmer Ken Rulon of Arcadia, Indiana, evaluated cover crop contribution on agronomic crop yields and his farm economics (Table 6). Rulon is one of five fifth-generation owners of a 6,100-acre family farm that is part of Rulon Enterprises LLC. Since 1991, the Rulons have practiced no-till (NT) farming, but they had adopted cover crops around 14 years earlier. Currently, most of their farming operation, about 6,100 acres with rented land, is in a continuous NT, cover-cropped corn-soybean rotation.

One of their key farm goals is to increase the soil organic matter (SOM) content while still maximizing economic profits and other services. Over the years of practicing NT cover crops, the SOM content in one of their fields has increased by an average range of 2.47 to 3.58%—a gain of 1.1% over a period of 14 years of conservation farming with corn production costs just over $3 a bushel (Table 6). The total long-term benefit is a $50.50 savings per acre. Over the last five years, the corn planted after cover crops yielded 30 bushels an acre more than the acres without any cover crops. By adopting cover crops in their NT corn-soybean systems, the Rulons had reduced input costs with an associated benefit of $74,620 for P and K fertilizers and $19,110 on N fertilizers. While the total cover cost was only $118,071, the cover crop benefit was $418,430. Based on their conservation farming practices with NT cover crops, the net economic return was $300,359 with a return on investment (ROI) of 254% over the years (Table 6). It is expected that a higher SOM (via carbon sequestration) improves soil quality properties by lowering summer soil temperatures with an associated benefit of conserving moisture, which facilitates better root growth for crops' nutrient and water uptakes. The impact was more pronounced during the drought years.

Table 6 Rulon enterprises LLC cover crop benefits, Fall 2017 (*Source*: Looker, 2018).

Itemised Benefits	Cost/Acre ($)	Total Acreage	Total Benefit ($)
Fertiliser saved: P & K (20 lbs. @ $0.38 + 30 lbs. $225)	$14.4	5,200	$74,620
Fertilizer saved: N (35 lbs./ac.: 200 vs. 165)	$7.4	2,600	$19,110
Corn yield (4 years × 64 Strips; Plot data; 7.1 bu. @ $4)	$28.4	2,600	$73,840
Soybean yield increase (1.95 bu. @ $10)	$19.5	2,600	$50,700
Total annual benefit	$42.0		$218,270
Drought tolerance (2004–2017: 30 bu. Every 5th yr = 6 bu. @ $4)	$24.0	2,600	$62,400
Carbon content (5.35 bu. per 0.1% of soil organic matter × 50% = 2.7 bu. @ $4)	$10.8	5,200	$56,160
Soil erosion reduction (2 tons/acre @ $4)	$8.0	5,200	$41,600
CSP program payment ($40,000)	$7.7	5,200	$40,000
Total long-term benefit	$50.5		$200,160
Total cover crop benefit			$418,430
Total cover crop cost			$118,071
Net economic return			$300,359
ROI = 254%	Net profit/ac. planted		$57.8

4. Cover Crop Economics Tool

In recent years, the USDA-NRCS has developed the 'Cover Crop Economics Tool' for clientele. The Cover Crop Economics Tool is a user-friendly economic assessment tool to assess the costs and benefits of incorporating cover crops into crop rotation. The tool assesses both short-term and long-term expected costs and benefits. A snapshot of the tool is as follows:

(https://www.nrcs.usda.gov/wps/portal/nrcs/detail/null/?cid=nrcseprd385825#:~:text=The%20Cover%20Crop%20Economics%20Tool,term%20expected%20costs%20and%20benefits)

Cover Crop Economics - Short Term Analysis

The Short Term analysis assesses the immediate cost and benefits. After completing of the short term analysis, an option is available to expand that information to a long term analysis.

Please refer to the "Instructions" worksheet for more detailed guidance on using the tool and entering data.

To get started with a new model, select the current rotation length and then select the "Start Model" button. Enter/edit information in the white boxes. To open an existing default scenario, select the "Defaults" button and follow the instructions provided.

Defaults

Select the length of the current rotation (1-5):

Example: for continuous corn select 1 Year, corn/beans select 2 Years, corn/wheat/double crop beans select 2 Years, corn, beans, wheat select 3 Years, etc.

Start Model

5. Conclusion

Cover crops are beneficial crops generally grown in the off-season between cash crops. They provide diverse and multifaceted benefits over time to improve soil's biological, chemical and physical properties associated with soil quality. Moreover, cover crops act as a living biological mulch or provide a substantial amount of organic residues to act as surface mulch to cushion against rainfall impact and control soil erosion; accumulate SOM; increase water infiltration and moisture storage; provide food, shelter and support beneficial insects; improve soil compaction and drainage; produce and recycle nutrients and suppress weeds and soil-borne diseases. With an improvement in soil quality, long-term cover cropping significantly reduces input costs, such as on fertilizers, herbicides and other chemicals; increases crop yields and ultimately improves farm profitability with greater ecosystem services.

References

Blanco-Canqui, H., Shaver, T.M., Lindquist, J.L., Shapiro, C.A., Elmore, R.W., Francis, C.A. and Hergert, G.W. (2015). Cover crops and ecosystem services: insights from studies in temperate soils. Agric. Ecosyst. Environ., 211: 1–9.

Clark, A. (2007). Managing cover crops profitably, 3rd edition, Sustainable Agriculture Network Handbook Series, USDA-SARE Beltsville, MD, USA.

Hamzaev, A.X., Astanakulov, T.E., Ganiev, I.M., Ibragimov, G.A., Oripov, M.A. and Islam, K.R. (2007). Cover crops impacts on irrigated soil quality and potato production in Uzbekistan. pp. 349–360. *In*: Lal, R., Suleimenov, B.A., Stewart, B.A. Hansen, D.O. and Doraiswamy, P. (eds.). Climate Change and Terrestrial Carbon Sequestration in Central Asia. Taylor & Francis/Balkema, The Netherlands.

Hoorman, J.J. (2020). Cover Crops Economics, a weekly news article, Hoorman Soil Health Services (Website: hoormansoilhealth.com).

Looker, D. (2018). The economics of cover crops, Successful Farming. https://www.agriculture.com/crops/cover-crops/the-economics-of-cover-crops.

Myers, R., Weber, A. and Tellatin, S. (2019). Cover crop economics opportunities to improve your bottom line in row crops, Ag Innovative Series Technical Bulletin, pp. 1–9. www.sare.org/cover-cropeconomics.

Nordblom, T.L., Hutchings, T.R., Hayes, R.C., Li, G.D. and Finlayson, J.D. (2017). Does establishing alfalfa under a cover crop increase farm financial risk? Crop Pasture Sci., 68: 1149–1157.

Varco, J.J., Spurlock, S.R. and Sanabria-Garro, O.R. (1999). Profitability and nitrogen rate optimisation associated with winter cover management in no-tillage cotton. J. Prod. Agric., 2: 91–95.

Index

A

actinomycetes 301
Aggregate stability 257, 258, 260, 261, 265
Agroecosystem 254
allelopathy 70, 72, 73, 75, 76, 78
arylsulfatase 302

B

Biochemical 84, 88, 91
Biocontrol 84, 94
Biodiversity 84, 87, 88, 92, 93, 125–127, 132, 139, 142,
 170, 173, 175–177, 179, 185, 191–193, 196
branch-and-bounds method 61, 62

C

Carbon dioxide 281, 282, 284
carbon management 174
climate change 1–3, 9
combinatorial optimization 61, 62
competition 70, 75–78
Conservation tillage 136, 139
Cool-season grass 102
Cover Crop 28–36, 58, 66, 69–79, 99–101, 103, 104,
 106–108, 112–120, 209–221, 227, 229, 230–243,
 268–276, 309–318
Crimper roller 34–36
crop production 209, 211

D

dehydrogenase 302, 303
Denitrification 210, 211, 213–221
diversity 299–302
Dribbling 151, 156

E

Economics 309–313, 316, 317
ecosystem 269, 274
ecosystem services 170, 172, 174–177, 179–181, 192,
 196, 197
Enzymes 125, 127–130, 139, 142
Erosion 254, 260–263, 265, 268–271, 273, 274, 276

F

Fluorinated Gases 282
Fodder 108, 109, 118
Forbs 153, 158

Fruit 147–150, 152–157, 159, 161–163
fungi 300, 301, 304–306

G

genetic algorithms 62–64, 66, 67
Ghana 41, 43–45, 47–50
Green Manure 42, 47, 48, 154, 158, 159, 161, 163
Green manuring 2

I

Irrigation 44, 45, 49

L

Labile carbon 287, 289, 301, 303, 305
Land degradation 1, 2
Leaching 209, 211, 218–220
Legume 45–48, 51–53, 100, 101, 103–108, 112–116, 118,
 149, 150, 152, 153, 155, 158, 160, 161, 163
legume crops 16, 17, 19, 23–26
Livestock 99–103, 107, 108, 113–120

M

Methane 281, 282
Microbial community 299, 300, 302
Microorganisms 289
Mighty mustard 91
Mineralization 210–216, 219–221
Mixed cropping 159

N

nematodes 84, 85, 88, 91–93, 227–229, 233–236, 239,
 241–243
Nitrogen 209–213, 215, 218, 220, 221
Nitrous Oxide 281, 282, 287, 289
no-tillage 178, 180, 185, 187, 189, 190, 193, 195, 197, 287
Nutrient cycling 126, 127, 129, 132, 141
nutrient recycling 18, 20

O

Orchard 147–163
Organic farming 28, 32, 33
Organic matter 125, 127, 128, 131, 135, 136, 139

P

Penetration resistance 256, 257
Porosity 256, 258–260

Profitability 312, 313, 318
protease activity 302, 303

R

Reduced input 316

S

Sod 149–157, 163
Soil 280, 282–291
soil and water quality 4
soil biology 227, 229, 230, 243
Soil compaction 7, 8, 10, 12, 256–258, 260
Soil Conservation 154
Soil fertility 41, 42, 47, 50–52
soil foodweb 227–229, 238, 240–243
Soil functionality 125
soil functions 171, 174, 196, 197
soil health 99, 100, 119, 170, 171, 175, 180, 192, 193,
    195–197, 309, 310, 313, 315, 316
soil microbial biomass 300, 301, 304

soil nutrients 271
soil organic matter 17, 170–172, 178, 190, 194, 256
soil quality 16, 20, 22
Species selection 28–32, 35
Sustainable agriculture 16, 17, 269, 270

T

Termination 29, 31, 33–36
Trapping 92
Turfing 151, 156

U

urease 302, 303

W

water infiltration 259–262
water quality 268–270, 273–276
weed suppression 69, 70, 73, 75, 77, 78
Weeding 154